Future index	$I_e = I_b\left(1+\frac{r}{100}\right)^n$	(5.41)
Learning	$T_u = KN^s$	(6.6)
Learning rate	$\log\Phi = s \log 2$	(6.8)
Average unit time with learning	$T_a \cong \frac{1}{(1+s)}KN^s$	(6.12)
Learning slope	$s = \frac{\log T_i - \log T_j}{\log N_i - \log N_j}$	(6.13)
Power law and sizing CER with index and independent cost	$C = C_r\left(\frac{Q_c}{Q_r}\right)^m \frac{I_c}{I_r} + C_i$	(6.15)
Factor	$C = (C_e + f_i C_e)(f_I + 1)$	(6.20)
Range	$E(C_i) = \frac{L + 4M + H}{6}$	(6.23)
Machining time	$t_m = \frac{L}{fN} = \frac{L\pi D}{12Vf}$ or	(7.2)
	$t_m = \frac{L\pi D_c}{12Vn_t f_t}$	(7.17)
Drilling time	$t_m = Lf_{dt}$	(7.5)
Taylor tool life	$VT^n = K$	(7.6)
Machining unit cost	$C_u = \sum\left[C_o t_h + \frac{t_m}{T}(C_{pt} + C_o t_c) + C_o t_m\right]$	(7.9)
Minimum cost cutting speed	$V_{\min} = \frac{K}{\left[\left(\frac{1}{n}-1\right)\left(\frac{C_o t_c + C_{pt}}{C_o}\right)\right]^n}$	(7.10)
Maximum production cutting speed	$V_{\max} = \frac{K}{\left[\left(\frac{1}{n}-1\right)t_c\right]^n}$	(7.12)
Cutting length	$L = L_s + L_a + L_d + L_{ot}$	(7.14)
Batch time	$\text{Lot hours} = SU_b + N \times H_b$	(7.20)
Operation cost	$C_{tbo} = \sum \text{lot hours}_i \times PHC_i$	(7.21)
Part cost per unit	$C_u = \sum_i^n PHC_i\left(\frac{SU_b}{N} + H_b\right)_i + C_{dm} + C_t$	(8.3)
Break-even with learning	$P = KN_{be}^s$	(8.10)

Cost Analysis and Estimating for Engineering and Management

Cost Analysis and Estimating for Engineering and Management

Phillip F. Ostwald
Timothy S. McLaren

Upper Saddle River, NJ 07458

Library of Congress Cataloging-in-Publication Data

Ostwald, Phillip F., 1931-
Cost analysis and estimating for engineering and management /Phillip F. Ostwald, Timothy S. McLaren.–1st ed.
p. cm.
Includes bibliographical references and index.
ISBN 0-13-142127-1
1. Engineering–Estimates. I. McLaren, Timothy S. II. Title.

TA 183. 082 2003
658.1'55–dc21 2003053580

Vice President and Editorial Director, ECS: *Marcia J. Horton*
Aquisitions Editor: *Dorothy Marrero*
Editorial Assistant: *Brian Hoehl*
Vice President and Director of Production and Manufacturing, ESM: *David W. Riccardi*
Executive Managing Editor: *Vince O'Brien*
Managing Editor: *David A. George*
Production Editor: *Kevin Bradley*
Director of Creative Services: *Paul Belfanti*
Creative Director: *Carole Anson*
Art Director: *Jayne Conte*
Cover Designer: *Bruce Kenselaar*
Art Editor: *Greg Dulles*
Manufacturing Manager: *Trudy Pisciotti*
Manufacturing Buyer: *Lynda Castillo*
Marketing Manager: *Holly Stark*

About the Cover: The cover shows a sculpture called *Imaginamachina*, which stands between the Discovery Learning Center and the Integrated Teaching and Learning Laboratory, located on the campus of the University of Colorado at Boulder. The word "Imaginamachina" was originated by artist David M. Griggs, Denver, Colorado. He envisions that "imagination," "magic," "animation," and "machina" are seen in the word. The Latin derivative of the word "machina" means "to devise, plan, and plot artfully" an appropriate engineering objective. We use *imaginamachina* for the first time in this book. Sculpture by David M. Griggs, Denver, Colorado. Photographed by Casey A. Cass, University of Colorado at Boulder.

Pearson Prentice Hall
Pearson Education, Inc.
Upper Saddle River, NJ 07458

Printed in the United States of America

10 9 8 7 6 5 4 3 2 1

ISBN 0-13-142127-1

Pearson Education Ltd., *London*
Pearson Education Australia Pty. Ltd., *Sydney*
Pearson Education Singapore, Pte. Ltd.
Pearson Education North Asia Ltd., *Hong Kong*
Pearson Education Canada, Inc., *Toronto*
Pearson Educación de Mexico, S.A. de C.V.
Pearson Education—Japan, *Tokyo*
Pearson Education Malaysia, Pte. Ltd.
Pearson Education, Inc., *Upper Saddle River, New Jersey*

To Doris
and to Mark, Phillip, and Lynne

To Penny

Contents

Preface

This first edition of *Cost Analysis and Estimating for Engineering and Management* provides the latest principles and techniques for the evaluation of engineering design. The theme for the book begins with four chapters devoted to an analysis of labor, material, accounting, and forecasting. In the next four chapters estimating is developed, and methods, operations, and product chapters are given. With those chapters understood, attention moves to Chapters 9 and 10, "Cost Analysis and Engineering Economy." Chapter 11, "The Enterprise, Entrepreneurship, and Imaginamachina," concludes the book, and it introduces principles that deal with bringing inventions to the marketplace. Wise and calculated risk taking for the entrepreneur (read engineer and manager) are important to the broader understanding of engineering for students. The organization of this book develops these principles in a systematic way.

With increasing importance of design over rote skills in contemporary engineering courses, this book can be used for a variety of teaching situations: for lecture only, for lecture with a laboratory menu, or for professional mentoring with business, and developed field trips. Courses that connect to on-line live or delayed video instruction can use this book, as the authors have personal experience with these delivery modes. Furthermore, lifelong learning programs for the professional in either formal or informal settings can use the book.

Academic requirements for this book/course may vary, and we believe that the book is suitable for a number of teaching approaches. The book has been written to appeal to engineering/management/technology settings. The student needs a mathematical maturity of algebra and introductory calculus. Typically, this book is used in the later college periods, and sometimes it coincides with the capstone course or other summary courses that occur in the final semesters. It is also suitable for graduate level courses in engineering/technology and management.

The instructor will notice Internet requirements that search for information and apply it in practical context. We provide Internet addresses for numerous assignments. (Regrettably, these addresses may change from time to time. Fortunately, many students are adept at finding their own way around the Internet.) In the interactive environment of teaching, this book is a part of modern courseware. Word processing and spreadsheet skills are assumed, and some CAD ability is always helpful. The student must have access to a computer, and system requirements would be typical of more advanced personal or college Pentium computers.

Various academic levels, either undergraduate or graduate, and backgrounds are appropriate and the instructor will find that this book is fitting for a variety of teaching styles. The authors have attempted to involve the instructor in the leadership of many exercises, calling on you, the instructor, to localize the assignments to your needs.

The book has more material than can be covered in one semester or quarter, and thus chapters can be chosen to meet the objectives of each class. Chapter order can be adjusted. For example, if the students already have an understanding of statistics, then Chapter 5 material can be excluded. Other sections can be dropped depending on student preparation and course objectives. Now and then the term "optional" is used with sectional material, and the instructor can either appropriately overlook that section or include it for enriching purposes. The instructor will find that the book is versatile.

This book has a range of difficulty for Questions for Discussion, Problems, Challenge Problems, Practical Applications, and Case Studies. Throughout the book, the authors have attempted to give the instructor the opportunity for outcomes-evaluation of student work with these many exercises.

There are 128 Questions for Discussion in the 11 chapters. They are qualitative and require back reading and a response of a few sentences for a thoughtful reply.

We believe cost analysis and estimating to be a problem-solving activity; therefore, many of the 245 Problems and 65 Challenge Problems request computations or sketches. Whenever the student is asked to set up and solve open-ended problems, much learning occurs. Indeed, some problems may have several appropriate solutions, and that depends on the assumptions and the route for the solution. This paradigm is instructive in a broader engineering context.

The problems have varying levels of difficulty. We want the Problems and the Challenge Problems to be tractable, either with calculator or spreadsheet, where the emphasis is on teaching concepts. It is not our desire to cause excessive computation, which is the nature of cost analysis and estimating problems. Thus, this book ignores software data and encyclopedias that are found on the Internet for estimating designs. Those software applications restrict the learning of principles. Nor do we give much attention to the minutia of extensive design practices, as those temporal trade details can be learned on the job, if necessary.

There is an end-of-chapter section that we call the Practical Application. The purpose of the Practical Application is to uncouple the student from books, libraries, and the classroom. As will be seen throughout the book, Practical Applications introduce the student to experiences in the real world. For example, it encourages field trips and communication with engineers, technologists, and management professionals. The instructor will appreciate this experiential approach, allowing him or her to use Practical Applications in exciting ways.

The end-of-chapter Case Studies are open ended, perhaps having several solutions. Students are often disturbed by this peculiarity, but instructors recognize cost analysis and estimating courses are unlike calculus courses with their singularity of correct answers.

The book contains 21 Picture Lessons. They describe important historical contributions of engineering. It is essential that students have an appreciation of the grand heritage and the remarkable two centuries of technological achievement of our profession. Selection of some of the Picture Lessons was from "The 20th Century's

Greatest Engineering Achievements," a collection identified by the National Academy of Engineering.

For the instructor, a comprehensive Solution's Manual and CD is available. Additional PowerPoint helps are included. This CD can be requested from the Prentice Hall college representative or from Dr. Timothy McLaren.

The authors are grateful to many people. Their advice and information has made this a much better book. For in writing a book of this magnitude, the authors are aware that friends and colleagues are hidden, but they are very important advisers. We are indebted to the following: Lawrence E. Carlson and Ross Corotis of the University of Colorado, Boulder; Rodney Ehlers, Boulder, Colorado; Stephen Burish; Boulder, Colorado; Lynne E. Lyell, Fort Collins, Colorado; Charles W. Stirk, Susannah Ferguson, and Qin Liu of CostVision, Boulder, Colorado; Michael Usrey, Boulder, Colorado; Edward Lyell, Adams State College, Alamosa, Colorado; Donald E. Forkner, Storage Technology Corporation, Louisville, Colorado; Mark Ostwald, Fish and Wildlife Service, Lacey, Washington; Mark Willcoxon, Coors Engineering, Golden, Colorado; Kurt Mackes, Colorado State University, Fort Collins, Colorado; Roger Eiss, Vancouver, Washington; Kevin Kilty, Vancouver, Washington; and Jack Swearengen, Santa Rosa, California.

The names used in the Problems and Case Studies are of real people, and they are mentioned because of our sincere regard for their contribution and friendship.

PHILLIP F. OSTWALD
TIMOTHY S. MCLAREN

Chapter 1

Importance

Study in cost analysis and estimating starts with simple observations. There are three important reasons for learning these principles:

- Professional enrichment
- Engineering management
- Enterprise profit

If there are lessons from the twentieth century, then there is a need for engineering students to know how business scores are kept and how to interpret the results. Although engineers are educated in mathematics and science and in their specific disciplines, little learning focuses on the money consequences that flow from engineering and management decisions. Thoughtful study of the material presented in this book will enable an aspiring engineer and manager to measure the cost of design, improve productivity, and develop a strategy for enterprise success.

Our free enterprise system means that there is free choice about how, when, and with whom a buyer and seller will conduct business. It also means that the marketplace is based on free and open competition. Enrichment, management, and profit are forerunners to success in this environment. Managing on instinct will not provide this knowledge.

Engineering is undertaken to satisfy a complex set of requirements. Along with the performance, form, safety, and function of the design, cost is one of the most important numbers. The journey now begins.

1.1 ENGINEERING AND DESIGN

Engineering and design are not separate functions; rather design is the creative part of engineering. We define design as follows: *Every design is a new combination of preexisting knowledge that satisfies an economic want.* It is the "new combination" that requires the creativity, while the "economic want" provides the motivation and the ultimate financial purpose for the enterprise. The driving forces for design are knowledge and wants because each alone is insufficient. Without wants, no problems exist; without knowledge, no problems can be solved.

There are six key steps in the design process, as illustrated in Figure 1.1:

- Identification and definition of the problem
- Development of concepts
- Engineering models
- Evaluation
- Design
- Implementation

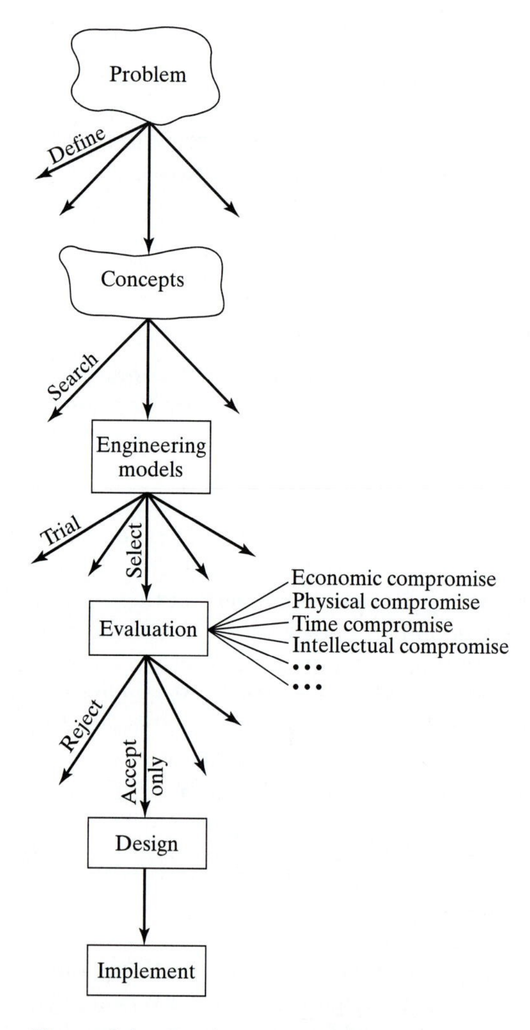

Figure 1.1 Engineering design process.

PICTURE LESSON The Steam Engine, Behemoth for the Industrial Revolution

The Industrial Revolution required energy. Drive shafts and belts had to be turned by steam engines, those behemoths that connected to the pulleys of machine tools, pumps, and dynamos, to the fans used for air-blowing Bessemer steel, and to other power-consuming designs that were flooding the factories and mills of that time. It was the steam engine that early on propelled the sinews of the Industrial Revolution, which started at about the time of the American Civil War.

The engraving shows a 40 HP compound steam engine with two cylinders, 19 in. and 33 in. dia with a 48 in. stroke. The flywheel is 16 ft dia, and weighs 19 tons. The date of the engraving is ca. 1890.

James Watt (1736–1819), a Scottish instrument maker and inventor of the steam engine, contributed substantially to the Industrial Revolution. While repairing a Newcomen steam engine in 1764, Watt was impressed by its waste of steam. In 1765 Watt came upon a solution—the separate condenser, his first and greatest invention. Watt realized that the loss of latent heat (heat involved in changing the phase of a substance—solid to liquid, liquid to vapor, etc.) was the worst defect of the Newcomen engine, and condensation must happen in a chamber distinct from the cylinder but still connected to it. He entered into partnership with John Roebuck, the founder of the Carron Works in 1768, who urged him to make an engine. After having made a small test engine, the following year Watt took out the famous patent for "A New Invented Method of Lessening the Consumption of Steam and Fuel in Fire Engines."

In 1775, after Watt's patent was extended by an act of Parliament, he and Matthew Boulton began a partnership that lasted 25 years. Boulton's financial support made possible rapid progress with the engine. In 1776 two engines were installed, one for pumping water in a Staffordshire colliery, the other for blowing air into the furnaces of John Wilkinson, the famous ironmaster.

During the next five years, until 1781, Watt installed and supervised numerous pumping engines for the copper and tin mines, which had the purpose of reducing fuel costs. Watt, who was no businessman, was obliged to endure keen bargaining to obtain adequate royalties on the new engines. By 1780 he was doing well financially, though Boulton still had problems raising capital. In the following year Boulton, foreseeing a new market in the corn, malt, and cotton mills, urged Watt to invent a rotary motion for the steam engine to replace the reciprocating action of the original. He did this in 1781 with his so-called sun-and-planet gear—a shaft producing two revolutions for each cycle of the engine. In 1782, at the height of his inventive powers, he patented the double-acting engine, in which the piston pushed as well as pulled. The engine required a new method of rigidly connecting the piston to the beam. He solved this problem in 1784 with his invention of the parallel motion—an arrangement of connected rods that guided the piston rod in a perpendicular motion—which he described as "one of the most ingenious, simple pieces of mechanism I have contrived." Four years later his application of the centrifugal governor for automatic control of the speed of the engine (at Boulton's suggestion), and in 1790 his invention of a pressure gauge, virtually completed the Watt engine.

In later years demand increased for his engine from paper mills, iron mills, distilleries, and canals and waterworks. By 1790 Watt was a wealthy man, having received £76,000 in royalties on his patents in 11 years. The steam engine did not absorb all his attention, however. Watt experimented on the strength of materials, and he was often involved in legal proceedings to protect his patents. In 1785 he and Boulton were elected fellows of the Royal Society of London, a great honor.

It should be noted that a precise sequential process is not intended. The design procedure is a combination of parallel activities and feedback loops progressing from the problem through implementation. Most design processes are carried out by one or more engineers, perhaps joined into a team, and individuals from other disciplines, all of them contributing in the areas of their strengths.

Problem. To get the design process started, someone must recognize the need for something. Marketing can determine the need or want for a new or improved service or product; engineering can require a better or faster product or production system, and on and on. Presumably, these needs are worth the spending of money. Once identified, the need becomes the economic want that drives the design process. After the problem is identified, it must be described.

The initial description of a problem is a vague statement addressing some want. It is necessary to transform this vague statement into an enlarged picture. Starting with a primitive problem description, information—technical or nontechnical—costs, and other data are superficially gathered to give form to the problem. Suppliers, customers, competitors, standardization groups, safety and patent releases, and laboratories are sources for ideas.

You may wonder why all the fuss over a simple problem statement. A thoughtful and reasoned statement specifying the problem will lead to a more efficient result. Remember that if the problem description is wrong, the solution will be wrong too. The formation of a problem statement is guided by questions—for example, Does this fit the company's needs, interests, and abilities? Are the people connected with the problem capable of carrying it to completion, or can suitable people be hired?

The problem definition step will conclude with a detailed specification of the requirements for the product. These requirements will include the physical limitations (size, weight), the performance parameters (load capacity, speed and range of operation, what the product needs to do), forecast quantities, and some idea of the size of the economic want—a cost target.

Concept. The stage is set for the search of concepts after a problem with subsidiary restrictions has been defined. This is the step in the process where the "new combination of existing knowledge" takes place. The beginning of the quest for concepts is searching, learning, and recognizing the collection of the existing knowledge; it is not yet the application. A timely and fortunate search may uncover unapplied principles. When we consider that about a million scientific and technical articles are published annually, it should be clear that a complete listing (ignoring study for the moment) of all information even within a narrow field is a hopeless cause. Nonetheless, the chance of finding the basic idea for a new development may be found through patent disclosures, new texts, or journals in the field. Instead of the unobtainable goal of completeness, there are other goals capable of being achieved in the search. Knowing where to start and when to stop in the search for concepts are the lessons of experience.

Once the requirements and relevant knowledge have been collected and assimilated, the creativity (the new combination) takes place. Creative thinking is an acquired skill that can be developed but not easily taught. It is also one thing that is uniquely human. As yet no machines or other creatures have been capable of creative thought. The concept generation itself should be unrestricted by rationale. The concepts, however wild, should be captured as they present themselves. Various techniques have been developed to aid in the generation of concepts; one of the most popular is brainstorming. Once the concepts have been captured, they can then be analyzed, embellished, revised and refined until a few promising concepts have been detailed.

Engineering Models. The engineering model involves the application of creditable concepts uncovered earlier. The formation of this model may range from a casual back-of-envelope model to a complicated physical shape. The formation of a model is an engineering trait and distinguishes the engineering pattern of thought. Models, whether experimental or rational, permit manipulation for theories or testing. Laboratory testing provides numerical answers by using physical mockups. Data are obtained, results are noted, and conclusions are stated. An engineer chooses the cheapest of the methods to state and understand a model.

The word *model* has many meanings. We can begin by defining it as a representation to explain an aspect of something. We are seldom able to manipulate reality because it may be either (1) impossible or (2) uneconomical. The market in a free economy is an illustration of the former; a nuclear reactor for electrical generation illustrates the latter. Prediction of reality through mathematical abstraction is the reason we use models. Engineering applies those ideas to scale larger problems. For example, an analyst may be unable to comprehend an actual system and have limited powers of perspective; a model may enable a satisfactory understanding. Discovery of pertinent variables, rejection or confirmation of prototypes, and comparison to a standard are reasons to use modeling.

We segregate models according to physical, schematic, and mathematical notions. All three are used in the design process. The scale of abstraction proceeds from physical models to mathematical formulations.

Physical models involve change of scale. A globe looks like the earth, for instance. A schematic model results when one set of properties represents a second set of properties. The model may or may not have a look-alike appearance to the real-world situation. Coding processes may be used, such as the chalkboard demonstration of a football play appearing as crosses and circles. Hydraulic systems are beneficial in understanding electrical systems and vice versa. Organization flow-process charts are other examples of schematic models. Schematic models capture the critical features and ignore the unimportant. Mathematical models operate with numbers and symbols in their imitation of relationships. Although those models are more difficult to comprehend, they are the most general. In an approximate way, mathematical models explain the real situation. They are desirable in engineering because they are easy to manipulate. It is customary to manipulate mathematical models according to the conventional rules of mathematics.

There are precautions that should be kept in mind in modeling. Models should be flexible to permit repeated applications. Mathematical manipulation should be simple. Arithmetic is preferred over algebra. Algebra is preferred over calculus. Calculus is preferred over vectors. Computerized spreadsheets may be sufficient. Simple models are sometimes preferred because of salability to management, and increased confidence results from their use.

Evaluation. Eventually, the engineering model reaches the evaluation point. A compromise forced on the engineering by economics, physical laws, social mores, ignorance, and faulty human judgment discolors evaluation. Even moral questions may be debated. The wisest person cannot foresee all the future effects of a design, but it is bold to ask. Others may cooperate at this point: The stylist may abridge and direct the progress, and the manager or marketing person may foresee other problems. But here is an appropriate place to stop, pause, and evaluate.

After the problem is wisely stated—couched in the design engineering model—estimating ideas are considered, an engineering cost model is selected, evaluation trials are started—shortcomings with time, money, staff, programs, or information impede the model. Finally, the cost estimate is completed. Actually, the optimum evaluation of engineering models is a continuous process. Experience suggests that it takes a long time to pass from the definition stage through the design stage.

Questions exist when performing evaluations: Does the design satisfy the requirements? What is the total cost of developing this design up to and including the sales promotion? What is the profit of the total investment during the first few years of production and sales? How long will it take for the initial investment to be returned? Does the new product coincide with the abilities and experience of the company? A number of factors are noted when considering these questions. Experience, study, and a questioning attitude are required traits for evaluation.

Above all, whatever the final results of the evaluation, something must be done; a service provided or a product manufactured. The choices include the manufacturability or serviceability of the concepts. This needs to be included as part of the evaluation process.

As can be seen in Figure 1.1, the ultimate product cost is largely determined during the initial design stages. Labor in the factory amounts to a small portion of the overall product cost (on the order of 15%, depending on the product)—which means that even if engineering is perfect at reducing the labor costs to nothing, they can only reduce the product cost by a small percentage. The major product cost is for purchased material and components (around 50%) and company overhead accounts for another 30% of the whole. The design process also influences the overhead amounts, and the design has a large impact on the material purchased. Careful and thoughtful designing can save significantly or cost considerably in terms of the engineering, production, and other infrastructure costs.

Design. Design is the process that executes the plan and gives it its shape. In fulfilling functional requirements, design involves, for example, computing, drafting, checking, and specifying as well as answering, questions: How do we build this? How will it work? By emphasizing design as a point in the design process, we do not intend to overinflate its importance. Nonetheless, a greater proportion of the designer's time is tied to designing. Engineering cost analysis provides one measure for this compliance.

Implementation. The final proof of the design and the cost analysis comes from the implementation of the design. No design process is complete until it has been put into action.

Clearly, before implementation takes place, a decision to proceed is necessary. That decision is based on answering several questions: Does the design satisfy the requirements? Will this enterprise yield a profit? To answer this last question, engineering management turns to the process of economic evaluation for the answer.

1.2 ECONOMIC EVALUATION

A key part of the definition for design is that it must "satisfy an economic want." Obviously, this implies a cost-conscious customer. A boundary is placed on the degree of "want." This limit is not an exact number, though it can be a range that will size the potential market—the higher the price, the fewer customers. The enterprise must satisfy the economic want with a design that can be produced at a cost below the price a customer—either commercial or individual—is willing to pay.

Engineering and marketing discover the products and the price. Among the possible routes of this development, they may first propose a product and then determine the potential market, or less likely, they may opportune the market and then propose a product. Market analysis is an important business practice that assists the many decisions involved in conceiving, designing, and producing a product. The study of marketing is important, but it is beyond the scope of this book.

Product development is a lengthy process and requires decisions, many of them involving economic evaluation. The steps that form the engineering design process are as follows:

- Designs or concepts are transformed into a cost analysis problem definition.
- Estimating models are proposed.
- Cost model is selected.
- Information is gathered.
- Cost-to-target trials are proffered.
- Cost estimates are completed.

The detail and accuracy of the estimates depend on the amount of information available at the time the evaluation is prepared. Early in the design process the information is limited and the estimates are preliminary. These numbers may be used to decide on continuation of the project or to compare various design concepts. As the design continues, the detail and quality of the information improves until the designs are complete. Cost analyses become more precise and are used for the authorization to proceed, price setting, production scheduling, and monitoring the production process.

The economic evaluation process has other purposes besides finding dollar costs and decision making for products. Who are the clients for this information?

1.2.1 Who Uses Economic Evaluation?

In the parlance of this text, economic evaluation includes—in addition to estimates of dollar or other monetary units—recorded actual data and other denominations of quantity, such as hours for labor, volume or weight for raw materials, and quantity of parts. In this expanded sense, many functions within the company take advantage of these evaluations. This extends from the decisions on long-term strategies to the day-to-day operations.

Engineering and manufacturing have the majority of the responsibility for design and production of the products. Thus these functions are the primary users and contributors to the economic evaluations. This does not mean that other functions are not parties to the evaluations. Sales and marketing are in charge of obtaining orders and selling the company's products. They are interested in obtaining the lowest—or a lower—cost for manufacturing the product. They want estimates in a timely fashion for their customers' requests for quotations.

The accounting department supplies information for the development of estimates and exploits estimate data to prepare operating budgets and financial performance predictions. Accounting may prepare portions of the estimate, such as overhead calculations and calculations involving interest rates and return on investment. Discussion of these topics will be reserved for later chapters.

Material management (receiving, stock room, internal transportation, shipping), quality assurance and test functions need to know the expected amount of material to be processed, and also to give costs for their operations.

1.2.2 Reasons for Economic Evaluation

Estimates for new products or changes to existing products are utilized by the enterprise to decide on continuing the product development at various stages. To conduct an economic evaluation for a product, engineering needs answers for three fundamental questions: (1) What are the performances, functions and features of the product? (2) What does it look like? and (3) How many? The reasons for the first question is obvious.

The reasons for the second question are less so. The methods of manufacture depend on the quantity of the production run and the design. For example, a metal part can be machined from a block of raw material, or it could be designed differently and be cast. Since casting has a tooling cost to form the pattern, there is a minimum quantity to recover this nonrecurring fixed cost. The machining process is more expensive for each unit, but if only a few parts are required it may be the lower cost method. The influencing factor is the anticipated quantity.

Cost estimates are helpful in deciding whether to proceed with the project, or which design alternatives to follow. As design progresses, the enterprise makes use of the estimates for other purposes. The following factors must be considered:

- *Cost to manufacture*: The product cost is important to other decisions, which may include long- and short-term capital planning, expansion of facilities, or the need to acquire additional equipment to manufacture the new or revised product.
- *Profit prediction*: Management will be analyzing a "make or buy" choice for all or portions of the product. Indeed, there will be an examination of whether to produce the produce domestically or elsewhere in the world. Furthermore, potential "angel investors" may be intrigued with the prospect of investment in the enterprise and the analysis is vital for their consideration.
- *Labor requirements*: If domestic manufacturing is the choice, managers will need to know labor skills, tooling requirements, start-up costs, and provisions for supplies and training. In the planning of labor, there may be a need to hire and train new employees.
- *Time*: Portions of the evaluation will give information for scheduling, both internally and with suppliers. Deliveries of subcontracted or raw material will flow from this information.
- *Control of operations*: Supervisors require estimates of how much time is necessary to produce the product in order to determine if the operations are efficient.
- *Improvement*: If operation improvements are anticipated, estimates are needed to see if the proposed methods will result in savings.

- *Budgets*: Engineering work with budgets. These are control tools that allow management to plan and monitor performance.
- *Equipment justification*: New products and proposed method improvements and replacement for older equipment require an economic justification to see if and how quickly the acquisition costs can be recovered. Estimates of the costs for production using the present equipment or process must be compared with the estimates for the costs for using the proposed plan. These methods are introduced in later chapters.

Countless other occurrences of economic evaluations occur throughout the enterprise on a daily basis. The opportunities for estimates range from the simple "how soon will that part be finished" to "what would happen if we built a new plant." As we shall see shortly, engineering cost and estimating tactics are driven by the strategies of the enterprise, and they are important.

1.3 STRATEGIES FOR THE ENTERPRISE

Strategies provide general direction for the whole organization toward common goals. Depending on the product and business strategies, the purposes of the functional organizations are greatly affected. Top-level business strategies flow down and are applied to engineering and design. Decisions are then made consistent with these strategies. For example, the firm's lofty ideal may be "earn a net profit after taxes of 4.2% for next fiscal year, incrementally 0.1% above the current year's performance."

There are overarching principles that the strategies need to address. One of the oldest ways to differentiate company situations is by diversification in the products or services offered and the quantities involved. Probably the most familiar for engineering is mass production, where very large quantities of a few very similar products are produced. Examples of this are automobiles and other consumer products such as television sets or clothing.

Other classifications in this quantity category are batch processes where the operations required are similar but the products change considerably within a range and the quantities are small. Traditional machine shops are examples of this business.

Finally, there are products that consist of a small quantity of a very specialized product—for example, the parts of the International Space Station. Similar analogies can be developed for the service enterprises. Once a production or service facility has been engineered to function in one of these situations, it is difficult and expensive to change the "spots of the tiger."

How a company competes in the market is a modern way to differentiate corporate strategies. Figure 1.2 offers a visualization of the basic principles. The design process discussed above results in committed and incurred costs.

Consider the following factors:

- *Design uniqueness or superiority*: There are products where a company will have a unique design advantage, which is usually based on performance or features. These are special products—very-low 10^{-9} torr. pressure instrument and sensor, which is inert to potential hydrogen atmosphere explosions—that are overlooked by competition or are protected by patents that prevent product incursion from other firms. The firm's strategy when flowing down to engineering may be

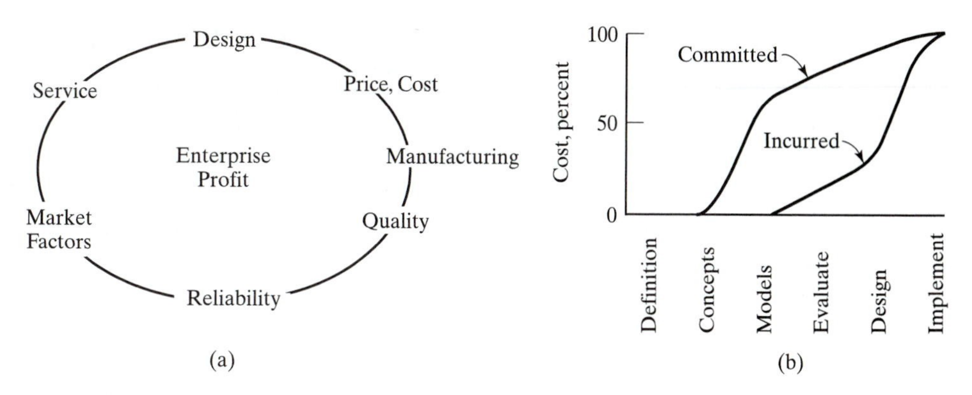

Figure 1.2 Perspective of engineering cost analysis and estimating. (a) strategies for the enterprise, and (b) design process with committed and incurred costs.

stated as "redesign very-low pressure instrument to sustain design superiority and market uniqueness by reducing part count by 7 components."

- *Price and cost*: Price and cost matter. Most companies compete on the basis of price of their products. Simply, they attempt to sell more by pricing less, or by reducing cost while keeping price constant, thus maintaining marginal profit of the product. This places pressure on engineering to perform the economic analyses. Price competition is of great importance for most consumers. If price is less than cost, then the company must take steps to reduce cost or abandon the product for long-term survival.

The firm's strategy when flowing down to pricing and engineering cost analysis and estimating functions may be stated as "confirm initial *cost to target* estimate of very-low pressure instrument which reduces cost by 8.1%, while permitting no-price increase in product."

- *Quality*: Companies compete on quality. For example, metrology instruments, which make precise hydrogen gas pressure measurements, need to be high in quality. In addition, consumers are bombarded by advertisements claiming advantages in quality. The firm's strategy when flowing down to the quality engineering activity may be stated as "understand consumer requirements on quality, and translate those requirements into achievable design specifications, which will be limited to no more than a 0.3% increase in product cost."
- *Reliability*: Still other products or services must be reliable, and perform without fail whenever used. Products and services involved in life-threatening situations—explosive gas environments for hydrogen fuel cell engines, for example—must perform with perfect reliability. The firm's strategy when flowing down to the reliability engineering activity may be stated as "ensure mean-time-between-failure parameter to minimize risk-insurance cost of very-low pressure instrument within range of ±0.02%."
- *Innovation and first to market*: A company may compete on the basis of innovation and being first to market. This is important in high-tech fields. Software firms recognize the importance of this strategy. The firm's strategy when flowing down

to engineering and marketing functions may be stated as "perform competitive analysis of other companies' designs, qualify their advantages and shortcomings, and determine their price positioning with respect to our very-low pressure instrument, and have the results available in the first quarter of the fiscal year." Another form for categorizing the business strategies for competitive enterprises is on the basis of how the products are sold.

- *Inventory, service, and supply to customer*: Companies can manufacture products for stock, where what is made is warehoused and then sold off the shelf or directly via the Internet or through a supply chain. This logistic activity is for retail and consumer goods, or it may be for original equipment manufacturers, or other technical suppliers.

The firm's strategy when flowing down to inventory management may be stated as "maintain lean manufacturing principles while holding inventory storage costs to 0.07% of full product cost."

- *Custom design or manufacturing*: These companies make products to order, such as machine shops, or they can assemble products such as personal computers to order. In some cases, a single company is involved in both. However, machining or assembly requires variation in the production facilities. Frequently these companies are important suppliers and they link to product-producing companies. These companies work closely with their customers, often providing design services whenever their components are involved in their customer's product. Some product-producing companies tie these businesses to their own success and make concessions in order to keep quality suppliers.[1]

  The firm's strategy when flowing down to the purchasing department may be stated as "sustain a supplier source, especially in the transducer design-and-build businesses for the very-low pressure instrument, and commit resources to enable the supplier to buy and build fixtures and tooling to allow the machining of the 0.21 ± 0.03 mm holes."
- *Commodity products*: Commodities—products that are provided by many suppliers and in large quantities—are made to standard specifications. Frequently, these products have little, if any, ongoing engineering support or design. These products are simply made without much technical attention. Strategies create differentiation in design and the economic analyses. Cost estimates for high production items must be accurate because any error is multiplied by the large production quantity. High quality and high reliability products will usually be made in smaller quantities, and they normally command higher prices but still allow little leeway for cost estimating errors.

[1]Linking a supplier to an auto manufacturer is an example of working closely together. The ordinary business model of auto builder and supplier is to have two or more suppliers able to provide the same component—say, brake drums to the auto builder. In the event of failure of one supplier to provide the brake drum, for whatever reason, there is the second supplier who is able to satisfy the need. In an extraordinary concession, the auto builder chose to have one supplier, but with the agreement of lower cost of the brake drum for several guaranteed years of the contract.

1.3.1 Traditional Business

By definition, traditional businesses deal with products and services that have been around for a long time and have changed little. The many examples would include appliances, automobiles, furniture, and raw materials such as lumber, steel, and aluminum and components such as bolts, hinges, and gears. These traditional products evolve slowly and their processes change little. The product designs tend to involve incremental changes and more often than not the design proceeds in the old "over-the-wall" sequential process, where design precedes the manufacturing process, as there is no concurrent effort.

There is a history of actual costs for these products, and changes are small deviations from previous conditions. Although cost analysis in the traditional businesses can be exact and straightforward to conduct, it is inscrutable to the uninformed and offers learning opportunities for the engineering student. Interestingly, the cost of many consumer products, such as consumer power tools, has declined over the last 25 years, while features and performance have improved. In contrast, the consumer-use birth-death phase of products such as household appliances may have declined. The student may want to ponder reasons for these observations.

1.3.2 High-Tech Business

Companies involved with high-tech products are in dynamic situations. Conditions and requirements change rapidly, and companies need to be quick to get a footing in the market—and they have to recover their cost and profit during the short life before another product makes them obsolete. Examples of these kinds of products are computers, video games, and other consumer electronics like audio and video reproduction devices. Think about the revolution of vinyl records to CDs to MP3 and VCRs to DVDs. The rapid changes require speedy times to market and company operations that are flexible and move swiftly.

To allow for a process that facilitates rapid and flexible response to changes, the design for new products must undergo *concurrent engineering*—an engineering practice that is almost essential for high-tech products. Multidisciplinary teams are formed to work on the product from conceptual development through design and implementation. As much work as possible is done in parallel or concurrently. The concepts and designs are reviewed as they are being created, instead of waiting until after they are completed. Implementation planning takes place during design and the facilities and materials must be ready to go as soon as the designs are finalized.

Cost analysis in this environment is challenging. The most innovative of the new products cannot rely largely on past experience and records. The cost analyses must be developed in a short period, with incomplete and changing information. The experience and judgment of engineering become important. Fortunately, if the enterprise reaches the market first, or close to the beginning, the profit margins are somewhat wider and thus there is some tolerance for the cost estimates. Of course, it must be pointed out that this tolerance does not permit less than quality cost analysis. Once operations begin, the costs can be more accurately determined and improvements in the design and processes can increase productivity and decrease costs.

PICTURE LESSON Electricity

It was during the 1881 Paris Exhibition of Electricity that Thomas A. Edison and his illumination lamps and dynamo became a sensation. The apparatus—the engraving shows the largest dynamo ever constructed at that time—was shipped from Menlo Park, New Jersey, to the Palais de l'Industrie. The exhibition featured the competition between gas illumination, which some thought was superior, and the electrical Edison carbon lamps.

This dynamo was to provide the electricity for some 1000 carbon filament lights. Electricity was the excitement of the times, and the steam engine, prime mover of 125 horsepower, constituted a direct-current machine of enormous size. Without the dynamo, electricity to power the lamps was not commercially possible. The system was designed to provide the current to light a small district of New York City.

Steam power, controlled by a fly ball governor, caused an eccentric crank to turn an armature. The electricians (now called electrical engineers) conceded that the Edison armature was the most ingeniously constructed piece. Cylindrical in form, the hollow shaft had straight copper bars attached to the cylinder without any wires, with the bars insulated by brown paper. The armature revolved at 350 rpm, and through a complex arrangement, commuted an electromotive force of 103 volts. The magnetic field was produced by cylindrical electromagnets, which were fixed in a horizontal position and induced the magnetism for two solid iron horse collars through which the armature rotated.

The filaments of the "light bulb" were carbonized cotton rather than the tungsten material of today, and the globe contained a noncorrosive gas rather than a vacuum. The electrical lamp competed with candles, oil lamps, and gas street lamps, which were all basically tiny fires. Edison's design of the dynamo and the electric lamp offered mass production possibilities. Within just 10 years after the Paris exhibition, there were 1300 lighting companies in America.

1.4 INFORMATION

It is necessary to have information of various kinds before engineering is able to estimate the economic want of a design. Two extremes in the type of information exist. Visualize the case where virtually no information is available—where we presume that it is unlikely that an estimate can be made. At the other extreme, assume that all data are available, which implies that the money has been spent for the design and an estimate is unnecessary. Whenever the data are all available, the process of after-the-fact cost analysis is likely to be conducted by accounting, not engineering. Engineering works in a cost data and design setting where the information is not fully disclosed and thus the reason for estimating unknown information.

We separate the accounting function from engineering cost analysis field on the principle that the accountant deals with cost quantities that have been spent and consistently recorded. Engineering attempts to forecast cost quantities that have not been spent for a design. Cost accounting and engineering cost analysis are specializations that work together and have much in common.

Some cost data are historical. Accounting reports are historical because they consist of the transactions recorded through cost-controlling accounts that may be kept in a ledger system. Money is expended, and materials, labor, services, and expenses (such as power or heat) are received. Specific accounting procedures must be provided for recording the acquisition and disposition of materials, for the recording and use of labor, and for their distribution. The internal function of cost accounting as it relates to our interest is discussed in Chapter 4.

Some data are measured. Engineering may find that work measurement methods give information that is amenable to estimation, either in time or dollar dimensions. Material quantities calculated from CAD files and drawings and specifications are a form of measured data.

Finally, some data are policy and have the property of being fixed for engineering purposes. The origins of policy data are varied and are accepted as factual and often unchallengeable by engineering. Union-management wage settlements are examples of policy data. Budgets and restrictions for legislation ranging from municipal to national laws dictate codes of conduct and cost. The federal government requires a social security tax from the employer for the purpose of providing old-age benefits. An unemployment compensation tax, sometimes called Federal Unemployment Insurance Cost, is collected by the states to provide funds to compensate workers during periods of unemployment. Such data may emanate from internal departments within the organization, official government sources, international agencies, trade associations, trade unions, sampling organizations, or any office that gathers and divulges design and economic information.

Accounting is a major source of information, but there are other departments internal to the organization that provide information. The personnel department, charged with the handling of employees, interprets the union contract (where unions exist), conducts labor contract negotiations, and keeps personnel records regarding wages and fringe costs.

Operating departments—whether, for example, in manufacturing or services—are the producing organizations and are concerned with doing. The supervisor or department manager knows the operating details at that moment. Frequently, he or she is a direct source for information. The supervisor may often assist in obtaining data on special forms that report extraordinary costs of process equipment, staffing, efficiency, scrap, repairs, or downtime. Sometimes he or she is the oracle for a "guesstimate" on operations with which he or she is familiar.

The purchasing department in many organizations is responsible for spending money for materials. Some companies believe that this department is responsible for the outside manufacturer. The purchasing department knows about purchasing and shipping regulations and can be a frequent source of information.

The contribution of sales and marketing is apparent in the pricing of products—for instance, market demand, sales, consumer analysis, advertising, brand loyalty, and market testing. A variety of basic economic facts and trends are available from the U.S. government. The Bureau of Labor Statistics (BLS) provides elements of cost on the prices of materials and labor.

Data are found from manufacturers' agents and jobbers, who, although they promote special interests, are willing to release information given to them by their clients. Trade associations, subsidized by groups of businesses sharing a common need, are typical organizations that publish data for engineering cost analysis.

Various methods for constructing economic analyses are discussed throughout this book. In some cases the analysis can be developed using actual record data. In most cases, however, engineering will use accumulated experience and skill to develop the analysis using estimates in place of actual data. The amount of information and time available to work on the analysis as well as the intended use for the end result will determine much about the estimate. Quick estimates can be made in a few minutes, with little more than conceptual information about the product. The resulting estimate may be "in the ballpark" and be unreliable for a contract. Full and complete estimates for establishing contractual cost, product prices, and the like are generally not possible until later in the design process and require time and skill for the analysis.

1.5 DOMESTIC AND INTERNATIONAL BUSINESS

From the beginning of the Industrial Revolution through the first half of the twentieth century, the United States produced many of the products needed within the country. Because of our many natural resources, there was little need for imports. With the exception of the production of war materials during the world wars, the United States did not participate to any extent in international business until the later half of the twentieth century. Because of the availability of land and resources and the relatively small population available as a workforce, the development of an engineering technology that relied on machines rather than laborers was needed to satisfy the country's requirements for products. The term "Yankee ingenuity" was applied to our ability to out-produce the rest of the world on a per capita basis.

Supplying war materials overseas during the world wars and materials for rebuilding war-damaged lives and industry enlightened U.S. manufacturers to international markets for their goods. These new markets facilitated the postwar growth of the U.S. economy and during the 1950s helped fuel a significant increase in our standard of living. The rebuilt factories in Europe and Japan then began to look for and supply international markets in competition with U.S. industry, and they even began to compete within the domestic market. U.S. manufacturers were able to remain competitive because of improvements in their manufacturing technology.

Those that study economics understand the evolution of American industry and the market conditions and productivity that allowed the United States to become dominant in world manufacturing with highly paid workers. During the second half of the last century, other countries caught up with and in some cases surpassed the United States in manufacturing technology. Their lower-paid workforce and established engineering systems enabled them to compete successfully with the United States, even after taking into account transoceanic transportation and import costs. This led to "globalization" of the production and marketing of goods and services. In addition, companies have merged across international borders, resulting in complex global organizations having facilities in many countries.

When we consider the impact of importing products from abroad, we must recognize the vast extent of a worldwide market. During the third quarter of the twentieth century, the United States was a net exporter to the world, but in the current period that situation has reversed itself. The United States now imports more manufactured product than it exports, by approximately 7 percent. Positioned in the center of the international arena of business, the United States must formulate a policy for future engineering and management strategy that will prepare for designing, manufacturing, and cost estimating in a global and domestic economy without borders. Trade treaties indicate that seamless movements of product across international borders will expand. This theme is considered in various chapters in this book, but especially in Chapter 8.

The net result of these world economic factors is that everyone involved in engineering, production, and commerce must think and conduct business within an international framework. Foreign markets and competitors must be considered in every business decision and design. Many U.S. products are already produced in SI units, and it will not be much longer before most product specifications are metric. People in most other countries of the world are used to considering cultural differences in their business dealings, and they are comfortable converting among various currencies. To be successful in the future, engineers in the United States need to acquire and hone these abilities.

We have learned to convert between units of physical measure using accepted conversion factors (e.g., 2.54 cm per 1 in.). These conversion factors are sometimes called constants because they do not change. Working with different monetary systems requires the same arithmetic skills, but the conversion rates are no longer constant. The value of one currency with respect to another varies on an almost continuous basis, depending on the currency exchange markets. This makes conducting business somewhat of a gamble. If you plan to buy a machine from Japan that will be delivered, and paid for in six months, you cannot be sure what your cost will be due to the fluctuations in the conversion rate between the dollar and the yen.

Consider the example of a company that will purchase machinery from Germany. We assume that the purchaser is in the United States and has the choice of currency for buying and paying for the machine. The company may agree to buy in euros or dollars.

If the purchase is in dollars, then it may appear that the foreign exchange has avoided a problem, but do not forget that the problem exists for the seller of the machine. The seller will be in receipt of dollars and will sell them for euros. On the other hand, if the machine is invoiced in euros, then the purchaser will buy euros before paying for the machine. A foreign exchange transaction is involved either way. It is axiomatic that movement of goods or services across a frontier causes a foreign transaction. Exporters and importers think in terms of their own national currency.

The prevailing foreign exchange rates between any two countries are the prices at which the currency of one country will sell in the currency of the other. Table 1.1 is a sample of foreign exchange rates. For instance, it takes $ U.S. 1.4608 to purchase £1, or you can buy £0.6846 with $ U.S. 1. Free and uncontrolled foreign exchange rates fluctuate daily.

TABLE 1.1 Illustrative Exchange Rates for $1 U.S.*

	Euro (€)	British Pound (£)	Japanese Yen (¥)	Canadian Dollar ($)
Spot	1.0732	0.6846	124.34	1.5326
6 Month	1.0817	0.6904	123.08	1.4906

*Information can be accessed online at http://www.ny.frb.org/pihome/statistics/forex10.shtml (U.S. Federal Reserve Bank).

A distinction is also seen in Table 1.1 between spot and future exchange, say six months from the date of the spot value. When an importer purchases spot exchange, delivery is actually taken for a definite amount of foreign exchange at the time of the purchase for which he or she pays the rate then quoted for the particular bill of exchange. When the importer purchases a future exchange contract, the importer agrees to purchase a given amount of exchange on a fixed date in the future or within a fixed period to pay for it at the rate specified in the future contract. This future rate may be higher or lower than the spot rate. Current exchange rates can be obtained from larger banks, business sections of the newspapers or from the Internet.[2]

Example: Conversion of Material for English Currency and Units

A material is priced currently at $3.65/lb. Find the equivalent spot market in British pounds and SI.

$$\left(3.65\frac{\$}{lbm}\right)\left(0.6846\frac{£}{\$}\right)\left(\frac{1}{0.4535}\frac{lbm}{kg}\right) = £5.51/\text{kg}$$

German material is currently priced at €485/m^3. Determine equivalent value 6 months from now in U.S. dollars and customary dimensional units.

[2]Public sources for exchange rates as found in these Web sites: http://markets.usatoday.com/custom/usatoday-com/html-investor-currency.asp; http://www.xe.com/ucc/; http://www.x-rates.com/.

$$\left(485\frac{€}{m^3}\right)\left(\frac{1}{1.0817}\frac{\$}{€}\right)\left(0.02831\frac{m^3}{ft^3}\right) = \$12.6933/ft^3$$

Observe that the trailing decimals on the right side of the equations are increased because of the exact nature of currency conversion.

A fluctuating exchange rate means that from day-to-day the amount of a foreign currency that can be obtained for a given amount of dollars changes. Some days a dollar can buy more of a foreign currency (strong dollar) and some days less (weak dollar). These variations certainly make international commerce challenging to track but also have various impacts on the costs of doing business. A strong dollar means we can purchase foreign made materials and equipment at a lower cost in U.S. dollars, but at the same time a strong dollar makes the goods that we want to sell in foreign markets more expensive in their currencies. The reverse is true for a weak dollar. A strong dollar means that a U.S. manufacturer must cut prices (provide discounts) to compete internationally.

In the larger picture of the U.S. economy, a short-term weakening of the dollar means an increase in exports and growth of the economy. However, in the long term, a weak dollar leads to inflation and higher interest rates. This translates to higher trade and federal budget deficits and discourages foreign investment in the United States. On the other hand, a strong dollar means short-term price cuts to compete internationally, but overall it is best for the U.S. economy. This means that to compete globally companies must be able to produce their products at lower costs.

1.6 INTERNATIONAL SYSTEM OF UNITS

The student of engineering can no longer be uninformed about worldwide systems of units. Understanding of dimensional conversions is required for international trade.

Le Systeme International d'Unites, known worldwide as SI, is a modernized metric system and incorporates many advanced unit concepts. With SI it is possible to have a simplified, coherent, decimalized, and absolute system of measuring units. Because it will be a long time before the United States deals exclusively with SI units in engineering, we must use a mixture of SI and U.S. customary units.

The SI system has seven base units (meter, kilogram, second, ampere, Kelvin, candela, and mole), two supplementary units (radian, steradian), and additional derived units. The list of derived units within the SI system is extensive.

The SI units for force, energy, and power are the same regardless if the design is mechanical, electrical, hydraulic, or chemical. Confusion is often found in the U.S. customary system of using both pounds force and pounds mass, but this is avoided in SI. The subject matter in this text deals with labor, measured in units of time that are the same in SI and the U.S. systems. Chapter 3 studies material cost and makes use of length, units of area, and volume and density.

The SI system has a series of approved prefixes and symbols for decimal multiples. An abbreviated table is shown in the back endpapers of this book. Be careful with capitalization of SI symbols to avoid confusion: for example, K is Kelvin but k is kilo.

Also note that we use the U.S. practice of commas to separate multiples of 1000; SI uses a space instead of a comma. This table is much abbreviated and contains only those conversion factors relative to material estimating. Also note that conversion factors for compound units (e.g., density or mass/volume) are not given. Conversions involving these other units can be found using the given values, a calculator, and dimensional analysis. Conversion factors are found in other reference books.[3]

Calculating tools (pocket calculators, computers) provide numerical results to many places to the right of the decimal point. The usefulness of all of those decimal places is limited by the decimal places of the input data and the type of calculation.

Example: Rules for Engineering or Cost Analysis Data

Specific rules are observed when engineering data are added, subtracted, multiplied, or divided. The first number is reported in millions, the second in thousands, and the third number is in units, as shown in (a):

(a)		(b)	
	163,000,000		163,000,000
	217,885,000		217,900,000
	96,432,768		96,400,000
	477,317,768		477,300,000

If those numbers were pure engineering data, then the numbers would first be rounded to one more significant digit to the right than that of the least accurate number and the sum given in example (b) and then rounded to 477,000,000. However, if the numbers are pure cost data (i.e., dollars), then the overriding choice depends on final use of the information. The rule for monetary calculations adopts example (a) as preferred but would accept example (b) if an approximation is all that is required.

The rule for multiplication and division of engineering data states that the product or quotient must contain no more significant digits than are contained in the number with the fewest significant digits used in multiplication or division. The difference between this and the rule for addition and subtraction should be noted. The last rule requires rounding digits that lie to the right of the last significant digit in the least accurate number.

Multiplication: $113.2 \times 1.43 = 161.876$ rounds to 162

Division: $113.2 \div 1.43 = 79.16$ rounds to 79.2

Addition: $113.2 + 1.43 = 114.63$ rounds to 114.6

Subtraction: $113.2 - 1.43 = 111.77$ rounds to 111.8

The product and quotient are limited to three significant digits because 1.43 contains only three significant digits. The rounded answers in the addition and subtraction examples contain four significant digits.

Numbers that are exact counts are treated as though they consist of an infinite number of significant digits. When a count is used in computation with a measurement,

[3]See *The Metric Practice Guide* published by the American Society for Testing and Materials (ASTM), 1916 Race St., Philadelphia, PA 19103.

the number of significant digits is the same as the number of significant digits in the measurement. If a count of 113 is multiplied by a value of 1.43, then the product is \$161.59. However, if 1.43 were a rough value accurate only to the nearest 10th and, hence, contained only two significant digits, then the product would be 160.

Rules for rounding are the same as for estimating and engineering practice. If the digit discarded is exactly 5, then the last digit retained should be rounded upward if it is odd or not if it is even. For example, 4.365 when rounded to three digits becomes 4.36, and 4.355 would be rounded to 4.36 if rounded to three digits.

Example: Conversions for Customary and SI Units

Convert the following numbers to SI:

12.52 ft to meters

17.2 ft^3 to cubic meters

5.15 lb-m to kilograms

2.005 in. to millimeters

2.4637 in. to millimeters

Using the rule of precision of the original measurements, $12.52 \times 3.048 \times 10^{-1} = 3.8161$, becomes 3.82 m; $17.2 \times 2.831 \times 10^{-2} = 0.48693$ becomes 0.487 m^3; $5.15 \times 4.53 \times 10^{-1} = 2.3330$ becomes 2.33 kg; $2.005 \times 2.54 \times 10^1 = 50.927$ becomes 50.93 mm, because a two-place decimal for the millimeters has the same degree of accuracy as contained in the original inch measurement; and $2.4637 \times 2.54 \times 10^1 = 62.5780$, which when rounded to three decimal places becomes 62.578 mm because of similar degrees of accuracy implied by a measurement.

1.7 A LOOK AT THE BOOK

This book provides the kinds of thinking that are found in engineering cost work. An effort has been made to assimilate theories and practices that are broadly attractive to all engineering students, whether they are, or are to be, employed in research, development, design, production, construction, sales, or management. Though engineering cost practices vary among the several fields of technology, the principles tendered in this book do not.

For engineering cost analysis, we follow the design in a logical manner to provide a scheme of estimating. The designs, whether for an operation, product, project, or system, provide the identifying feature. Formats, procedures, and ramifications vary for those types of designs. Methods of estimating and analysis do not. Figure 1.3 provides a look at the organization of the book into the chapters. The chapters are arranged to follow the development of an estimate from the parts to the whole. Some chapters contain material that discusses "tools" that are used in the development of the estimates.

The engineering student who brings to his or her job an understanding of the economic consequence of design is valued in industry, business, and research. She or he is prepared to become an engineer, designer, project engineer, supervisor of engineering

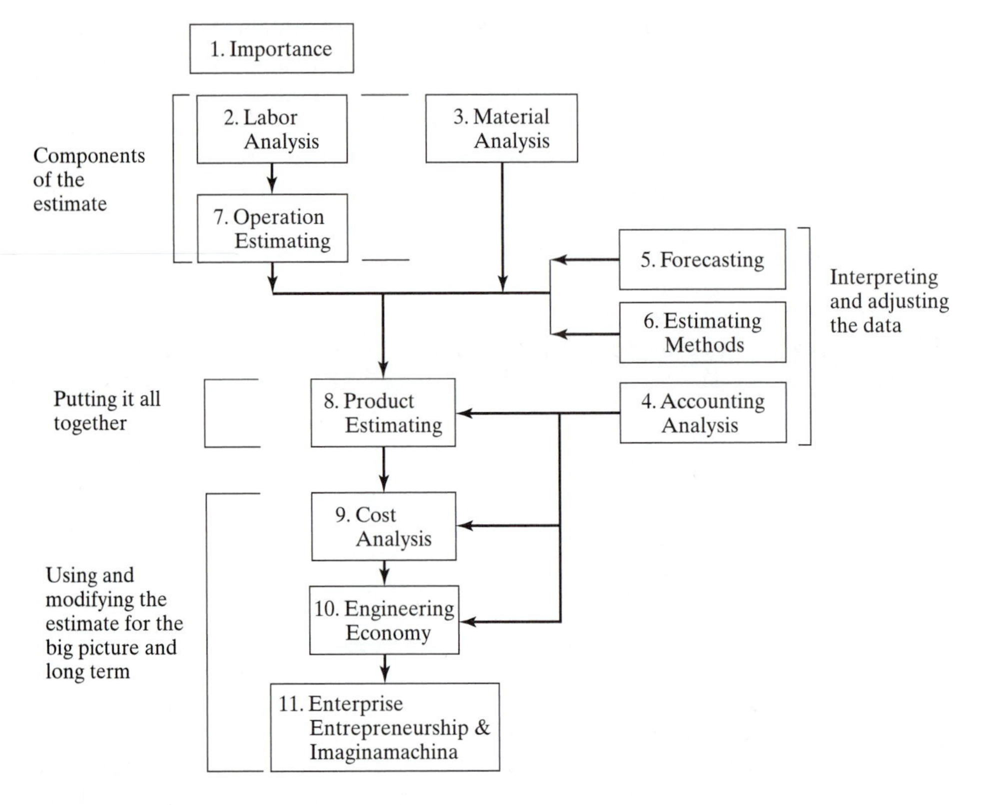

Figure 1.3 Chapter organization.

activities, or a business person, working closely with design, development, and the research team.

Manufacturing and construction are essentially a "value added" process. A service adds value by changing some condition. Labor provided by the workers changes material into some desired product or produces a service. In order to make a profit the value added must exceed the cost of affecting the change or providing the service. Chapter 2 describes how the cost of labor is calculated.

Engineering and others engaged in estimating product costs must consider the costs of the materials and components that are converted into the company's products. The cost of other materials used to facilitate the conversion process must also be included. Chapter 3 describes how to estimate the quantity and cost of the materials used to manufacture the designed products.

Chapter 4 deals with aspects of accounting that are of particular usefulness in cost analysis for products and services. Accounting is the means of analyzing the money transactions of business. Accountants prepare balance sheets, statements of income, and information to aid the control of cost, and these facts are essential for the determination of overhead. Engineering deals with costs and designs before the spending of money; cost accounting records the cash flow facts.

Business forecasting comprises the prediction of prices of material, availability and cost of labor, market demand, and costs of manufacturing. The usual approach to business forecasting involves extrapolation of past data into the future. Forecasting is a small but important part of the professional field of engineering cost analysis. Chapter 5 considers the basic statistical and indexing methods used for forecasting.

Even though engineering designs differ greatly, estimating methods are remarkably similar. Chapter 6 introduces general estimating methods. These procedures are discussed and their advantages and shortcomings are noted. Methods depend on experience and judgment to mathematics.

The common sense of operation estimating is, in a substantive measure, a talent for breaking a task into essential elements. A design formulated sufficiently for this analysis is available, and engineering uses cost equations and engineering performance data that model the details of the design. Thousands of labor and material operations exist. Regrettably, the explanation in Chapter 7 extends only to a few, using the metal machining processes to illustrate the methods of operation estimating.

A product estimate is prepared in response to a request. Engineering will obtain the necessary data, calculate component labor and material costs, and compile the results into a total product manufacturing cost. Finally a price for the product is established. Chapter 8 describes the various elements that make up the product cost and demonstrates how to assemble a complete product estimate.

Chapter 9 presents methods for applying cost analysis to engineering designs. Cost analysis is performed during the engineering and manufacturing "tradeoff" period. These tradeoffs happen early in important engineering products—perhaps 60 to 80 percent of the consequential cost/tradeoffs are made before and during design. The analysis is special for each design. However, there are some principles that are helpful, and these are described.

Chapter 10 provides a useful discipline for making choices among engineering alternatives, where the project is dominated by an investment. The units for the decision-making evaluation are in terms of money, but the added ingredient at this point is that some amounts of money are future revenues or costs. Their timing influences the choice of the alternative. This is where the concept of interest—the time value of money—comes into play.

Chapter 11 introduces concepts about creativity and the nuts and bolts for starting a business or augmenting the opportunities for an existing business. This chapter provides some thoughts about economic evaluation in the context of the enterprise, entrepreneurship, and imaginamachina. Topics include designing for profit, preparation of the enterprise plan, and financing with stocks, bonds, or credit.

SUMMARY

There are three important reasons for learning the principles given in this book: professional enrichment, engineering management, and enterprise profit. The remainder of the book concentrates on those simple orders.

We have described the environment where cost analysis takes place, first in terms of the product design process, then in terms relating to the various business strategies and finally in a global environment. The need for and uses of cost analyses were introduced, and some discussion focused on the sources for data needed for analyses.

Finally, some computational material relating to conversion between different systems of measurement was presented. The traditional dimensional conversions, especially length and mass dimensions, will be used for computing cost estimates and are required for global commerce. Monetary conversions and the fact that the conversion factors constantly change were discussed.

QUESTIONS FOR DISCUSSION

1. Define the following terms:

Design	Historical information
Domestic business	Measured information
Economic want	Policy information
Engineering cost analysis	Profit
Engineering model	Product estimate
Estimating	SI
Exchange transaction	Spot exchange

2. Prepare a list of career opportunities in engineering cost analysis from the Internet.
3. How does competition and failure of the enterprise interact with principles of cost analysis? Discuss.
4. How would you define profit? Discuss fully. What are the consequences of negative profit (loss)? How can profit be made appealing to the individual?
5. List positive and negative results of failure in business. Should governments prevent business failure? Does company size and political power affect your answer?
6. What distinguishes the act of cost analysis and estimating? Will this activity ever become a science?
7. How do economic laws and physical laws differ? Are the well-ordered cause-and-effect relationships separable in business and engineering fields?
8. Relate the role of cost analysis to design engineering. Contrast these roles.
9. What do exchange rates between countries reflect? Why do they fluctuate?
10. Do you agree that the dimension "dollar" is as important as other engineering dimensions?
11. Distinguish between product and technology competition. Which affects your life more?

PROBLEMS

Convert the following eight problems from U.S. customary units to SI units or from SI to U.S. units. Show correct abbreviations.

1.1 (a) 17 ft^2, 2.4 ft^2, 450 $in.^2$, 5000 ft^2 to $meters^2$ (m^2)
(b) 0.15 $in.^2$, 0.035 $in.^2$, 20.61 $in.^2$ to $millimeters^2$ (mm^2)

1.2 (a) 18 pounds force (lb-f) to Newtons (N)
(b) 180,000 lb-f, (with appropriate prefix for newtons)
(c) 1.8×10^6 lb-f to newtons, (with appropriate prefix for newtons)

1.3 (a) 18 ft, 2.0 ft to meters (m)
(b) 10 in., 0.01 in., 100 in., 0.00015 in. to millimeters (mm)

1.4 (a) 15 ounce mass (ozm) to kilograms (kg)
(b) 25 tons to kilograms (kg)
(c) 1400 kg to pounds mass (lb-m)

1.5 (a) 15 pounds-mass/ft^3 (lbm/ft^3 = density) to kilogram/meter^3 (kg/m^3)
(b) 180 kg/m^3 to lbm/ft^3

1.6 (a) 180 feet/minute (fpm), 500 fpm to meters/second (m/s)
(b) 180 inches/second (in./s), 1855 in./s to meters/min (m/min)

1.7 (a) 0.37 ft^3, 125 ft^3, 700 ft^3 to meters^3 (m^3)
(b) 0.01 in.^{3}, 12 in.^{3}, 150 in^3 to millimeters^3 (mm^3)
(c) 1000 yards^3 to m^3
(d) 250 mm^3, 80 mm^3, 1500 mm^3 to in.^{3}

1.8 (a) 800 ft^3/min, 65 ft^3/sec to m^3/s
(b) 1000 m^3/s, 75 m^3/s to ft^3/min

Use Table 1.1 for Problems 1.9 to 1.12.

1.9 A casting costs U.S. \$17.50 per unit. Determine the spot value of the casting

(a) in euros
(b) in British pounds
(c) in Canadian dollars

1.10 If a catalyst is worth \$85 per gallon in the United States, calculate an equivalent value

(a) in euros and metric units
(b) in yen and SI

1.11 International export opportunities may send materials from one country to another and back to the originating country because of labor cost or a technology advantage. An electronic product is transported to Japan, and a value of 12,434¥ is added.

(a) What is the U.S. value?
(b) If a value of \$110 is estimated for equivalent U.S. work, then what is the exchange rate that is indifferent to the decision? (*Hint:* Indifference means that the decision can go either way, or is indeterminate.)
(c) For the work to remain in the United States, must the exchange rate increase or decrease relative to the indifference rate?

1.12 An American businesswoman travels from New York to four countries. She will start with U.S. currency and exchange her dollars in each country by using that country's prevailing exchange rate. In each country she will buy the next airfare and will incur business expenses. Travel and expense budget, expressed in the currency of the country, are as follows:

U.S. to Canada	Canada to England	England to France	France to Japan	Japan to U.S.
US $875	Can $2200	£1600	€3500	250,000¥

What is the minimum amount of U.S. cash will she need in New York?

CHALLENGE PROBLEMS

1.13 A small country is preparing for its anniversary, and a politician wants 2500 busts of the president for distribution. A copper alloy will be used, but the weight of the bust depends on the copper alloy. When the master of the bust is submerged in water, it displaces 0.00098 m^3 of water. An 80 to 20% copper-zinc alloy requires $2.50 per kilogram for copper and $2.07 per kilogram for zinc. The densities of copper and zinc are 8906 kg/m^3 and 7144 kg/m^3 respectively. Your billing price is figured at seven times the metal cost. The country's currency is renolas, figured at a current exchange rate of 14.3 units/U.S. $1.

(a) Find the price for 2500 units in renolas and U.S. dollars.
(b) Convert the values to U.S. customary units and repeat.

1.14 A U.S. contractor will design and fabricate equipment for a fossil-fuel power plant, and supervise its installation in country X. Designing and fabricating are performed in the United States, while installation is in country X using the national labor of country X. The deal is agreed to at base time zero and will be adjusted for inflation within country X and exchange rate. The contract at time zero is as follows:

Time	U.S. Material	U.S. Cost for Installation in Country X
0	$6 million	$2 million

Engineering believes that the rate of inflation will be 10% and 25% annually in the United States and country X respectively. The exchange rate of country X to the U.S. base is 1:1, 2:1, and 3:1 for years 1, 2, and 3. U.S. material cash flow is spread evenly over the first two years while erection will occur during year 3 in country X. The contract requires that the full lump sum is paid at year-end 3, adjusted for inflation and exchange rate. (*Hint:* Assume that the inflation and exchange rates are independent.)

(a) If the U.S. material costs are spread uniformly over the first two years, then roughly what is the total amount spent including inflation? (*Hint:* Consider inflation to increase at a compound amount, and that money spent at the start of the year will inflate for the entire year while money spent at the end of the year will not inflate for that year. Thus, consider average inflation and the time period.)
(b) If the contractor will pay the national labor in that country's currency during year 3, what is the cost in U.S. dollars? How many country X currency units are roughly expected?
(c) What approximate sum of U. S. dollars is due to the contractor at year-end 3? What is that in country X currency?

1.15 A reactor vessel can be built in Germany, England, or Japan and shipped to country Y for integration in a refinery. Quotes are received from contractors and ratioed to the U.S. estimate in terms of U.S. currency by using Table 1.1 and spot exchange rates. According to the contract, payment is transacted between the two parties on the day the reactor reaches the port of entry of country Y. The exchange rate for the scheduled arrival is, however, different from that listed by the table and is shown as a relative change to the table.

	Germany	England	Japan
Ratio of subcontract bid to U.S. estimate	0.9992	1.0065	1.0062
Exchange rate on arrival compared to Table 1.1	+0.3%	−0.5%	+0.01%

The U.S. estimate for the reactor is $100 million.

(a) In which country should the vessel be built, and what is that cost in U.S. currency?
(b) What is the dollar penalty if the next-lowest country is selected?
(c) Where does nonnumerical judgment enter the decision process?

PRACTICAL APPLICATION

The Internet's educational value for engineering cost analysis and estimating prompts this practical application. You have already developed cyber skills, which can enhance your learning in the course you are taking. From time to time assignments are given to examine the Internet for information that adds to the instructional value of this book.

You are to study several Internet search engines and hunt for Websites that associate closely with the course objectives that your instructor will identify.[4] Once you have located useful sites and information, transmit this information to another student (or to the instructor or to your college department, or to your company, if you work for a business that might be interested). There are many other Internet home pages for you to find. Be sure that you verify that the sites you select are reputable. (The Internet is not like printed publications, in that there are no editors to ensure the validity of the posted material.)

Conference proceedings, papers, abstracts, patent searches, governmental information and a great deal more are candidates for this searching. Your instructor will add more objectives.

The transmission of the information can be to the intranet email system, or to the mailto function on the college's or the instructor's or your companion's home page on the Internet. Your instructor will discuss the desired information, a plan for your work, and the rules on grading.

[4]These associations closely identify with professional engineering activities: American Society of Mechanical Engineers, Institute of Electrical and Electronic Engineers, International Association for Management of Technology, Society of Manufacturing Engineers, Institute of Industrial Engineers, American Association of Cost Engineers International, Association for Information Management, and other professional organizations and engineering cost consultants.

CASE STUDY: PROFESSOR JAIRO MUÑOZ

"Good morning, Professor Muñoz. I'm Rusty, and I'm in your eight o'clock engineering cost class."

"Uh huh," replies the professor without looking up from the desk. "What can I do for you, Rusty?"

"Well, it's like this. I'm not sure that I belong in your class." Rusty smiles and continues as the professor looks up. "Oh, it's not you, Professor, but I really want to be a (*make your own selection*) and it's unclear what this course will do for me later on."

"Yes, go on," says Professor Muñoz, leaning forward in his chair. "What are you looking for?"

"I do want a career-oriented program, but will cost analysis and estimating help me in that direction?" Rusty looks at the professor expectantly.

"That's an important question, Rusty. There are many, many considerations."

Help the professor answer Rusty's question. Consider the following:

- Call and talk to an engineer involved in cost analysis.
- Check the newspaper want ads, trade journals, technical magazines, and your college career office for employment opportunities.
- Determine the educational requirements and the experiences needed for these careers.
- Consider tangential opportunities that require cost analysis. Are these positions in the management area?
- List the job titles and their relationship to engineering cost analysis. Do these situations change after one year, five years, or ten years?
- Which courses—basic and engineering science, mathematics, humanistic and social science, technical elective, or design—should you emphasize to reach your goal?
- What do you enjoy doing?

Once the investigative part of the case study is concluded, write a prospectus for an internship that you may submit to a company. Provide a copy of the prospectus to your instructor.

Chapter 2

Labor Analysis

Labor is the underpinning for engineering cost analysis and estimating. A serious examination of this topic is important to the student, as there are cultural and business reasons for identifying labor as the initial factor for study. Because of its historical development for over a century, labor in North America is analyzed in a certain way.

Engineering is interested in productivity, and procedures that identify, describe, measure, and then model labor are crucial. This chapter, for the most part, deals with definitions and measurement of labor in order to assemble later analysis and estimates. The student or engineer may not be involved with these procedures, but to overlook their role does a disservice to the broader understanding needed for later chapters.

2.1 BACKGROUND

Labor has received intensive study over the years, especially direct labor, and many recording, measuring, and controlling schemes exist in an effort to understand labor. Labor can be considered in a number of ways, for instance:

- Political-historical understanding
- Union influences
- Professional education
- Attendance- or performance-based payment schemes
- Direct-indirect labor classification

The range of the political-historical understanding of labor is vast. There are many threads that we could follow, but we focus on capitalism and socialism for this first interpretation. In 1867 Karl Marx published *Das Kapital*, his critique of political economy.[1] His writing came at a turning point in history, as the agriculture home-craft era evolved into the engineering industrial era. How can a book titled "Kapital" or "Capital" be concerned with labor? Marx understood (as did Adam Smith and others

[1]*Das Kapital, Kritik der Politischen Oekonomie*, by Karl Marx, 1867.

too) that equipment and capital transfers its worth, along with labor value, to a product. If the product is textiles or iron forgings, machinery,[2] either in performance or capacity, continues to replace labor, and as a proportion of the absolute or relative cost of a product, labor value declines. There is an economic deficit of labor when compared with the machinery replacement of labor, and this substitution increases the profit that an article is able to receive. Comparative break-even cost analysis between the simple tools and labor of the home-craft era and the labor and equipment of the industrial age gave the advantage to the engineering of industrial plants.

The roots of our country's trade unions extend deep into the early history of America. Primitive unions which were then called guilds, made their appearance, often temporary, in various cities along the Atlantic seaboard. In the early years of the nineteenth century, efforts by unions to improve the workers' conditions—through either negotiation or strike action—became more frequent. As ineffective as these first efforts to organize may have been, they reflected the need of working people for economic and legal protection from exploiting employers. The application of the steam engine to operate machinery was developing a factory system not much different from that in England, which produced misery and slums. Starting in the 1830s and accelerating rapidly during the Civil War, the factory system accounted for an ever-growing share of American production. It also produced great wealth for a few, but grinding poverty for many.

There were many leaders of the union movement. Individuals such as Samuel Gompers founded The American Federation of Labor in 1886. "The various trades have been affected by the introduction of machinery, the subdivision of labor, the use of women's and children's labor, and the lack of an apprentice system—so that the skilled trades were rapidly sinking to the level of pauper labor," the AFL declared. "To protect the skilled labor of America from being reduced to beggary and to sustain the standard of American workmanship and skill, the trades unions of America have been established."

In 1935 John L. Lewis led the creation of the Congress of Industrial Organizations (CIO) to carry on the effort for industrial unionism. Industrial unions organized an entire industry regardless of skill, in short, forming unions of unskilled workers. The CIO conducted organizing campaigns, and over the next several years brought industrial unionism to large sectors of basic American industry. They were united into the AFL-CIO in 1955. The merger brought about the virtual elimination of jurisdictional disputes between unions, which had plagued the labor movement and alienated public sympathy in earlier years.

The early years of the industrial era saw plants where workdays of 10 to 12 hours were not unknown. Remember that electrification and Edison's incandescent lighting systems did not become known until the 1890s or so. Plants before then were naturally lighted through sawtooth construction of roofs having windows to allow as much light as possible into the working area. Oil or gas lamps were not efficient enough to allow second or third shifts, and the industrial barons of that time encouraged utilization of

[2]Marx reports on the production rate of a hot forging press of small hammers, as 700 strokes per minute, a reminder that even today, this output is extraordinarily high. This author wonders if the observation is correct, as press strokes per minute are different than units per minute, which is more plausible with a multiple cavity forging die.

PICTURE LESSON Child Labor

Lewis Hine, Library of Congress

Early in the Industrial Revolution, child labor was endemic, and America's army of child laborers had been growing steadily. In the first years of the twentieth century, more than two million American boys and girls under the age of 16 years of age were a regular part of the workforce. Many would work 12 hours per day for pitiful wages and in unhealthy and hazardous mills, factories, mines, cotton fields, and canneries. The nation's economy was expanding. Poor immigrant families had a need for income, and children were chattel for survival. Many families were trapped into subsistence living and children provided income, however meager it was.

Thousands of young boys descended into dangerous coal mines, as depicted in the photograph of a sooty-faced boy in West Virginia after his shift below ground. A typical job for "breaker boys" in coal mines would be to separate slate from coal as the coal would pass underneath (giving them bleeding fingers), then cast the nonburning material aside. In Pennsylvania, thousands of 14- and 15-year-old boys were employed legally, at the same time that thousands of younger boys, some of them nine and ten, worked illegally. The state's child-labor law was almost useless, since it required no binding proof of age.

Small girls tended noisy machines in spinning mills, and after working long hours, they were kept awake by having cold water thrown in their faces. In canneries, mothers and children worked side by side shucking oysters, among other tasks. In Colorado, the stoop labor of picking sugar beets involved itinerant families with five- and six-year-old children working from sunrise to sunset during harvest season.

In the early 1900s, many Americans were calling child labor "child slavery" and were demanding an end to it. They argued that long hours of work deprived children of an education and robbed them of their chance for a better future. Child labor only promised a future of illiteracy, poverty, and misery. The concern was not with children who worked at odd jobs after school or did chores around the house or family farm, or youngsters working as trainees and apprentices, learning skills they would use for the rest of their lives.

Lewis Hine, a New York City schoolteacher and photographer, was an early reformer of this period. He knew that a picture could tell a powerful story. As an investigative photographer for the National Child Labor Committee, he scoured the country aiming his camera at underage workers. Not always welcomed into the working scene, Hine used clandestine means for taking his pictures. His equipment was a simple box-type 5 × 7 inch view camera, bulb shutter, glass plate negatives, and magnesium flash powder for illumination. With that technology, the pictures generated the sympathy that caused the country to act and to create the child labor laws that we have today.

the plant by long shifts. Over several decades of union negotiation and the continued electrification and modernization of industrial plants, the workweek typically matured to a five-day eight-hour day period, which introduced the three-shift operation.

In recent decades there has been a steady decline in union membership and influence. One reason is that unions raised workers' wages. Consequentially, many union-made products became so expensive that sales were lost to foreign competitors, and continued improvement in technology made our economy less reliant in the types of industrial jobs that tended to be union strongholds.

Artificial or not, the pressure from labor unions with the prospect of the interruption of business and the increase in labor costs, encouraged removal or reduction of labor costs. Engineering's perceived response was improvement in productivity and continued reduction of labor content from product cost.

Industrialization gave impetus to the necessity of specially educated employees for these businesses. Congress created the Morrell Land Grant Act and agriculture and mechanic arts colleges were directed to improve these arts for the specific good of the country.[3] Indeed, professional programs developed, separating themselves from liberal arts. Universities and colleges developed engineering, business, and economics programs. These specializations went their own direction and gave identity and definitions to the labor content of work. The professions of accounting, economics, and engineering collaborated only to the degree where there was common interest. Accounting defined the cost of the work of production to meet the requirements of the business documents of the profit and loss statement and the balance sheet. Economics carved

[3]In 1862 engineering education was jump-started by the Morrell Land Grant Act, and signed by President Abraham Lincoln, accelerating the growth of the Industrial Revolution, and from that beginning more than 2400 accredited engineering, technology and applied science programs have emerged.

out its own niche and concerned itself about political economy of national and local governments and their interaction with business. Engineering was involved in the productivity of the industrial plant and its operations, and the economy of equipment and labor. From the beginning of these professional distinctions, the understanding and interpretation of labor diverged.

For example, payment of wages may be based on attendance or performance. In the early years of the industrial era, it was soon realized that the workers had control of the output of machinery, and if their performance could be increased, the cost of labor and machinery would decline. Thus, the concept of incentives or piecework was born, and the American culture accommodated to this "science" of production rationalization.[4]

Engineering activities defined the production processes and the terms direct labor and indirect labor were coordinated with accounting activities. Of particular interest to the engineer is *direct labor*, or labor that is directly involved with the changes to the material during manufacturing. Examples of direct labor are assemblers, machine operators, and painters. *Indirect labor* in manufacturing is shop labor that supports direct labor, such as stockroom clerks, first level supervisors, forklift operators, material handlers and quality inspectors. Indirect labor and its costs are usually covered as a factor of direct labor or through overhead. Stated differently, direct labor "touches" the design, while indirect labor supports that effort.

Labor content is an important component of most manufacturing operations. The percentage of cost attributed to labor in an operation has decreased as computer-based automation has increased, but on the average, direct labor still contributes approximately 15% to the cost of producing a product. The percentage of labor cost certainly will vary as to the type of product and the operations involved in producing the product.

One only needs to review the impact of labor costs in the United States. Following World War II, manufacturing was mostly conducted in the United States for its own domestic needs. But labor-intensive operations were sent to Japan because of its lower cost, and then to Korea, Taiwan, Brazil, and Mexico, (although in recent years Mexico has become more expensive than China or Malaysia).[5] The reality of finding the lower cost of labor means that there are no national boundaries, and productivity of labor is frequently a deciding factor as to domestic or international manufacturing. U.S. companies are accustomed to outsourcing product manufacturing, and sometimes development and engineering, to Asia and other low-cost countries.

The two major costs of a product are material (discussed in Chapter 3) and time utilized in the facilities, on the equipment, and by the workers. Workers are paid a wage for the hours that they work (dollars per hour) and equipment has an operating cost (capital cost, maintenance, consumables, etc.) which is usually expressed in dollars per hour. It is important then to have some knowledge of how much time is required to

[4]Congress noticed that in 1912 a paradigm shift had occurred in engineering and management and labeled the principles of this teaching and philosophy as "scientific management."

[5]In other parts of the industrialized world, cost analysis and estimating are not practiced as extensively as it is in North America. A typical offshore approach is to find the North American price or cost, and then expect that their "on-shore" delivered cost will be less. Of course, analysis is done, but negotiation between the parties becomes more important.

manufacture a product to determine its cost. It is also necessary to have an idea about the amount of time an operation should take in order to monitor and control the productivity of the work center.

For new products and changes to existing products, it is desirable to have an estimate of the expected cost and time for manufacturing. This topic will be covered in this and later chapters. For products currently in manufacture, there are two ways to determine the manufacturing time: (1) record the time charged to a specific product derived from accounting or shop records and (2) measure the time actually required performing the manufacturing operations. Accounting records are generally not specific enough to identify labor time for individual products, operations, and/or manufacturing areas. These records are available only after the production is over and are thus not helpful. Nor do these records separate out nonproductive time. The values obtained represent the overall span time (clock hours) during which the product was being (or supposed to be) acted upon. If an operator charged one day (8 hours) to a job, then that is what is reported to accounting. However, the fact that the same operator spent 2 hours that day waiting on machine maintenance and another hour waiting for material is not apparent. In this latter case the actual time spent working on the job is only 5 hours.

For products not currently in production or for new products, other procedures become necessary to determine the labor contribution, and there are several chapters that deal with this important need.

Labor and its cost, as the above discussion argues, can be considered from several views. Culturally speaking, the view given in this book is a North American perspective. The impact of political-historic, wage motivation schemes, distinctiveness of labor, and professional segregation of engineering, accounting, and economics fields, guide a confluence of thoughts and procedures. Table 2.1 identifies definitions for this labor culture.

To determine time-costs for any design, it is necessary to obtain accurate measurements. This can be achieved using actual time records, but those would need to be fairly detailed and reported honestly. The historical and accepted method for determining job times is to perform some type of work measurement. Two work measurement methods are discussed in Sections 2.3.2 and 2.3.3: time study and work sampling.

Many times the engineer is only concerned with operation times. The cost is computed from the times provided to engineering by some other company departments such as shop or accounting. However, if the wage rates are known to engineering, the computation of cost for a given time is straightforward and simple:

$$\text{Labor cost} = \text{time} \times \text{wage} \tag{2.1}$$

Units for the wage and time estimate must be dimensionally compatible. If the time is in hours per unit, and the wage is expressed in dollars per hour, then the labor cost will be dollars per unit. The wage may be the amount that the worker sees in the paycheck (gross pay, before the worker's deductions) or it may include all or part of the company costs for fringe benefits (insurance, retirement, vacations, etc.)

Engineering desires dimensions for labor, as much as that is possible, and we now turn to study schemes for finding time measurements.

TABLE 2.1 Definitions of Labor for Engineering Cost and Time Analysis

Term	Definition
Actual time	Time reported for work, which may include delays, idle time, and inefficiency, as well as efficient effort, also called clock time.
Allowance	An adjustment to the normal time providing for personal needs, fatigue, and delays inherent in the work.
Allowed time	Normal time increased by appropriate allowances (see standard time); also, reported man-hours adjusted because of work differences or judgment.
Avoidable delay	Interruption in the work that caused additional time that could have been avoided or minimized by the worker or equipment by using better skill or judgment.
Constant element	An element whose normal time is unaffected by various independent effects on the element.
Continuous timing	A method of time study where the total elapsed time from the start of the study is recorded at the end of each element.
Cycle	Total time of elements from start to finish in a repetitive operation.
Delay allowance	One part of the allowance included in standard time for interruptions or delays beyond the operator's ability for prevention.
Element	Subpart of an operation separated for timing and analysis; beginning and ending points are described, and the element is the smallest part of an operation observed by time study.
Elemental breakdown	Description of the elements of an operation in a measurable sequence.
Fatigue allowance	One part of the allowance caused by physiological reduction in ability to do work, sometimes included in the standard time.
Foreign element	Unrelated to the operation and removed from the time study and standard time.
Frequency of occurrence	Number of times an element occurs per operation or cycle (can be fractional).
Idle time	Time interval in which the operator, equipment, or both are not performing useful work.
Incentive	Financial methods motivating a worker to exceed a standard.
Labor hour	Unit of measure representing 1 person working for 1 hour. Formerly called man-hour; see also *person-hour*.
Machine interference	Idle machine time occurring as a result of operator attention on another machine in multiple-machine work.
Machine or process time	Time required by a machine or process.
Normal time	An element or operation time found by multiplying the average observed time by a rating factor.
Observed time	Time observed on the electronic or stopwatch clock and recorded on the time study sheet during the measurement process.
Operation	Designated and described work subject to work measurement, estimating, and reporting.
Personal allowance	One part of the allowance included in the standard time for personal needs that occur throughout the working day.
Person-hour	Unit of measure representing 1 person working for 1 hour. More commonly called *labor hour*.
Rating factor	Involves comparing the performance of the operator under observation by using experience or other bench marks; additionally, a numerical factor is noted for the elements or cycle; 100% is normal, and rating factors less than or greater than normal indicate slower or faster performance.
Regular element	Elements that occur once in every cycle.
Snap back	Method of timing that records the elemental time at the end point of the element.
Snap observation	An observation made virtually instantaneous as to the state of the operation (i.e., idle, working, or the nature of the element).
Standard time	Sums of rated elements that have been increased for allowances.
Variable element	An element whose time depends on one or more dependent effects.

2.2 THE MYTHICAL MAN-HOUR

The ever present and the most popular metric that gives an estimate of work time has been called the "man hour," or *labor hour*. This basic unit is defined as one worker working for one hour. Examples of labor hour units may be given in many ways:

Number of welder labor hours per inch of welding
Number of assembler labor hours to complete a subassembly
Number of machinist labor hours to fabricate a complex shaft

Literally, thousands of labor hour units are found throughout engineering.

Though the number of working days vary in a month or year owing to holidays, vacation, sick leave, leap year, and so on, for cost analysis the *labor year* is defined as working 52 weeks with the popular 40 hours in each week, or 2,080 hours. A *labor month* is 173.3 hours per month, $((40 \times 52)/12 = 173.3)$. Labor seconds or labor weeks are uncommon units and are discouraged in cost analysis application.

Many estimates end up missing the mark because of several misconceptions. We confuse effort or hours at work with progress, leading to the assumption that labor hours and clock hours are interchangeable. While cost does vary as the product of the number of men or women and labor hours, progress does not. For example, if 20 labor hours are proposed for a task, this could mean one worker will consume 20 hours, or a crew of two will consume 10 hours. This divisibility is an accepted but an unproved hypothesis, and sometimes will not work in practice. Furthermore, there is a linkage between consecutive tasks in manufacturing. There is a philosophy of estimating that pretends that the sum of the individual labor hour estimates will equal a total for the job without any slippage of the labor hour constants or without any idle or wasted time. As an example, many times machine setup changes between jobs have labor times that vary depending on the preceding job. Thus the efficiency of the machine setups is dependent on the scheduling sequence of jobs on the machine. In addition, labor hours for the many tasks are assumed to be independent, which is fallacious reasoning as there are interface dependencies. A worker on one flow line assembly task may be required to wait for the completion of preceding work, thus consuming labor hours without producing any work.

With this background now covered, we return to the simple Equation (2.1), where the terms "time" and "wage" are introduced in a larger context. The student will want to study these amplifications because of their importance, even though he or she may not be involved directly in these activities.

2.3 TIME

Through most of the twentieth century, engineering management philosophy was that worker productivity could be improved by financial work incentives. The *incentives* were based on performance relative to a standard amount of work. Simply put, workers were assigned an amount of work for a given period (i.e., a work day or job lot) and if they exceeded the standard amount, they received additional wages. While this system was attractive to workers in that they could earn additional earnings, there remained issues over a fair and reasonable standard. Obviously, management wanted the standards set to maximize the work obtained, but fairly, and the workers wanted standards that would yield higher pay for lesser amounts of work, but fairly. This situation led to the refinement of sophisticated methods to determining accurate and fair standard rates for the work involved.

Toward the end of the last century, management philosophy had mostly turned away from a purely incentive-based concept. It was realized that workers were not solely motivated by money (of course, a base level of satisfaction was necessary) and that other forms of encouragement were needed to improve performance. Despite the form of management system, it is necessary to know how much time a job should take, or standard time, to monitor and control performance.[6] In addition, the times determined for performing jobs are used to estimate future work costs and quantify the effects of product and process improvements. As management philosophy has evolved, it is still necessary to measure work time in order to establish work standards.

2.3.1 Ergonomics

Ergonomics is the study of the interrelationship of humans, in both physical and psychological aspects, and the equipment and tools needed to perform some operation. The practice of ergonomics is divided among the psychological, medical, and engineering sciences. One aspect of ergonomics is closely involved with safety and the guidelines issued by Occupational Safety and Health Administration (OSHA). In a strictly economic sense, the cost of providing a safe work environment can be assumed to be less than the cost of workplace injuries. Ergonomics recently has taken on additional importance and come under media and legal observation in the area of repetitive motion injuries. These physical problems occur in many areas of work—from the office to the shipping dock—and are a result of constantly repeating the same set of motions over extended portions of the workday and for a long period of time. Rules and legislation are still evolving, but engineers involved in workplace operations should always keep ergonomics in mind when designing manufacturing processes and setting up work motions.[7]

Measured time, by its nature, involves jobs that are currently being performed. There are two commonly used methods for measuring work: time study and work sampling.

2.3.2 Fundamentals of Time Study

For our purpose here, it is not necessary to delve into the historical background of time study. Suffice it to say that around the beginning of the twentieth century, Frederick W. Taylor was the founder of time and motion study and that Frank and Lillian Gilbreth were two leading pioneers. Because of competition in any industry, it is necessary to know cost factors and how much a proposed engineering change decreases or increases cost. The finding of these cost comparisons is possible by time study, which is the backbone of some cost systems.

Time study is actually composed of two parts: (1) the analysis of an operation to optimize it by eliminating unnecessary elements and determining better and faster

[6]An oversimplified philosophy in improving productivity is to identify TITMOP (time is the measure of productivity) as a rule of management. Using this attitude, it is easy to see how a "time" metric can be adopted.
[7]Additional current information can be found on the Internet at www.osha-slc.gov/ergonomics/standards.html and at www.ergoweb.com.

methods of performing the operation in order to standardize methods, equipment, and conditions and (2) then, and only then, the determination by measurement of the number of standard hours required for an average worker or for the equipment (as it can be automatic without the assistance of an operator) to perform the optimized job. Time study has the following advantages that are useful in industry:

- System for estimating cost
- Justification for methods improvement
- Reduction of operation costs
- Improvement of engineering design
- Information for managing productivity

The basic time study process measures the time of industrial operations. It needs to be noted that our discussion does not develop skills for the individual but is given for appreciation purposes. The student will see that this emphasis is on the analysis of time measurements once those measurements are concluded.

At its heyday the stopwatch procedure was arguably involved in some 75% of industrial operations but now is suspected to be in 10% of operations, and mostly labor-intensive work. But its role in gathering time for production remains significant, though in a more informal setting. Many businesses in the United States depend on time study for information used in cost analysis for monitoring productivity and estimating costs.

Over the years, refinements have established a procedure for "taking" a time study, and it goes like this:

1. Conduct analysis of methods and improve, if possible.
2. Record equipment, layout of the process, and operator information.
3. Separate the operation into elements.
4. Using a timer, record the time consumed by each element as it occurs for each repetition.
5. Rate the pace or tempo at which various elements of work are performed.
6. Convert rated elements into normal time.
7. Determine the allowances.
8. Calculate the standard time using the allowances.
9. Express the standard in common units of production.

The equipment for taking a time study is simple: a clipboard and an electronic timer or decimal minute or hour stopwatch. Video camcorders can be useful, giving the pictures and the time necessary for the work, and are prominent in this analysis.

The first and most important phase of taking a time study is the preparation. Is the job ready for timing? The time study technician resolves this question by answering the following questions:

1. Is the proper tooling being used, and is it laid out correctly?
2. Is the material laid out in an efficient manner?

3. Are proper machines or tools being used?
4. Is the quality of the finished part or operation up to the inspection standards?
5. Is the motion pattern employed the most economical that can be devised now?

The second phase of the time study is to record the information on the time study form (see the following example). The recording of significant information starts with a plan view sketch of the process layout. Only the barest of detail is shown for the process, but information such as distances must be included as shown in Figure 2.1.

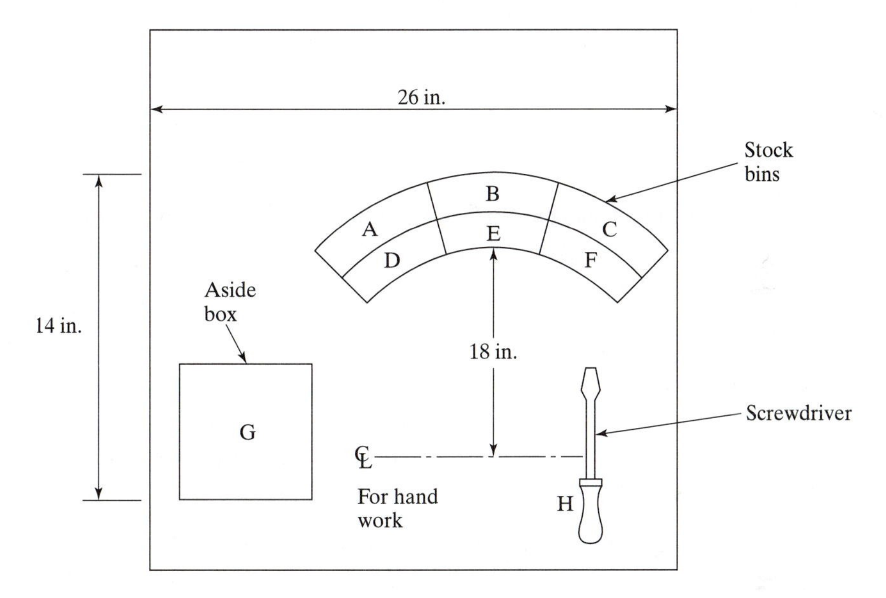

Figure 2.1 Layout of an assembly bench.

Certainly, drawings are available if additional information becomes necessary. After the title block information is recorded, the elements of work must be identified and written down in sequence. When breaking down the operation, the technician keeps in mind that the elements should be as short as possible but long enough for timing accuracy. The elements must also have definite and easily identifiable beginning and end points. Wherever possible, manual operator time is separated from machine time.

Elements that are constant (or nearly so) should be separated from variable elements. The time for *variable elements* depends on some other attribute of the operation, such as a length, the number of parts involved, or some other identifiable factor. Times for these elements will then need to be variable based on the factor. The time for the operation will then include a constant plus the variable factor time. An example for welding would be $5.23 + 0.45\,x$ where x = number of inches of weld.

Example:

Consider an example of an automatic placement process of integrated circuits onto a printed circuit board. Operator attendance is necessary for surveillance of machine operational problems, but during normal operations the times required for the elements are controlled by the machine cycle. Normally, costs for the machine time outweigh the operator hourly cost. But the observation and control of the process is necessary for quality assurance of the printed circuit board.

The machine is composed of robot equipment that loads and inverts bare printed circuit boards, dispense epoxy on the board at all locations where components will be placed, position and insert components, component testing, and placement inspection systems, an ultraviolet curing module, and a stacker for the completed boards. Computers control the machine operation and the testing and inspection. The UV module surface cures the epoxy, which minimizes the movement of the components until soldering.

At this point the time study technician watches the details of the operation and separates the operation into elements. Some processes have manual-controlled elements, but this example is automatic. Trial and error of the selection and definition of the elements is expected because short-time elements that have clearly defined start and stop points are important. In Figure 2.2, for example, element A is "Inload PCB to machine." In a similar way the other 12 elements, B through M, are described. There may be some experimental timing to get a feeling for the length of the time for the elements. Coding can be given to foreign elements if they occur unintentionally during a cycle. Foreign elements (those items normally not part of the operation) are removed from time studies later. In this example there are no foreign elements.

Example:

Stopwatch timing can be done by one of two methods, continuous or snap back. In a *continuous* study the watch runs without stopping, with the first element of the first repetition beginning at zero. The end time for each element is then recorded as the watch runs. The actual time for each element must be calculated by subtracting the beginning time (end time of the previous element) from the end time. In the *snap-back method,* the end time for each element is observed, the watch is reset to zero, and the observed time is recorded. This is the case for the example described here.

Time studies for repetitive operations may be recorded in seconds, minutes, or hours. The time study of Figure 2.2 is in decimal minutes. The number of readings necessary for a good average is a matter of judgment and depends on the consistency of the cycle time.

During the time study for manual operations, the observer judges the effort of the operator in performing the elements and operation. This is called *rating*. The observer watches the performance or speed as motions, elements, and operations are made and compares this mentally to a speed that an operator should be expected to work under normal circumstances and for extended periods. The operator's effort, which is reflected in the rating factor, includes intangibles such as health, interest in work, skill, and speed. Operators may intentionally, or unintentionally, react to the fact that they are being observed and depending on their interpretation of the eventual effects of the study, try to influence the outcome by speeding up or slowing down their work.

If the operator's effort is considered in excess of normal (100%), then the rating factor is greater than 1. On the other hand, if the effort is considered less than 100%, then the rating factor is less than 1. This rating may be recorded during or at

Elements	Inload PCB to machine	Invert & Position PCB	Move to lower level	Getting below epoxy station	Epoxying the PCB	Getting below comp. placement	Placing components on PCB	Reaching machine vision	Sensing Defects	Waiting in queue	Going through oven	Reaching loader	Inserting PCB and change slot	
	A	B	C	D	E	F	G	H	I	J	K	L	M	
Line														Cycle time
1	0.025	0.05	0.083	0.1	0.87	0.1	0.62	0.1	0.72	0.87	0.083	0.05	0.034	3.705
2	0.025	0.05	0.083	0.1	0.62	0.1	0.7	0.1	0.87	0.75	0.083	0.05	0.034	3.565
3	0.025	0.05	0.083	0.1	0.7	0.1	0.72	0.1	0.75	0.67	0.083	0.05	0.034	3.465
4	0.025	0.05	0.083	0.1	0.72	0.1	0.87	0.1	0.67	0.7	0.083	0.05	0.034	3.585
5	0.025	0.05	0.083	0.1	0.87	0.1	0.75	0.1	0.7	0.72	0.083	0.05	0.034	3.665
													SUM =	17.985

Summary

Total time	0.125	0.025	0.415	0.5	3.76	0.5	3.65	0.5	3.7	3.7	0.415	0.075	0.17
Number of reading	5	5	5	5	5	5	5	5	5	5	5	5	5
Average reading	0.025	0.05	0.083	0.1	0.75	0.1	0.73	0.1	0.74	0.74	0.083	0.05	0.034
Frequency	One out of One												
Average Time	0.025	0.05	0.083	0.1	0.75	0.1	0.73	0.1	0.74	0.74	0.083	0.05	0.034

Average cycle time $= \frac{17.985}{5}$

$= 3.597$ min

Cycle rating factor 1.00

Normal cycle time = 3.597

Allowances:

Personal : 5%
Fatigue : 5
Delay : 5
Total : 15%

Standard time per unit : 4.232

Figure 2.2 Time study worksheet to manually place electronic components on printed circuit board.

the conclusion of the study. Though the rating process may be criticized as arbitrary, observers can be trained to be quite accurate at evaluating performance. Despite the criticism, it is evident that rating is necessary for effective cost analysis. For the preceding example, the operator did not influence the speed of the automatic equipment, and thus the rating factor is one. The shop portion of the time study is concluded when the rating factor is posted, and the observer then returns to his or her desk to complete the analysis.

The next step is to analyze the time study. After concluding the shop portion of the time study, and back in the office, subtractions are made for each element, if the time study was continuously timed. A spreadsheet program assists with the office analysis of the time study. Since snap-back timing was used in the example, the subtractions are not required.

The columns and rows for each element are totaled and divided by the number of observations to obtain the average of the readings. In this time study, the elements are regular because they occur once for each cycle. It is possible that an element may occur several times or fractionally for each cycle or be irregular. This fraction of the noncyclic elements is entered in the frequency row. Once the element is multiplied by the performance-rating factor, we get *normal time*, which is the time for an average qualified person working at a normal pace. With the performance rating for the previous example of 1, the normal time is 3.597 minutes per unit.

After determining the normal cycle time, job allowances must be accounted for. During a time study, all foreign and irregular elements or interruptions for personal and unavoidable delays are purged. Thus a time study is a picture of regular work elements only. But there are other legitimate needs for an operator to sustain work throughout the shift. Time for *personal and functional body needs* are added, (about 4 to 5% is the U.S. customary value).

Fatigue—physiological body wear reducing the ability to do work—is another factor included within allowances. It is not "tiredness," because that is expected and can be accommodated with the rating factor. Fatigue is not subject to precise measurement. For instance, hot, heavy, dirty work such as forging of steel billets has a higher allowance than light assembly work in a controlled, air-conditioned factory. Some practitioners believe that fatigue does not exist in most factory environments. Fatigue allowance percentages are determined by practice or union negotiation or common sense. Fatigue allowances may vary from 0 to 25%, depending on average weight handled, percentage of time under load, repetitive or nonrepetitive work, sitting or standing, and cycle time.

Time to accommodate operator interruptions beyond their control is required for tool breakage, out of parts, supervisor instructions, and so on, and is considered. The usual range for *delays* is 2 to 8%.

The total of these allowances—personal, fatigue, and delay—are sometimes referred to as PF&D allowances. A typical PF&D allowance is 15%, but it can vary from 8 to 35%. Typically the allocation of the allowance in cost analysis procedure is as a percentage of a 480-minute (8 hour) shift. Because productive time in the workday is inversely proportional to the amount of PF&D allowance, the allowance should be expressed as a percentage of the total workday. To obtain a precise PF&D allowance factor, we divide the total workday (100%) by the productive day expressed as a percentage of the workday,

$$F_a = \frac{100\%}{100\% - \text{PF\&D}} \tag{2.2}$$

where F_a = allowance multiplier for PF&D

PF&D = personal, fatigue, and delay allowance, percentage

For example, assume that 72 minutes of the 480-minute workday are spent in PF&D activities, which means that allowances amount to 15%. Converting this allowance to a percentage of the 408-minute productive day results in a multiplier of

$$F_a = \frac{100\%}{100\% - 15\%} = 1.176$$

Apply the allowance by multiplying the normal time by the allowance multiplier. If the rated productive time is 408 minutes, then a job standard of 408 minutes × 1.176 = 480 minutes (a full day's work). The job standard is computed as

$$T_s = T_n \times F_a \tag{2.3}$$

where T_s = standard time for an operation per unit

T_n = normal cycle time per unit

An alternative common expression is

$$T_s = T_n(1 + \text{PF\&D}) \tag{2.4}$$

where the PF&D allowance is expressed as a decimal. It is not as accurate in definition but is often used for simplicity. Observe that for the electronic printed circuit example and Equation (2.3) we have 3.597 × 1.176 = 4.232 standard minute per unit. If the simplified alternative equation (2.4) were used, then the standard would be 3.597 × 1.15 = 4.137.

Equation (2.3) did not state whether the time was minutes, hours, or person-days. That depends on the dimension of the time measurement (i.e., week, day, hour, minute, or second). Though the basic approach is unaffected by the nature of the time units, it is customary in production work to use minutes or hours. If minutes are the dimension of the measurement, then minutes are converted to pieces or units of production per time period by a reciprocal relationship. If pieces per hour are desired, then

$$\text{Pieces per hour} = \frac{60}{T_s} \tag{2.5}$$

where T_s = standard minutes per unit

Various firms prefer expressions such as pieces per hour, units per minute, or packages per week. In some businesses it is common to express the rate per dozen or gross. However the expression may be written, the dimension units per time is not as preferred as hour per unit or units for cost analysis work. The conversion is handled by

$$H_s = T_s \times \frac{N}{60} \tag{2.6}$$

where H_s = hours per quantity of units

$$T_s = \text{standard minutes per unit}$$
$$N = \text{specified quantity of units}$$

The value N may be 1, 10, 100, 1000, etc., depending on plant practice.

Example:

Consider the time study example of the assembly of a small seven-component electrical product. The product consists of one back plate, one mounting ear bracket, two contacts, one faceplate, and two screws.

The present method, where assembly fixtures are not used, is examined for possible improvements. The first step to record significant information starts with a plan view sketch of the bench layout. Observe in Figure 2.1 that only the barest of detail is shown because the precise location of the screwdriver or stack bins is not critical.

At this point the observer watches the details of the operation and separates the operation into elements. The time data record sheet that has been prepared is shown in Figure 2.3, where the eight elements are abbreviated. Each line number is for a particular cycle.

Time studies for repetitive operations may be recorded in seconds, minutes, or hours. The continuous time method time study is shown in Figure 2.3 in decimal minutes. Once time recording is under way, the observer time-marks the end of each element for each line under the column R (reading).

During the time study, the observer judges the effort of the operator in performing the elements and operation. A single rating factor for the study is entered for each element in the row for the element rating factor. The rating factor for the elements and the operation are not necessarily equally weighted. This rating may occur at the conclusion of the study.

The next step analyzes the time study. Subtractions are made for each element and posted under column T (time). For instance, on line 1 the time for element 4 (R = 0.21) is found by subtracting the reading for element 3 (R = 0.16) to give an element time of 0.05, which is written under column T4. This subtraction is made for all T columns and cycles. If snap-back timing was used, then the subtraction is unnecessary.

Next we total the T column for each element and divide by the number of observations to obtain the average of the readings. In this time study, the elements are regular because they occur once for each cycle. The average element times are multiplied by the element rating factors, to obtain normal time. This entry for each column T is made in the bottom row, "Element Normal Time," and the row sum equals 0.489.

Observe the sum of 7.36 arising from the total under 8R. The average cycle time gives 7.36/15 = 0.491, which is marked on the form. When multiplied by the 1.05 cycle rating factor, the normal cycle time is 0.515 (=0.491 × 1.05), which compares to the 0.489. That these two values are not identical is not surprising because the rating factor is a trained but arbitrary value.

Finally, the normal time is multiplied by the PF&D factor of 1.176 for a 15% allowance. The standard time per unit is thus 0.606 minutes.

So far, we have discussed time study to find measures of time. There are other ways to find "time" for the fundamental Equation (2.1). We next study work sampling, a method, that makes sampling observations of large workforce populations over a period of days, weeks, or months.

Time study observation

Date *Sept. 23.*
Time start *8 : 30 AM*
Time stop *9 : 45 AM*
Elapsed time *1 : 15*

Sheet no. *1*
No. sheets *1*

Elements	Pick up back plate		Mount frame to back plate		Contacts in back plate		Contacts in back plate		Face plate over back plate		Place screw in back plate		Place screw in back plate		Tighten screw and aside			
Number	1		2		3		4		5		6		7		8			
Notes / Line	T	R	T	R	T	R	T	R	T	R	T	R	T	R	T	R	T	R
1	.03	.03	.05	.08	.08	.16	.05	.21	.05	.26	.09	.35	.04	.39	.21	.60		
2	.03	.03	.03	.06	.05	.11	.03	.14	.06	.20	.04	.24	.05	.29	.17	.46		
3	.02	.02	.04	.06	.04	.10	.04	.14	.08	.22	.04	.26	.05	.31	.16	.47		
4	.02	.02	.04	.06	.04	.10	.04	.14	.06	.20	.04	.24	.05	.29	.16	.45		
5	.02	.02	.05	.07	.05	.12	.03	.15	.05	.20	.04	.34	.05	.29	.15	.44		
6	.02	.02	.03	.05	.05	.10	.04	.14	.04	.18	.06	.24	.04	.28	.15	.43		
7	.02	.02	.03	.05	.05	.10	.05	.15	.06	.21	.04	.25	.06	.31	.16	.47		
8	.04	.04	.03	.07	.05	.12	.05	.17	.04	.21	.06	.27	.08	.35	.14	.49		
9	.02	.02	.04	.06	.05	.11	.04	.15	.05	.20	.04	.24	.08	.32	.15	.47		
10	.04	.04	.04	.08	.04	.12	.04	.16	.05	.21	.06	.27	.05	.32	.16	.48		
11	.02	.02	.05	.07	.07	.14	.07	.21	.05	.26	.04	.30	.09	.39	.14	.53		
12	.03	.03	.04	.07	.08	.15	.07	.22	.05	.27	.04	.31	.04	.35	.21	.56		
13	.02	.02	.03	.05	.06	.11	.03	.14	.08	.22	.05	.27	.05	.32	.15	.47		
14	.03	.03	.04	.07	.02	.09	.06	.15	.12	.27	.06	.33	.05	.38	.18	.56		
15	.03	.03	.02	.05	.06	.11	.03	.14	.08	.22	.05	.27	.05	.32	.16	.48		
16													Sum = 7.36					

Summary

	1	2	3	4	5	6	7	8	
Total time	.39	.52	.79	.67	.92	.75	.83	2.45	
No. of readings	15	15	15	15	15	15	15	15	
Average of readings	.026	.035	.053	.045	.061	.050	.055	.163	
Frequency	One	out	of	one					
Average time	.026	.035	.053	.045	.061	.050	.055	.163	
Element rating factor	1.05	1.00	1.00	.95	.90	1.00	1.00	1.05	
Element normal time	.027	.035	.053	.043	.055	.050	.055	.171	

Sum = 0.489

Foreign elements

SYM	R	T	Description
A			
B			
C			
D			
E			
F			
G			
H			
I			

Av. cycle time $\frac{7.36}{15}$ = .491
Cycle rating factor 1.05
Normal cycle time 0.515

Percent allowances

Pers.	Fat.	Delay	Total
5%	5%	5%	15%

Std. time per unit 0.606
Pieces per hour 99
Std. hours per 100 1.010

Allowances in minutes

Pers.	Fat.	Delay	Total
.008	.008	.008	.024

Avail. prod. min. per hr. ____
Pieces per hour ____
Std. hrs per 100 ____

Figure 2.3 Time study worksheet for manual assembly.

2.3.3 Fundamentals of Work Sampling

Work sampling is a counting and statistical technique for gathering information about large workforce populations. Its objectives are to measure labor and efficiency and equipment and utilization, and their costs, and to help identify and make adjustments for improved cost performance. In comparison to time study, work sampling can do the following:

- Be applied to situations where multiple workers and/or machines are involved
- Study operations spread over a larger area
- Study work with long cycle times
- Require less time devoted to the study and cost less
- Be less disruptive to the workers being observed
- Avoid bias caused by worker awareness of being observed

However, work sampling has at least two disadvantages:

- Inefficient for studying single worker or machine situations
- Not feasible for breakdown and study of brief operations

When used for the right application and when the results are used for general purposes, such as estimating costs for future jobs, scheduling, determining labor requirements, and monitoring and managing performance and productivity, work sampling is a helpful tool.

Work sampling is a counting method for quantitative analysis of the proportions of a span of time that are spent on preidentified activities. The principles of probability and statistics are used. In a statistical study, a random sample is taken that is deemed representative of the distribution for a future population.

In time study with full-time observations of elements, the entire span of work is observed. But in work sampling, only discrete observations are made—for example, using sample photographs of the work. Can "still" photographs of a work task taken at many instants in time be indicative of the total work? The answer is yes, if the number of the photographs is sufficient, if they are taken at random times, and if they are descriptive of the work, and the voids in observations are insignificant relative to the durations of the work activities.

A work-sampling study consists of a number of field observations that pertain to the specific activities of the person(s) and equipment at *random* intervals. Those observations are classified into predefined categories directly related to the work situation. The technician instantaneously makes tally marks, such as "working," "idle," or "absent," by walking around the job site and noting the state of the workers and/or equipment during the course of the work-sampling study.

A variety of technical methods are available to provide the observations. For instance, the sampling observations can be made by a remote video camera. Later, the videotape is analyzed as if a technician were making the direct observations by walking through the production area.

The key to accuracy is the number of observations, for which the requirements may vary. One survey may require very broad areas to be investigated, in which case relatively few observations are required for meaningful results. On the other hand, many thousands of observations may be needed to establish engineering standards. Four thousand observations will provide more reliable results than 400. To determine the number of observations necessary, the engineer specifies the desired accuracy for the results. The method for determining the number of observations will be discussed shortly.

Because work sampling is a statistical technique, the laws of probability must be followed to obtain an accurate estimate. In this approach, an event such as "equipment working" or "idle" is instantly tallied. For this "snap" or instantaneous observation, mathematicians define a binomial expression, where the mean of the binomial distribution is equal to Np_i, with N equaling the total number of observations, and p_i the probability or relative frequency of event i occurring. The variance of this binomial distribution is equal to $Np_i(1 - p_i)$.

As N becomes large, the binomial distribution approaches the normal distribution. Because work-sampling studies involve large sample sizes, the normal distribution is an adequate approximation of the binomial distribution. The purpose for this approximation is that the normal distribution is well understood and mathematically easier to work with than the binomial distribution.

In work sampling we take a sample of size N observations and obtain a proportion of the observations for the events p', which is an estimate of the actual proportion p_i, or

$$p'_i = \frac{N_i}{N} \tag{2.7}$$

where p'_i = observed proportion of occurrence of event i, decimal

N_i = instantaneous observations of event i, number

N = total number of random observations

As shown in textbooks on probability, the standard error of a sample proportion for a binomial distribution is given by

$$\sigma_{p'} = \left[\frac{p'(1 - p')}{N}\right]^{1/2} \tag{2.8}$$

where $\sigma_{p'}$ = standard deviation of the proportion of the binomial sampling distribution, number

Bias and errors occur in any sampling procedure. This results in a deviation between the estimated p' and p, or the true value. Now we need to ask two questions:

1. How close to the actual value does our sample need to be or what is the interval?
2. How sure do we want to be that the sample data are in fact within the desired interval, or what is the confidence level?

A tolerable maximum sampling error in terms of a *confidence interval* I commensurate with the nature and importance of the study can be preestablished. For instance, if $p' = 62\%$, and if a maximum interval of 4% is desired, then $I = 4\%$ for 60 to 64%, or 62% ± 2% where $I/2 = 2\%$. The *relative accuracy* is defined as $I/2p' \times 100$, is $0.02/0.62 \times 100 = 3.2\%$.

The confidence and interval may be understood by examining Figure 2.4. For a 90% confidence level, which is usual for work-sampling studies, the factor $Z = 1.645$ is obtained from a table of values of the standard normal distribution function (see Appendix 1).

The total area under a normal curve is 100%, and the opportunity for a sampling value p' to fall within the two tails is given in Appendix 1 and is equal to

$$\pm Z \propto 2 \times (\text{probability from the table for a given } Z) \tag{2.9}$$

The table in Appendix 1 provides values for the area to the right (positive) side of the center value. In order to find the values for both the + and − sides, the table value must be multiplied by 2. For instance, if Z ranges from −1.645 to +1.645, then the probability that the true value p will be between the limits is $2 \times 0.45 = 0.90$, while the probability that the true value will be outside the limits is 1 – 0.90 or 10%. The value 0.4500 is determined from Appendix 1. Note the Z value is familiar as the multiplier in front of the standard deviation—for example, 3σ or 6σ. Some values of Z

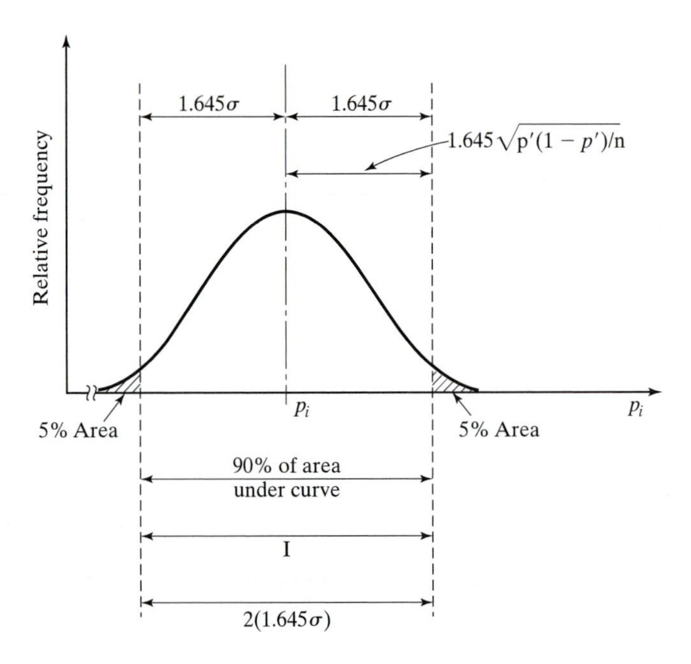

Figure 2.4 Normal curve showing a 90% area under the curve and within the two tails.

corresponding to confidence areas follow:

Area Between Limits (%)	$-Z$ to $+Z$	Area Outside Limits (%)
68	± 1.000	32
90	± 1.645	10
95	± 1.960	5
99	± 2.576	1

The sampling interval is determined by

$$I = 2Z\left[\frac{p_i'(1 - p_i')}{N}\right]^{1/2} \tag{2.10}$$

where I = interval, decimal

Z = value of the standard normal distribution function for a chosen confidence, number

We expect that the true value of p falls within the range $p' \pm 1.645\sigma_{p'}$ approximately 90% of the time. In other words, if p is the true percentage of the work estimate, then the estimate will fall outside $p' \pm 1.645\sigma_{p'}$ only about 10 times in 100 owing to chance or to sampling errors alone.

Equation (2.10) may be solved for the sample size when the other factors are either assumed or known, as

$$N_i = \frac{4Z^2 p_i'(1 - p_i')}{I^2} \tag{2.11}$$

Equation (2.11) may be solved for each N_i of the several elements of the study. Then the maximum N_i from all of the events i is chosen as N for the overall work-sampling study.

A preliminary work-sampling study helps to uncover problems before the major study is started. The study sells the idea, and, importantly, the percentage of observations falling into each activity gives a useful, albeit rough, estimate of the universe percentages. We might find that a proposed element is not significant enough to consider separately, especially if we have little control over that task. But during the sampling we might find that some other event deserves a separate identification.

After the preliminary study is known, and with the job elements defined, the next step deals with calculating the number of observations for the full work-sampling study. This number depends on the percentage of observations in each element and the size of the desired confidence interval. For example, we might estimate that the first element should occur somewhere between 23 and 27%, with a 90% confidence interval or a plus-minus tolerance of 2% or less is desirable.

Our next step is to spread the observations equally among the days for the study and then randomly within the working day, excluding breaks and lunch periods. Random times can be obtained from scientific calculators, spreadsheet programs or tables of random numbers. We leave it to the student to suggest the steps for this procedure. The snap observations will then commence.

As the sampling study progresses, it will be helpful to generate a control chart, similar to a *p*-chart used in statistical process control (SPC), to monitor the critical sample elements. After several days of observations, it is possible to calculate control limits for the element observations and then to monitor subsequent data using SPC techniques. These topics can be studied in engineering statistical texts.

Once the observation portion of the study is completed, it is possible to compute labor cost. A model using these data is

$$H_s = \frac{(N_i/N)HR(1 + \text{PF\&D})}{N_p} \tag{2.12}$$

where H_s = standard man-hours per job element i

N_i = number of event i observations

H = total man-hours worked during study

R = rating factor, decimal

PF&D = allowance, decimal

N_p = work units accomplished during period of observing this event

The H_s values are multiplied by appropriate cost rates for labor and equipment, such as productive hour cost, a topic studied in Chapter 4.

Example: Work Sampling for Integrated Circuit Operation

As part of integrated circuit (IC) fabrication, a process known as photolithography provides a pattern of material on a silicon wafer surface. Part of the photolithography process consists of applying a thin uniform thickness of a photosensitive polymer on the surface of the wafer using a spinning technique, where liquid polymer with a controlled viscosity is dropped onto the center of a spinning wafer. For a small prototype IC fabrication facility without automation for this process, the technicians are occupied by the elements shown in Table 2.2. A preliminary work sampling study of 3 coating stations (with 1 operator at

TABLE 2.2 Preliminary Observation Proportions

Element	Description	% of Time
1	Preparation of the photopolymer	10
2	Setup and filling of the spinning machine	15
3	Handling of wafers	25
4	Loading/unloading spinning machine	20
5	Monitoring spin cycle	12
6	Cleanup	8
7	Coating inspection	3
8	Idle	7
	Total	100

each) for one 12-hour shift (IC fabrication is a 24/7 operation) yielded a total of 180 observations. The preliminary proportions for the elements are shown in Table 2.2.

Our first estimate of sample size is based on preliminary percentages, and Table 2.3 indicates the desired tolerances. The sample size required for each element is first calculated using Equation (2.11) with the results shown in Table 2.3. The largest sample size will control the study; in this case we need about 1534 total observations (the preliminary 180 observations may be included as part of the whole sampling). Note that for this example, element 3, which is not the largest proportion, determines the number of observations needed.

TABLE 2.3 Finding Sample Size for Photolithography Study by Using 90% Confidence Interval

Element	Rough Proportion, %, p'	Desired Interval, I	Observations for Job Element, N_i	Relative Accuracy, $\frac{I}{2p'} \times 100(\%)$	90% Confidence Interval
1. Preparation of the photopolymer	10	0.03	1083	15	0.085–0.115
2. Setup and filling of the spinning machine	15	0.03	1534	10	0.135–0.165
3. Handling of wafers	25	0.04	1269	8	0.23–0.27
4. Loading/unloading spinning machine	20	0.04	1083	10	0.18–0.22
5. Monitoring the spin cycle	12	0.03	1270	12.5	0.105–0.135
6. Cleanup	8	0.02	1211	12.5	0.07–0.09
7. Coating inspection	3	0.02	479	33.3	0.02–0.04
8. Idle	7	0.02	1071	14.3	0.06–0.08

In retrospect it becomes possible to determine the magnitude of the sampling error after the study is underway or concluded. Assume that for element 4, loading/unloading spinning machine, partially through the study when $N = 875$, a total of $N_2 = 166$ tally marks are indicated, and $p' = 0.19$. This would give an interval of

$$I = 2(1.645)\left[\frac{0.19 \times 0.81}{875}\right]^{1/2} = 0.0436$$

We say that the true value lies within 0.19 ± 0.022 with a probability of 90%. Observe that the lower value 0.168 ($= 0.19 - 0.022$) does not lie in the interval of element 2 (0.18, 0.22) as given in Table 2.3. The student may want to explain this observation in terms of sampling errors, interval, and probability.

We continue the example by showing the results of the completed study. The work sampling study of 3 spin coating stations (with 1 operator at each), for one 12-hour shift per day, was conducted over a 10-day period with the results shown in Table 2.4. A total of 6000 wafers were coated during this study that included 1550 observations.

The time required for each element has been determined by multiplying the proportion by the total time covered by the sampling. If the number of wafers coated divides these times, than a representation for the standard times can be computed.

We now determine total time observed, observed time for each operation, and standard time (assume a performance rating of 100%) per wafer and per "boat" of 50 wafers.

TABLE 2.4 Results of Photolithography Work Sampling Study

Element Number	Description	Observation	% of Observations	Time (hr)
1	Preparation of the photopolymer	146	9.4	33.8
2	Setup and filling of the spinning machine	248	16.0	57.6
3	Handling of wafers	398	25.7	92.5
4	Loading/unloading spinning machine	296	19.1	68.8
5	Monitoring the spin cycle	176	11.4	41.0
6	Cleanup	110	7.1	25.6
7	Coating inspection	64	4.1	14.8
8	Idle	112	7.2	25.9
	Total	1550	100.0	360

Note that idle times are not included in the standard calculations. Total observed time = 12 hours per shift × 3 machines × 10 sampling days = 360 hours of total observation time. An allowance of 20% is provided to include entering and exiting the clean room, which requires dressing and undressing in clean environment attire, breaks and mid-shift meal. Then PF&D = 100/80 = 1.25. The standard output per wafer = (360 − 25.9)/6000 × 1.25 = 0.0696 hr per wafer = 4.18 min per wafer and 0.0696 × 50 = 3.48 hr per boat.

Examination of Table 2.4 shows that the most significant element is the handling of the wafer, at 25.7%. At this point, a series of questions can be asked about this work, and efforts to improve the element can be considered.

Work sampling is a tool that provides broad opportunities for analysis of workforce populations or equipment, making it possible to learn facts about some production activity in an easy, fast, and economical way. In summary, the following considerations should be kept in mind for work-sampling analysis:

- Explain and sell the work-sampling method before putting it to use.
- Isolate individual studies to homogeneous groups of equipment, jobs, or activities.
- Use as large a sample size as is practical, economical, and timely.
- Observe the data at random times.
- Take the observations over reasonably long periods, (e.g., 2 weeks or more), although rigid rules must bend with the situation and design of the study.

2.3.4 Labor-Hour Reports

It may not be wise to measure work times by the above methods. Jobs that are performed infrequently, long-task durations, crew work, professional jobs (especially those requiring intellectual activities,) and indirect labor do not lend themselves to traditional work-study methods. In these cases the only work time data that may be available are time card records and specific *labor-hour reports*. These records provide only actual time consumed in the performance of the job, including time spent for personal needs and nonproductive time.

The objective is to determine the amount of time the job *should* take from the labor-hour reports. Hardly rocket science, the number of labor-hours one worker would require to complete the job, or *allowed time* is desired. The difference between the span time, which is the clock or calendar duration between the start and completion of the job, and the labor-hours can be due to multiple workers working simultaneously on the job, planned or unplanned delays, or workers dividing their time among several different jobs during the same span.

Most workers submit work time information (time card or computer entry) indicating start and stop times for each workday. Workers who are given several jobs during the same work period may register and indicate the amount or portion of the time devoted to each job. The supervisor or timekeeper may perform the data recording for the labor-hour reports.

Information can also be recorded in written form on job tickets (see Figure 2.5), on time cards or through computerized time recording systems that may utilize barcoded work orders. The recorded time and task data are later combined with engineering drawings, parts lists, and work instructions, and they are used to determine the allowed time for the tasks as broken down in the time recordings. The data can be adjusted for unproductive effort and for PF&D allowances.

Consider an example of a small contract welding shop where job tickets are used. Observe from Figure 2.5 that the items of interest are operation, elapsed time, and units completed. Note that this job ticket has no supervisor's approval, timekeeper's mark, or any external verification. In some situations, such as small job shops, there is little repetition of similar work, so the analyst may chose to purge or alter information, based on her or his experience and judgment. Reworking of the data is risky, but in analyzing historical data, the first step is to clean up the information as rationally as possible. If raw data are used as received, then other problems may arise.

Job tickets are collected for similar work, and a worksheet is devised for the data. From Figure 2.6 we can see that the welding operations are varied and the engineer identifies the general operation as "welding" rather than divide the processes into one of the many welding methods. Several thicknesses, welding methods, and two part numbers have been posted on the worksheet. The mixing of processes and handling time into one lump time value is not an immaterial detail. For this job shop, one may

TRITON JOB TICKET

Name Michael Blakely Employee no. 505 30 9709
Date May 13 Order no. 101
Part no. 6682 Part name Bracket
Time started 9:15 Time stopped 11:43 Quantity 3
Operation Weld 40 X 80 X 3/4 in. plate.
Submerged arc weld. Butt weld two fillets, 240 in.

Special instructions Automatic weld

Figure 2.5 Worksheet for typical job ticket.

Part no.	Material	Thickness × length	Welding method	Job ticket actual man-hours	Remarks	Allowed time man-hours
6682 (3 units)	Steel	3/4 × 240 in.	Submerged arc butt weld 2 fillets	2.47	Fillet 2 sides reduce by 1/2 for one side and 2 extra units	1.24
7216	Steel	1/2 × 84 in.	Shielded metal arc, 1 fillet	1.06	No change	1.06
8313	Steel	3/16, 1/4, 1/2, × 344 in.	Shielded metal arc	1.18	Increase by 1/4 for various Thickness	1.48

Figure 2.6 Worksheet for a labor-hour report for welding.

choose to reduce, leave as is, or increase the three job-ticket times to give *allowed time*. These labor-hours for the tasks can be used for estimating future similar work.

2.3.5 Other Methods for Determining Time

The "time" to estimate the time for a job may be limited. The job is small or of short duration and it is not feasible to spend more than a few minutes to estimate allowed hours. Indeed, for new product development and product changes, there may be insufficient product information to assemble detailed estimates. In these cases, comparative or simplified methods may suffice.

Similar Products. If a product exists that is similar to a new product, and there are actual or measured time data, or a good estimate for the existing product, then those can be used for determining the time for the new product (for example, see Section 6.2.3). It may be possible to break down the products into component parts or subassemblies that can be costed using existing data and then reassembled into the new product configuration. Here's an example: A new, faster PC (imagine that!) is being designed that will use a new motherboard and CPU. The cost of the new CPU has just been published. Given the specifications for the new motherboard, a 15% larger power supply and the current costs of the daughter cards (which will be used unchanged), hard drive, and case, an estimate for the new computer can be calculated. When the product is not easily modularized, an engineer may determine that the new product is akin to an existing product but is a percentage more or less than the existing product. In this case, the time data for the existing product is ratioed by the estimated percentage. Experience and judgment are necessary for these instances.

Estimating Tools. There are some quick estimating tools that can determine times for new or redesigned products. Knowledge about the design and the processes is needed and some experience and training are recommended. Chapters 6 and 7 give additional information.

Engineering Performance Data. Software and estimating data that is commercially available are able to estimate production. The development of engineering performance data is discussed in Section 6.3.2. See the references for a listing of the sources.[8] These computer programs address metal machining, assembly, presswork, and almost all of the production processes. In addition, some computer-aided design (CAD) programs have limited estimating capability.

Since the 1920s, breakdown of manual work elements into universal micromotions have been available. For example, one system, methods-time-measurement (MTM) is practiced in high-volume labor intensive industries. A typical micromotion element would be "reach" or "move."[9] For example, a work element, "load part into die," is broken down into 20 very small or so micromotions with detailed measurements of distances and descriptions of motions. A patented system then matches preassigned times for each micromotion. A time is eventually gathered for many of these micromotion patterns.

These systems are costly to acquire, and require skilled and trained technicians and significant effort to use. If a complete, detailed design and a process plan are available, the times obtained from these systems can reasonably be considered standard times. These systems are disappearing because of their cost to find the times for the products. Furthermore, these systems only cover the manual side of the operation. The processing part of the operation cannot be evaluated using these micromotion tables and must be determined by measurement, or from the supplier database.

In-House Systems. For certain manufacturing operations, in-house estimating systems are effective. Topics discussed later can help with the creation of these programs. A representative estimating system can be developed using a spreadsheet program. For example, printed circuit board assembly utilizes common operations (inserting DIP packages, axial and radial leaded devices, wires, soldering, etc.) although the quantities and configurations of placements vary. However, if the company has times for the common operations, then only a parts count and some board sizes are necessary to determine the board assembly time.

Experience and Judgment. Finally, for those jobs where there is little previous experience, for the one-time job, repairs, rework, retrofit, and for those situations when little time or funding are available for the estimate, it may fall on experience and "engineering judgment" to find the estimate of allowed time. Obviously, the estimate will be only in general terms and in broad time increments such as days or weeks. Sometimes this activity is called *guesstimating*. However, as long as the accuracy of the estimate is considered when quoting costs, it should be possible to obtain a satisfactory value for these situations. These methods are also useful in situations where a *rough*

[8]One reference is *AM Cost Estimator*, Fourth Ed., Phillip F. Ostwald (Penton Education Division, Penton Publishing: Cleveland, OH), 1988.
[9]Additional information can be found at www.MTM.org.

PICTURE LESSON F. Taylor's Scientific Management

F.W. Taylor Collections, Stevens Institute of Technology, Hoboken N.J.

Frederick W. Taylor (1856–1915), American inventor and engineer, was apprenticed to learn the trades of a pattern maker and machinist. Following this training, he started as a machine shop laborer and successively rose through the ranks to chief engineer at Midvale Steel Company. In 1881, at age 25, he introduced time study at the Midvale plant, which was subsequently adopted throughout the industrialized world.

Studying at night, Taylor earned a degree in mechanical engineering from Stevens Institute of Technology in 1883. Taylor might have benefited from a brilliant career as an inventor—he had more than 40 patents to his credit—but his interest led him to develop a "new profession," that of consulting engineer in management.

The American Society of Mechanical Engineers elected him as president in 1906, the same year as the publication of his famous book *On the Art of Cutting Metals and The Principles of Scientific Management*, which used time study as a rationalizing component, and was published commercially in 1911. These and other books increased Taylor's fame.

In 1912 he appeared before a special committee of the House of Representatives, which was investigating this new profession, and appropriately Taylor became known as the

Father of Scientific Management. Essentially, Taylor suggested that production efficiency in a factory could be greatly enhanced by close observation of the individual workers and elimination of waste time and motions in the operation. Though the Taylor system provoked resentment and opposition from labor when carried to its extremes, its value in rationalizing production was undeniable, and its impact on mass production techniques for the last century monumental.

order of magnitude (ROM) or *not to exceed (NTE)* estimate is all that is required. (See Section 6.2 for further discussion.)

Methods to determine "time" are complete. We now study wage and fringe rates, which must be represented in compatible units with the time rates and the general accounting and engineering cost analysis procedures. After the product of Equation (2.1) is taken, we have the direct labor cost.

2.4 WAGE AND FRINGE RATES

Labor costs, which are dollars paid for wages or salaries for work performed, are a major ingredient of an estimate. A wage is paid or received for work by the hour, day, or week, and can be expressed for a period of time, so that it becomes a wage rate—for example, $25.71 per hour. A salary is paid for a period of time, say a week, month, or year, and may not be precisely dependant on the hours worked.

Payrolls cover two classes of workers: first, management and general administrative support employees, who may or may not be on a salary basis; and, second, hourly labor. General administrative employees may be management, engineers, programmers, foremen, inspectors, or office personnel.

Hourly labor is sometimes classified for cost analysis as direct or indirect. *Direct labor* involves the transformation process that adds value and refers to employees who can be associated with a product directly, such as milling machine operators or assembly workers; that is, the term encompasses any work that can be preplanned or designated to a specific job or product. *Indirect labor* refers to workers who support the direct labor generally performing undesignated work, such as clerk-typists, janitors, stock room attendants, production control, maintenance personnel, industrial engineers, and superintendents. In an allocation sense of cost, their work and effort is usually for a variety of tasks, making it difficult to designate precisely what portion of their work contributes to the particular operation, product, project, or system. Indirect labor costs are handled by a process called "overhead," which is discussed in Section 4.9.

There are labor costs along with the wage rate that are known loosely as *fringes*—those costs to the company that are required for employees in addition to the wage paid including things like paid holidays, vacation days, and insurance benefits. It is necessary to understand their content and value to obtain a complete labor cost. Fringe benefits, which are related to wages and salaries, constitute as much as 30% of the actual cost incurred for labor. More extensive descriptions are included in

Section 2.4.2. Variety exists in the handling of wages and fringes. Note that a deduction from the employee alone (usually those items shown on the paycheck stub), is not explicitly identified for estimating because it is included in the employee wage rate.

To determine the labor cost once the design is known, the work is planned and job descriptions of employees are matched to the work. Those job descriptions indicate the skill, knowledge, and responsibility required of the worker. Each company will have occupational descriptions. The example below illustrates a job description for an assembler. The occupations are graded, and a company will have a pay scale that increases from the very simple occupation to the most difficult. Normally job descriptions and their wage rates are policy information available to engineering, especially for operation and product designs. In some cases, engineering is responsible only for determining the labor hours required and other company functions, such as finance or accounting, determine the dollar costs using the engineer's estimate of hours after applying the wage information.

Example: Job Description, Assembler

Assembler (bench assembler; floor assembler; jig assembler; line assembler; subassembler): Assembles and/or fits together parts to form complete units or subassemblies at a bench, conveyor line, or on the floor, depending on the size of the units and the organization of the production process. Work may include processing operations requiring the use of hand tools in scraping, chipping, and filing of parts to obtain a desired fit as well as power tools and special equipment when punching, riveting, soldering, or welding of parts.

Class A: Assembles parts into complete units or subassemblies that require fitting of parts and decisions regarding proper performance of any component part or the assembled unit. Work involves any combination of the following: assembling from drawings, blueprints, or other written specifications; assembling units composed of a variety of parts and/or subassemblies; assembling large units requiring careful fitting and adjusting of parts to obtain specified clearances; using a variety of hand and powered tools and precision measuring instruments.

Class B: Assembles parts into units or subassemblies in accordance with standard and prescribed procedures. Work involves any combination of the following: assembling a limited range of standard and familiar products composed of a number of small- or medium-sized parts requiring some fitting or adjusting; assembling large units that require little or no fitting of component parts; working under conditions where accurate performance and completion of work within set time limits are essential for subsequent assembling operations; using a limited variety of hand or powered tools.

Class C: Performs short-cycle, repetitive assembling operations. Work does not involve any fitting or decision making regarding proper performance of the component parts or assembling procedures.

Wages and fringe costs are considered by one of two methods: (1) wage only, and (2) wage and fringe combined, or the gross hourly cost. In the first method the fringe effects are collected in overhead, which will be described in Section 4.9. In some cases where machinery is involved, a composite rate of the operator's wages and some amount representing the hourly cost of the machine operation is used to determine the productive hour cost, which is discussed in Sections 4.9.5 and 8.4.1.

2.4.1 Wage-Only Method

In the *wage-only method*, only the wage paid directly to the worker is included in the labor cost determination. The costs of all of the fringes are accounted for in overhead and are added to the cost later in the process, as discussed in Chapters 4 and 8. Wage payment can be classified into two general groupings: those that pay for attendance and those that pay on performance. In time attendance wage plans, gross wages are figured easily. The time in attendance is multiplied by the rate. An engineer who earns $54,000 per year earns $4500 per month. If he or she starts or leaves within the month, the pay is prorated to the number of calendar or working days, depending on company policy.

The qualitative formula given by Equation (2.1) can be more formally expressed as

$$C_{dl} = H_a \times R_h \tag{2.13}$$

where C_{dl} = cost for direct labor effort, dollars

H_a = actual hours

R_h = rate per hour, dollars per hour

Naturally, R_h can be found from annual, monthly, or weekly pay scales. For a $4500 monthly scale, the weekly scale is $4500 \times 12/52 = \$1038.46$. There are 173.3 hours per month ($=52 \times 40 \times 1/12$), so the hourly scale is $25.96. Those calculations use the popular 40-hour week. In attendance-based plans (sometimes called *day work*), the worker is paid for the amount of time spent on the job.

Of course, the worker is interested in as large a wage as possible. The employer, on the other hand, is interested in the labor cost reduction. If the employer is able to encourage increased output from the worker, the employer might be willing to pay higher wages. In these performance plans, the worker's earnings are related to productive output and are called incentive or piece-rate plans. Management philosophy has been moving away from incentive wage plans in recent years, but because of its continued presence, it is described here. For a pure incentive rate plan, the cost is determined as

$$C_{dl} = N_p R_p \tag{2.14}$$

where N_p = number of pieces produced

R_p = standard rate per piece, dollars per unit or piece

If $R_p = \$0.1466$ per unit and $N_p = 128$ units, then $C_{dl} = \$18.76$, which is paid to the worker. Most incentive plans do not pay additional wages until 100% standard is reached. Below that level a *guaranteed wage* or *day work* is pledged. A formula expressing this relationship is given as

$$C_{dl} = H_a R_h + R_h(H_s N_p - H_a) \tag{2.15}$$

where H_s = standard hours per piece

This equation is valid as long as the standard is met. If the standard is not met, the second term becomes negative and C_{dl} will be below the guaranteed wage. So a second proviso must be met:

$$C_{dl} \geq H_a R_h \text{ or } H_s N_p - H_a \geq 0 \tag{2.16}$$

Example: Wage with Incentive

Assume that $H_a = 0.8$ hour, $R_h = 14.50$, $N_p = 128$, and $H_s = 1.010$ hour per 100 units. Then,

$$C_{dl} = 0.8 \times 14.5 + 14.5[(1.010/100) \times 128 - 0.8] = \$18.75$$

If only 70 units had been produced under the piece-rate plan during 0.8 hour, then pure incentive earnings would be \$10.26. Under the guaranteed 100% plan, earnings would be \$11.60 because $C_{dl} \geq H_a R_h$.

Earned hours are recorded for pay purposes when an incentive plan exists. Thus, for a job that is set in standard hours, and the worker may earn more dollars by producing the job in fewer actual or clock hours. The advantage of this approach is that the standard hour base is the same for each operation unit, whereas the actual hours for an operation unit vary with the worker's efficiency on that job. *Efficiency* is defined as the ratio of the task amount of work, or standard hours to hours actually taken in performing the work, is

$$E = N_p \frac{H_s}{H_a} \times 100 \tag{2.17}$$

where E = labor efficiency (or productivity), percent

In this example where $N_p = 128$ units, $E = (128 \times 1.010/100)/0.8 \times 100 = 162\%$, and for 70 units $E = 88\%$.

2.4.2 Gross Hourly Cost

In some cases the fringe costs are determined explicitly for estimating and are not included in overhead. When combined, wages and fringes are called *gross hourly cost (GHC)* and may be determined for production workers. GHC uses the principles of activity based costing, a practice described in Sections 4.9.6 and 8.4.2.

Federal and state laws and negotiated labor contracts regulate wages and salaries paid by employers. Two are prominent: the wage-hour law and the Walsh-Healy Act. Manufacturers engaged in interstate commerce are subject to the wage-hour law. The Walsh-Healy Act covers only companies having federal contracts. Both require the payment of wages at *time-and-a-half rates* for more than 40 hours in 1 week. Walsh-Healy adds the same requirement for more than 8 hours in 1 day. Some groups of workers are exempted from the wage-hour law, such as management and engineers. Workers with a 40-hour base rate of \$12 must be paid \$18 per hour for their overtime hours. The federal wage-hour law also specifies a minimum dollar per hour as the lowest paid wage. For example, the wage minimums as established by Congress constitute a floor for employed labor in most categories. Other labor laws exist of course, such as antidiscrimination, limiting work conditions for children, safety (e.g., OSHA) and sanitation laws, unemployment insurance, and social security and other taxes.

Contractual agreements may specify additional requirements, such as a number of holidays, time off, sick leave, and uniforms. In addition to wages earned by employees, the employer must pay the appropriate government agency or insurance carriers additional amounts. Time paid but not worked (holidays and vacations) is prorated annually over the actual hours worked by addition into the fringe costs.

A partial listing of fringe costs may include (1) legally required payments, such as payroll taxes and workers' compensation; (2) voluntary or required payments, such as group insurance and pension plans; (3) wash-up time, paid rest periods, and travel time; (4) payment for time not worked, holidays, vacations, and sick pay; and (5) profit-sharing payments, service awards, and payment to union stewards. Fringe costs depend on local situations and must be determined individually for each case.

The Social Security Act of 1935 requires that businesses pay a tax for retirement and medical benefits. Officially known as the Social Security tax, or *FICA* (*Federal Insurance Contribution Act*), this amounts to a percentage of a fixed sum of the gross earnings of an employee. The employee contributes an amount equal to the employer through a deduction from the wage paid. The rate and the base on which it is levied are subject to change by Congress and have marched steadily upward since the first law back in 1935. Initially, the employee paid 1% of the first $3000 earned and the employer paid an equal amount. For a recent year, the employee deduction was calculated as 7.65% of eligible income to a ceiling of $80,400 for FICA and 1.45% with no ceiling for Medicare. These deductions are changed whenever Congress finds it necessary. The employer pays the same amount, which is added to the fringe cost. The employee's share of the FICA tax is not a business cost, though it is clearly a reduction of the wages that the employee enjoys.

Workers' compensation is a levy against employers for continuation of income to employees in periods when they cannot work because of accidents occurring on the job. Payments may be made to the state compensation insurance fund or to state-approved insurance carriers. Employees are grouped into work types, and rates based on experience of risk incidence are established for each type. In the event of injury, the insurance carrier provides financial assistance to the injured person or to the survivors in the event of death.

The Social Security Act of 1935, also established a federal-state tax on wages for *unemployment insurance* Setting up the employment services was delegated to the states, which collect the major portion of the tax to operate their own unemployment offices. State legislation has to conform to the requirements of the federal act. An employee who is laid off through no fault of his or her own is entitled to unemployment "pay" for a limited period. The unemployment insurance pays for this involuntary employee cost. Built into the system is a merit-rating procedure to reward "good" employers (i.e., employers who operate their labor pool to prevent repetitive hirings and firings and who terminate employees only for sound economic reasons). When a business is established or enters the system, it pays the maximum state rate. If its employment record becomes stable, then the rate drops until the minimum rate is reached. The employer is rewarded for good employment practices. Not all employees, such as farm workers or domestic help, are covered.

Fringe benefits include expenditures, other than payroll taxes and workers' compensation insurance, that benefit employees individually or as a group. Because those benefits are not legally required, they may vary from employer to employer and be based on labor-market competitiveness, industry practice, management's attitude toward employees, management's social consciousness, or labor-management contract.

One such benefit is the portion of medical and dental insurance paid for by the employer. *Supplemental medical insurance* covers the employee and many times the family for all or part of ordinary illness, or could apply only to catastrophic illness. Usually, a schedule of payments by the employee and the employer is established, a percentage of present employees are required to initiate the plan, and all new personnel are included automatically. The company's portion of the premium needs to be considered for estimating. Medical insurance is a large expense and can amount to around $500 per month per employee ($2.89/hr). Employee contributions are not a business cost since they are included in the basic wage.

Other supplemental benefits are *life insurance* and *disability insurance*. Life insurance pays a contracted amount upon the death of the employee. Disability insurance pays a specified periodic (weekly or monthly) amount to replace the wages of an employee who is disabled and cannot work. A schedule is established showing the employee's and employer's portion of the premiums. In initiating this plan the personal histories of present employees are secured, and a stated percentage of the workforce must participate. All new personnel may be automatically included. The company's portion of the premiums needs to be included in the fringe amount.

Holiday and vacation pay is also considered a fringe benefit. Vacation policy varies as to length. *Sick pay* is similar to vacation pay. In cases where an employee is entitled to a specified number of paid sick days (or hours) per year and the time accumulates if not needed, then the cost is known and can be included in the fringe costs. In cases where sick pay is determined on some individual basis and if not needed is not charged at all, it may not be formally set up on the accounting books as such and is charged to overhead when taken. It can, however, be estimated on experience. In the example shown in Table 2.5, the employee receives 6 paid holidays and 10 vacation days each at the nominal 8 hours per day. In this example, no sick days were paid, however, if known a number of annual sick days (or hours) may be included. The total number of paid holiday, vacation and sick hours are multiplied by the wage rate and added to the annual costs of other fringe benefits.

Another fringe benefit is an *employee supplemental pension plan*, which provides retirement benefits in addition to social security. The plan may be self-funded, in which case the company agrees to invest contributions, or the plan may be funded by an outside agency such as an insurance company or a mutual stock fund company. In the latter case, the company makes the premium payments required under the plan. In a third arrangement, the plan may be fully funded by the company or partially by employees. Again, any employee contributions are not part of the fringe costs for the company.

Another fringe benefit is the *stock-purchase plan*, by which the company encourages employees to become shareholders in the company by allowing them to purchase stock at less than market value. The immediate benefit the employee receives is the difference between market price and the price paid.

TABLE 2.5 Method for the Calculation of Gross Hourly Cost

Job title: *Assembler*
Department: *Mechanical*
Future effective period: *Jan-July*
Wage: *$14.50/hour*
Shift: *1*
Name: *Spencer Tucker*
Clock no: *10017*
Subsidy: *10%*
Entitled vacation days: *10*
Holidays: *6*

	Item	Annual hours	Annual cost	Annual excess cost
1	Regular paid clock hours	2,080	$30,160	
2	Planned overtime hours	192	2,784	
3	Overtime cost at 50% of 2	—	1,392	$1,392
4	Subtotal	2,272	$34,336	
5	Holidays	48	696	696
6	Entitled vacation	80	1,160	1,160
7	Paid sick leave	—	—	
8	Expected nonchargeable hours, 5%	104	1,508	1,508
9	Subtotal (5 + 6 + 7 + 8)	232	$3,364	$3,364
10	Chargeable (4–9)	2,040	30,972	
11	Expected performance subsidy	204		
	10% × 10	—	3,097	3,097
12	Chargeable standard (10–11)	1,836	$27,875	
13	Nonhourly costs			
	FICA @ 0.076 × 4			2,610
	Workers' compensation, 3% × 15,000			450
	Unemployment insurance			300
	Supplemental medical insurance			550
	Supplemental pension			—
	Supplemental life insurance			—
	Union welfare			—
	Bonus, gifts			—
	Uniforms, tools, etc.			50
	Profit sharing			—
	Other			—
14	Subtotal of 13			$3,960
15	Excess costs (3 + 9 + 11 + 14)			$11,813
16	Excess hourly rate (15/10)			5.79
17	Wage			14.50
18	Effective gross hourly cost (16 + 17)			$20.29
19	Percentage increase in hourly base rate (16/18)			29%

Example: Gross Hourly Cost for Assembler

The effective gross hourly cost for an assembler (see the job description given in the example on page 58) is determined in Table 2.5. By knowing the job title and employee, it is possible to figure out entitlement for vacation, shift differential, and so on. The expected nonchargeable hours of 5%, line 8, include washups, which are not a part of overhead or not included in production allowances. Line 11, expected performance subsidy, may be used for award pay rates if provided by management policy. Thus an effective gross hourly cost of $20.29 is 29% greater than the basic wage. It is also possible to calculate gross hourly costs for a class of employees, using averages for entitled vacation, overtime, holidays, and all of the items listed in Table 2.5.

In many companies, each job classification has an assigned wage rate (either typical, average, or high), which may include the fringe costs, that is used in labor cost calculations by engineering or finance.

2.5 JOINT LABOR COST (OPTIONAL)

In some cases, work is performed on material that will later be split into separate products, or work is performed that produces different products, essentially at the same time. Think of wafers, which are eventually split into many integrated circuits. Before these wafers are divided into the unit IC, fabrication handles and processes the wafer as if it is one unit or a common unit. In the second case a worker could be tending one or more machines that are simultaneously producing different parts.

These situations involve joint labor costs. In the first case material may be worked through a process to prepare it for further work that will convert it to one or more of several different products. The joint labor performed before the split off will need to be proportioned in some fashion into the cost of the final product(s). In order to determine actual costs, the engineer dejoints, or "unitizes," the cost. Joint costs include not only labor but also material and overhead, which will be discussed later in Chapters 3 and 4.

One of the production causes for joint costs is *multiple-machine operation*. In certain kinds of work it is possible for one operator to tend a number of machines, depending, of course, on the amount of attention that each machine requires. If these machines are producing different products, then the operator's time or cost must somehow be equitably divided among the products.

The problem with having one worker tend multiple machines is that several machines may need attention at the same time. If a machine must wait for operator attention, the waiting is called *machine interference*. This situation is described by queuing theory. The problem arises, therefore, as to how many machines an operator should tend. Although queuing methods are more elegant and exact, simpler methods are preferred when estimating because this information is available. Machine interference exists whenever too many machines are assigned to one operator, but this may be less costly than operator idleness. Joint labor cost or time exists when an individual operator is involved in the following situations:

- Similar machines each producing the same output
- Similar machines each producing different output
- Dissimilar machines each producing different output

Crew or flow-line work is another example of joint labor cost. Crews are found frequently in manufacturing. In crew work the interest is to balance the conveyor workload such that there is no idle time or that all idle time is equal within the crew. Time balance is unlikely, so the entire crew time is subject to the time of the worker with the longest time. Estimating of crew work for flow line production is discussed in Chapter 7.

Dejointing of shared labor divides the time or labor cost among the different output products. The point of division is called the *split point*. Labor up to the split point is *joint*, while afterward it is called *unit*. In order to split the labor, some common metric must be used in order to find proportional relationships for the products. Common units of measure need to be used, such as pounds, number of units, inches, or feet. Using the common measurement units, a worker's output can be specified, for example, as 30% for product A, 20% for product B, and 50% for product C. Then the labor cost can be similarly proportioned.

Example: Dejointing of Labor Time for Left-and Right-Hand Product

Two camera brackets, left and right hand, are machined together. Loading and unloading are simultaneous. The left bracket requires special machining. Gross hourly costs are $20. A time study gives the following data:

Element	Bracket	Minute per Element
Loading	Both	0.20
Position and drill hole	Both	0.15
Rotate and position	Left	0.05
Index collet	Left	0.08
Counterbore	Left	0.21
Eject parts	Both	0.15
Clean jig	Both	0.10
Total		0.94

A dejointing solution by time would be as follows:

Elements	Left	Right
1, 2, 6, 7	0.30	0.30
3, 4, 5	0.34	
Total	0.64	0.30

$$\text{Cost for left bracket} = 0.64 \times 1/60 \times 20 = \$0.213/\text{unit}$$
$$\text{Cost for right bracket} = 0.30 \times 1/60 \times 20 = \$0.100/\text{unit}$$

We can also use the two units of output as the basis for dejointing to obtain $0.157 per unit ($= 0.94 \times 1/60 \times 20 \times 1/2$). This method of calculation is not as preferred.

This example splits the costs in a manner that most accurately reflects the actual as produced costs. However, costs can also be dejointed by market effects that stipulate that the cost be a function of price or number of units sold. Marketing strategy can affect allocation accuracy by falsely requiring that some products subsidize others. This policy says that selling price or market value is not proportional to processing costs and thus distorts the estimated cost. In the bracket example, suppose that the left bracket can be sold on a ratio of 4 to 3 units of dollar compared to the right bracket. We proportion the labor costs based on the ratios, 4/7 and 3/7 and compute the cost as

$$\text{Left bracket} = 0.94 \times 1/60 \times 20 \times 4/7 = \$0.179$$
$$\text{Right bracket} = 0.94 \times 1/60 \times 20 \times 3/7 = \$0.134$$

Note that in all methods, it is necessary that the total cost of a pair of brackets (one left and one right) is $0.313.

2.6 LEARNING

Reduction in the cycle time is expected as a worker continues to produce similar units. This trait is called *learning*. Job performance improvement is measured by the time per unit. The direct-labor cycle time declines as the number of units produced increases, although up to a point. This widely observed phenomena becomes predictable, and companies are able to measure job performance during training. The result of this is that the initial units will take longer to manufacture (cost more) than those produced once the job is learned well.

Look at Figure 2.7. Four new employees are hired under probationary terms. Employees 1 and 4 are relatively unskilled or slow in job speed. After a period of training, employee 1 is able to produce a unit of product in less than the standard time and is retained. Operator 4, perhaps having less dexterity, skill, or motor ability, is unable to achieve the standard. Operators 2 and 3 have initially greater skill or faster job speed, but only operator 3 achieves the output requirements. Operators 2 and 4 would continue training, or be reassigned or be terminated in some cases.

Though the learning phenomena is widely experienced by anyone who has performed manual and repetitive work, the application of the learning theory to cost analysis is broader than this limited application to direct labor and repetitive production. The learning improvement is sufficiently predictive and additional models are given in Sections 6.4.1, 8.4.3, and 8.5.

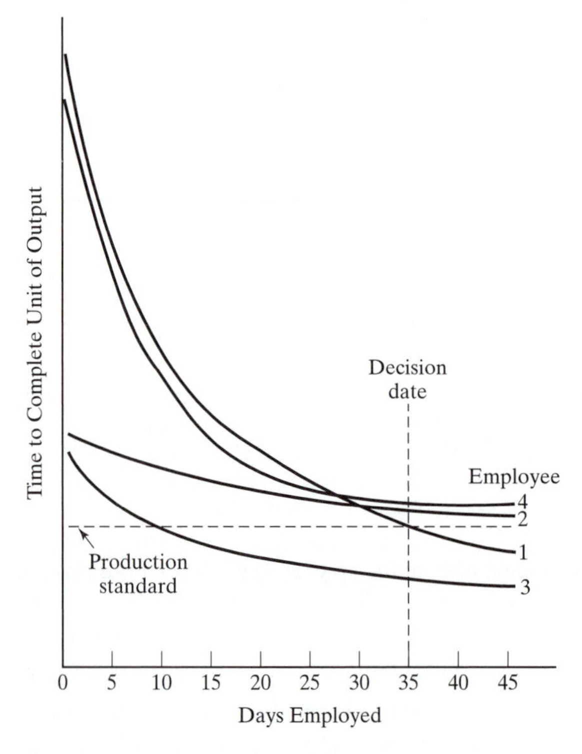

Figure 2.7 Learning application in operator training, performance, and retention.

SUMMARY

The cost of labor is a significant part of the cost of a product. Time and the hourly wage comprise this cost. While it is amount of time charged to production that becomes the actual cost, frequently it is the estimated future labor cost that is more important. Estimated labor cost becomes a part of the quotation, and it is this number that creates new business.

In order to assist management and provide accurate information for making estimates, knowledge of what an operation *should* cost is necessary. The usual ways to measure labor are time study, work sampling, and man-hour reports. The method must fit the situation that is being measured and the purpose for which the information will be applied. An engineer may not be involved in these measurements, but he or she needs to understand the implications of the measurement and their context.

The wage amount applied to the estimated or actual hours requires analysis. There are many components to the wage rate in addition to the amount actually paid to the employee. These include holiday, vacation, and sick pay; insurance premiums; and retirement, some of which are legally or contractually mandated.

The objective for these measurements and calculations is to estimate an *accurate* future cost of labor.

QUESTIONS FOR DISCUSSION

1. Define the following terms:

Allowance
Dejointing labor cost
Direct labor
Gross hourly cost
Historical time
Indirect labor
Interval
Job description
Job ticket
Labor-hour
Measured time
Normal time
Person-hour
Personal time
Pieces per hour
Rating
Standard
Supervisor's report
Time study
Work sampling

2. Why do we separate labor into direct and indirect categories?

3. What is the purpose of rating? A job is time studied and rated greater than 100% and the standard is used for a second worker. What does the fact that the original rating was greater than 100% mean to the second worker?

4. What are allowances intended for?

5. Why is an operation divided into elements for the time study?

6. Why do we use labor-hour reports? Name types of work that are appropriate.

7. Determine the current FICA rate and the base for the employee and employer. What are the political ramifications?

8. List mandatory and voluntary types of fringe costs.

9. Specify the nature of work that makes it joint. List allocators that dejoint work.

10. Supply a practical illustration of learning for direct labor.

11. Why is it essential that random times be used for work sampling?
12. Describe the situations where time study and work sampling would be used.
13. Why is the allowance multiplier different from the allowance factor?
14. Explain the difference between the interval and the confidence level used to calculate the sample size.
15. Describe the cultural and historical progress of labor in the United States.

PROBLEMS

2.1 (a) A machine has a cycle time of 1/2 second. What are the productions pieces per hour at 100% and 75% efficiency?
(b) The floor-to-floor time (meaning a complete productive cycle) is 31 seconds. Find the gross units per hour at 100% and 90% efficiency.

2.2 (a) If the hourly production is 11 units, find the hours per 1000 pieces.
(b) The production is 29.5 units per hour. Find hours per 100, 1000, and 10,000 units.
(c) Find pieces per hour and standard minutes per unit for 15.325 hours per 100 units.

2.3 A time study of an assembly operation is summarized:

Description	Frequency	Element Minutes	Rating Factor
Part A to 5 in. dowel	1/1	0.037	1.10
Part B to 5 in. subassembly	2/1	0.064	1.15
Short piece to base	1/1	0.089	1.20
Long piece to base	1/1	0.129	1.10
Grab pieces from tote box	1/5	0.185	1.10
Place on conveyor	1/1	0.087	1.10

Allowances for this work total 9.5%. Find the standard minutes per unit, pieces per hour, hours per 100 units, and labor cost per unit, where the wage rate is $14.75 per hour.

2.4 A mailroom prepares the company's advertising for mailing. A time study has been done on the job of enclosing material in envelopes. Develop the elemental normal time, cycle standard time, pieces per hour, and hours per 100 units, providing 15% for allowances, for the continuous stop watch data. (*Hint*: The time readings are in hundredths of a minute.)

	Cycle							
Description	1	2	3	4	5	6	7	Rating %
Get envelops	11		55		105		151	105
Get and fold premium	22	41	65	83	116	134	160	115
Insert premium in envelope and seal	29	48	73	97	123	141	182	95

2.5 A method has been engineered and an operator has been time studied. The method has produced an average overall time of 2.32 minutes by actual watch timing. The overall pace rating applied to the job was 125%. This company, which uses an incentive system, adds extra time into the production rates to cover personal time, fatigue, and unavoidable delays. The total allowance is 15%.

(a) Determine the rated time or normal minutes, the total allowed time or standard minutes, the standard hour rate for the job per 100 units, and the pieces per hour.
(b) The labor rate per hour is $15.30. What is the standard labor cost per unit?

2.6 Element time study observations that use the snap-back method are given as follows:

Element Description	1	2	3	4	5	Rating (%)	Allowance (%)
Place part in jig	0.03	0.08	0.05	0.04	0.06	117	11
Drill hole	0.31	0.38	0.37	0.39	0.33	100	5
Remove part, place on conveyor belt	0.07	0.08	0.09	0.12	0.11	93	13

(a) Find the average, normal and standard times for each element.
(b) Calculate the production per hour and hours per 100 units.
(c) By an older method, 0.834 minute of standard time was required to complete this same process. Calculate the increase in output in percent and savings in time in percent.

2.7 Review the time study summary below.

Element Description	Frequency/ Unit	Average Time	Rating (%)
Get part from conveyor	1/1	0.040	100
Get subassembly from conveyor	1/1	0.030	105
Connect parts to subassembly	2/1	0.055	110
Install 3 brackets	3/1	0.153	100
Assemble side 1	1/1	0.234	90
Mate side 2	1/1	0.183	90
Put on conveyor	1/1	0.032	100

(a) With an allowance factor of 15% to cover P (personal), F (fatigue), and D (delay), what are the normal minutes per piece, the standard minutes per unit, and the number of hours per 1000 pieces?
(b) If the performance against standard has averaged 120%, what is the incentive hourly if day work is paid at $15.75 per hour?

2.8 The raw continuous-timing data for a two-element punch press operation are provided. The times are in minutes.

Description	1	2	3	4	5
Handle part	0.01	0.04	0.08	0.11	0.14
Punch part	0.03	0.06	0.10	0.12	0.16

(a) What is the average time for each element? If the ratings are +8% for element 1 and −9% for element 2, what is the normal time? The PF&D allowance is 20%. What is the standard minute per piece for this operation? How many pieces per hour? How many hours per 100 units?
(b) If the PF&D allowance only applies to element 1, what are the standard minutes per unit? How many hours per 100 units?

2.9 Engineering aides prepare job tickets to track their daily work. After a sufficient period of data collection and classification, the data are summarized. A personal allowance of 15% is used for this work. (*Hint*: For simplification of calculation, ignore holiday and vacation days.)

Description	Average Minimum	Frequency/ Drawing
Post drawing numbers	7.0	1/1
Duplicate drawings	19.0	1/1
Correct computer entries	38.0	2/1
Phone calls	1.5	1/4
Update CAD changes	27.0	1/2

(a) Determine the standard labor hours per engineering drawing.
(b) A new product design is anticipated and a separate staff will be assembled. The new product design will probably result in 2500 drawings over a 1-year period. How many aides will be required, and at $25 per hour, how much should be budgeted for their work on the drawings for the new design?

2.10 The results from a working sampling study of a workforce is divided into 12 categories. This sample covers a span of 25 eight-hour days. What are the percentages and expected hours per category?

Category	Number of Observations
1	92
2	99
3	37
4	11
5	25
6	14
7	24
8	33
9	3
10	22
11	8
12	32
Total	400

2.11 The pediatrics department in a hospital is work sampled over a total of 608 hours. The results are shown below. There were a total of 1578 observations. Find the percentage of occurrences for each category, cumulative percentages, the element hours, and cumulative hours for each category.

Category	Count	Category	Count
Routine nursing	496	Other	79
Idle or wait	263	Feeding	52
Unit servicing	183	Bathing	22
Report	129	Elimination	11
Personal time	128	Transporting	8
Intervention	102	Housekeeping	7
Unable to sample	91	Ambulation	7

2.12 A work sampling study is taken of a department with the following information obtained: number of sampling days, 25; number of trips per day, 16; number of workers observed per trip, 3; and number of items sampled, 4. The observations counted for each of the items was, A−80; B−320; C−1600; and D−2800.

(a) How many labor-days and observations were sampled?
(b) What are the percentages and equivalent hours for the activities?
(c) For a confidence level of 90%, what is the relative accuracy for each item?

2.13 We want to determine the percentage of idle time of a machine shop by work sampling. A confidence level of 95% and a relative accuracy of ±5% is desired, where a rough estimate of 25% is suspected for idle time.

(a) How many observations are necessary?
(b) Assume that the relative accuracy is $\pm 2\frac{1}{2}\%$. How many observations are required now?
(c) What happens to the number of the observations required, as the relative accuracy becomes less?

2.14 (a) To get a 0.10 interval on work observed by work sampling that is estimated to require 70% of the worker's time, how many random observations will be required at the 95% confidence level? Repeat for 90%.
(b) If the average handling activity during a 20-day study period is 85% and the number of daily observations is 45, what is the interval allowed on each day's percent activity? Use 90%; repeat for 99%.
(c) Work sampling is to be used to measure the not-working time of a utility crew. A preliminary study shows that not-working time is likely to be around 35%. For a 90% confidence level and a desired relative accuracy of 5%, what is the number of observations required for this study? Compare to a 95% confidence level.

2.15 A shipping department that constructs wooden boxes for large switchgear has five direct-labor workers. A work-sampling study is undertaken, and the following observations of work elements are recorded over a 15-day, 8-hour period:

Description	Count
Set up and dismantle	312
Construct crates	264
Load switch gear in crates	204
Move materials	324
Idle	96

A rating factor of 90% is found. The number of switchgear shipped during this period is 26. This firm uses an allowance value of 10% for work of this kind. Average labor costs $18.75 per hour.

(a) Find the elemental costs.
(b) What is the standard labor cost per box?
(c) Estimate the actual cost.

2.16 An eight-person CAD department working with sizes A, B, and C drawings is sampled by a management consultant over a standard 4-week period. A chart summarizes the categories:

Item	Count
Drafting and tracing	778
Calculating	458
Checking prints	110
Classroom	125
Professional time off	172
Personal time, idle	270

During this period, 55 drawings (A = 20, B = 25, C = 10) were produced with a total payoff of $26,400. Let relative size A = 1, B = 2A, C = 2B.

(a) Given a performance rating of 1.0 and a PF&D factor calculated from the personal time, idle element, determine the standard time per drawing unit (A size) and standard times for the B and C sizes.
(b) Determine the hourly pay rate. A new order is estimated at A = 10, B = 30, and C = 25 drawings has been placed. What is the estimated cost of this order?

2.17 Find the effective gross hourly cost for drill press operators paid an average wage of $22 per hour. No overtime is planned. Company policy allows six paid holidays, and the average entitled vacation is 10 days. There is no nonchargeable time or performance subsidy. Sick leave is charged against vacation time. FICA is at current rate, and workers' compensation is at 2% of the first $20,000. The company pays $400 for unemployment insurance.

2.18 A tool and die worker is paid $26.50 per hour. The work year consists of 52 40-hour weeks with overtime schedules for 26 Saturdays. The company allows 8 paid holidays and 10 days of vacation. Four days of sick leave are budgeted and have historically been used. Expected nonchargeable hours are 2%. There is no subsidy for performance. Nonhourly costs include FICA taxes at current governmental rate, workers' compensation at 2% of regular wages up to $20,000, an accident insurance sum of $600, a major supplemental medical plan for $300, and unemployment insurance of $800.

(a) Find the effective gross hourly cost.
(b) What is the job cost if a job will require 10 man-hours with a productivity of 90%?

2.19 Determine a company's base annual cost for a top-grade worker. The base hourly rate is $15.75. Use the current FICA costs of the base. State unemployment compensation runs 2%. Health insurance premiums cost $40 per month. The company carries term life insurance that costs $45,000 per year for all employees (there are 60 employees). In addition, the company profit-sharing plan usually pays 5% of the base wage.

2.20 Find a typical hourly cost for direct labor. In this instance, the average workweek is 48 hours, of which the final 8 hours are premium at one and a half. The wage is $30.30 per hour. Each day a total of 20 minutes is permitted as a coffee break, and the final 15 minutes is cleanup. Annually, there are 2 weeks of paid vacation on the basis of 40 hours worked and five paid holidays. Taxes for FICA are at the current rate, and there is a 1.2% workers' compensation tax for the amount of $15,000. There is a requirement for $250 per month for company-paid medical and life insurance, and a $25 Christmas bonus.

2.21 A man worked 8 hours on incentive and nonincentive jobs. While on incentive, he completed 8 tasks, each with a 1-hour standard time. The nonincentive jobs took 2 actual hours. The incentive plan has a base rate of $13 per hour, though the man's wage rate for nonincentive tasks is $16 per hour. How much did the man earn?

2.22 (a) A worker produces 56 units on an incentive plan during the 40 hour week. Her base hourly rate is $15 per hour, and the standard for 1 unit of product is 1.10. What are her weekly earnings?

(b) Find the full employee cost for the product if the employee has a FICA rate of 7.60%, unemployment insurance costs the company 2.1%, workers' compensation rate is 2.5%, there are 10 days of yearly vacation, and $400 per month for medical insurance. Approximate the actual wage and fringe cost per hour and per unit. Assume a weekly wage for a 52 week year. Yearly vacation is at base rate of $15 per hour.

2.23 A worker is paid the day-work rate if she earns less than 100% incentive premium. The standard is 75 units per hour, and the operator completes 140 good parts in 1.4 hours. The rate is $25 per hour.

(a) Find the piece rate.

(b) What are the earnings and labor efficiency?

2.24 (a) A sheet metal operator is told that her standard is 1.875 hours per 100 units. If her wage is $16 per hour and she makes 800 units in 12 hours, what are her earnings?

(b) If the actual time is 16 hours, what is her guaranteed pay and efficiency for the 100% plan?

2.25 Efficiency of 125% for an operation is calculated. The standard is 10 hours per unit and the wage is $23.50 per hour. What is the expected cost for 210 units?

2.26 A molding operator tends two machines and plastic products A and B are produced. Machines A and B have a production rate of 400 and 250 strokes per hour. Molds A and B produce three or four buttons per stroke respectively. The operator wage is $11 per hour. Dejoint these labor costs when allocation is based on (a) number of machines, (b) product output of machines A and B, and (c) marketing believes that A to B value is 5:4.

2.27 A time study observes the operation of producing two parts, which are labeled 1 and 2. The elements and their standard minutes are given as follows:

Element	Part	Standard Minutes
Load part in vise	1	0.17
Load second part	2	0.08
Balance and tighten vise	1, 2	0.17
Start machine	1, 2	0.06
Machine	1, 2	1.17
Stop, unload	2	0.13
Start, retighten	1	0.15
Fine machining	1	0.83
Stop, unload	1	0.14
Clean vise	1, 2	0.16

The labor wage is $23 per hour. Dejoint the cost using allocation methods of (a) number of parts, (b) time required by part, and (c) potential sales price ratio of 3:2.

2.28 One operator controls four automatic machines. After those machines are set up they produce parts independent of the operator except for occasional inspection. A cam controls the unit time to make one piece and is 4 seconds at 100% efficiency. Ignore the setup time because it is small.

(a) If the actual efficiency is 85%, the operator controls four machines, and the labor wage rate is $15 per hour, then what is the dejointed labor cost per unit?
(b) Repeat for hours per 100 units.
(c) Repeat for dollars per 100 units.

2.29 A technician tests and repairs printed circuit boards. On the average, printed circuit boards require 10 minutes for testing and 12 minutes for testing and repairing, if necessary.

(a) If, during a labor-hour study, 18 units tested OK and 5 failed requiring repair, what proportion of the labor wage is due to testing and repair?
(b) If the gross labor rate is $21.75 per hour, what is the labor charge per unit for testing and repair? Use proportional methods to dejoint cost.

CHALLENGE PROBLEMS

2.30 The fire department is concerned about the speed of the crews assigned to the all-purpose trucks. The chief wants to know the time required after receiving the alarm before starting to fight the blaze. To answer the chief's questions, a time study of the activities of one crew on seven different alarms is conducted. On the observation sheet that follows, continuous (no reset) electronic watch readings in hundredths of minutes are recorded, indicating full minutes only when it changed.

	Alarm						
Crew element	1	2	3	4	5	6	7
Start timing	.00	1.51	2.33	3.24	4.12	5.04	5.91
Get dressed	.12	1.58	2.43	3.36	4.25	5.15	6.01
Board truck	.29	1.82	2.60	3.55	4.41	5.32	6.19
Start engine	.44	1.94	2.76	3.70	4.55	5.48	6.32
Drive to fire*							
Unload hoses	.64	2.00	3.02	3.89	4.78	5.63	6.50
Connect hoses	.89	2.33	3.24	4.12	5.04	5.91	6.75
Unload ladders	1.08						6.95
Position ladders	1.51						7.31
End timing for this alarm	1.51	2.33	3.24	4.12	5.04	5.91	7.31

*Watch stopped because of variable nature of distances.

(a) Determine the average time for the elements.
(b) If the crew is rated at 110% for all elements, find normal time in minutes per occurrence.
(c) Determine the cycle-time standard for a 20% allowance.

2.31 We are interested in finding the estimated direct-labor cost for the job description of industrial electrician. The year consists of 52 40-hour weeks, and overtime is seasonal for 12 weeks consisting of Saturday work of 8 hours. The contract allows for nine paid holidays and two weeks of vacation at regular time. Four days of sick leave are paid. Expected nonchargeable hours are 5%. A subsidy for performance will be 15%. Nonhourly costs include FICA taxes at the current federal rate, workers' compensation at 1% of regular wages up to $15,000, accident insurance sum of $200, major medical plans for $500, and a Christmas gift of $50. The hourly base is $27.10.

(a) Find the effective gross hourly cost.
(b) What is the job cost for this electrician if a job requires 25 man-hours?
(c) What is the loss or gain if efficiency will be 85% or 115% for the job?

2.32 A preliminary work sampling study is made of an assembly operation. The work elements and the results are shown in the table below. The desired interval for each of the elements is also shown.

Element	Number of Observations	Desired Interval (I)
1. Assemble components	31	0.04
2. Setup	8	0.03
3. Wait for material	6	0.02
4. Painting	2	0.02
5. Inspection	3	0.02

(a) For a 90% confidence interval, find the number of observations required for the complete study.
(b) At the completion of the whole work sampling study, the productive elements amounted to 83% of the time. The study took 4 weeks and observed a crew of 6. Given a PF&D of 20% and a performance rating of 0.95, find the standard time per part if 23 parts were completed during the study.

PRACTICAL APPLICATION

Perform a preliminary work sampling study to determine the amount of program and commercial time for prime-time television. Determine times for 60 observations at random intervals between 8:00 and 11:00 P.M. Carry out the observations on one night for a major network (ABC, NBC, or CBS). (*Hint:* The selection of random numbers can be found using a telephone directory, where the last four numbers are considered "practically" random. A random number table from a statistics book can also be used. Spreadsheet programs will have a random number generator. It will be necessary to think up an algorithm for this conversion of a number called "random" to the observing time.)

Make snap observations of the following categories: programming, commercial, promotion of other programs or local station and news update or news related promotions, which will include weather. (*Hint:* Remember the "snap" notion of instantaneous recording.)

Calculate the amount of actual program time per hour. Prepare your report considering night, network, observation times, summary table of observations (number of observations for each category), and observations outside of the specified categories along with the description. Talk about your coding of the observing times. Conclude with a discussion of the belief of your work sampling conclusions.

Your instructor will provide additional instructions and reporting guidelines.

CASE STUDY: THE ENDICOTT IRON FOUNDRY

"We can't make any profit on that job," said Dick, the foundry superintendent, to George, engineer for the Endicott Iron Foundry. "There's too much labor cost in it."

The company, like its competitors, has always estimated costs on a per pound basis for the delivered casting. Difficult castings are quoted at a higher price per pound than simple castings, but the difference in price (often based on the estimated cost) did not seem to be great enough to warrant the extra labor costs.

Dick suggests that the company is making little profit and sometimes even a loss on jobs that take considerable labor. What will happen if Endicott starts quoting higher prices for casting requiring extra labor? Should it recover the full cost? What ideas can you suggest to improve the estimates? Should the engineer depend on the knowledge of Dick as final? What constitutes a loss or profit for the estimate?

In your consideration of the case study, you may want to call foundries and inquire how they estimate future work.

Chapter 3

Material Analysis

Engineering is intimately involved with the transformation of material from one state to another more advanced state. Ores, minerals, and petrochemicals are a starting point, and basic materials of iron and steel, gasoline, polymers, and paint, for example, become a finished product. Engineering is also engaged with the conversion of materials from a semifinished condition, such as annealed cold-rolled steel bar or sheet metal stock to machined or formed components.

Moreover, engineering uses these components to design the processes to assemble an endless list of products—from airplanes, appliances, cars, cell phones, computers, generators, and satellites to tools, toothpicks, and toys.

Look around you. What you see are the products of the engineering profession. Research, design, manufacturing, and construction are keystone activities in this conversion of materials to products. The infrastructure of free-enterprise businesses, from large original equipment manufacturers to suppliers and small job shops, bring together these material conversions.

Yes, engineering is a business of converting materials. Students need to understand the importance and the fundamental engineering approach to material cost analysis, because you will be involved.

3.1 BACKGROUND

We define *materials* as the substance being altered. This may involve iron ore, coke, and limestone to a basic blast furnace industry for producing pig iron; steel ingots to a steel rolling mill for refining and rolling into strip and coil; tin sheet to a can producer manufacturing 12-ounce tins; and cases of 12 cans for a food processor for canning frozen food.

The scope of what constitutes materials depends on the situation. Materials have been purchased, not manufactured, by the plant that uses the materials. Thus sheet steel is a product from a rolling mill, but is a material to a sheet metal forming plant.

PICTURE LESSON The Capitol Dome Under Construction

Library of Congress (photo by Corbis).

President George Washington laid the cornerstone of the U.S. Capitol building in 1793 and construction proceeded steadily. Burned by the British in 1814, the building was rebuilt, redesigned, and refurbished several times and has been refreshed with American spirit and idealism and inhabited over the years by lively, colorful, and dedicated people. The U.S. Capitol inspires awe and appreciation.

As the country grew, and as the building wings expanded to accommodate the increasing number of elected senators and representatives, it became apparent during the 1850s that the dome, dwarfed by the magnitude of the enlarged Capitol, was not attractive. The Bulfinch dome (named for the leading architect during the period 1818–1829) was built of masonry and wood with copper sheathing. Later the architect Thomas Walter used design elements of the great masterpieces of that time—St. Peter's in Rome and St. Paul's in London—to produce the enlarged dome of today that is instantly recognizable.

The photograph included here shows the construction in May 1861, less than a month after President Abraham Lincoln was inaugurated the first time. Troops were garrisoned to protect the Capitol during the Civil War. Cast iron frames were constructed over the existing dome. Statuary for the Senate pediment is in front of the steps. Throughout the trying months during the summer of 1861, construction on the great cast iron dome was halted by the Army Corps of Engineers. But President Lincoln's personal determination prevailed, and construction forged ahead. The Statue of Freedom atop the Capitol dome was installed amid a proud 35-gun cannon salute (one for each state) on December 2, 1863, during Lincoln's first term.

My country, 'tis of Thee, Sweet Land of liberty, of Thee I sing:
Land where my fathers died, Land of the pilgrims' pride,
From ev'ry mountainside Let freedom ring.

Engineering operations add value to a material by converting it from a received material into a deliverable product. This conversion results from the following manufacturing process categories:

- Mixing, combining, separating, and refining
- Casting and molding
- Cutting, shaping, and forming
- Changing properties
- Joining and assembling
- Cleaning, painting, coating, and finishing
- Electronic fabrication and packaging
- Packaging

Manufacturing uses plant facilities, equipment, capital, overhead, and labor for the conversion of the purchased material. For most operations the material that is purchased from suppliers accounts for about 50% of the cost of the product. Roughly, direct labor adds about 15%. Thus, it is important to know the cost of the material.

Prior to purchase of the material, exact costs may not be known. But to predict the material cost, compare various designs and material specifications, and to monitor and control material expenditures, estimates of the material requirements and costs are necessary.

Manufacturing processes material through a series of operations until it reaches a specified level of completion suitable for delivery as a product to a customer. The customer can be a purchaser of the product external to the company, or if the product is being transferred to a different plant within the company, the customer is considered internal. A supplier delivers its product as material to a customer for further processing.

The product design shows the completed physical description of the finished product. The fact that the production of the product changes some procured material into the finished configuration implies that the procured material is in some way different from the design condition. This means that both the starting design—procured material and part description and the finished design—must be considered when estimating material costs. In most manufacturing processes, some amount of the starting material does not end up in the finished product. Material is lost to chips and to small, unusable pieces in cutting operations, extra material is required in casting and molding operations in the gating structures—sprues and runners—and in most operations there is material lost because of processing errors. The cost of all materials, whether it ends up being shipped out as product or in the scrap bin, must be recouped in the product price and thus must be considered in the cost estimates.

Component parts are usually purchased in discrete unit quantities with a known or easily computed cost. The design specifies these parts and a bill of material gives the required quantity. The cost for these parts can be determined from vendor catalogs or specific quotations, or for early estimates, a "ballpark" cost can be obtained.

Material, on the other hand, is usually sold in bulk quantity and or in shapes that do not match exactly what is required for the design. A simple example is plywood, which is sold in 4 × 8 ft sheets. The plywood pieces required for a product may be 3.1 × 5.2 ft, but a full size sheet must be purchased because suppliers are not producing this size. In many cases material is not sold in the same units of measure that are used in the design. Most fabricated metal products are sold by weight, but the design is in linear dimensions. Calculations are undertaken to determine the material costs for these types of materials.

In analyzing the product design, engineering determines the *quantity* of material required to a high degree of accuracy. But finding the *cost* per unit of the material is not always as certain or accurate. Fluctuation of prices adds to the apparent disorganization of material cost. Inflation or deflation of material cost is possible owing to complex interactions.

The approach in this chapter is to identify the type of material, find its quantity, and select a cost policy to eventually find the cost per unit.

3.2 MATERIAL

Product design specifies all materials that are a part of the product. Everything that is used to make a product is referred to as *material*, from the significant to the insignificant.

Items that become a part of the product are called *direct material*. Examples are scrap metal and ingots that are melted for castings, metal bars, rods, sheet and plate that are machined, and plastic pellets used for injection molding. These raw materials are further manufactured or fabricated, or otherwise processed.

Materials that are used for processing but do not become part of the end product are called *indirect materials*. Examples of indirect materials are sand for casting molds, flux, and gasses used in welding or heat-treating and cutting fluids.

In many cases material items are acquired complete and ready to assemble directly into the product. These items are called *components* and are generally specified by quantity (e.g., 10-32 × 1 in. hex head bolts, 26 each).

In manufacturing industries, engineering determines the cost of material per product unit after the itemization of the materials has been established. The estimate of materials involves calculation to include allowances for waste, short ends, and other losses. After those calculations are completed, often with the aid of data from catalogs on weights, allowances, and CAD, costs are found. Costs can be determined with varying degrees of accuracy from sources such as the following:

Previous transactions (actual costs from previous orders)
Material cost estimates from similar previous designs

Inquiries to suppliers

Formal quotations

Catalogs, printed or online from the Internet

Information from these sources can be faulty due to frequent price changes and negotiated price-volume breaks between the company's purchasing agents and the seller.

Efficient use of time in preparation of cost estimates should be considered. For example, small inexpensive parts such as nuts, bolts, and washers can be estimated using a multiplying factor correlated to some other direct materials. This can be done for standard purchased parts whose costs are insignificant.

The design documents include the CAD files and drawings, engineering bill of material and specifications. The bill of material (BOM), or parts list, accompanies the design drawings and CAD files. The BOM consists of an itemized list of the materials for a design. Table 3.1 presents two examples of BOMs. The BOM is discussed again in Chapter 8. In some industries the BOM has other names. In the food industry the list of materials is called a recipe (e.g., for making cookies or snack crackers) or a formula (e.g., the exact contents of a soft drink). The list may be prepared on a separate sheet, sometimes in a computer file that allows easy transfer among design, manufacturing, procurement and inventory control, or it may be lettered directly on the drawing. A BOM contains the part numbers or symbols, a descriptive title of each part, the quantity, material, and other information such as casting pattern, number, stock size of the materials, weight of parts, or volume of materials.

Engineering is conversant with the drawings and specifications and is already familiar with the details of the design and contract. At this point it begins the job of calling out material requirements.

Specifications are separate from, but augment the bill of material and the engineering drawings. Basically, specifications state the technical details about the item.

TABLE 3.1 Simple Bill of Material: Fabricated Metal Part and Assembly of Components

Part No.	Description	Quantity	Unit of Material
2388-665	6061-T6, plate, 3 in. thk, 4 $\times$ 6 in.	1	Ea
2388-647	1018 Cold rolled steel plate, $\frac{1}{2}$ in. $\times$ 5 $\times$ 5 in.	1	Ea
2388-556	1/8 in. $\times$ 2 $\times$ 2 in. angle	5	Ft

Item	Part No.	Description	Quantity	Unit of Material
1	127-2348	Screw, skt head cap 8–32 $\times$ 3/4	8	Ea
2	127-0028	Nut, 8-32	8	Ea
3	56-324	Bearing, TCF 29-36	2	Ea
4	45-778	Wire, hook up, 18 ga, solid, red	5	Ft
5	2388-665	Housing	1	Ea
6	2388-647	End plate	2	Ea
7	FL342	Oil, SAE 90, gear	2	Lb
8	2300-005	Transmission assembly	1	Ea

Specifications put into writing all relevant details about the material so that there is no misunderstanding between the specifier and the supplier. For example, the tires for a new automobile are called out on the BOM as 4 each, "white lettered," 215/75-R15 (which describes the tire size). The tire specifications will describe the tire construction (steel belted), load and speed rating, mileage rating, and tread type (mud and snow). Additionally, government coding is lettered on the tire. Specifications are used to discuss the proposed work so that bids can be compiled or quotations collected so that the correct material can be purchased. Specifications are also used as a guide or a book of rules during manufacturing.

In addition to the specification of the material, it is important that correct quantities such as units, weight, and volume are known. To have a correct cost such as 86 cents per pound of casting material and then improperly specify the number of pounds required for the casting is a serious flaw. To determine the quantities involved, engineering examines the CAD files and specifications and makes a material estimate based on the part design and the shape and size of the raw material. For example, an engineer will calculate the weight of steel in a casting, or the length of a piece of bar stock used to make one or more parts.

Direct materials are further divided into raw materials, components, subcontract items, and interdivisional transfer materials. These material types relate to each other and are used by a company as shown in Figure 3.1. *Raw materials* include fabricated, intermediate, or processed material in a form that will receive direct labor work in conversion to another design. *Components* are a class of materials normally not converted; rather, they are accepted in a manufactured state (tires used by an automobile assembly plant, integrated circuits used in the electronics industry, standard hardware items such as bolts, washers and nuts). Component materials may be a significant portion of total material cost and are considered separate from fabricated raw materials.

Subcontract items are parts, components, assemblies, intermediate materials, or equipment produced by a supplier or vendor in accordance with designs, specifications, or directions applicable only to the design being estimated. These items do not appear

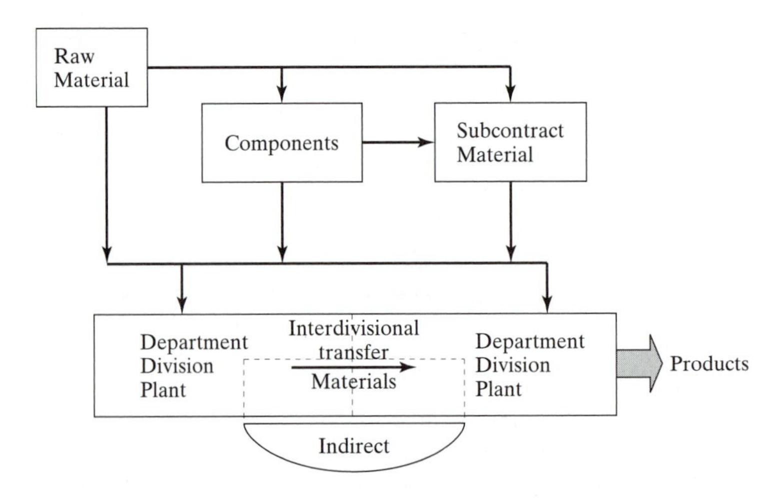

Figure 3.1 Material relationships.

in a supplier's catalog and are generally not sold to any other customers. Examples include castings, specially designed extruded shapes, weldments, and other parts made by some other company to a particular company's design.

Interdivisional transfer materials are materials sold or transferred between divisions, subsidiaries, or affiliates that have common ownership or control. Those sales are ordinarily handled on a cost, no-profit basis. Occasionally, other arrangements may be made for items regularly manufactured and openly and widely sold.

Some materials are used in the manufacture of products but are not specified explicitly on the BOM and may not actually become part of the product. These *indirect materials* consist of things that are required but that are handled in bulk and purchased as part of overhead expenses. Indirect materials are those things necessary for the conversion of direct materials and are not directly traceable to the design. Solder, welding rod, lubricating oil, process gasses (oxygen, acetylene), and perishable tooling costs are expenses that are indirect. Some materials can be classified as either indirect or direct. Fuel, such as natural gas, can also be a raw material (cracking refinery gases for the production of ethylene or for heat-treating furnaces). Convenience of the costing dictates whether it is simpler to classify some material costs as direct or indirect.

Other indirect costs include operating supplies such as shop rags, gloves, brooms, and small hand tools, which are too diverse and unimportant to be considered in the cost analysis. Company records are generally used to determine the costs for indirect materials. Overhead is the usual way to account for their costs.

3.3 SHAPE

Raw material is purchased to manufacture the design. How much material is necessary and what will it cost are two questions. Three different quantities of material are involved:

1. The amount of raw material needed to make the design—which is always greater than or equal to the amount of material in the design
2. The amount of material theoretically composing the design, called part shape
3. The difference between these two, which is the loss during manufacturing processing

Shape describes the amount of material in the design. Shape implies mass (or weight), volume, area, length, count, or some other convenient engineering dimensional unit. The shape of the design should be determined using units compatible with, or easily converted to, the units used for pricing and purchase of the raw material.

Since estimating is in advance of production, an actual part is unavailable to allow measurement of the amount of material. If a part did exist, its weight could be determined. Similarly, the volume of some parts can be determined, for instance, by submersing them in water and measuring the volume displaced. As a reminder, assemblies of different materials will not be successful for this procedure. Other physical measurements are possible to find part properties.

Engineering makes two calculations: (1) finding the shape and (2) optimizing the shape with raw material. For parts that are in the design stage and yet to be made, shape is found by calculations. For simple parts the dimensions obtained from the designs—either paper or monitor—are used to calculate the volume. An algebra known as mensuration—a branch of mathematics dealing with the determination of length, area, or volume—has many special equations to aid the calculation. For some designs, engineering breaks the part into simpler sections and computes the volumes of these smaller portions and then adds or subtracts the resulting pieces to obtain the shape (see Figure 3.2[a]). The density of the material converts the calculated volume into weight. If one of the many popular computer-aided drawing packages is used to create the design, the program calculates volume.

Seldom do we see simple parts like Figure 3.2(a). Furthermore, if the design is an assembly of two or more dissimilar parts, the problem is more complicated. Different materials and densities make the calculation procedure more interesting. We leave those considerations for the student.

In addition to the ideal shape, the amount of material required to manufacture the part is necessary, and frequently is more. The ideal shape requirements are found from the drawings. The amount of material to manufacture the part often depends on a particular process and the arrangement of the part dimensions to the raw material dimensions.

This can be seen in the arrangement of the design on a purchased flat, glazed, and hardened pure platinum sheet, 10×15 cm, as shown in Figure 3.2(b). A rectangular part, 2.1×4.8 cm, is required. Shearing will separate the sheet first into strips and then into blanks. The dashed lines show shearing. The shearing process separates the material without any loss, unlike a sawing procedure, which will lose material because of the kerf of the diamond blade.

If the parts are arranged as in Figure 3.2(b), then two strips per sheet and seven blanks per strip for 14 blanks per sheet can be produced from one sheet of material. If the parts are arranged in the other direction as shown in (c), then only 12 parts can be obtained from a sheet. Engineering gives inordinate but necessary efforts to optimize shape on raw material, because of the importance of the efficiency of the shape with respect to raw material cost.

This example may seem like an obvious choice, but situations can be more complex. The arrangement of the parts in the example shown in Figure 3.2(b) may be determined by other requirements besides simple parts per sheet. The part orientation can be dictated by design or manufacturing. For example, there could be a requirement for matching the grain direction, which results from rolling the sheet or because there is some limitation in the processing as the material only fits into the processing equipment one way. In addition, manufacturing often requires extra material around the part to allow for processing.

Solid material that will be machined to remove unwanted material to reveal the design shape is usually procured in specific sizes. There are many examples of this, such as material obtained in bars, sheets, plates, or coils. These stock sizes are usually available in dimensions and can be purchased only in unit quantities, even though the costs may be quoted in some other customary units ($/lb, $/ft, $/each, etc.) Engineering must determine the amount of material to purchase in order to manufacture the parts, taking into account both the shape of the individual parts and the quantity to be made. The cost of the total amount of material procured for the production order

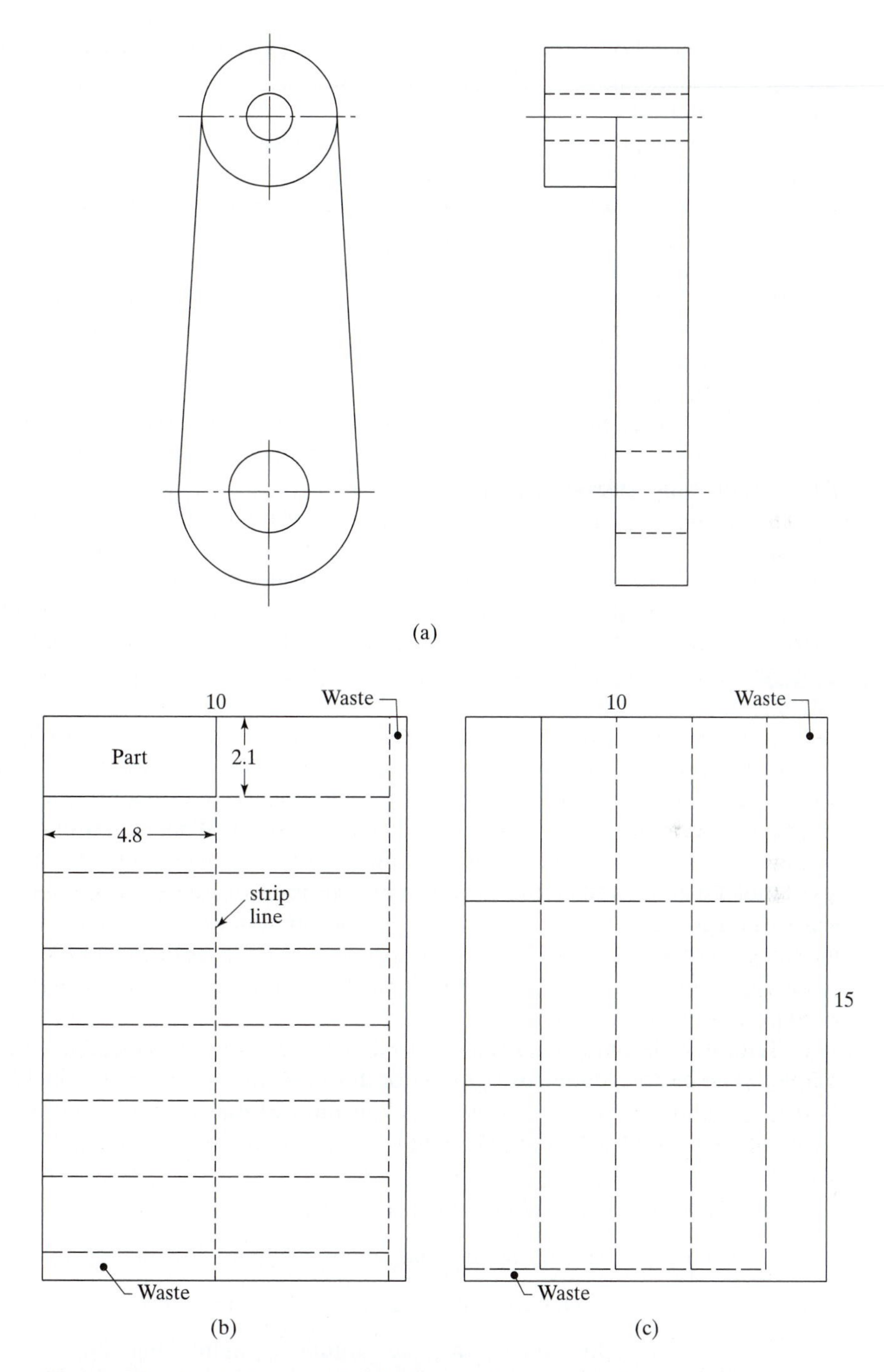

Figure 3.2 Typical material calculations for cost analysis: (a) finding volume and weight by mensuration calculations; (b) optimizing the efficiency of raw material into part shape; and (c) suboptimal layout of blank on sheet giving only 12 units.

is then apportioned over the number of parts actually produced. In the case of cast or molded parts, manufacturing adds material to the shape to accommodate sprues, runners, risers, draft, and cores.

The part shape is the desired minimum amount of material. As has been explained, the actual amount of material required to make the part is larger. The difference between the two is given the term *waste*. The waste is composed of chips, which are cut away during machining, sprues, and runners cut off after casting, and the trimmed edges during stamping. Waste is determined by the design in combination with the raw material dimensions and the manufacturing processes and is generally constant for a specific part. An interesting word, *offal*, is sometimes found in machining operations, and means waste, but its original meaning was the drop-off waste of a butchered animal.

A great deal of engineering effort is concerned with the reduction or elimination of waste. Despite this effort, waste can sometimes amount to as much as 15% or more. Then too there are processes that are waste free.

In addition to waste, manufacturing processes generate *scrap*—faulty material caused by human mistake or equipment malfunction. For example, mislocation of drilled holes is a shop mistake. However, if the mistakes are caused by the designer's error (a hole position was dimensioned incorrectly), then the scrap is the fault of engineering. The allowance for scrap can vary and is usually controlled by monitoring and management. A scrap allowance factor is usually determined by past experience from actual process records.

Another loss is *shrinkage*—the loss of material because of theft or physical laws. Originally, shrinkage dealt with volumetric reduction of lumber owing to drying, but now we mean the economic effects owing to deterioration of materials due to aging, oxidation, chemical reaction, natural spoilage, and so on, if reduction in quantity and quality will occur. For example, polymer raw material may have a limited shelf life, and, if not molded and cured before the onset of deterioration, there is economic, not physical, shrinkage. Rusting of ferrous materials is a common form of shrinkage. Food commodities have serious problems with shrinkage losses, both economic and in the reduction of flavor or moisture. As with waste, a shrinkage factor is usually available from process records.

The total amount of material required to manufacture the part, called the starting shape S_s, can be determined by figuring for the losses in addition to the final part shape S_f. For the additive type processes—molding and casting—the amount of starting material can be computed by the following:

$$S_s = (S_f + L_1) \times (1 + L_2 + L_3) \tag{3.1}$$

where S_s = starting shape, magnitude in engineering dimension

S_f = final part shape, magnitude in engineering dimension

L_1 = loss due to waste, magnitude in engineering dimension

L_2 = loss due to scrap, allowance, decimal

L_3 = loss due to shrinkage, allowance, decimal

Example: Gray Iron Casting, an Additive Process for Finding Shape

A gray iron casting is being designed. The CAD system calculates a volume for the casting of 350 in.^{3}, and the foundry engineer estimates that the gating system (sprue, runner, and riser), and pattern draft will require approximately 75 in.^{3} of additional material. In addition, the scrap allowance for gray iron casting with the green-sand process is 15%. A shrinkage allowance for stack loss of iron due to repeated cupola remelting is estimated to be 5%. How much material is needed for the casting? The amount of material to be costed S_s is

$$S_s = (350 + 75) \times (1 + 0.15 + 0.05)$$
$$= 425 \times 1.2 = 510 \text{ in.}^3$$

If the density of gray iron is 0.26 lb/in.^{3}, then the weight is 132.6 lb.

For the subtractive processes—machining, presswork, and many more—the number of parts of S_f that can be obtained from a single unit of procured material, parts per sheet or bar, must first be determined. Then the number of units of purchased material needed for the quantity of finished parts can be calculated. Unusable material required for clamping, the material lost separating parts, and dressing ends must be included. A single, simple equation like (3.1) does not exist for all possibilities. However, the following is useful in determining the amount of starting material:

1. Obtain the finished part shape S_f and quantity.
2. Determine the specification of the raw material. This consists of the dimensions of the specified or supplied material.[1]
3. Determine cutoff, holding and clamping, and die stamping material requirements.
4. Layout the parts for the cut out or stamping of the raw material.
5. Determine the number of parts that can be made from each unit of raw material, n_1.
6. Divide the total number of parts required by n_1 to obtain the number of units of raw material.
7. Compute the total amount of material in units compatible with the pricing scheme—pound for $/lb.

To compute the cost of material for the part, all of the material—both comprising the part shape and the removed material—must be considered. The volume for costing is computed from the raw material dimensions. In addition, allowances for scrap and shrinkage must be included.

Example: Bar Stock Component from 6-ft Raw Material, a Subtractive Process

A $1\frac{7}{16}$ in. OD carbon steel bar has a finish length of $1\frac{1}{8}$ in. Historical experience indicates that scrap L_2 is 1% for these operations. Knowledge of the process suggests that waste is composed of a 1/8 in. cutoff tool width, and a facing length of 1/64 in. is required

[1]The American Standards for Testing Materials (www.ASTM.org) is one of several agencies that officially promulgate standards and give extensive details about the material.

for accurate dimension. The lathe selected for this turning is limited to a 6 ft bar length. The lathe collet requires a gripping length that reduces the quantity produced from the bar stock by one.

Thus each unit will require $1\frac{1}{8} + 1/8 + 1/64 = 1.266$ in. of length, and the exact number of pieces will be $72/1.266 = 56.9$. Reducing this to 55 units allows for an integral number of parts and material for last-part gripping. The amount of starting material S_s that is apportioned to each part is $72/55 = 1.31$ in. In addition to this, the 1% scrap allowance is included by $1.31 \times (1 + 0.01) = 1.32$ in. This is slightly larger than the part length including the cutting allowances and larger than the finished part length S_f of 1.125 in.

We can calculate the percentage of waste using the length dimensions. The percentage of the total raw material remaining in the finished part is $55 \times 1.125/72 \times 100 = 85.9\%$, so the waste L_1 is $100 - 85.9 = 14.1\%$. There is no shrinkage, $L_3 = 0$, for this part.

Waste in this example includes grip length, short ends, cutoff, and facing material. We assumed a finished bar stock diameter equal to the original diameter. If there were chip removal on the circumference, then our volume calculation would find the exact final amount plus those losses. Based on the above information, the volume of raw material required per part is computed:

$$S_s = \pi \times (1\tfrac{7}{16})^2/4 \times 1.32 = 2.142 \text{ in.}^3$$

Example: Panels from Sheet Stock

Metal desktop panels 30×60 in. are to be sheared from 0.0625 in. thick sheets that are 72×120 in., which are sold by the pound. Experience shows a scrap rate of 5% and no shrinkage. What is the material quantity that should be used for calculating the material cost?

Four top panels can be sheared from one sheet. So, for 4 pieces $S_s = 72 \times 120 \times 0.0625 = 540 \text{ in.}^3$, and material required to for each top panel (including the design waste) is $540/4 = 135 \text{ in.}^3$

If we assume that the build quantity is sufficiently high to apportion out the scrap, then the purchased amount of material required for each top is $135 \times (1 + 0.05) = 141.75 \text{ in.}^3$

The waste can be calculated per panel as $= S_s - S_f$. Continuing, $S_f = 30 \times 60 \times 0.0625 = 112.5 \text{ in.}^3$; and waste $= (540/4) - 112.5 = 22.5 \text{ in}^3$. The waste L_1 is found by dividing the waste amount by S_f, or $L_1 = 22.5/112.5 \times 100 = 20\%$.

Efficiency of the material conversion is always important. Engineering computes this metric by

$$E_s = \frac{S_f}{S_s} \times 100 \tag{3.2}$$

where E_s = shape yield, percent

Example: Finding Shape for Stamping Fabrication from Strip Stock Material

Stamping products require unique shape calculations. Stamping fabrication of sheet metal parts begins with coil stock or with shearing and blanking operations from sheets

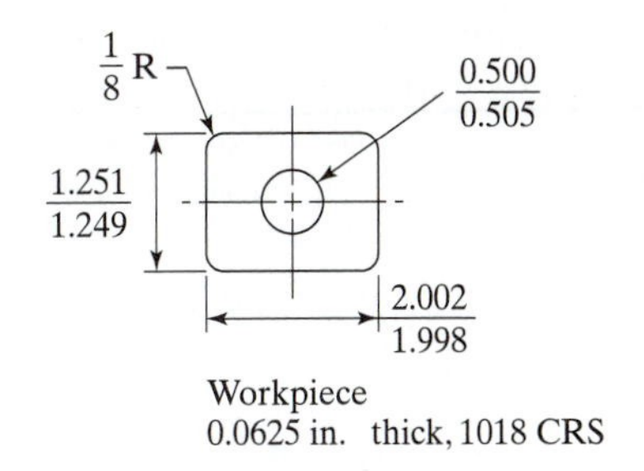

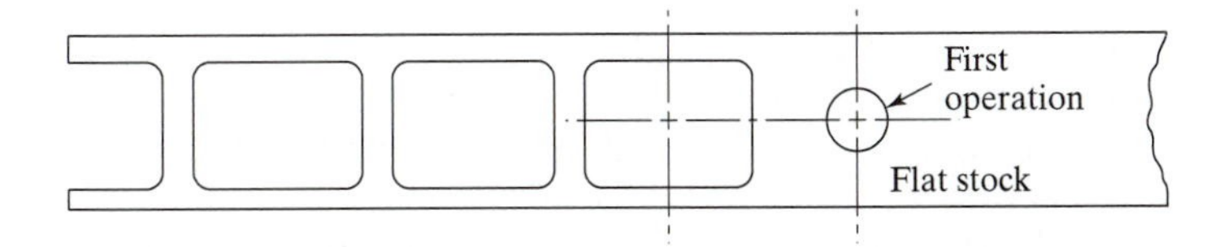

Figure 3.3 Sheet metal stamping.

or strips. Study the work piece shown in Figure 3.3. This design is fabricated from rigid strip stock, though coil stock is a possibility. Since quantities are low, however, coil raw material is unadvisable.

The die for this stamping operation is two station: (1) the point where the first station punches a hole in the strip, and (2) the point where the second station blanks out the work piece. The material is supplied as strips. The strip is wider and longer than the work piece because the strip is pulled by the advancing mechanism of the die. After the workpiece is separated, a skeleton remains, which is waste. Because of the stamping fabrication requirements, the practical distance between blanks is limited to a minimum of 0.75 × thickness, and the margins between the edge of the strip and the blank must be at least 0.90 × thickness for each side.

The width of the strip is calculated as $1.251 + 2 \times 0.9 \times 0.0625 = 1.3635$ in. In sheet metal manufacturing, the dimension of the part that is along the length of the strip plus the spacing allowance is called the "advance." For this example, the advance is $2.002 + (0.75 \times 0.0625) = 2.0489$ in. If $L_1 = 0.25\%$ and $L_3 = 0$, and if we ignore the corner radius and the round slug, then $S_s = 1.3635 \times 2.0489 \times 0.0625 \times (1 + 0.0025) = 0.175$ in.^{3} Notice that the dimension includes the plus side of the tolerance, rather than any nominal dimension. This practice is conservatively admitting that the material usage will be maximum.

Now the density for AISI 1018 cold rolled steel is 0.28 lb/in.^{3} and then the weight per unit = 0.049 lb. For this stamping example (and where the four corner radii and the slug have been included in the calculation), the shape yield can be calculated based on the part area as

$$S_f = (1.250 \times 2.000) - (0.25^2 - \pi 0.125^2) - (\pi 0.25^2) = 2.2902 \text{ in.}^2$$

$$E_s = \frac{2.2902}{(1.3625 \times 2.0489)} \times (1 + 0.0025) \times 100 = 82.2\%$$

Now we consider the popular 12-oz beverage can, which is perhaps the most advanced application of all presswork operations. For presswork technology, it may be argued that the beverage can is the most ubiquitous metal container ever made, being used for countless sodas, beers, and other drinks. Some container plants produce 25 million units per day. In terms of total presswork production quantity ever manufactured, perhaps only drawn bullet-casings or coins exceed the popular can.

Example: Calculating the Shape Requirements for the 12-oz Beverage Can from Aluminum Coil-Stock Material

Beverage containers, those handy cans that seem to be everywhere, are amazing in many ways. With annual worldwide production in the billions of units, engineering is relentless in its pursuit for lowering costs for the container. Such cost is mostly material, and cost per unit and yield of the process are the driving metrics.

The modern-day aluminum can was first developed in the 1960s as a 5-oz extruded design. The process later morphed into punching a circle blank from soft aluminum coil stock, then drawing an intermediate cup, ironing and drawing the cup to its height, and trimming off the excess waste from the open end.

An example of material estimating is given for the 12-oz beverage can, which is composed of the body, top, and pull ring. We consider only the container body here, which is blanked from 3004-0 aluminum coils and an intermediate cup is formed without any significant change in wall thickness. The cup is drawn in a horizontal drawing machine, and metal is squeezed to a sidewall thickness of 0.0055 in. The bottom thickness remains unchanged. The can is trimmed to the final height to give an even edge for later rolling to the top. Four blanks are punched out simultaneously for the 5.3176-in. advance in the die, and the coil is repeatedly advanced and punched in this manner, leaving a skeleton of waste, still in coil form.

Figure 3.4 illustrates a "4-out" from a 19.231 in. wide aluminum coil. The aluminum supplier will provide these engineered coils because of the extreme demand for this material.

The dimension of 5.3176 in. shown in Figure 3.4 gives four blanks for each advance of the coil. Over the years, the number of blanks per advance—which increases the yield of the operation—has been increasing. Parenthetically, as the width of the coil stock increases, the blanking, cupping, ironing, and drawing operations require bigger press tonnages and larger dies, so the engineering decision to increase the width is complex. It involves many cost analysis and estimating choices for material, equipment, plant, material moving and handling and the like.

Table 3.2 gives a summary of the calculations. Line 1 provides the volume in a 12-oz can body, and line 2 indicates the strip volume of advance. Dividing by 4 gives the can body volume. Line 3 provides the metal efficiency of the can body to strip. The blank, 5.2476-in. OD, also has losses in drawing to the final can, and line 5 indicates a yield of 86.3%. Line 6 indicates the cost per pound. The density of the material is 0.0982 lb/in.^{3} Line 8 indicates the salvage value for the waste, and line 9 shows the recovery on a per can basis. Finally, the net cost per can body is calculated on line 11 as $0.0384. The serious student may want to calculate a "five-out" (see Problem 3.7) and make comparisons to the "four-out."

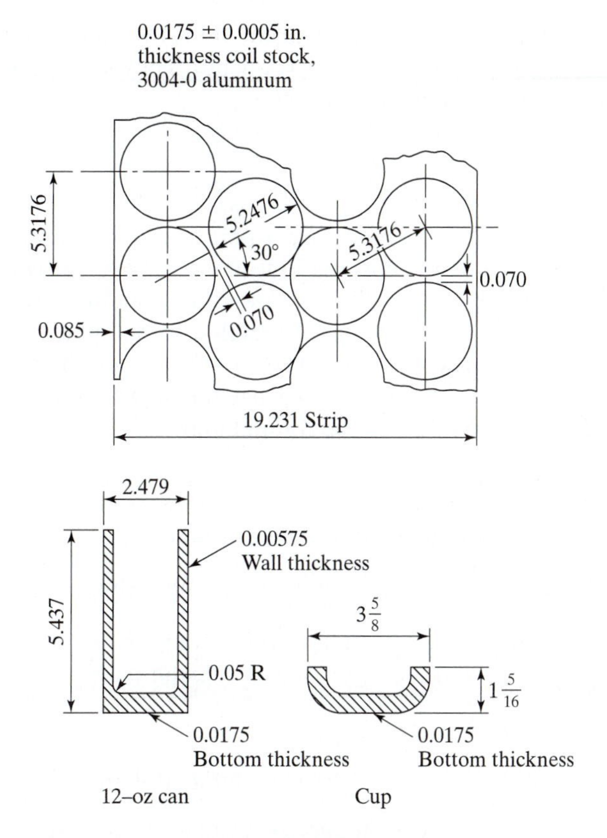

Figure 3.4 Engineering design layout of popular 12-oz beverage can body.

TABLE 3.2 Calculations for Direct Material Cost for a 12-oz Can

Line	Property	Calculation	Result
1	Final metal in 12-oz can body		0.3266 in.^{3}
2	Strip volume per can body	$5.3176 \times 19.231 \times 0.0175/4$	0.4474 in.^{3}
3	Metal efficiency	0.3266/0.4474	73.0 %
4	Metal volume in 5.2476 in. blank		0.3785 in.^{3}
5	Can body to blank efficiency	0.3266/0.3785	86.3 %
6	Cost of metal		\$1.0017/lb
7	Cost of can body	0.4474×0.0982 lb/in.^{3} $\times 1.0017$	\$0.0441
8	Waste salvage value		\$0.483/lb
9	Salvage of waste per can body	$19.231 \times 5.3176 \times 0.0175(1 - 0.73) \times 0.0982 \times 0.4830/4$	\$0.0057
10	Metal cost in can body	$0.3266 \times 0.0982 \times 1.0017$	\$0.0321
11	Net cost per can body in strip	$0.0441 - 0.0057$	\$0.0384
12	Yield	$0.0321/0.0384 \times 100$	83.6 %

We have provided several examples of calculations for simple designs. Material shape calculations are unique, and geometry, production knowledge, starting material, costs, and yields are important to the tabulation.

3.4 COST

The cost of the material to manufacture a part can be determined from the starting shape S_s by multiplying by the cost per unit of measure of the raw material. For some materials, the scrap and waste, called *salvage,* can be sold to a recycler or the original supplier or it can sometimes be reused/recycled in manufacturing process. In these cases, the value of the salvage can offset part of the cost of the material. Conversely, if the manufacturer must pay to have the salvage taken away, this cost adds to the cost of the material. The salvage value can be computed by

$$V_s = (S_s - S_f) \times C_s \tag{3.3}$$

where V_s = salvage value, dollars per unit

C_s = salvage value, units compatible with shape dimensions

C_s is negative if there is a charge to have the salvage material hauled away. If the salvage cost is significant, then the material cost calculation is

$$C_{dm} = S_s C_{ms} - V_s \tag{3.4}$$

where C_{dm} = cost of direct material, dollars per unit

C_{ms} = cost of purchasing material, dollars per unit

We presume that S_s and C_{ms} are in compatible units. If both sides of the equation are multiplied by the number of pieces produced N_p, then we have total or lot cost, depending on the nature of N_p.

It is possible to compute the overall efficiency of the economic conversion of raw materials per part by using the equation

$$E_m = \frac{S_f C_{ms} - V_s}{S_s C_{ms}} \times 100 \tag{3.5}$$

where E_{ms} = material cost yield, %

Example: Material Cost with Salvage Value

Laminations for a 14,200 kva transformer core are blanked and formed from a special steel provided in coil stock. The electric-arc steel supplier sells this material for \$0.925/lb. The fabricator waste is 3.25% due to design; manufacturing contributes 0.4% scrap and there is a 0.5% shrinkage loss because of rust and irregular edging during storage. New material costs \$0.925/lb; the steel supplier purchases salvage for \$0.07/lb. A new transformer is designed with a core weight S_f of 9255 lb. Determine the material cost:

$$S_s = 9255(1 + 0.03255 + 0.004 + 0.005) = 9639 \text{ lb and}$$

$$V_s = (9639 - 9255) \times 0.07 = \$26.89.$$

Finally, $C_{dm} = 9639 \times 0.925 - 26.89 = \$8889.19$ per transformer.

3.5 MATERIAL COST POLICIES

Once the amount of material is determined, material cost analysis proceeds to the second step of finding the cost of that material. This is not simple because there may be multiple materials—with different prices—that fulfill the requirements, prices fluctuate with quantity and time of purchase, and companies negotiate various contracts that add to the puzzlement of determining an accurate cost value. Furthermore, a manufacturing firm or job shop may not order materials until the job is won. This means that the cost of materials will not be known exactly until commitments have been made.

In view of the difficulties of picking a cost for material, engineering fashions policies to allow consistency and accuracy for these procedures. That is what this section is about—understanding the principles and practices.

Engineering or purchasing needs to provide a cost estimate for design material. Complications in material cost finding arise because of several reasons:

- Specification and magnitude of choices
- Commodity materials
- Special or common design
- Quotation method
- Inventory methods

3.5.1 Specification

Suppose that there is a design, that specifies "steel,"[2] and a price is required. A call is made to a steel supplier, and, surprisingly, a steel warehouse is perplexed when given the simple request for a material cost for the alloy steel. The choice of their response extends to many thousand, and nationally there will be millions of selections.

Common steel contains 0.18% of carbon, the most popular alloying element, which is customarily referred to as AISI 1018 steel.[3] This low carbon grade steel can be provided in specified shapes—flat, hex, round, square, rectangle, sheet, bands, coil, angles, strip, bar, tubing, plate, wire, blooms, billets, ingots, structural shapes and more—and in hot or cold rolled condition (and if in cold rolled, then in polished or oiled) and it may be annealed or tempered or to a designated maximum hardness.[4] Materials are called "merchant quality"—with little qualification—or are issued with a certificate of quality.

[2]For iron and steel prices, see the following addresses: www.iss.org/resources/producers.htm; www.onlinemetals.com (some aluminum too); www.armstrongtools.com; www.metalprices.com.

[3]AISI is American Institute of Steel and Iron.

[4]Technically, the steel AISI 1018 is not a hardenable grade: thus for hardness, annealing, and other heat treatment requirements, a steel having more carbon and other alloying elements (i.e., AISI 1045) would be called out.

PICTURE LESSON The Bessemer Process: The First Patent to Make Steel

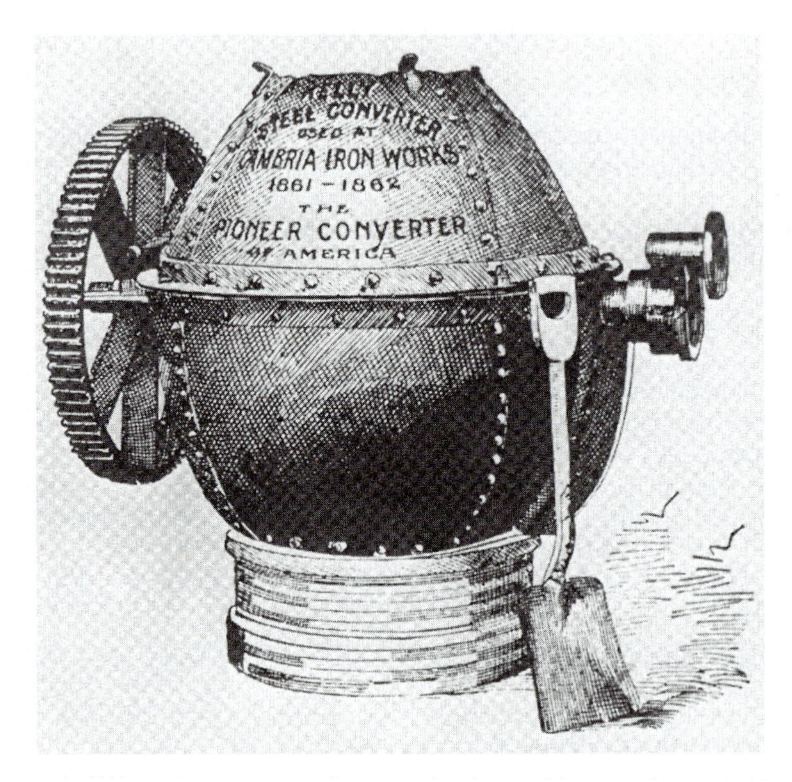

Smithsonian Institution.

The first method discovered for mass-producing steel is named after Sir Henry Bessemer, an English engineer, but the process evolved from the contributions of many investigators. It was apparently conceived independently and almost concurrently by Bessemer and by William Kelly, a businessman-scientist from Pittsburgh, Pennsylvania. As early as 1847, Kelly began experiments aimed at developing a revolutionary means of removing impurities from pig iron by an air blast. Kelly theorized that the air injected into the molten iron would supply oxygen to react with the impurities, converting them into oxides separable as slag, but also that the heat evolved in these reactions would increase the temperature of the mass, keeping it from solidifying during the operation. After several failures, he succeeded in proving his theory and rapidly producing steel ingots. The photograph shown here of the engraving of Kelly's "pot" is the one used at the Cambria Iron Works, a very early iron plant. Air was blown into the bottom to burn out the impurities such as sulfur and to "convert" the molten iron into low carbon steel.

In 1856 Bessemer developed and patented the same process. Whereas Kelly had been unable to perfect the process owing to a lack of financial resources, Bessemer was able to develop it into a commercial success. Another engineer found that adding an alloy of carbon, manganese, and iron after the air-blowing was complete restored the carbon content of the steel while neutralizing the effect of remaining impurities, notably sulfur. With improvements and reliability, the end result was a means of mass-producing steel. The resultant volume of low-cost steel in Britain and the United States soon revolutionized building construction and provided steel to replace the brittle cast iron in railroad rails, for example. The discovery was important to the Industrial Revolution, which was taking place at that time. Even today, more tons of steel are produced than all other ferrous and nonferrous metals.

The invention of the open-hearth process in the late 1860s eventually outstripped that of the Bessemer process. Today this has yielded to oxygen steel making with the basic oxygen furnace, which blows in pure oxygen, and is a derivative of the Bessemer process. The impressive process starts with scrap, about 30% of the total charge. Molten pig iron is poured into the mouth of the vessel, and a water-cooled oxygen-carrying lance is positioned above the bath. With oxygen blowing over the surface, ignition starts immediately and the temperature climbs close to iron's boiling point of about 3000 F. Carbon, manganese, and silicon are oxidized. Fluxes are added and impurities are removed as slag, and the tap-to-tap time to produce 300 tons of high-grade steel alloy is about 45 minutes, a huge step from Kelly's tiny converter.

At this point, the length of the raw material may be specified—say, 20, 16, 12 or 6 ft or a random length. There may be dimensional tolerances that are large, standard, or tight, and these may depend on the shape, condition, and so on. The stock may be turned, ground, and polished. The material may be banded to keep the stock together or it may be delivered loose. After that opportunity, the material may be paper wrapped and protected, or bundled in a box.

Quantity considerations are important. If only a 16 ft bar is available, and one and one-half bars are necessary, then is there a cutting charge to prepare the one-half bar? Is the inquiry for a few or for many units? All of this makes a difference in the cost a steel supplier will quote.

Is the request for AISI 1018 steel of one bar, where a 16-ft bar and $1\frac{1}{2}$ in. dia is 120. 3 lb, or does the request fall into other weight blocks for which there are price breaks? Indeed, the quantity may be so large that the supplier can quote "base quantity" rates, which begin at 10,000 lb, and are a special order to a steel mill on a new-order request. Cost-per-pound values do not reduce once the base quantity is exceeded.

Is the request so unusual that only a steel supplier some distance from the plant can fill the specification? Delivery costs or FOB (free on board, meaning that the material is loaded onto a truck free) is a factor.

The combinations of the response ranges in the thousands for a simple 1018 steel, and they vary every day. The magnitude of choices is large. The design requirement for "steel" is unsatisfactory, and it needs to be clear and precise—a specification following the rules of a national or international standard is necessary.[5]

[5]See the information provided by the American Standards for Testing materials (www.ASTM.org; www.iso.ch/iso/eng/ISOoline.frontpage).

3.5.2 Policies for Evaluating Commodity Materials

Some purchased materials are *commodities*.[6] For our understanding here, these material can be resistors,[7] printed circuit boards, nuts, bolts, and items that are available to a standard specification by a number of suppliers. Indeed, they are widely obtainable in very large quantities, and the maker of the commodity will not know the specific end user. Furthermore, engineering may not provide a CAD file or drawing for the commodity but will rather use a *call-out* for the item on the drawing or the BOM. Commonly stocked materials—such as those sold from a catalog, or items that are widely used and are produced by a number of individual manufacturers—have an ongoing market price. In this case, market competition establishes reasonable prices. Engineering uses those market values as they determine a cost. Commodity materials are costed by engineering or purchasing.

For the most part, engineering is concerned with new or revised designs that are special and with bringing such designs to a marketable state. These designs are proprietary to the firm, and they are the beginning point for cost analysis.

3.5.3 Policies for Evaluating Contractual Materials

Analysis of direct material cost is separated into contractual and inventory methods. The contractual method, used for subcontract and special order materials, implies that a buyer-vendor arrangement exists, and engineering solicits a material quotation for the specific design. Complex and special items that have limited application and few suppliers are often subject to wide price variation. Greater care is required to estimate those values.

For large orders or special stocked material, the *quotation and cost method* is the most widely used one to determine material cost. In this method, the cost of the material is established between engineering and the supplier. The price is considered fixed, subject to the guarantees of the mutually agreed-to contract.

Quotations can be solicited orally if insufficient time is available or if the purchases are relatively small in value. Written solicitations should generally be tendered when special specifications are involved, when a large number of items is included in a single proposed procurement, or when obtaining oral quotations is not considered economical or legal. In most cases, facsimile copies can be considered the same as written communication and for some companies, e-mail quotes are valid. Involved in the wording of the tendered quotation are statements about the design and specifications, terms or conditions, delivery date, and price. The obligation to contract at fair and reasonable prices does not diminish as one moves down the scale from multimillion-dollar contracts for system acquisition to the dime and quarter prices for nuts and bolts.

A subset of the quotation and cost method is the *quote or price-in-effect* (QPE) method, which is a collaborative legal agreement between the buyer and seller. As usually established, the contract allows for adjustment to the original price should the seller incur material costs in excess of those estimated. If the seller's material cost falls

[6]An example of commodity prices may be examined at www.gs.com/gsci/#components.

[7]For electronic components, see www.futurlec.com/components.html and www.partminer.com.

below that which was estimated in the contract, then the buyer agrees to the original price and adjustment is unnecessary. If at the time of delivery of the product the seller can prove that the material costs incurred to the seller escalated above the original estimated value, then the buyer will make up the difference by use of a formula or through negotiation. The QPE method may seemingly avoid the troublesome problem of making accurate material analysis and forecasts. However, competition may not allow the luxury of this type of contract.

The QPE method becomes more popular during periods of inflation, as the supplier wants protection against price increases in the company's own supply chain. Should suppliers have these price increases, they desire to pass them along, down the ladder, and eventually to the customer.

The engineering policy choice for contractual methods is simple. The value of the contract is adopted for the material cost. But when a company employs an open inventory to stock its materials for the manufacture of the product, the policy choice of material cost is not so evident.

3.5.4 Policies for Evaluating Materials from Inventory

When materials are purchased specifically and are identifiable with a contract, the actual (or estimated) purchase cost is used as the estimate for the design, and the contractual methods as described earlier are appropriate. However, when materials are issued from inventory for manufacturing the special design, it is difficult to find the value because inventories are extensive and constantly changing. Most manufacturers carry raw material inventories, and material costs are affected by the method used to evaluate inventory. Frequently, accounting has established the rules for the evaluation of the inventories, and these methods are sometimes incompatible with the needs of engineering.

Costs vary according to the method and market conditions. Often the materials are consolidated for many designs (a company may standardize on a certain grade and size of steel for shafts). Furthermore, materials are purchased at different times and in various quantities. Quantity discounts are normally offered as a business practice, and lower prices can be obtained when several design requirements are consolidated into a single purchase order.

Consider the representative inventory situation shown in Figure 3.5. Units of a material are plotted against time periods. The units of measure are not important. Time can be in days, weeks, or months, and the material units can be in pounds, count, or any other convenient measure that is representative of the procurement measure. In Figure 3.5 the time scale is divided into past and future by the present point where the estimate is made ($t = 0$ and shown by E, the time of the engineering estimate). Point D is the delivery time of the transformed raw material, now a product, to the customer.

For an undesignated material, the plotted functions show the following:

1. The amount of the material going into the inventory supply at time t, $TS(t)$, with the plot showing the cumulative total of the material received.
2. The amount of material issued from inventory at time t, $TU(t)$ for all jobs, with the figure showing the cumulative total of the material released.

3. The amount of material used by a single job, J, at time t, JJ(t) showing the total material used.

The vertical difference between plots TS(t) and TU(t) at a given time period indicates the current amount of inventory on hand. For example, at period $t = -1$, TS($t = -1$) = 10 and TU($t = -1$) = 6, the current inventory = 4. Immediately after this point, there is a jump in the total supply by 6 units.

The JJ(t) requirement and all other jobs in progress (not shown) total up to the issues for period t on the TU(t) curve. With this as background, recognize the following:

$$\text{TS}(t) = \text{total supply of material in inventory at time } t\text{, units}$$

$$\text{TU}(t) = \text{total usage of material for all jobs at time } t\text{, units}$$

$$\text{JJ}(t) = \text{total usage for job J at time } t\text{, units}$$

To the right of period 0, each of the functions represents inventory projections. The step sizes at each of the integer times represent the amount of material going into or out of inventory for that time period. To illustrate, at time 3 the estimated inventory receipts TS(3) will be 2 units while 2 units for TU(3) are expected to be issued, 1 of which goes to job J. A total of 3 units, TS(3)−TU(3), remain in inventory at time 3.

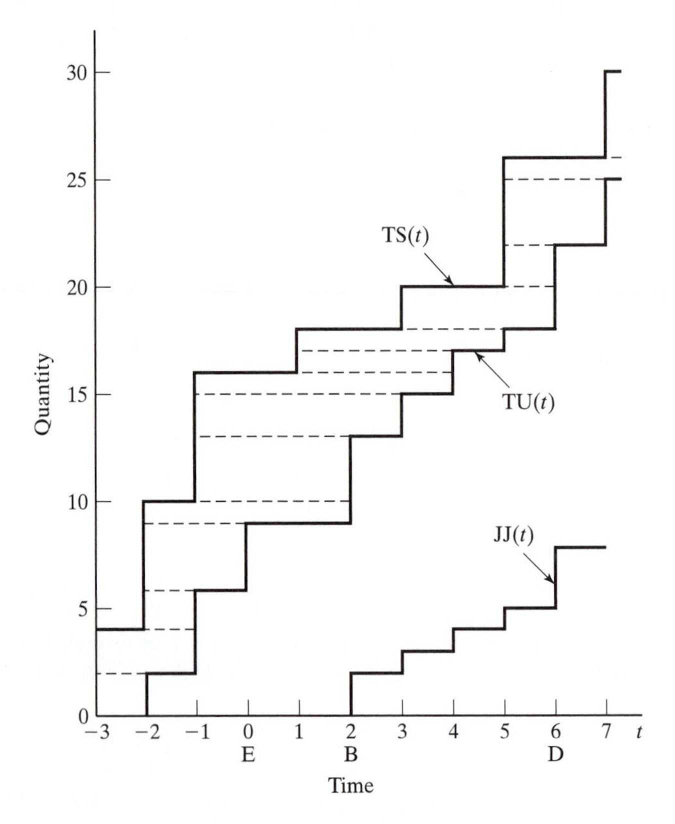

Figure 3.5 Inventory purchases, usage for several designs, and usage for job J.

The inventory supply curve TS could follow a general review and reorder discipline, such as MRP (material requirements planning) or JIT (just-in-time) ordering policy. But those principles do not influence our present discussion.

For this discussion, a fixed lead time of one period is assumed for all orders. What does this mean? Once a reorder issue is made and the supplier is informed of the order, it takes the supplier one period to deliver the material.

Our current customer is called JJ. What is the cost of the material that will fill out the estimate for JJ? Now the JJ order is for 8 units. The estimate is made at period E = 0. The JJ step-function consists of 8 units with initial inventory withdrawal beginning at period 2 (B) when manufacturing begins. The last inventory withdrawal occurs at period 6 (D) when the 8 units are delivered to the customer.

The horizontal dashed lines in Figure 3.5 show the connection between the material used and the period it was received. Notice that 4 units were issued at $t = 2$ as total usage, and 1 of the 4 units came from an inventory supply at $t = -2$, while 3 units were supplied at $t = -1$ period. This information will be useful in several of the inventory costing methods covered in the upcoming discussion.

Our discussion now turns to historical and forecast costs of the material. Table 3.3 displays the historical and forecast unit costs our firm faces in purchasing the material shown in Figure 3.5.

The costs given for E = 0 and earlier are historical—that is, we have actual costs and they are given in the table. For periods $t = 1, 2, \ldots,$ the values are forecast, say by using the statistical methods of Chapter 5.

Price breaks are available for various lot quantities. Costs in any row show quantity effects on the cost, as the quantity increases, costs decline. Table 3.3 reflects the period and the quantity purchased. For instance, if 1 or 2 units were purchased 3 periods back ($t = -3$) from the current time of the estimate (E = 0), then each unit cost

TABLE 3.3 Historical and Forecast Unit Costs for Quantity and Period

Period	Quantity			
	1–2	3–4	5–6	7–8
	Historical Costs			
−4	10.00	8.00	7.00	6.50
−3	10.70	8.60	7.50	7.00
−2	11.10	8.95	7.80	7.20
−1	11.25	9.00	7.90	7.30
0	12.00	9.65	8.40	7.85
	Forecast Costs			
1	12.95	10.01	8.73	8.12
2	12.97	10.44	9.10	8.46
3	13.52	10.89	9.49	8.83
4	14.10	11.36	9.89	9.20
5	14.69	11.84	10.32	9.60
6	15.31	12.35	10.75	10.01

\$10.70. If 3 or 4 units were purchased in a lot, then the unit cost is \$8.60. If 2 periods in the future ($t = 2$), 5 or 6 units are to be purchased, then each unit will cost \$9.10.

Inflation effects are indicated by a column entry—as older purchases proceed to newer purchases, cost increases. At $t = -4$ the cost for 1 or 2 units is \$10.00 per unit, but at period 6 and for the same quantity, the cost is \$15.31.

The stage is set with Figure 3.5 and Table 3.3, and we now study six different policies for the evaluation of material cost per unit where the firm uses inventory to build its products:

- Original
- Last
- Current
- Lead-time replacement
- Delivery
- Money-out-of-pocket

Original Cost Policy. The original cost policy assumes that materials are used in the order received and establishes as a cost estimate the unit cost of the oldest material sustained in inventory. In Figure 3.5 the oldest material in stock at time E = 0 came from a lot of six purchased at time −2. Table 3.3 shows the unit cost as \$7.80. Commonly called FIFO (first in, first out), this method has been popularized by the accounting profession for inventory valuation purposes.

Last Cost Policy. The last cost policy assumes that the latest materials purchased are the first to be used and establishes as a cost estimate the unit cost of the most recent material in inventory. In Figure 3.5 the most recent material in stock at time E came from a lot of six purchased at time −1. Table 3.3 shows the unit cost is \$7.90. This method, also called LIFO (last in, first out), is also frequently used in accounting.

Current Cost Policy. In the method known as current cost, a unit cost at period E = 0 is used as the value for the estimate and thus is time coincident to the preparation of the estimate. If price breaks exist, then a lot size must be determined to arrive at an estimate. Remember, inventory need not be added every period. Rules may be arbitrarily established—for example, that the entire requirement will be purchased if no other quantity is known for that period. If the purchase lot is assumed to be 8 at time 0 (even though no material purchases are made in Figure 3.5 at $t = 0$), then the estimate from Table 3.3 is \$7.85 per unit.

Lead-time Replacement Policy. The lead-time replacement method adopts as an estimate the replacement cost of the first lot of the material at the time it arrives. This is commonly called the NIFO (next in, first out) method, even though this label is an obvious anomaly. Once the job is won, material is ordered, and material specifically ordered for job J arrives at time 2. Given a lead-time requirement of one period, the order for 2 units will be placed at time 1, for a unit cost estimate from Table 3.3 of \$12.95. This method, used by engineering where material renewals are significant, is for bidding situations where there is a delay between submission of the estimate and knowing if the bid wins.

Delivery Policy. The point in time when the production order is delivered to the customer establishes the policy for the delivery cost method. As in the other methods, if any quantity is to be purchased at time D, then we use that. Since there were no scheduled purchases for inventory during period 6, assume a quantity perhaps equal to the delivery, and use that as the value of the material estimate. In Figure 3.5 the delivery time is in period 6, and, if we assume a lot size of 8, then the unit cost is taken as \$10.01.

Money-Out-of-Pocket Policy. Money-out-of-pocket (MOOP) methods refer to a general class of techniques where the overriding philosophy has the estimate reflect future and actual material expenditure. If the material is purchased in a single lot, then the estimate will simply be the unit cost of the lot. If the material is taken from inventory, then the estimate will be the original purchase cost of the material. If the material is to be purchased in the future, then a forecast of the unit price will be used. The MOOP method for material supplied from inventory is illustrated in the example that follows below.

Figure 3.6 presents a closer look at inventory transactions for the first five periods than that shown in Figure 3.5. At $t = 2$, the total usage for Job J and another job amounts to 4 units. One of the 4 units was purchased at the $t = -2$ period, while 3 units were purchased at the $t = -1$ period. The cost of the total usage at period 2 is \$31.50 (= 7.80 + 3 × 7.90) and the weighted average = 7.88 (= 31.50/4).

In Figure 3.5, the material for job J is taken from a number of past and future inventory purchases. The MOOP policy is a weighted calculation for forecasted usage of material in as accurate a manner as possible. In general, suppose that at time t we withdraw $n(t)$ lots of the material in inventory that were purchased at times t_j where $j = 1,2, \cdots, n(t)$. At time t consider the following:

Define $f(t_j)$ = fraction of material inventory coming from the time t_j lot, decimal

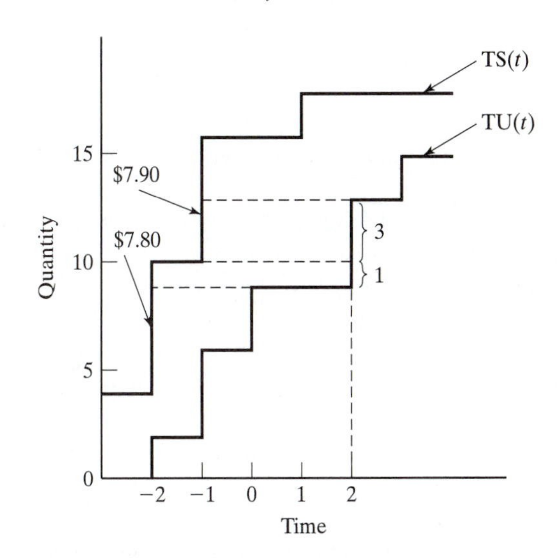

Figure 3.6 Closer look at inventory transactions for the first six periods from Figure 3.5.

$p(t_j)$ = unit cost of t_j lot, dollars per unit

$d_j(t)$ = number of units of material used for job J at time t

$$\text{then } C_{MOOP} = \frac{\sum_t \sum_{j=1}^{n(t)} p(t_j) f(t_j) d_J(t)}{\sum_t d_J(t)} \tag{3.6}$$

where C_{MOOP} = MOOP cost of material, dollars per unit

j = period number, time

$n(t)$ = number of lots of material in inventory purchased at times t_j, $j = 1,2, \cdots, n(t)$

Example: Money-Out-of-Pocket-Policy

Figure 3.6 shows an enlargement of the first 5 periods of inventory transactions from Figure 3.5. To repeat, look at the material withdrawn from inventory in period 2, TU(2). This material is from two purchases, one from the 6 units purchased in period −2 and 3 units from the 6 purchased in period −1. Referring to Table 3.3, the cost of the total usage at period 2 is 7.80 + (3 × 7.90) = \$31.50 for an average cost of \$7.88 per unit.

Referring to Figure 3.6, job J will use 2 of the units from the inventory withdrawals in period 2. According to the MOOP method, these 2 units will cost \$7.88 × 2 = \$15.76. For the complete job J, material is taken from a number of past and future inventory purchases.

Observe Figure 3.6 and the data in Table 3.3. The dashed horizontal lines of the figure relate the time of inventory purchase to time of usage, and we see that of the 4 units total usage of period 2, 1 unit can be considered as purchased at −2 time for \$7.80 per unit and 3 units as purchased at −1 time for \$7.90 per unit. The dashed horizontal lines of the figure relate the time of inventory purchase to time of usage. Under MOOP, the 2 units of job J usage for the second period would be costed as $(7.80 \times \frac{1}{4} + 7.90 \times \frac{3}{4}) \times 2 = \$15.75$. The remaining periods of production for job J (t = 3,4,5, and 6) are similarly costed as $(7.90) \times 1 + (7.90 \times \frac{1}{2} + 12.95 \times \frac{1}{2}) \times 1 + (12.95) \times 1 + (13.52 \times \frac{1}{2} + 10.32 \times \frac{1}{2}) \times 3 = \$67.04$. Total lot cost for job J is \$82.79, and for the lot quantity of 8, the MOOP unit cost is \$10.35 (= 82.79/8).

The calculation of the total MOOP cost of the material in job J is summarized using Equation (3.6) as

$$\left[\frac{(7.80 \times 1) + (7.90 \times 3)}{4} \times 2\right] + \left[\frac{7.90 \times 2}{2} \times 1\right] + \left[\frac{(7.90 \times 1) + (12.95 \times 1)}{2} \times 1\right]$$

$$+ \left[\frac{12.95 \times 1}{1} \times 1\right] + \left[\frac{(13.52 \times 2) + (10.32 \times 2)}{4} \times 3\right] = \$82.79$$

In addition, the MOOP unit cost for 8 units in job J is \$10.35 (= 82.79/8).

TABLE 3.4 Unit Cost of Material Issued from Inventory for the Different Policies

Policy	Unit Cost
Original cost (FIFO)	7.80
Last cost (LIFO)	7.90
Current cost	7.85
Lead-time replacement (NIFO)	12.95
Cost at delivery	10.01
Money-out-of-pocket (MOOP)	10.35

Table 3.4 summarizes the estimated unit costs of raw material computed using the different policies for evaluating material cost. It is easy to see that there is a large variation among the different methods. Depending on the business opportunity, tolerance of the company for absorbing loss, level of competition for the order, and other considerations, engineering may want to adopt a policy that is most representative of the true cost than the other methods for cost analysis. Generally, the procedure is determined by company policy.

It would be impulsive on our part to suggest that these policies can be simply implemented into practical business situations. The observant student may have noticed a number of assumptions for all policies. Some astute companies have developed algorithms to find those inventory cost per unit values that meet the overarching principle of money-out-of-pocket, which is the preferred objective. But it is the important theme that needs to be stressed again—engineering wishes to estimate future materials costs that will be verified as money-out-of-pocket when the estimate becomes an actual cost.

3.6 JOINT MATERIAL COST (OPTIONAL)

During the manufacture of some products, other products are produced as a consequence, either deliberately or not. Sometimes this is as simple as selling your salvage to be used by another manufacturer—for example, waste from a meat packing plant for use in pet food production. At other times, additional effort may be required to yield a useful secondary product. When there are secondary products, it is necessary to determine the true joint material cost of all of the components. Although joint labor cost (discussed in Chapter 2) is exclusively labor, joint material is usually confounded with elements of material and labor costs.

A frequent task is the preparation of a detailed cost analysis of a manufacturing process dejointing the cost of common material into separate finished products. Traditional approaches to joint costs of material involve concepts that serve accounting or marketing needs and do not reflect the true or actual costs for engineering.

Joint materials are those materials that result from the processing of a singular raw material supply. Joint materials are intermingled up to the point at which the materials are divided into separable units. The point of division is called the *split point.*

Material and labor costs up to the split point are referred to as joint costs, but afterward are called unit cost.

The key element in this definition is the concept of the singular raw material, which by virtue of a processing step, becomes two or more discrete products. For example, the processing of raw milk into ice cream and skim milk illustrates the notion of singular raw material into two discrete products that will be individually marketed. Another example is the rough log, which on sawing and milling, becomes first and second grade lumber, wood chips, and sawdust. A molded plastic part where the die has several cavities, possibly for different parts, is a common production joint-cost problem. Similar situations occur frequently in the chemical, petroleum, integrated circuits, food, metallurgical, and timber industries. Those industries include processes to create marketable products of what were essentially byproducts of basic processing steps. The problem in processes that result in multiple products is the tracing of the cost contribution of raw material and labor to the individual final products. Often the choice of allocators is not clear, product lines are not direct, or other factors appear to prevent cost traceability.

A distinction is necessary between distributing and conversion types of joint costs. In the distributing process, the material is divided up among the various products, but it enters the next stage for each product essentially the same material. An example of this is steel fabrication. The steel starts out in the ingot form and is rolled down to some intermediate shape that can then be further formed into bars, rods, wire, and other structural shapes. The starting material is distributed into all of these cases, but the material remains essentially the same.

Another example of the distributing type of joint costs is illustrated by the multiple-cavity die problem because plastic pellets are intermingled prior to the molding operation. The characteristic of the plastic is not altered (in a joint-cost sense), and cost traceability is valid albeit complicated.

For conversion processes, the resulting material is significantly changed in more than just physical shape from the starting material. Conversion usually takes place in processing industries where a quantity of raw material is converted into new products or material. An example is separation and liquefaction of gases, where air is subjected to complex thermodynamic processing (compression and chilling) with different elemental gases coming off as the temperature is decreased. Splitting of raw material sometimes requires that the essential cost nature of the material be changed, resulting in two or more discrete products or materials with differing characteristics and physical measures or values. Figure 3.7 describes this distinction between distributing and conversion joint costs.

Marketing practice may require that some products subsidize others because the selling price or the market value of a product is not necessarily proportional to the processing costs. An accounting or marketing policy may distort true costs. For example, slight variations in the production of integrated circuit microprocessors can result in a variation in the maximum operating speed of the processor. The slower processors must be sold at a discount, so a premium is charged for the faster processors, even though the cost of production is the same for all of the processors. The allocation of joint cost on the basis of quantity can result in invalid cost data and arises when products are valued for inventory purposes at levels higher or lower than their actual

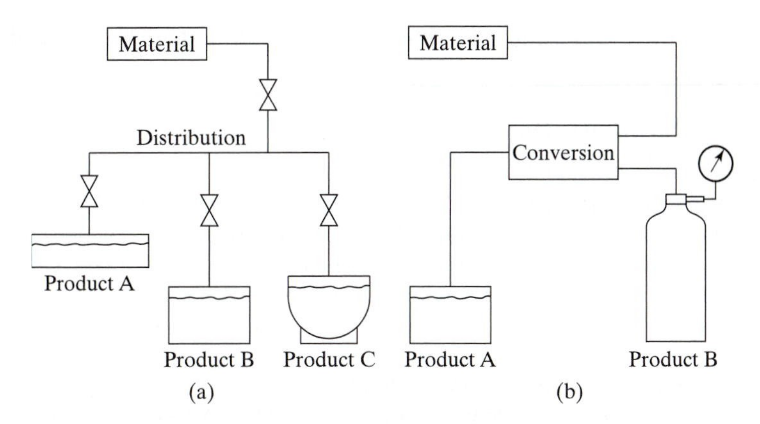

Figure 3.7 Two classes of joint cost: (a) material being "distributed" and (b) material being "converted" into two or more products.

production costs. A similar situation occurs when marketing policy dictates price levels that reflect market-driven selling prices, not manufacturing costs.

In the example of a rough log subject to a splitting process, the production of first-grade lumber is associated with the production of lesser-grade lumber and sawdust. The processing investment to finish and market secondary lumber products may be subsidized by profits from first-grade lumber production. Profitability decisions on specific products cannot be made on the basis of accounting data, which may cover a group of products, and may not reflect the true costs of an individual product. This leads to the need for an exact cost estimate that examines joint product costs.

Now consider units of measure commonly used to prorate or allocate costs. We select a unit of value per unit of reference to separate joint costs into unit cost. Commonly, this is a dollar value per unit of measure: \$/ft^2, \$/lb, \$/kW, \$/hr, \$/product unit. Fundamental reference units that are readily measured are preferred. The simpler the measured unit, the clearer its use becomes and the more accurate are the conclusions drawn from the analysis.

Allocators can be classified as follows:

- Physical measure: geometry, weight, shape
- Energy: Btu, kW
- Time: second, year, labor hour
- Units of finished product: each, 100 units

The unit of reference needs to be defined for traceability of costs to avoid changing the unit through a process. For example, if pound is used in the material stage, pound is convenient as a finished product rather than square foot. This becomes important with split converting where split products may be in different physical states. In the example of timber processing, the lumber produced has units of board feet (a volume measure), while the wood chips and sawdust cannot be conveniently measured in volume due to packing density, so are measured by weight. Unnecessary unit conversion should be avoided. For example, rather than defining the products of a petroleum cracking process as gallon and barrel, a common unit such as pound should be used throughout the process.

Example: Dejointing Lumber Production

Production of lumber starts when trees are harvested in the forest. The trees are sold by the landowner and are felled, limbed, cut to length, and loaded for transport to the lumber mill. The logs have a joint material value, as each log is sawed into various commercial sizes.

Imagine a tree that grows precisely as a 4 × 4 in. square for a height of 8 ft and bark and limbs do not reduce the rough initial size. This means that a 4 × 4 in. × 8 ft parallelepiped can be sawed from the tree. Under these idealized conditions, the yield approaches 100% and there is no joint cost of material, as one unit of input raw material becomes one unit of output product.

Imagine a tree that is a right circular cone where the lower base and upper base have as their radius r_2 and r_0. Telephone poles are examples of right circular cones, and these Douglas fir poles have a high yield for this application. But if this geometrical shape (as shown in Figure 3.8) is to be sawed into a variety of commercial shapes, then what is the yield and joint cost of product?

This is an important question, and the lumber industry considers this an imperative requirement for profit and survival—and optimum solutions are always desired. One solution is external laser mapping of the product, which is geometrically analyzed for the greatest yield for the variety of product that the mill is organized for.

Consider this first geometrical solution. The volume of the geometrical form of Figure 3.8 is given by

$$V = \frac{1}{3}\pi L_2(r_2^2 + r_0^2 + r_0 r_2)$$

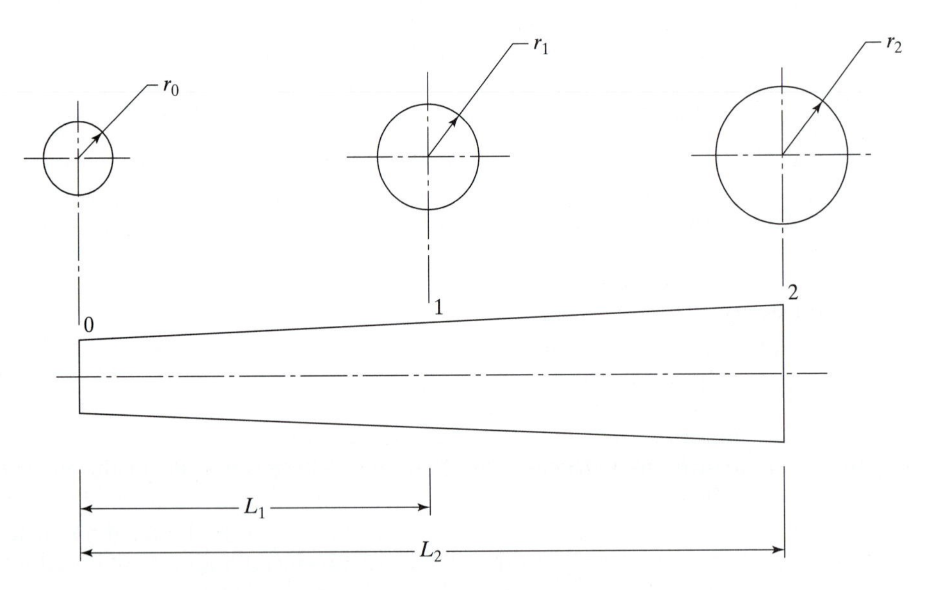

Figure 3.8 Model of tree idealized as right circular cone.

where

r_0 = radius of smaller end, in.

r_1 = radius of intermediate location, in.

r_2 = radius of larger end, in.

L_1 = length of location 1, ft

L_2 = length of tree, ft

A simplified approach begins by saying that a 4 in. square × 8 ft product can be can sawed from a $r_0 = 2.83$ in. [= $(2^2 + 2^2)^{1/2}$] and $r_1 = 4.24$ in. [= $(3^2 + 3^2)^{1/2}$] where L_1 = 8 ft from a right circular cone. This provides a yield of 40%, which is not very good. For a log that is 16 ft long, and for $r_2 = 5.66$ in. [= $(4^2 + 4^2)^{1/2}$] then for the 4 in. square and the 6 × 6 in. square green lumber, the yield increases to 44%, still poor.

An algorithm to find the yield is possible, but each log must be geometrically evaluated, which is not realistic. In addition, multiple dimension lumber can be sawed from the logs. In industry a real-time laser model scans each log individually for the greatest yield and sawing pattern, which we do not discuss here.

Another method is to evaluate the raw material for joint cost of product output with historical values. For example, approximately 60% of the log by volume becomes dimension lumber. The dimension lumber is broken down into two product grades: construction and economy. The split between construction and economy is 80 and 20%. There is a price advantage to construction grade. The price for the finished construction grade lumber is $0.35 per bd ft and for economy grade it is $0.10 per bd ft.

Twenty-five percent of the material from the log ends up as chips that are sold for pulp, which is sold at $40 per green ton. The remaining 15% of the material is bark and sawdust, which is used to fuel the kiln for drying of lumber.

Virtually all of the material that leaves the forest is utilized. To remain profitable, all material delivered to the mill must be used, and the value of the joint cost of product must be maximized. Besides the dimension lumber, there is pulp that is marketable. How should the joint cost of the original log be distributed between the two products? How should the heating value of waste product be considered?

Consider this solution. Begin with a green Douglas fir that is 13 in. OD at the small end and 20 ft long, and the minimum volume of the log is 18.4 ft^3. Cost for the purchased log to the mill is $35, and unit cost is $1.90/ft^3.

Fir specie has an approximate density of 59 lb/ft^3, so the weight of the log is 1088 lb. About half of the weight at this stage is water. The lumber is kiln dried before it is sold to reduce the water content, and thus decreasing the weight and transport cost and improving quality. The pulp chips are sold by green weight (with the normal water content) and the bark and sawdust replace other conventional fuels to produce utility heat for drying.

Since the weight of the material changes throughout the process, it is more consistent to use the volume unit of measure. There are 12 bd ft per ft^3. In this case the volume of the chips and sawdust are calculated "in the log" at the original density and not when it is piled loosely in a container where the density would be significantly different.

Examine the table below to see the calculations and the finding of proportional cost and income.

Dejointed Products	Calculation	Cost	Board Feet	Price per Board Foot	Income
Lumber	35 × 0.60	$21.00			
Pulp	35 × 0.25	8.75			
Fuel	35 × 0.15	5.25			
Construction grade	18.4 × 12 × 0.60 × 0.80		106.0	$0.35	$37.10
Economy grade	18.4 × 12 × 0.60 × 0.20		26.5	0.10	2.65
Pulp	18.4 × 0.25 × 59 × (1/2000) = 0.136 ton				40 × 0.136 = 5.44
Total		$35.00			$45.19

While the overall $45.18 income exceeds the $35 cost, notice that economy grade lumber and pulp were discounted below cost, and their income did not recover their proportion of original cost.

The primary product is the product that forms the financial and physical justification for a company or process to exist. All other products are secondary products regardless of their value and would not exist were it not for the production of the primary product. Pulp is a secondary product, but the opportunity makes the primary products less costly.

Converting industries, unlike the manufacturing, the fabrication, and the durable-goods industries, deal with this kind of joint-cost problem. Further discussion on this topic is outside the scope of this book.

SUMMARY

Products are made of materials that are classified as "direct"—raw materials, standard commercial items, subcontract items, and interdivisional transfer items—and "indirect"—materials required to make the product but that do not become a permanent part of the product. Roughly, material costs account for 50% or so of the product cost. Thus, care is necessary for the material cost estimates. The cost of making the material estimates should be in proper proportion to the material cost itself. Some direct material, because of the difficulty of estimating, may be analyzed as indirect and included in overhead, although, for accuracy, direct costs are preferred.

The amount of material that ends up in the completed part will be less than the amount purchased, but it is the amount procured that we must pay for. The material requirements are estimated to include the actual part shape plus the waste, scrap, and shrinkage.

Once the amount of material is determined, it is a simple calculation to determine the product material cost using the material cost rate. Contractual arrangements and inventory schemes affect the method in which the material cost rate is found.

When material is maintained in a common inventory, there can be difficulties in determining the exact cost of the material issued for a particular job. Factors such as quantity discounts, time of placement of the order, and purchase agreements cause variation in the cost of the same material held in inventory. Methods for estimating the cost of the material from inventory were explained.

QUESTIONS FOR DISCUSSION

1. Define the following terms:

Current cost	MOOP
Dejointing	Quote or price-in-effect (QPE) method
Direct materials	Shape
FIFO	Shrinkage
Indirect materials	Specification
Joint products	Subcontract materials
LIFO	Waste
Mensuration	Yield

2. Discuss the complications of materials as they relate to engineering design.

3. Define material in terms of "alteration."

4. Divide direct materials into categories. Discuss.

5. Give some typical engineering units for shape.

6. What are the advantages of the MOOP method over other inventory methods?

7. List allocators for joint material costs.

8. Three future trend lines (constant value, deflation, and inflation futures) are assumed for the cost of unit raw material and are shown in Figure Q8. The moment of the engineering estimate is located at time E, exchange of the money between you (the buyer) and the seller is at the buy time B, and the delivery of the transformed product to your customer, following its manufacture by a value-added operation, at time D. Write a policy statement defining the point in time (either E, B, or D) for the selection of a cost value for an engineering product estimate. Does your recommendation of which value to select differ for the three sketches?

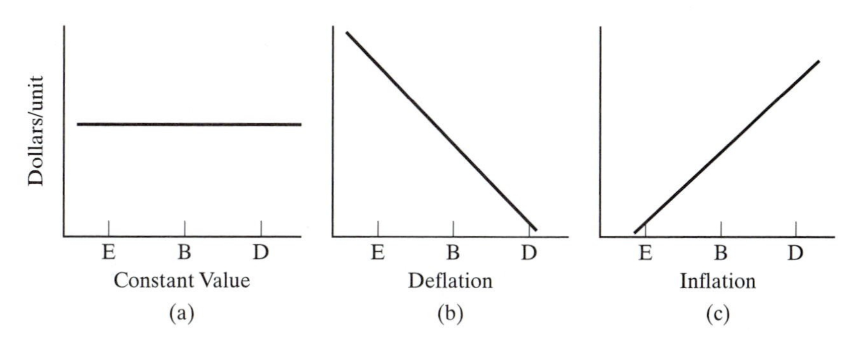

Figure Q.8

PROBLEMS

3.1 A round shaft 6.5 in. OD has a specified length of 31.675 in. The facing dimension necessary for a smooth end finish is 1/16 in. The width of the lathe cutoff tool is 3/16 in. The length of the raw bar stock is 12 ft. The lathe collet requires a minimum of 4 in. for gripping the last part. The material costs $0.95/lb and has a density of 0.29 lb/in^3.

(a) Determine the number of pieces that can be removed from the raw bar.
(b) What is the unit cost for the raw material?
(c) What is the value of material that is lost to waste, given that waste material has a salvage value of 10% of the original value?
(d) Find the shape yield.

3.2 Examine the 2024-T4 aluminum shaft in Figure P3.2. Raw material is purchased to match the outside dimensions. The bar stock for this part is supplied in 12 ft lengths. The density is 0.0975 lb/in.^{3} and the cost is $1.20/lb. Assume and state allowances for facing, cutoff, and collet grip length for the last part. (*Hint*: Ignore the radius and tapered end in the hole.)

(a) Determine the number of parts per bar.
(b) Estimate the cost of the raw material per part.
(c) Find the approximate shape yield.

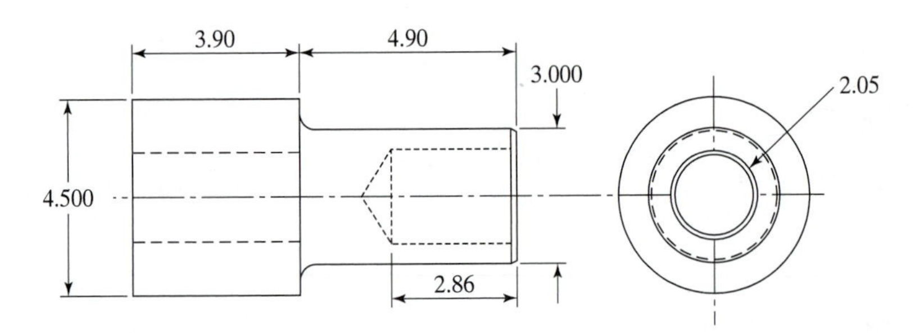

Figure P3.2

3.3 (a) Determine the theoretical and actual material required to produce the part, shown in Figure P3.3. Raw stock is supplied in 0.875 in. diameter. A lathe cutoff tool width is

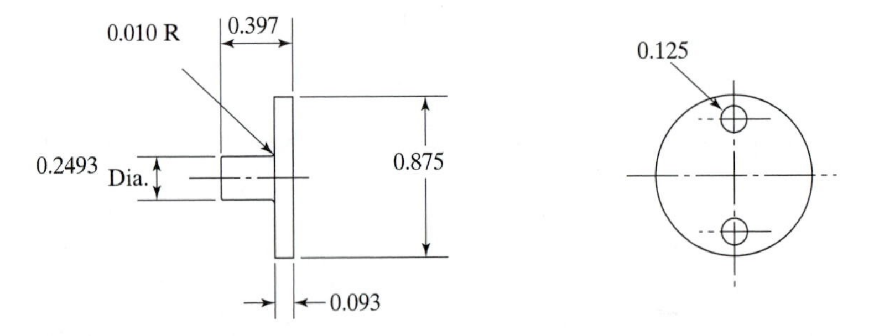

Figure P3.3

0.125 in. A 0.015 in. stock allowance is necessary for facing the small diameter end of the part. Allow 4% of shrinkage and bar end losses. The raw material weighs 2.05 lb/ft.

(b) Find the shape yield.

(c) What is the unit cost for this hot-rolled steel stock, which costs $0.88/lb?

3.4 In the integrated circuit (IC) fabrication process, individual circuit chips are rectangular in shape. Multiple chips are produced on round silicon wafers. The chips are laid out in a rectangular array on the wafer with a saw clearance (called a *street*) in between, which is used in the subsequent dicing process where the chips are cut apart. Square IC chips, 5 mm on a side, will be fabricated on a 200 mm diameter wafer. The streets are 0.1 mm wide.

(a) Calculate the number of chips that can be laid out on the wafer. (*Hint*: For simplicity, assume that all of the chips are laid out in a rectangular area inscribed within the circular wafer. For the more challenging problem, the semirounded areas outside of the inscribed rectangle can be utilized for additional chips.)

(b) Based on the area, determine the amount of waste.

(c) Assume a scrap allowance of 10% of the chips and no shrinkage. Determine the material yield.

3.5 A V-block is manufactured of cast iron. The finish dimensions are shown in Figure P3.5. Because hot metal occupies a greater volume than cold, a wood pattern is made using a shrink rule for the green sand casting. A 4% allowance on the finished shape is given to cover extra stock for finishing all faces, and a 1.5% draft allowance is made on the vertical sides for pattern removal. The as-cast material has a density of 0.26 lb/in.^{3} and costs $0.12/lb.

(a) Calculate the volume of the raw casting.

(b) Determine the cost of the raw casting.

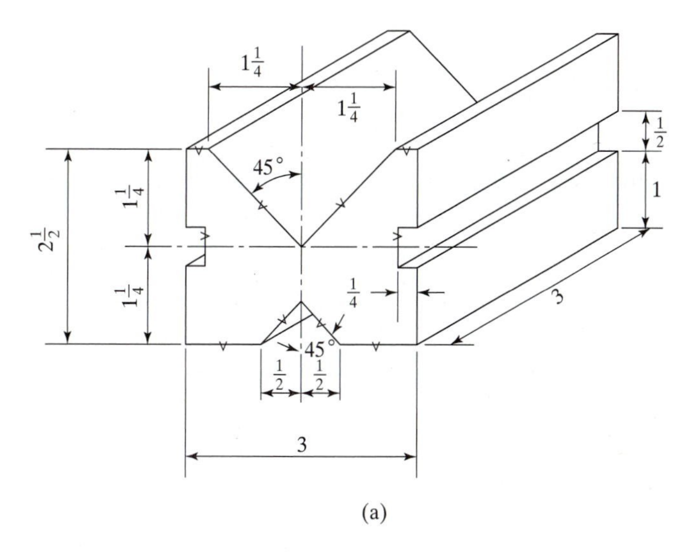

Figure P3.5

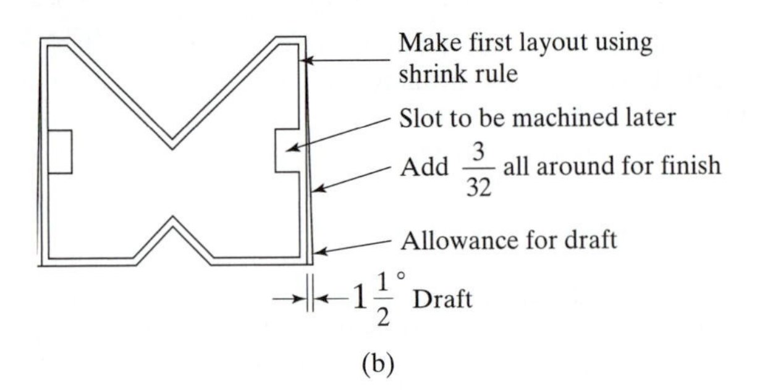

Figure P3.5 (*Continued*)

3.6 Circular blanks are to be stamped from strip stock. The company can buy the 0.125 in. thick cold-rolled steel in either of two widths. The stamping patterns for each size strip are shown in Figure P3.6. Both sizes of material are sold for \$1.14/lb and have a density of 0.278 lb/in^3. Choose the material size to use for this job by determining the following:

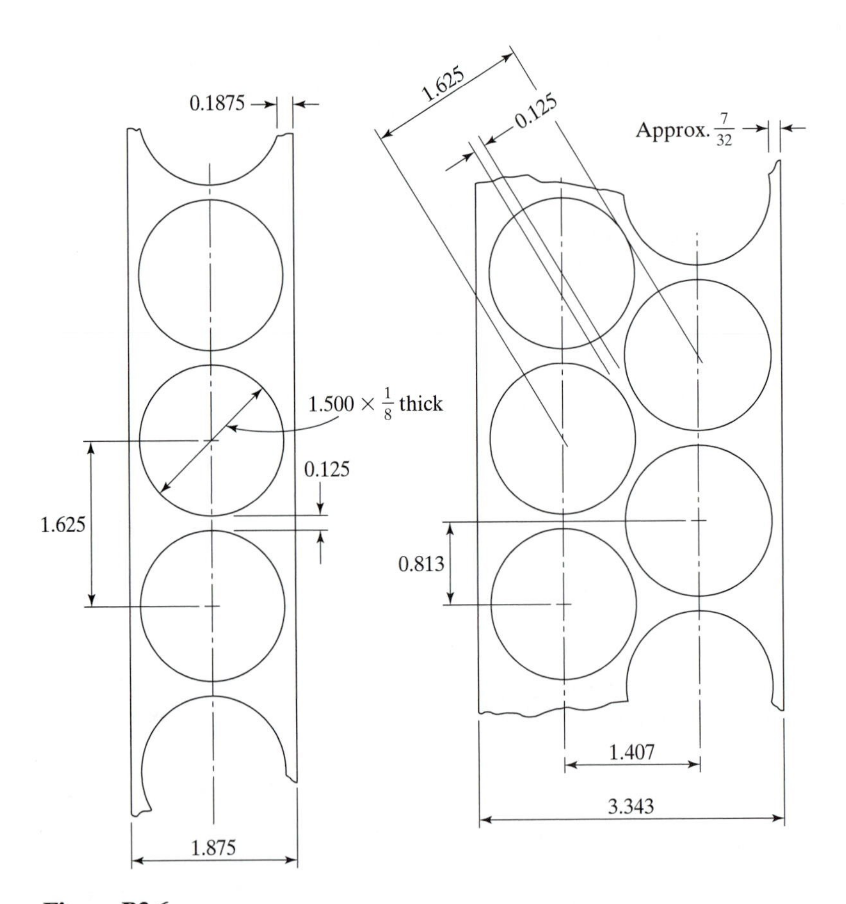

Figure P3.6

(a) Estimate the material cost per blank for each strip size.
(b) Assume there is an additional 5% overall blanking loss. Find the total loss per part for each strip size. Determine the shape yield.
(c) The salvage recovery value is 10% of the original material value. Find the net cost per piece for each strip size.
(d) Which size strip is preferred?

3.7 A 12-oz beverage can is composed of the body, top, and ring. The container body is blanked from 3004-H19 aluminum coils with the layout given by Figure P3.7. An intermediate cup is then formed without any significant change in thickness. The cup is then drawn to a sidewall thickness of 0.0055 in. The bottom thickness remains unchanged. The can is trimmed to a final height of 5.437 in. to give an even edge for later rolling to the lid. The coil stock costs \$1.0728/lb and has a density of 0.0981 lb/in.^{3} Recovered waste is sold at \$0.530/lb.

(a) What is the volume of metal in the trimmed can? (Ignore the 0.05 in radius.)
(b) Find the yield of the can body to blank.
(c) Find the amount of metal from the strip used for each can.
(d) Determine the shape yield.
(e) What is the cost of the material from the strip used for each can?

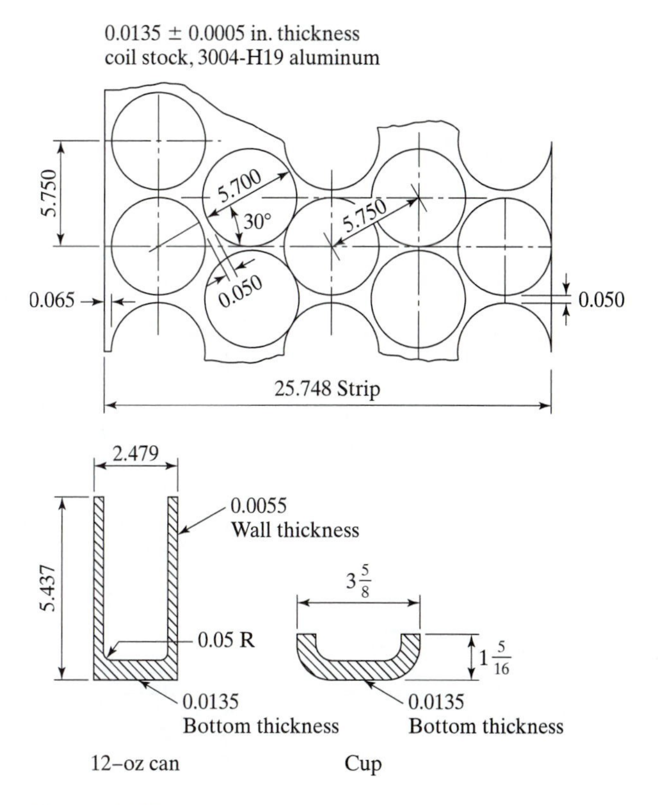

Figure P3.7

(f) What is the recovered value from the sale of the recovered waste per can?
(g) Estimate the net cost of material per can body.

3.8 Look again at Table 3.3 and Figure 3.5 in this chapter. Find lead-time replacement, delivery, and MOOP unit material costs for the following adjustments to Figure 3.5.

(a) For just the JJ(t) changes in the table below.
(b) For TU(t) changes.
(c) For TS(t) changes.
(d) For the combined changes.

Period	JJ(t)	TU(t)	TS(t)
2	1	2	2
3	2	3	0
4	0	1	2
5	1	3	4
6	4	4	8

3.9 The inventory plan for a material is given by Figure P3.9. Use Table 3.3 from the chapter and find the following unit costs for the six units of stock used for JJ:

(a) Original
(b) Last
(c) Current
(d) Lead-time replacement
(e) Delivery
(f) Money-out-of-pocket

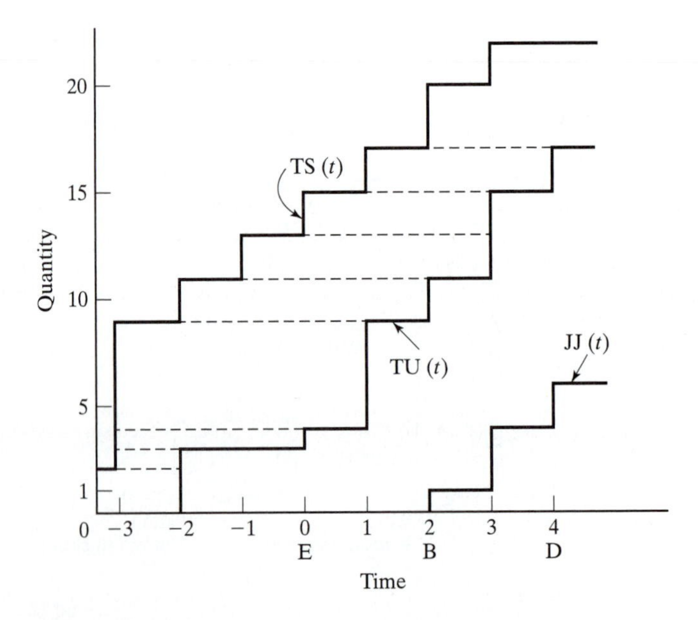

Figure P3.9

3.10 A flange is forged from C-1020 steel bar 1.5 in. OD Each bar is 7-9/16 in. long and will produce 4 flanges. The bar includes a 1 in. long tong hold for all 4 flanges which is later trimmed off as waste. The material cost is \$1.70/lb and has a density of 0.29 lb/in.^{3} Waste and scrap are sold at 10% of original value. A labor crew consists of a hammer-man (wage = \$21.75/hour) and a helper (wage = \$19.65/hour). Each member of the crew performs different elements of the operation, and their joint output is 0.54 hour per 100 flanges.

(a) Find the material yield.
(b) Find the unit material cost, adjusted for the sold salvage material.
(c) Determine the labor cost per unit.
(d) Estimate the total unit cost.

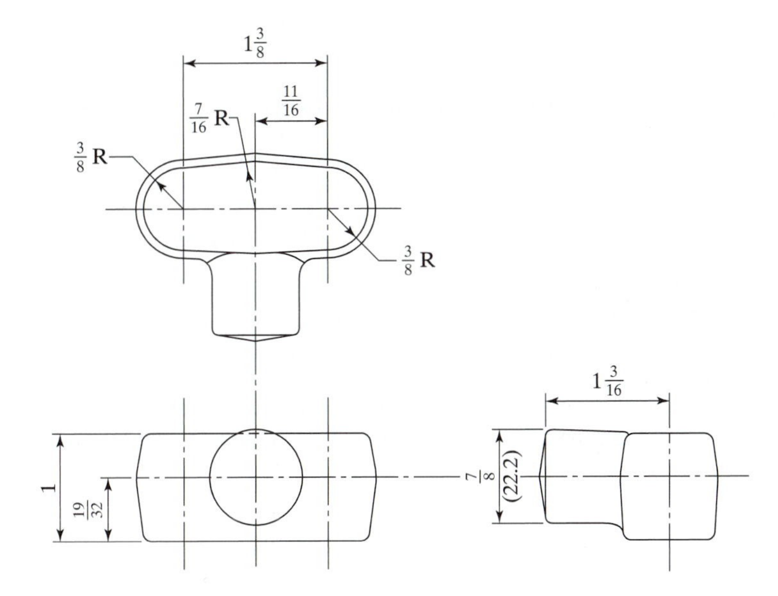

Figure P3.10

3.11 A back plate is used as a "snap-in" cover for a mating cassette case. The back plate is molded of polystyrene resin and weighs 0.02 lb/piece. There is no waste. Each shot fills 8 cavities. The high impact polystyrene costs \$2.20/lb. and has a density of 0.0365 lb/in^3. Production time is 8 seconds per shot. The molding machine operator has a gross hourly direct wage of \$28.80/hour. The operator monitors 3 machines simultaneously with the back plates made on one of them. Nothing is know about the operations on the other two machines.

(a) Expressing the information on a per 100 pieces basis, find the direct material cost, direct labor cost, and total cost per 100 pieces.
(b) A Canadian source has been found for the polystyrene resin at a cost of Can\$7.39/kg due to a favorable exchange rate of Can\$1.4785 to U.S.\$1.00. Should the company buy the Canadian material?

3.12 Integrated circuits (ICs) are fabricated on 12 in. diameter silicon wafers that cost \$75 each. For an IC design, 441 chips can be patterned onto a single wafer. A normal production batch consists of a "boat" of 50 wafers. Six process-monitoring wafers are included at the start of each batch and are removed one at a time at various points during the processing to check on the quality of the processing. At the end of the processing, each IC is tested on the wafers. The cost for processing a boat of wafers through fabrication is estimated to be $15,000. From the latest batch, an average of 25 ICs per wafer failed.

(a) What is the cost of a good IC?
(b) The process yield is calculated by:

$$Y = (N_g/N_s) \times 100$$

where Y = percent yield
N_g = number of good parts
N_s = number of parts starting processing

Find the yield for this process.

(c) The company wants to increase the yield to 99%. How many failed chips per wafer are allowed? What is the cost of an IC when this yield is reached? Explain what it means to have a fraction of an IC per wafer as a failure.

CHALLENGE PROBLEMS

3.13 Find the part cost for the following casting. A foundry is to cast the motor cylinder shown in Figure P3.13. Compute the volume of the casting, the overall volume of metal required, and the cost of making the casting based on the following factors: From experience, the shop yield is 54.5% and there is a metal loss of 10%. The furnace and overhead are $0.06 per pound of poured metal. The cost of the metal charged is $0.12 per pound. Forty percent of the charge is remelted metal that is valued at $0.08 per pound. The density of the

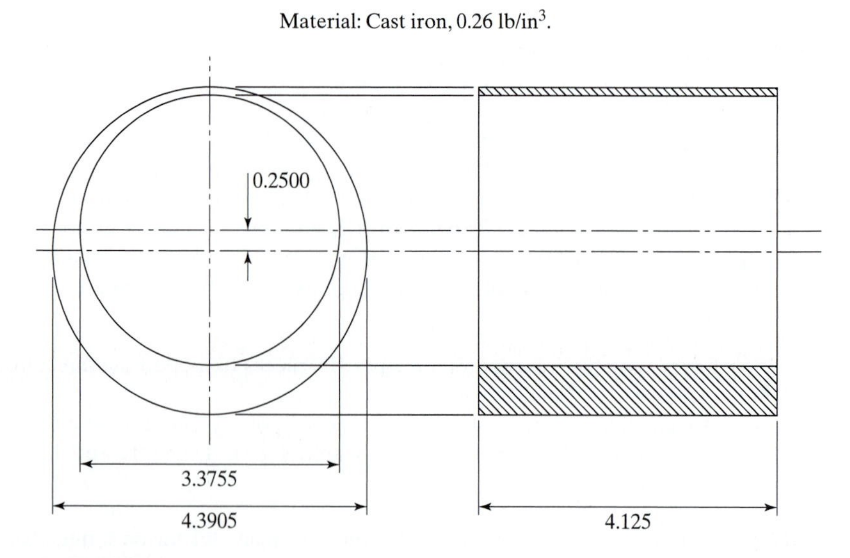

Figure P3.13

cast iron is 0.26 lb/in.^{3} Allow 1/8 in. of stock for all machined surfaces (assume all surfaces are machined).

(*Hints:* The pouring weight is finished weight per shop yield. The remelted weight is the pouring weight multiplied by remelt factor.)

3.14 A plastic material is blended, pigmented, and injection molded. The part is molded in a three-cavity mold and the die has runners that connect the pieces to the sprue. The essential features of the simple product, which is called a button, include runners and a sprue and are given by Figure P3.14.

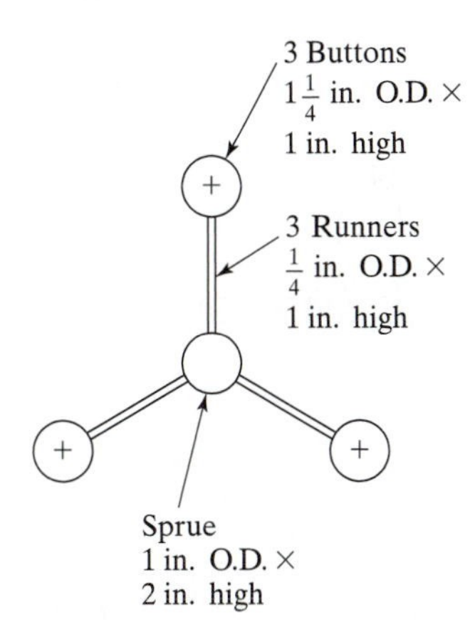

Figure P3.14

The buttons are $1\frac{1}{4}$ in. OD × 1 in. high. Only the button is used. The runners and the sprue are trimmed and considered waste; they cannot be reground and reused because of the degrading of the material during the high-temperature process. The raw material is a flourlike polymer, which when heated in the injection molding machine becomes fluid and flows into the cavities of the die. The material costs \$2.50/lb and has a density of 0.0275 lb/in.^{3} Dejoint the common cost for one injection molding shot, and find the cost of one button. (*Hint:* The following calculations may be useful.)

Property	
Volume for 3 buttons	3.6816 in.^{3}
Weight for 3 buttons	0.1012 lb
Sprue	1.5708 in.^{3}, 0.04312 lb
Runners	0.1473 in.^{3}, 0.00041 lb

3.15 The inventory plan for a material is given by Figure P3.15. Using Table 3.3 from the chapter, find the following unit costs for the 8 units of stock used for JJ:

(a) Original
(b) Last
(c) Current
(d) Lead-time replacement
(e) Delivery
(f) Money-out-of-pocket

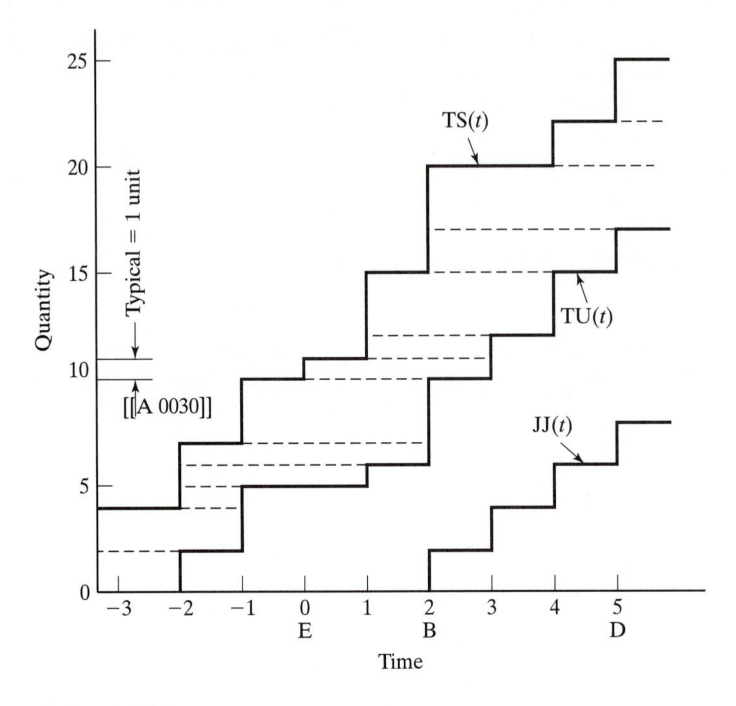

Figure P3.15

3.16 A computer part is molded two at a time from clear polycarbonate plastic. A partially dimensioned sketch (Figure P3.16) provides the part size along with the sprue and runners for filling the mold. The sprue and runners will be removed from the parts and recycled.

(a) Compute the volume and weight for 1 part and for the complete shot consisting of 2 parts, 2 runners, and the sprue. The material density is 0.0404 lb/in.^{3}
(b) What is the yield of part material to total material?
(c) The cost of this material is \$6.20/lb and waste is recovered at 15% of the original value. Find the net cost per part, including the waste recovery.

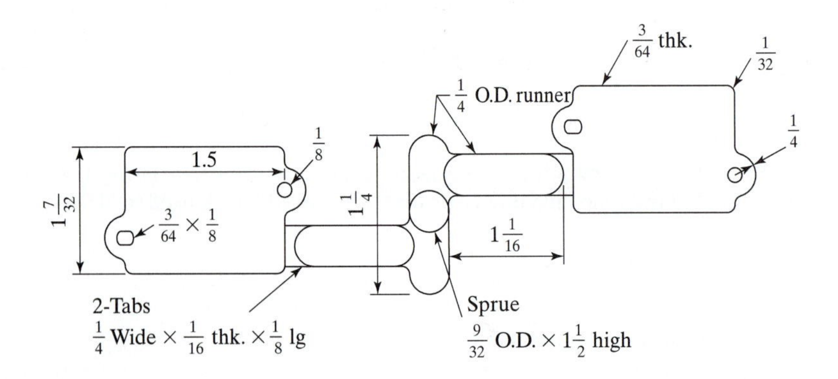

Figure P3.16

(d) The cycle time for one operator running one injection-molding machine is 45 seconds per shot. The "detab" time is 10 seconds for both parts and is performed during the molding cycle. For a gross labor rate of $33 per hour, find the labor cost per part.
(e) What is the total cost per part?

3.17 Compute the material cost for the part shown in Figure P3.17 for a lot size of 500 units. The material is cold rolled steel bar 0.5 $\times$ 1.25 in., 10 ft length, density 0.26 lb/in.^{3}, and costs $1.27/lb. Lengths less than 3 ft are scrapped without salvage value, and lengths over 3 ft are restocked without charge. Sawing is the operation of choice and results in a 0.125 in. saw kerf per part.

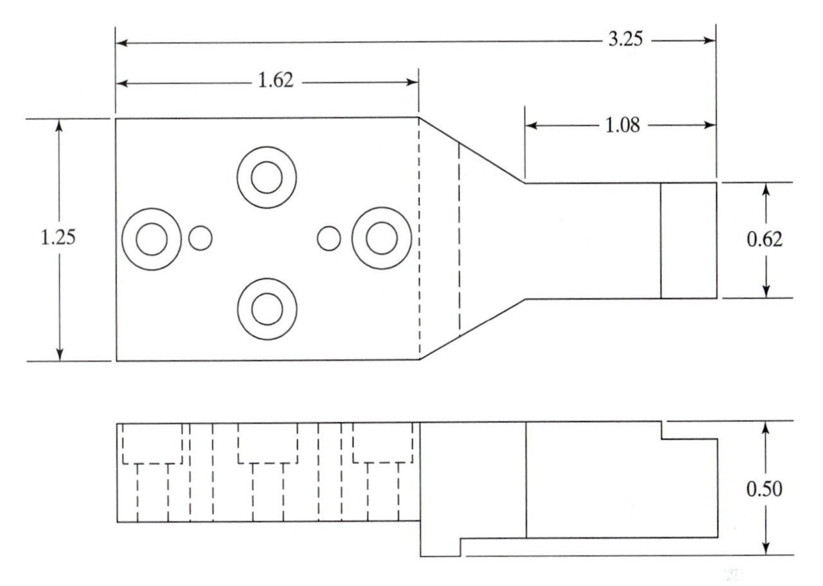

Figure P3.17

3.18 To make wafers for the integrated circuit industry, extremely pure silicon is melted and allowed to solidify into a single crystal (starting with a seed crystal attached to a drawing wire) in a reactor. The resulting crystal is roughly cylindrical in shape (see Figure P3.18) and is called a "boule." After the crystal is grown, the seed and tail ends are removed and then the outside diameter is ground to obtain a nearly perfect cylinder of single crystal silicon. The resulting cylinder is then sawed into individual wafers and the wafer surfaces are polished.

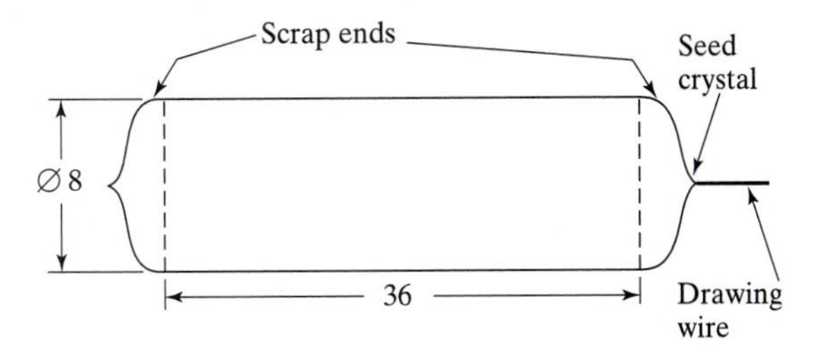

Figure P3.18

An 8 in. diameter × 36 in. long silicon cylinder is obtained by removing 125 in.^{3} from each end of the boule (which can be sold for a salvage value of \$2.25/lb) and then the process grinds 0.25 in. off of the diameter. Wafers 0.031 in. thick are obtained after sawing the cylinder with a 0.012 in. thick saw blade and polishing 0.0008 in. off of each face. Determine the following:

(a) What is the number of wafers obtained from one boule?
(b) What is the volume and weight? (*Hint:* The density of Si is 0.084 lb/in.^{3} of the total finished wafers, the total starting boule and waste.)
(c) An additional loss of 33% of the boule results from the melted silica (Si) remaining in the crucible, which can be sold for salvage. Raw Si material has an original purchase cost of \$110/lb. There is a per boule cost of \$500 for indirect and overhead materials. Determine the material cost per wafer and the overall material yield.
(d) It requires 24 hours cycle time to grow a boule. During this time, 3 operators (costing \$25/hr each for wage and benefits) attend 8 machines (with staggered cycles so only one reactor is being started or finished at a time). Determine the per wafer cost of the labor and the total finished material cost per wafer.

3.19 A purchased bar is 12 ft long and 2 in. OD. The drawing specifies a design that is 16 in. long in final dimension, but engineering requires a facing dimension of 1/16 in. and a cutoff of 3/8 in. The gripping for the last piece by the collet is 6 in. Scrap historically is 1% and shrinkage is $\frac{1}{2}$%. The cost per pound is \$0.75 and the density is 0.283 lb/in.^{3} A total of 260 final units are required. How many bars are required? What is the yield? Find the material cost for each piece. The labor for performing this job is paid at \$18.50 per hour and the production rate is 1.825 hr per 100 pieces. Analyze this job for direct costs.

3.20 Sheet, cold-rolled carbon steel, 0.1345 × 36 × 96 in. with a density of 0.29 lb/in.^{3}, is sold for \$84.88 per 100 lb in the United States. A Canadian company that has adopted SI, desires to buy 5 sheets today. Typical exchange rates in U.S. dollars are spot at 0.8378 and one year at 0.8506. (*Hint:* The SI conversion factor is 0.4535 kg/lb.) Each sheet produces 2 units of product with a waste loss of 10%. There is no shrinkage or scrap. Salvage is figured at 5% of the original value.

(a) What is the quoted rate cost to the customer for the material, and the total cost for the purchase of the lot if the transportation costs U.S.\$10 per shipment from Detroit to the customer in Windsor?
(b) What is the Canadian material cost per unit if the product is to be produced in Windsor?

3.21 A cold-rolled sheet is 36 × 96 in. in size. The sheet is commercial grade, oiled, No. 10 thickness (0.1345 in.) and it costs \$85/each free on board (FOB) at the customer's site. A manufacturer plans to first shear strips and then shear blanks from the strips. A strip is either 36 or 96 in. long. The shearing process cuts along a straight line and loses no material, although there may be a trim loss (drop-off) after shearing of strips and blanks. The placing of the strips along the sheet and the choice of strip dimension followed by the selection of the blanking dimension along the strip can affect yield from the sheet. Parts with dimensions of 5 × 10 in. are to be made.

What is the maximum number of parts that can be made from a sheet? What is the maximum shape yield? What is the lowest cost of material per part if the salvage is sold back at 10% of original material cost? Assume an infinite quantity run, and develop your answers with respect to one sheet.

3.22 A plant produces printed wiring assemblies (PWA) using an automated line. At the end of the line, the PWAs are tested. Currently, the line is producing PWAs with a yield of 85%. The cost of a completed PWA, good or bad, is $55. Some of the initially failed assemblies can be reworked to passing. For the current design, 65% of the failed assemblies can be fixed. A model that reflects this situation is

$$N_g = N_s(Y) + N_s(1 - Y)(R)$$

where R = PWA units repairable, percent and other variables are given in Problem 3.12

(a) Find the total number of acceptable units from a batch of 500.
(b) If the cost of repair is $10 per unit, what is the unit cost of the delivered units?
(c) If the customer's order is for 500 units, how many PWAs should the company start?

PRACTICAL APPLICATION

Great volatility is expected in raw material cost. These costs vary day to day even from the same supplier for identical material requirements. Material delivered to a plant site or FOB at the dock of the supplier can affect the cost. Finding these material costs will expose you to real-world circumstances.

Your practical application is to find a cost that you would pay for this material if you ordered it for delivery to your location today. (*Hint*: You will find this information by calling or visiting a local material supplier, recent catalogs, or the Internet.)

In your report, give the conditions for the quotation. Be careful to note the specifications. Cite the source of your information. (Remember you are asking for a "courtesy" quote, so be respectful for their efforts.)

(a) Steel bar stock, C-1020, $1\frac{1}{2}$ in. diameter, sufficient for 1000 flanges. (*Hint*: Use Figure P3.10.)
(b) Aluminum bar stock, 2024-T, 4.500 in. diameter, 12 ft length, sufficient for 500 quantity. (*Hint*: Use Figure P3.2.)
(c) Material that your instructor will specify.

CASE STUDY: DESIGN FOR RUNNER SYSTEM

"What counts in this design is minimum plastic volume in the runner system" Sarah, the die designer for General Plastics, mutters to herself. Sarah knows that for this plastic mold design it will be impractical to reuse the scrap because the plastic part will be colored and thus have little value as salvage. The part to be molded is roughly 25 mm in diameter and 10 mm thick, and has novelty impressions on the surfaces.

Recently hired in this job, Sarah has learned that full-round runners are preferred. They have a minimum surface-to-volume ratio, thus reducing heat loss and pressure drop. Balanced runner systems permit uniformity of mass flow from the sprue to the cavities, because the cavities are at an equal distance from the sprue. Main runners adjacent to the sprue are larger than the secondary runners.

Design	Runner Section	Diameter (mm)	Section Length (mm)
a	1	5	25
	2	6	25
b	3	5	12
	4	6	75
	5	8	25
c	6	5	8
	7	6	100
	8	10	175

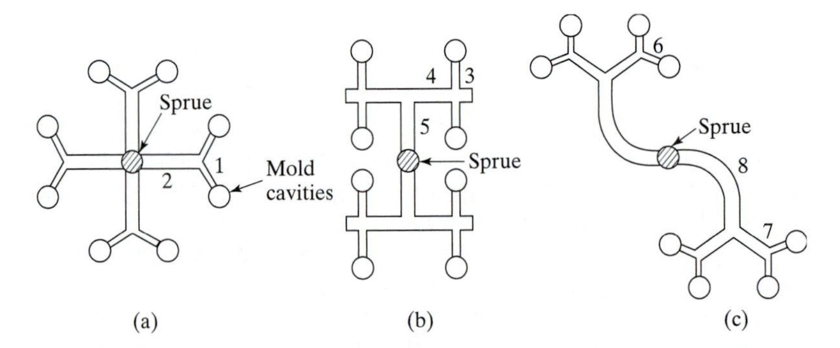

Figure C3.1 Runner system designs: (a) star, (b) "H", and (c) sweep.

Sarah has designed three configurations (Figure C3.1) and will select the one that uses the least amount of runner material. The cycle time factor is not critical, because the three arrangements provide identical numbers of parts per shot. With the design dimensions shown in the table above, determine which arrangement has a minimum of material in the runner system.

If the plastic costs \$1.20/kg and the density is 1050 kg/m^3, then what is the cost per part including the prorated loss per unit? Determine the shape efficiency where shape efficiency equals to material in parts per material in shot.

Chapter 4

Accounting Analysis

Accounting specialties fall into six categories: general, public, auditing, tax, government, and cost. Cost and tax accounting are more important to engineering than are the others. Cost accounting emphasizes accounting for costs, particularly the cost of using productive assets. Tax accounting, because of myriad laws and practices, includes the preparation of tax returns and, importantly, the consideration of the tax consequences on business transactions.

Accounting is the means of analyzing the money transactions of business. Accountants prepare balance sheets, statements of income, and information to aid the control of cost, and these facts are essential for the determination of overhead. Although cost-accounting data may have been carefully collected and arranged to suit the primary purposes of accounting, these raw data are incompatible with analysis, estimating, and pricing maneuvers.

The relationship between engineering and accounting is an active one, for both deal with much of the same information. On the one hand, engineering works with costs and designs before the spending of money; on the other hand, cost accounting records the cash flow facts. Roughly, engineering looks ahead and accounting looks back, and both are necessary for successful cost analysis.

This chapter exposes the student to basic accounting fundamentals. Business transactions, the income and balance-sheet statements, and overhead are the significant lessons for learning. We stress the understanding of these important documents, not their preparation, that being less important. In addition, engineering managers are responsible for budgets, and these are prepared oftentimes using accounting rules.

Why should engineering students be interested in accounting? If there is a curiosity in engineering cost analysis and management as a career goal, then the lessons of this chapter are imperative.

4.1 BUSINESS TRANSACTIONS

The business transaction is composed of two elements that are reported in a financial record. This duality leads to double-entry bookkeeping, a practice several centuries old.[1] Tested and found true, principles of double-entry bookkeeping have changed little, even though the overall growth of business has complicated professional accounting. The essential practices show similarity to the earliest commercial records.

In double-entry bookkeeping the business transactions are collected in records called *accounts*. The simplest form is the *T-account*—a graphical presentation of a transaction. The recording of a business transaction is an *entry*, as shown by Figure 4.1(a). An entry on the left side of the account is called a *debit (Dr.)*, and an entry on the right side is a *credit (Cr.)*. The terms debit and credit, when used for the bookkeeping of transactions, have several meanings. For example, debits increase assets and expense accounts; credits increase capital, liability, and revenue accounts. Double-entry accounting is basic to balance-sheet and income-statement presentation.

The T-account is an artifice used for textbook presentation. In practice, a more complete description supplies columns for additional data, as shown in Figure 4.1(b). The *columnar style* provides space for the date of each entry, description, folio (F), or a cross-reference to another record. This arrangement is suited to computer data processing. If the status of the account is desired, then it is only necessary to total the debit and credit entries and show the balance on the larger side. This process of summing and finding the larger amount is called *footing*.

The evidence for any entry is found in the vouchers, invoices, receipts, bills, sales tickets, checks, and the many documents relating to a transaction. With this evidence, the original entry is made to a journal that contains the chronological record of the transactions. The information is summarized in Figure 4.2. In practice, the journals are files, computer hard drives, floppy disks, or other media, which we collectively call *journals*. A single journal may suffice for entries for a small business or owner. For the larger business, however, many types of journals exist, such as cash, sales, purchase, and general journals. This recording in journals is termed *journalizing*.

The next step involves the transferring of journal entries to appropriate accounts in a *ledger*—a group of accounts. Perhaps one page of the ledger is used for each account. *Posting* is the term applied to the process of transferring the debit and credit items from the journal to the ledger account. As each item is posted, the number of the

T-account

Debit (entry on left-hand side)	Credit (entry on right-hand side)

(a)

Account title

Date	Description	F	$	Date	Description	F	$

(b)

Figure 4.1 Illustrative forms for the entry of debit and credit: (a) T-account, and (b) columnar style.

[1]The beginning of rational economic analysis is sometimes linked to the work of the Italian Benedetto Cotrugli, who published in 1458 the first known work on double-entry bookkeeping. The revolutionary nature of double-entry bookkeeping paved the way for understanding business transactions aimed at achieving profit, which then commingled commerce and profit.

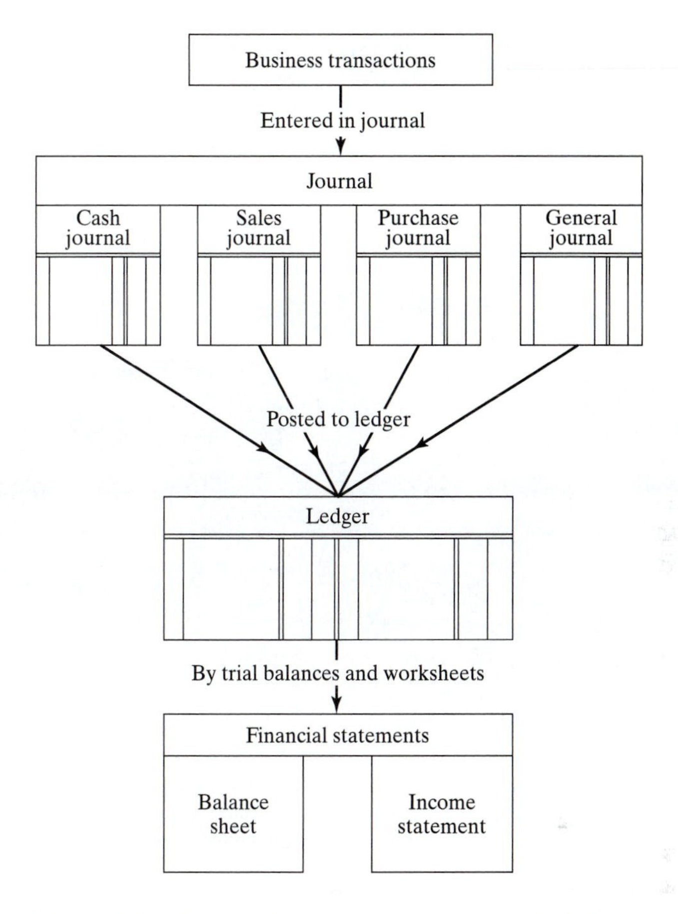

Figure 4.2 Flow of the raw transactions to the balance sheet and the income statement.

ledger account is indicated in the journal folio. Similarly, the journal page number is placed in the folio column of the account of the ledger.

The flow from the original recording of business transactions to the eventual development of the balance sheet and income statement is shown in Figure 4.2. All of these maneuvers are computerized for a faster response. From time to time, the accountant adjusts the accounting of the business to present the financial situation accurately.

4.2 CONVENTIONS

Accounting uses many important conventions, which the student needs to understand. Those that are studied include the following:

- Money measurement
- Fundamental accounting equivalence
- Conservatism
- Consistency

PICTURE LESSON The Era of the New York Central Railway

Fig. 2. The "De Witt Clinton;" New York Central Railway, 1831.

The third locomotive that was built in America had two inclined cylinders 5 in. OD with a 16 in. stroke. The two axles were coupled with wheels 4 ft 6 in. OD, of turned and finished spokes, which were "let into the cast iron rims." The engine weighed 3 tons. The boiler was copper tubular and burned anthracite coal. The train would travel 30 miles per hour with three to five cars.

Train travel was the equivalent of going to the moon during the 1830s in the United States. The building of locomotives was in its infancy in 1831 when the DeWitt Clinton (named after an early governor of New York) was inaugurated. The papers reported "The first journey to Schenectady and back was accomplished successfully, and without other incident than frightening horses and burning most of the umbrellas of the party by sparks from the tunnel."

The New York Central Railroad Company, one of the major American railroads that connected the East Coast with the interior, was founded in 1853. It was a consolidation of 10 small railroads that paralleled the Erie Canal between Albany and Buffalo; the earliest was the Mohawk and Hudson, New York State's first railway, which opened in 1831. The New York Central's moving spirit was Erastus Corning, who for 20 years had been president of the Utica and Schenectady, one of the consolidated railroads. He served as president of the New York Central until 1864. In 1867 Cornelius Vanderbilt won control, after beating down the Central's stock, and combined it with his New York and Hudson railroads running from Manhattan to Albany.

Vanderbilt and his son William acquired other railroads, until the Central system had 10,000 miles of track linking New York with Boston, Montreal, Chicago, and St. Louis.

After World War II, the New York Central began to decline. Efforts to merge with its chief competitor, the also ailing Pennsylvania Railroad Company, culminated in 1968 with the creation of the Penn Central Transportation Company—a merger that later included the New York, New Haven, and Hartford Railroad, in 1969. The new colossus had 21,000 miles of track. Its creators hoped to achieve a division of labor, sending freight to New York and New England north along the New York Central's water-level route while the Pennsylvania main tracks served the industrial needs of Philadelphia, Baltimore, and the Delaware and Schuylkill valleys.

The merger failed, however, and the new road was forced into bankruptcy in 1970. Passenger services were taken over by the federally established National Railroad Passenger Corporation (Amtrak) in 1971. The company's other railroad assets were merged with five other lines in Consolidated Rail Corporation (Conrail) in 1976, although the New York–Washington route was later transferred to Amtrak.

- Cash and accrual accounting
- Relevance

A *money measurement* fundamental requires that business transactions be recorded only with money. This practice expresses many different situations in common units. A fabricating plant can be compared to a construction company. Moreover, designs denominated in various currencies, such as dollars, euros, or pesos can be algebraically manipulated to guage the company's performance. Because of the importance of the money-measurement fundamental, it is universally adopted, even though it must ignore technology expertise, engineering skill, prominence of design, and a company's reputation. While these are important features, they are difficult to demarcate in money units.

An *accounting fundamental equivalence* insists that business transactions be recorded in double-entry fashion and that total debits must *equal or balance* total credits. This leads to the important accounting equation, where assets equal liabilities plus owner's equity, or modeled differently in Equation (4.1),

$$\begin{aligned} \text{Assets} &= \text{liabilities} + \text{owner's equities} \\ \text{Assets} &= \text{liabilities} + \text{net worth} \end{aligned} \tag{4.1}$$

Both creditors and owners have claim to equities. The creditor has first rights, leaving the remaining rights to the owners. An owner's equity or net worth or proprietorship is the ownership interest in the business assets.

A convention known as *conservatism* encourages the recording of financial data as the lower of possible values. For example, in the evaluation of finished goods inventory (products held by the company for sale) we could value the inventory as either the cost to the business for production or the market value. Conservatism dictates the lower value. In public disclosure for the annual financial report, the statement is made according to "the cost or market value, whichever is lower" to reflect this convention.

The *consistency* convention states that business transactions, if accounted for in one way, are recorded and accounted for that way in the future. Companies obtain discounts for prompt payments of materials. One firm may use the discounts to reduce the cost of the materials. Another firm may take discounts and record them as income realized on prompt payment. The consistency convention adopts one of those two methods and persists in its use in succeeding periods. This convention discourages a business from manipulating its figures to reflect favorable conditions on one occasion and then, as convenient, changing its approach.

The *going-concern fundamental* implies that the business is operated in a prudent and rational way. This policy, it is assumed, perpetuates the business over an extended period. This going-concern assumption is one justification for the depreciation principles, discussed later.

A *business-entity fundamental* is a simple notion that accounting transactions of a business are for the sake of the business rather than for individuals. If an owner withdraws money from the cash box, then the owner is richer but the business has less. The accountant records the effect of this transaction on the business and ignores the effect on the owner.

A *cost fundamental* recognizes that cost does not necessarily equal value. Though it is possible to list the assets in preparing a balance sheet, their value is determined by several methods. Consider a car for sale. Early in the day, a potential customer asks what the owner will sell it for. Later, a tax assessor asks the same question. Though the answer might honestly differ, the point is that "value" is subjective. Even so, there are two other ways to evaluate business assets. Market value can be obtained, but this depends on the purchaser's needs. It is possible to find several evaluations of the worth of an asset, which returns the selection of value to subjective reasoning. We can value assets on the cost of their replacement. A replacement approach leads to a range rather than a single value. Market and replacement value lead to confusing choices, but it is possible to determine the value of the asset as given by the *original* cost. The receipt for the payment of the asset is a document of record and demonstrates cost, albeit not value necessarily. The primary appeal of the cost fundamental is objectivity and expediency. When coupled with the going-concern principle, it is assumed that the asset is used to advance the business.

If *cash accounting* is used, then income is recorded when the cash is received and expense is recognized when cash is paid out. The cash method is used by small companies and individuals for personal and family records. Cash accounting is not allowed by the Internal Revenue Service for larger businesses or owners.

The *accrual fundamental* is concerned with the majority of business and is generally unfamiliar to individuals. In accrual, income is recorded when it is earned, whether the payment is received during the period or not. Expenses incurred in earning the income are recorded as expense whether or not payment has been made during that period. The profit-and-loss statement includes incomes earned during the period covered by the statement and expenses concurrent to the same period. Many businesses prefer accrual accounting because it matches revenues to the expenses in a specific period. The time of collection of the income or payment of the expense is not a factor.

A convention of *relevance* permits the accountant to use judgment in handling certain expenses. Business transactions that affect income are measured by increases or decreases in net worth or services. Moneys received increase net worth and are called *revenues*. Costs that a business incurs to provide the design, goods, or services decrease net worth and are called *expenses*. If a transaction is irrelevant to the financial results of the business, then there is discretion as to how and when to record that event. If a piece of paper and pencil have been consumed, then, theoretically, the paper and pencil become an expense. However, this penny-watching is unwise, because average monthly office-supply expenses are a common-sense way of handling this transaction.

4.3 CHART OF ACCOUNTS

Accounts are established for various business transactions. A formal *chart of accounts* is grouped into five principal categories:

- Assets
- Liability
- Net worth
- Revenue
- Expenses

For instance, a total sales account may lead to a specific customer sales account. Every item of financial information on the balance sheet and income statement has an account. A chart of accounts is necessary for a business. A typical chart or listing of accounts intended for a manufacturing company is illustrated in Table 4.1.

TABLE 4.1 Illustrative Chart of Accounts for an Accrual Type of Business

- *Assets*
  - Current Assets
    - Cash in bank
    - Petty cash
    - Notes receivable, customers
    - Inventories
    - Accounts receivable
    - Prepaid insurance, taxes, interest
    - Supplies
  - Fixed assets
    - Land
    - Buildings, equipment
    - Accumulated depreciation
- *Liabilities*
  - Current liabilities
    - Accounts payable
    - Notes payable
    - Accrued wages payable
    - Accrued interest payable
    - Accrued taxes
    - Deferred rent income
    - Dividends payable
  - Fixed liabilities
    - Mortgage payable
    - Bond payable
- *Net Worth*
  - Capital stock, preferred
  - Capital stock, common
  - Retained earnings
- *Revenue*
  - Sales
  - Interest earned
  - Dividends received
  - Sale of waste and scrap
- *Expenses*
  - Manufacturing costs
    - Purchase of materials
    - Salaries and wages
    - Heat, light, water
    - Telephone
    - Depreciation
    - Direct labor
    - Indirect labor
    - Insurance
    - Repairs and maintenance
    - Manufacturing supplies used
  - Selling expenses
    - Salaries and commissions
    - Advertising and samples
  - General and administrative expenses
    - Salaries
    - Traveling expenses
    - Telephone, postage
    - Supplies
- *Taxes, All Kinds*

Assets. The *assets* of a business are those things of dollar value that it owns—tangibles such as land, buildings, equipment, or inventory, or intangibles such as trademarks, designs, and patents. Assets may be segregated into current assets, fixed assets, and intangible assets.

Current assets have three inventory accounts representing raw materials, in-process goods, and completed products. *Fixed assets* include office equipment, manufacturing equipment, and buildings, less accumulated depreciation. Land is an asset that does not depreciate. In retail and manufacturing businesses, inventories are owned by a company and are a tangible asset with dollar value. Assets are capable of providing future benefits; otherwise, they are expense.

For an example of intangible assets, an engineering firm designs new methods through its research and development department and obtains patents on the procedures. The cost of research, engineering, and testing leading to the development of a new procedure may be significant and in theory could be treated as an asset. However, under present financial tax accounting, those costs are expenses when incurred. Because many research projects may be underfoot at the same time, cost can be incurred over a period of years. Some firms treat engineering costs as a part of current operating expenses. Patents and copyrights are not recorded as assets if developed within the firm. If they are purchased or acquired through merger, then they are shown as assets.

Liabilities. The *liabilities* of a business are the debts the firm owes. The category of liabilities is frequently broken down into current and long-term debts. Current liabilities are less than or equal to 1 year. Some of the more common items of business liabilities are (1) accounts payable (debts of the firm to creditors for materials and services received), (2) bank loans (amounts the firm owes to banks for money borrowed), and (3) mortgage payable (debt to investors for money loaned to the business on the security of its real estate or equipment).

Net Worth. The *net worth* of a business is the ownership interest in the firm's net assets. In certain accounting situations, proprietorship and capital are considered synonymous with net worth. In a simple case the net worth of a business consists of its capital stock and the retained earnings:

Example: Net Worth of Business

Net Worth	
Capital stock	$45,000
Retained earnings	4,000
Total	$49,000

Broadly speaking, capital stock is the portion of the net worth paid in by the owners. Surplus, another term used within net worth, is that portion that the owners paid for stock. Retained earnings refer to the accumulated profits and losses of the firm.

A company issues capital stock, which is divided into units of ownership called *shares*, and the owners of the company are referred to as *shareholders*. The ownership of a shareholder in the net worth of the company is related to the number of shares he or she owns. The surplus of a company increases as the company earns profit and decreases as the company incurs losses or distributes the profits among the shareholders as dividends. If the losses and dividends of a company since inception exceed its profits, then a negative profit or a deficit within retained earnings results.

Income and Expense in Business. *Revenues* are generated by sales before the deduction of cost. *Expenses* represent the costs of doing business. Income is received from the sale of jobs, products, or materials. Expenses include salaries, advertising, power and light, telephone, rent, insurance, and interest. The profit-and-loss statement of a business is a summary of its incomes and expenses for a stated period. If the statement discloses a net profit or loss, then the change represents an increment or decrement in the retained earnings during the period, arising from the incomes and expenses of a business, and it is carried to the net worth section of the balance sheet.

Gross income is the difference between income and expense. Once taxes are removed, we have *net profits*. Because there are so many types of taxes, our discussion on taxes is simplistic, and we refer to taxes as "all kinds." The interested student is referred to other sources for detailed information.

4.4 STRUCTURE OF ACCOUNTS

An expanded accounting equation using T-accounts is given as

$$\frac{\text{Assets}}{+|-} = \frac{\text{liabilities}}{-|+} + \frac{\text{networth}}{-|+} + \frac{\text{revenue}}{-|+} - \frac{\text{expenses}}{+|-} \tag{4.2}$$

This is sometimes referred to as the *financial and operating equation*, because it has plus and minus signs for each class of T-accounts. The plus and minus signs show increases and decreases and are summarized as follows:

Debit Indicates	Credit Indicates
Asset increase	Asset decrease
Liability decrease	Liability increase
Net worth decrease	Net worth increase
Revenue decrease	Revenue increase
Expense increase	Expense decrease

Consider an example for General Manufacturing Company. The integrated example flows from the transactions to the noting of the effect on the T-accounts, to a trial balance, and finally to the profit-and-loss and balance-sheet statements. Our simplified approach starts with the initial capitalization of the firm, leading to its first reporting of financial documents.

Each transaction is worded to suggest a dual recording of the transaction. The numbered entry connects to the affected accounts. Consider transaction 1 in Table 4.2.

TABLE 4.2 Transaction Effects for General Manufacturing Company

Transaction	Accounts Affected	Type of Account	Effect of Transaction: On Account	Is Recorded by a Debit of	Is Recorded by a Credit of
1. General Manufacturing is founded and $50,000 paid for capital stock	Cash	Asset	Increase	$50,000	
	Capital stock	Net worth	Increase		$50,000
2. Business buys materials from S. W. Specthrie on account, $10,000	Inventory	Asset	Increase	10,000	
	Account payable	Liability	Increase		10,000
3. Pay monthly rent on shop, $1500	Rent	Expense	Increase	1,500	
	Cash	Asset	Decrease		1,500
4. Pay S. W. Specthrie on account, $4,000	Account payable	Liability	Decrease	4,000	
	Cash	Asset	Decrease		4,000
5. Sell to P. Hall on account, $15,000	Account receivable	Asset	Increase	15,000	
	Sales	Income	Increase		15,000
6. Pay salaries, $2,850	Salaries	Expense	Increase	2,850	
	Cash	Asset	Decrease		2,850
7. Retire for cash $5,000 of capital stock	Capital stock	Net worth	Decrease	5,000	
	Cash	Asset	Decrease		5,000
8. Collect $2,000 from P. Hall	Cash	Asset	Increase	2,000	
	Account receivable	Asset	Decrease		2,000
9. Buy equipment for $3,000 cash	Equipment	Asset	Increase	3,000	
	Cash	Asset	Decrease		3,000
10. Receive $500 rebate on month's rent	Cash	Asset	Increase	500	
	Rent	Expense	Decrease		500
11. P. Hall returns $4,000 of material for credit	Sales	Income	Decrease	4,000	
	Accounts receivable	Asset	Decrease		4,000
12. Pay advertising bill, $800	Advertising	Expense	Increase	800	
	Cash	Asset	Decrease		800
13. Buy on credit $12,000 computer from Englewood and Co.	Computer	Asset	Increase	12,000	
	Account payable	Liability	Increase		12,000
14. Pay insurance premium, $500	Insurance	Expense	Increase	500	
	Cash	Asset	Decrease		500
15. Take depreciation charge, $600	Depreciation	Expense	Increase	600	
	Accumulated depreciation	Asset contra	Increase		600
16. Pay taxes, $1,250	Taxes	Expense	Increase	1,250	
	Cash	Asset	Decrease		1,250
Total				$113,000	$113,000

A cash account is an asset type of account, and capital stock is a net worth account. The effect of the cash on the asset is to increase the T-account by the recording of a debit to the Cash T-account, and net worth is increased with the entry of $50,000 to the credit side of the T-account. Each of the transactions follows the financial and operating equation.

Notice that the transactions of Table 4.2 are hooked to the T-accounts of Table 4.3. By using those rules for applying debit and credit, we record the transaction into a T-account. The identity of the account is described further, such as A or asset. In the T-accounts presented in the Table 4.3, the numbers in parentheses next to the dollars amount in relate to the transaction numbers shown in Table 4.2. For example, the "(1)" preceding the debit of $50,000 in the Cash T-account indicates the transaction presented in Table 4.2. The transaction deals with the founding of General Manufacturing and the $50,000 which has been paid for capital stock. The dual entry is in the Capital stock T-account, where $50,000 is recorded as a credit and is noted with a 1 next to the $50,000.

In Table 4.2, one transaction affects two accounts—for example, the transaction of the founding of General Manufacturing. Both cash and capital stock are influenced, and those accounts are identified into asset and net worth types. The effect of the transaction is to either increase or decrease the account, which is further explained by a recording of either a debit or a credit.

Because of the dual effect of a transaction, the record of the transaction must be balanced, and for every debit, there must be a credit. It is not necessary that there be the same number of debit and credit items in any T-account. Table 4.3 is a record of transactions

TABLE 4.3 T-Accounts for Transactions of General Manufacturing Company, Table 4.2

Cash (A)				Accumulated depreciation (A)				Insurance (E)			
(1)	$50,000	$1,500	(3)			($600)	(15)	(14)	$500		
(8)	2,000	4,000	(4)								
(10)	500	2,850	(6)								
		5,000	(7)								
		3,000	(9)								
		800	(12)								
		500	(14)								
		1,250	(16)								
Inventory (A)				**Accounts payable (L)**				**Depreciation expense (E)**			
(2)	$10,000			(4)	$4,000	$10,000	(2)	(15)	$600		
						12,000	(13)				
Accounts receivable (A)				**Sales (I)**				**Rent (E)**			
(5)	$15,000	$2,000	(8)	(11)	$4,000	$15,000	(5)	(3)	$1,500	$ 500	(10)
		4,000	(11)								
Computer (A)				**Salaries (E)**				**Capital stock (NW)**			
(13)	$12,000			(6)	$2,850			(7)	$5,000	$50,000	(1)
Equipment (A)				**Advertising (E)**				**Taxes**			
(9)	$3,000			(12)	$ 800			(16)	$1,250		

that affect the amounts of assets, liabilities, net worth, income, and expenses of a business. Any business transaction may be recorded in terms of equal debit and credit elements based on the increase or decrease effect. Purchases are considered an expense because they are an offsetting cost to income sales. That portion of purchases remaining unsold at the close of the period is termed inventory and is classed as an asset.

An open account has either a debit or a credit balance. A closed account has debit and credit of equal amount and, therefore, has no balance.

	Rent			Capital stock	
	$8,000	$2,000		$5,000	50,000
		6,000			
Balance	0	0	Balance		45,000

The Rent account is a closed account, because the sum of the debits equals the sum of the credits, which in this case is zero. This is not the case for Capital stock, which is open, or $45,000.

Each journal entry provides for equal debits and credits, which are posted to the ledger accounts. If the posting is accurate, then the ledger must have equal debits and credits. Additionally, the sum of the debit ledger account balances must equal the sum of the credit ledger account balances. This equality is periodically tested by a *trial balance*, which is a list of the open ledger accounts as of a stated date. The trial balance shows the debit or credit balance of each account.

The account groups in the trial balance are broadly divided into those used to prepare the balance sheet and the income and expense statement. A few accounts contain both balance-sheet and profit-statement elements and are called *mixed*. Those are separated into the two components during worksheet analysis.

Note that for the Cash T-account in Table 4.3, the cash balance is a debit of $33,600, which is entered in the trial balance under debit and the Cash account in Table 4.4. In a similar way all Table 4.3 T-accounts are footed and their debit or credit balance entered under the trial balance column in Table 4.4

The periodic trial balance of the ledger provides reasonable proof of the arithmetic accuracy of journalizing, posting, and ledger account balancing. The ledger lists account balances from which the balance sheet and income statement are later prepared. In most businesses the trial balance is performed after the end of the month.

A worksheet example is given in Table 4.4, which continues the development given in Tables 4.2 and 4.3. Various account titles are listed, and the balances from the ledger are posted in the Trial Balance column as either debit or credit. Those entries are carried over to either the income-statement or balance-sheet columns.

Generally, a worksheet adjusts the accounts for accrued expenses, accrued incomes, deferred expenses, depreciation, and bad debts. Once those adjusting entries are disposed, the next step concludes the trial balance by using a worksheet. Note that in Table 4.4 the first trial balance shows the effect of the ledger accounts. Those in turn are separated and extended horizontally into profit-and-loss and balance-sheet entries. The worksheet must balance. If it does not, then an error of some kind is indicated, and must be found.

TABLE 4.4 Trial Balance for General Manufacturing Company

	Trial Balance		Income Statement		Balance Sheet	
Account	Dr	Cr	Dr	Cr	Dr	Cr
Cash	$33,600				$33,600	
Inventory	10,000				10,000	
Accounts receivable	9,000				9,000	
Computer	12,000				12,000	
Equipment	3,000				3,000	
Accumulated depreciation	(600)				(600)	
Accounts payable		$18,000				$18,000
Sales		11,000		$11,000		
Salaries	2,850		$2,850			
Advertising	800		800			
Insurance	500		500			
Depreciation expense	600		600			
Rent	1,000		1,000			
Capital stock		45,000				45,000
Taxes	1,250		1,250			
Total	$74,000	$74,000	$7,000	$11,000	$67,000	$63,000
Profit to retained earnings						4,000
Total Assets and Total Liabilities and Net Worth						$67,000

4.5 UNDERSTANDING THE BALANCE-SHEET STATEMENT

The *balance sheet* is a tabular presentation of Equation (4.1)—the important accounting equation—which summarizes the assets, liabilities, and net worth at a point in time, what might be called a snapshot. The information for the balance sheet is obtained from the worksheet. Balance sheets always follow a standard form, usually the complete name of the firm, followed by the title "Balance Sheet," and the date of the evaluation.

Example: A Balance Sheet for XYZ Manufacturing Company

A balance sheet is shown for XYZ Manufacturing Company and highlights the accounting equation terms.

XYZ Manufacturing Company
Balance Sheet
May 31, 20xx

Assets		= Liabilities	
Cash	$15,000	Bank loan	$15,000
Inventory	10,000	Mortgage	5,000
Land	15,000		
Building and equipment	40,000		
		+ Net worth	
		Capital stock	45,000
		Retained earnings	5,000
	$80,000		$80,000

TABLE 4.5 Balance Sheet for General Manufacturing Company

General Manufacturing Company
Balance Sheet
June 30, 20xx

Assets			Liabililities and Net Worth		
Current assets			Current liabilities		$18,000
Cash		$33,600	Net worth		
Accounts receivable		9,000	Capital stock	$45,000	
Inventory		10,000	Retained earnings	4,000	
Fixed assets					
Equipment	$3,000				
Less accumulated depreciation	(600)				
Computer	12,000				
Subtotal fixed assets		14,400			49,000
Total assets		$67,000	Total liabilities and net worth		$67,000

The General Manufacturing Company balance sheet is given in Table 4.5. Important points about the balance sheet are the length of time covered, handling of the accumulated depreciation, and the asset and liability groups disclosed. Note that the closing date shown is the end of June. The balance sheet does not provide any hint about what the assets, liabilities, and net worth were for any date before or subsequent to June 30.

Balance-sheet assets are not valued on the same basis. Cash, customer receivables, and inventories are valued at cost or net realizable cash value according to the conservatism convention. Land is valued at the amount originally paid for it, and depreciable fixed assets are valued at original cost less the accumulated depreciation. The liabilities are valued at the cash amount required to liquidate at the time of their maturity date. The net worth is a conglomerate value, because it represents the difference between assets and total liabilities.

4.6 UNDERSTANDING THE PROFIT-AND-LOSS STATEMENT

The statement of earnings of a company, known either as the *profit-and-loss* or *income-and-expense statement*, is a summary of its incomes and expenses for a stated period of time. The net profit or loss represents the net change in net worth during the reporting period arising from business incomes and expenses. This usually covers the period between balance-sheet statements and will provide details about the difference between two balance sheets.

Definition of Profit. *Profit* represents the excess of revenue over cost. It is an accounting approximation of the earnings of a company after taxes, cash, and accrued expenses (representing the costs of doing business), and certain tax-deductible noncash expenses, such as depreciation, which are deducted. Sometimes it is termed *net profit. Loss* represents the excess of cost over bid—for example a manufacturing job costing $1,800,000 with a bid of $1,700,000, results in a loss of $100,000.

Example: Effect of Net Worth on Profit-and-Loss Statement

The following balance sheet describes the effect on business net worth of profit or loss:

JD Manufacturing
Balance Sheet
May 31, 20xx

Assets		= Liabilities	
Equipment	$400,000	Bank loan	$100,000
Job A material	800,000	Accounts payable	200,000
Job B material	600,000		
		+ Net worth	
		Capital stock	1,500,000
	$1,800,000		$1,800,000

If JD Manufacturing sells Job A material for $1,000,000 cash, its balance sheet changes as follows:

JD Manufacturing
Balance Sheet
June 30, 20xx

Assets		= Liabilities	
Equipment	$400,000	Bank loan	$100,000
Cash	1,000,000	Accounts payable	200,000
Job B material	600,000		
		+ Net worth	
		Capital stock	1,500,000
		Retained earnings	200,000
	$2,000,000		$2,000,000

Net worth increases $200,000. If the business sells the asset Job B material for $500,000 cash, then its balance sheet looks like this:

JD Manufacturing
Balance Sheet
July 31, 20xx

Assets		= Liabilities	
Cash	$1,500,000	Bank loan	$100,000
Equipment	400,000	Accounts payable	200,000
		+ Net worth	
		Capital stock	1,500,000
		Retained earnings	100,000
	$1,900,000		$1,900,000

The $600,000 Job B material is replaced by $500,000 cash, and the net assets and the net worth are decreased $100,000. Notice that profits increase the net worth because profits increase the net assets, and losses decrease the net worth because losses decrease the net assets.

Continuing on with the worksheet and balance sheet, as developed previously for General Manufacturing Company, its profit-and-loss (P&L) statement is given in Table 4.6.

TABLE 4.6 Profit-and-Loss Statement for General Manufacturing Company

General Manufacturing Company
Profit-and-Loss Statement
June 30, 20xx

Income		
Product SALES		$11,000
Expenses		
Salaries	$2,850	
Rent	1,000	
Advertising	800	
Insurance	500	
Depreciation	600	
Total		5,750
Gross profits		$5,250
Taxes at 23.8%		1,250
Profit (to retained earnings)		$4,000

These P&L statements should be studied for their heading, income and expense groupings, and length of time and dates covered. Certainly, profits depend on the time of earnings. Net sales measure the net revenue from sales, and allowances for sales returns, freight out, and sales discounts are deducted from gross sales. Cost of goods sold covers the expense of the products sold to the customer. If the cost of goods sold is a gross value, then freight in and purchase discounts may reduce the value.

Operating expenses list recurring usual and necessary costs for conducting the business. Miscellaneous income and expense arise from interest and discounts and other small items of revenue and expense that are unrelated to the major business thrust. Except for depreciation the income statement items result from current-period transactions. *Depreciation* is an *allowable noncash tax expense* and reduces total income. Administrative expenses are found in almost all companies and cover the cost of managing the company; they may also include heat, power, rent, insurance, accounting, engineering, and legal costs. Income taxes, or the provision for income taxes, are an item reducing business income and are identified separately.

The accounting documents of the profit-and-loss statement and the balance sheet are very important for engineering, but other subjects are necessary for study too. Depreciation cost study by itself can be considered. Though less important to the central accounting documents, it is useful for engineering cost calculations for product and project cost analysis and estimating.

4.7 DEPRECIATION

The understanding of depreciation is important for students because of its impact on business and for engineering cost analysis. For example, the accounting statements, cash flow analysis, budgets, overhead calculations and taxation effects are influenced by depreciation practices. Manufacturing is a capital-intensive business and depreciation is a significant topic. We study the following topics related to depreciation:

- Background
- Purpose for depreciation
- Property classification for depreciation methods
- Methods of calculating annual depreciation amount

4.7.1 Background

Initially, capital money is spent to acquire assets that contribute to revenue over long periods. Examples of fixed assets include plants, equipment, computers, trucks, and furniture. Depreciation is applied to fixed assets, not current spending or expenses. Depreciation is charged against assets, and as a result of this charge it is also treated as a current expense and a reduction of current income.

The cost of manufacturing equipment, for example, is a positive factor in generating revenue. In the original practice, depreciation was recognized as a decrease in the value of property and the following factors contributing to this decline in market value:

- Physical wear and tear resulting from ordinary operation
- Functional factors such as inadequacy and obsolescence
- Governmental and political actions

Physical and life declining factors for plants and equipment—for example, wear, corrosion, retirement, and sudden death syndrome, are commonplace and exert an influence on economic efficiency and safety. Well known is the maintenance cost that increases with age.

Owing to technological progress, obsolescence is a frequent situation, where the asset is retired not because it is worn out but because it is outmoded. Superior efficiency of a later design is one reason that compels new equipment designs for the market. Sometimes alterations of the design through research and development techniques make for immediate obsolescence.

Inadequacy frequently occurs when the asset, although neither worn out nor obsolete, is unable to meet the demands made upon it. An electric power company installs a small hydroelectric generator and after several years finds that the demands of an expanding community placed on the generator exceed availability of hydroelectric power to meet peak loads.

Governmental laws may prevent manufacturing operations, such as loss of raw material sources or laws prohibiting waste disposal. These disposal laws limit product design because of hazardous materials that are a part of the product.

While these wear out and decline-in-utility considerations were important originally, it was the development of federal tax laws in the last century that significantly altered the business practice of depreciation. Today, taxation effects for business dominate the generalized mechanics of depreciation calculations and some of the thinking in cost analysis.

4.7.2 Purpose of Depreciation

Depreciation is a process of allocating an amount of money over the recovery life of an asset in a systematic manner. Furthermore, it is an accounting charge that provides for recovery of the money that initially purchased the physical assets.

The depreciation charge is not a cash expenditure. The actual cash outlay takes place when the asset is acquired. Depreciation charges are the assignments of that initial cost over the recovery life of the asset and do not involve a periodic disbursement of cash. Thus depreciation is a noncash deduction to the income statement. There are no interest charges or any recognition of a changing dollar value because of inflation during this period of time of depreciation.

Depreciation is used for the calculation of income taxes and there is no real money transaction happening. It is an important part of overhead calculation, however. Eventually, we will see that the depreciation is removed from the income statement as an expense item but is then added back for cash flow analysis purposes after the taxes are calculated.

When the rate at which the asset is depreciated increases, the depreciation charge does not increase the outflow of cash. In fact, the opposite reaction happens, because the depreciation charges reduce taxable income and the outflow of cash for taxes. The initial investment is a prepaid operating cost that is expensed or allocated to an operating expense account, typically in overhead.

Because federal laws require income measurement, it is necessary that an appropriate portion of the cost of the equipment be charged to or matched with each dollar of revenue resulting from the income generated by its activity. The amount of such cost, matched with revenue, during any one period is the estimated amount of the cost that expires during the period. Thus a fixed asset, which will not last forever, has its useful value exhausted over a period of time.

Engineering estimates are required for the value of the asset, erection (if necessary), and operating capital and costs for consideration of new projects. In every project there are certain costs that may possibly be tax expensed immediately. These would include expenditures for nonphysical assets, some physical assets of extremely short life, and certain installation and startup expenses.

The question of *life* is an important matter. Some businesses are concerned about economic life with little regard for physical life, while others, public utilities for example, are restricted to earning a specified amount on capital invested, and life takes on another meaning. The economic-life estimate of an asset is affected by tax laws and the actions of Congress.

4.7.3 Property Classification for Depreciation Methods

Before calculating depreciation, it is necessary to understand property or asset classification. A depreciable property is used in the business or held for the obtaining of revenue and has a useful life longer than one year. The property will wear out, decay, become obsolete, or lose value from natural causes. Thus, the value of land is not depreciable.[2]

[2]However, in mining operations, the removal of natural resources results in a decrease of the value of the land. This removal is called *depletion*.

Depreciable property may be tangible or intangible, personal or real. *Tangible property* such as buildings and equipment is seen or touched. *Intangible property*—designs, patents, and copyrights for example—can be amortized if its useful life can be found. *Real property* is land and generally anything attached to, growing on, or erected on the land. Because land has an indeterminate life, it cannot be depreciated. Personal property, which does not include real estate, includes equipment, machinery, or vehicles, for example. Engineering deals with tangible personal property and tangible real property.

Tax laws affect the practice of depreciation. The Internal Revenue Service (IRS) provides various definitions. *Recovery property* is subject to the allowance for depreciation. A recovery period, a prescribed length of time, is designated for recovery property:

- Three-year property has a life of 4 years or less and is used in connection with production and tooling.
- Five-year property excludes 3-, 10-, or 15-year public utility property. Examples include machinery and equipment not used in research and development, autos, and light trucks.
- Ten-year property is public utility property with a class life of more than 16 years but less than 20 years. An example could be railroad tank cars.

A firm may choose an alternative method to determine depreciation by using a longer recovery period. The following table is typical.

Recovery Period	Optional Recovery Period, $N(k)$
3	3, 5, or 12
5	5, 12, or 25
10	10, 25 or 35
15	15, 35, or 45

4.7.4 Methods of Calculating Annual Depreciation Amount

The annual depreciation amount is calculated by several means, but we consider only three of them:

- Straight line with and without salvage
- Units of production
- Accelerated cost recovery

Because depreciation is a money amount, it is found as follows:

$$D_j = P(j) \times P \tag{4.3}$$

where D_j = depreciation in jth year for specified property, dollars

$P(j)$ = percentage for year j for specified property class

P = cost of asset, dollars

Straight-Line Method. The simplest method—a method that is popular worldwide and is occasionally used in this book—is the *straight-line method*:

$$P(j) = \frac{1}{N(k)} \tag{4.4}$$

where $N(k)$ = recovery period, years

The straight-line method of Equation (4.4) assumes that the salvage value is zero. The $P(j)$ is a constant for the straight-line method.

Example: Straight-Line Method of Depreciation

Assume a $100,000 asset, which for physical wear-and-tear reasons we choose to depreciate over a 5-year life. The asset will decline to a zero value at the end of the fifth year. The amount of annual depreciation is calculated as follows.

Year	Cost	Straight-Line Depreciation, %, $P(j)$	Book Value at Year Beginning	Yearly Depreciation, D_j
0	$100,000		$100,000	
1		20	100,000	$20,000
2		20	80,000	20,000
3		20	60,000	20,000
4		20	40,000	20,000
5		20	20,000	20,000
				Total $100,000

Book value is ordinarily taken to mean the original cost of the asset less any amounts that have been charged as depreciation. Book value should not be confused with salvage. The word "book" implies the accounting book.

For general understanding, a *salvage value* is often associated with straight-line depreciation, and the straight-line method is modified as follows:

$$D_j = \frac{1}{N(k)}(P - F_s) \tag{4.5}$$

where F_s = future salvage value of property, dollars

Example: Straight Line with Salvage Method of Depreciation

Reconsider the investment of $100,000 with a $10,000 salvage at the end of 5 years. The yearly depreciation is found as follows:

Year	Cost Less Salvage	Straight-Line Depreciation, $P(j)$, %	Book Value at Year Beginning	Yearly Depreciation, D_j
0	$90,000		$100,000	
1		20	100,000	$18,000
2		20	82,000	18,000
3		20	64,000	18,000
4		20	46,000	18,000
5		20	28,000	18,000
				Total $90,000

Although appealing because of their simplicity and their modeling of the physical life principle, the straight line with and without salvage and the units of production methods are not the most frequently used methods in business. That distinction belongs to the accelerated methods. The notion of "acceleration" means that the investment dollars that originally purchased the asset are returned sooner to the company than the straight line methods. The overarching principle of depreciation changed from forecasting "wear-out" and life to one of governmental laws dictating the return of the investment. Accelerated methods also provide tax advantages, or reduced taxes in the earlier years of the asset life, to the owner of the asset.

Units of Production. A second method of depreciation is based on the premise that an asset wears out exclusively as demands are placed on it. Called the *units-of-production method*, the computation is given as

$$D_{up} = \frac{1}{\sum N(i)}(P - F_s)N(i) \tag{4.6}$$

where $=$ D_{up} units-of-production depreciation, dollars

$N(i)$ $=$ units for ith year

This method has the advantage that expense varies directly with operation activity. Retirement in those cases tends to be a function of use. An estimate of total lifetime production is necessary.

Example: Unit-of-Production Method of Depreciation

The example involving $100,000 cost and $10,000 salvage value is estimated to have 200 units of output over the 5-year period. The depreciation annual charge is found as follows.

Year	Cost Less Salvage	Units of Production, $N(i)$	Book Value at Year Beginning	Yearly Depreciation, D_{up}
0	$90,000		$100,000	
1		15	100,000	$6,750
2		45	93,250	20,250
3		50	73,000	22,500
4		55	50,500	24,750
5		35	25,750	15,750
		200		Total $90,000

The units-of-production method is not widely used because it is difficult to estimate the production output of the asset.

The government alters the tax laws and frequently affects the depreciation.[3] Changes in the laws governing income taxes and codes are enacted periodically, about every other year. In recent decades the federal government coined the term *modified accelerated cost recovery system* (MACRS). Useful life, recovery period, and salvage value are terms that are carefully defined.

There are advantages for business if the capital cost for the asset is quickly returned, and the straight-line method suffers seriously in that each year a uniform amount is depreciated. This is unlike the actual declining life values of equipment, which is greater in the early years and less in senior years.

The advantages of accelerated methods of depreciation are compatible with the logic that the earning power of an asset is created during its early service rather than later, where upkeep costs tend to increase progressively with age.[4] Accelerated methods offer a measure of protection against an unanticipated contingency such as excessive maintenance, and they return the investment more quickly and simultaneously, decreasing the book value at the same rate. A high book value would tend to deter the disposing of unsuitable equipment even when the need for replacement is pressing. Rapid reduction of book values, provided the owner or business overlooks the tax benefits from capital loss, leaves the owner or business more free to dispose of inefficient and unsatisfactory equipment.

Accelerated Cost Recovery. A method called *accelerated cost recovery* uses Equation (4.3), given previously. The values of $P(j)$ are established by the IRS code. Typical values of $P(j)$ are given for years 3, 5, 10, and 15 and are shown in Table 4.7,

[3]The student may be interested in the home page of the Internal Revenue Service, where many current details are given. The address is www.IRS.gov.

[4]The important 1981 federal act featured a dramatic departure in the treatment of business outlays for plant and equipment—i.e., capital cost recovery or tax depreciation. Heretofore, capital cost recovery had attempted roughly to follow a concept known as economic depreciation, which refers to the decline in the market value of a producing asset over a specified period of time. The 1981 act explicitly displaced the notion of economic depreciation, instituting instead the Accelerated Cost Recovery System which greatly reduced the disincentive facing business investment and ultimately prepared the way for the subsequent boom in capital formation

TABLE 4.7 Illustrative Schedule for Modified Accelerated Cost Recovery System

	Percentage Depreciation			
Recovery Year	3-Year	5-Year	10-Year	15-Year Public Utility
1	33	20	10	7
2	45	32	18	12
3	22	24	16	12
4		16	14	11
5		8	12	10
6			10	9
7			8	8
8			6	7
9			4	6
10			2	5
11				4
12				3
13				3
14				2
15				1

which is presented for illustrative purposes, as actual IRS schedules are different. Note that the sum of each column is 100% and that values change each year. If percentage values were equal (i.e., 20% for the 5-year property class), then the depreciation would be straight line. The method in Equation (4.3) disregards any expected salvage value. Many laws are connected with the use of this table. Sale or purchase during the tax year, averaging midyear or month conventions, and other methods of calculation are left to the interested student willing to read other texts and the IRS code.

Example: Modified Accelerated Cost Recovery Method of Depreciation

Consider a 5-year recovery property having a cost of $100,000. The property will be sold at the end of the sixth year. Find the annual depreciation charge to the income statement.

Year	Cost	MACRS Depreciation, % $P(j)$	Book Value at Year Beginning	Yearly Depreciation, D_j
0	$100,000		$100,000	
1		20	100,000	$20,000
2		32	80,000	32,000
3		24	48,000	24,000
4		16	24,000	16,000
5		8	8,000	8,000
				Total $100,000

In most businesses the amount shown in the depreciation account does not appear as cash. If a special fund is set aside specifically for this purpose, then it will be called a *sinking fund*. To create a fund of this nature suggests that a fund is actually invested outside the company to earn interest. However, interest rates found on the outside are hopefully less than the earning rate enjoyed by the company. Usually, it is wiser to employ the money for operations. The amount equal to the depreciation will appear as other assets, such as working capital, raw materials, materials in storage, or even cash. When it becomes necessary to buy new equipment, management must convert physical assets into cash (unless sufficient cash is on hand), or draw on existing profit to pay for the new equipment. Other financial means for raising money for business purposes are introduced in Chapter 11.

The appearance of accumulated depreciation on the balance sheet, as with other assets, represents capital retained in the business, ostensibly for the ultimate replacement of the capital asset being depreciated. Accumulated depreciation is also known as a *contra account*, whose balance reduces the value of an asset, which is shown at cost. The value of the contra account is enclosed in parentheses, indicating a negative value. For example, in Tables 4.2 through 4.5 the contra account is ($600).

Example: Cash Flow Effects on the Income Statement with Accelerated Depreciation

How do we show the effects of accelerated depreciation on the income statement? Now reconsider Table 4.6, which is the profit-and-loss statement for General Manufacturing Company. In that example, the first year depreciation amount is shown as $600. If the allowable depreciation amount is $850, how does that affect the income statement? Repeating the information, and with the changes we have the following:

	$600 Depreciation		$850 Depreciation	
Income				
Product income		$11,000		$11,000
Expenses				
Salaries	$2,850		$2,850	
Rent	1,000		1,000	
Advertising	800		800	
Insurance	500		500	
Depreciation	600		850	
Total		5,750		6,000
Gross profits		$5,250		$5,000
Taxes at 23.8%		1,250		1,190
Net profit after taxes		$4,000		$3,810
Cash flow = net profit + depreciation		$4600		$4660

The greater depreciation amount of $850 reduces the gross profits to $5000, as the total expenses increases. This in turn reduces net profits, but the effect of the acceleration is to increase the cash flow from $4600 to $4660.

Factors in any depreciation model are subject to estimation: salvage value and, particularly, life. If the estimates prove faulty, then it is possible to retire an investment

PICTURE LESSON Eli Whitney: Inventor of the Milling Machine and the American System of Mass Production

Smithsonian Institution, Photo No. 73–6431, and New Haven Historical Society.

Eli Whitney, (1765–1825), a mechanical engineer and manufacturer, is best remembered as the inventor of the cotton gin, but his most important contribution dealt with the concept of mass production of interchangeable parts.

Whitney graduated from Yale College in 1792, where he learned many of the new concepts and experiments in science and the applied arts, as technology was then called. After graduation, Whitney saw that a machine to clean the green-seed cotton could make the South prosperous and make its inventor rich. He constructed a crude model and was able to perfect his machine, securing a patent in 1794. He and a partner went into business manufacturing and servicing the new gins. However, the unwillingness of the planters to pay the service costs and the ease with which the gins could be pirated put the partners out of business by 1797.

When Congress refused to renew the patent, Whitney concluded that "an invention can be so valuable as to be worthless to the inventor." He never patented his later inventions, one of which was a milling machine. The photograph included here shows Whitney's milling machine, where a leather belt is attached to a pulley at the left end, which rotates a cutter mounted to the right spindle.

Whitney learned much from his experience. He redirected his mechanical and entrepreneurial talents to other projects in which his system for manufacturing gins was applicable. The government, threatened by war with France in 1797, solicited 40,000 muskets from private contractors, because the two national armories had produced only 1000 muskets in three years. Like the government armories, the manufacturers used the conventional method whereby a skilled workman fabricated a complete musket by machining, assembling, and fitting each part. Every weapon was unique, and if a part broke, its replacement had to be made to order.

Whitney broke with this tradition with a plan to supply 10,000 muskets in two years. He designed machine tools by which an unskilled workman made only a particular part that conformed precisely, as precision was then measured, to a model. The sum of such parts was a musket. Any part would fit any musket of that design. He had grasped the concept of interchangeable parts. "The tools which I contemplate to make," he explained, "are similar to an engraving on copper plate from which may be taken a great number of impressions perceptibly alike."

Eleven years passed before Whitney delivered his 10,000 muskets. He constantly had to plead for time while struggling against unforeseen obstacles, such as epidemics and delays in supplies, to create a new system of production. Finally, before President-elect Thomas Jefferson, he demonstrated the result of his system: from piles of disassembled muskets they picked parts at random and assembled complete muskets. They were the witnesses at the inauguration of the American system of mass production.

before its capital is recovered. In those circumstances where net income received is less than the amount invested, an unrecovered balance remains. This unrecovered balance is referred to as *sunk cost*, which is the difference between the amount invested in an asset and the net worth recovered by services and income resulting from the employment of the asset.

As an illustration of sunk cost, consider a capital investment of $100,000 to be recovered in 5 years with a residual $10,000 salvage, or $90,000 depreciation. Based on straight-line depreciation with salvage, the amount invested and to be recovered per year will be $18,000. As a result of excessive use, the equipment is sold after 3 years for $40,000 and has actually consumed $60,000 in 3 years or $20,000 per year on the average. The sunk cost is equal to the difference in the actual depreciation and the depreciation charge, in this case $60,000 − $54,000 = $6,000. Stated yet another way, sunk cost is the estimated depreciation value (or book value) minus the realized salvage of the asset. Sunk cost is unaffected by decisions of the future and must be faced realistically. Sunk cost is a concept considered again in Chapters 9 and 10.

We have illustrated depreciation for one piece of equipment. In actual situations, the firm will have many depreciable assets, and they will not follow these individual calculations of depreciation. Group schemes are available that simplify these methods. Multiple-component and additions of assets are numeric complications left to the experts.

Furthermore, terms such as surtax, credits, accelerated write-off, inflation or recession, depletion, obsolescence, and new capital spending confuse any discussion about depreciation. Various accounting terms such as contra-depreciation, reserves for depreciation, allowances for depreciation, and amortization and retirement enlarge the uncertainty. We do not study these advanced topics, and leave their discussion to other textbooks and specialists. However, they are important in the practice of cost analysis.

Depreciation is important, yet it is only important in the larger context of accounting documents, and for budgeting, which is studied next.

4.8 BUDGETING

Budgeting is a frequent engineering assignment. A budget is a written plan covering the activities of a business unit for a definite future time. The process of budgeting draws from many sources of information, and as a detailed plan the budget is the first step in finding overhead. The chart of accounts, discussed earlier, is helpful for budget preparation. Indeed, it may serve as a template for a general business plan, and an abbreviated chart is useful for this goal.

Budgets deal with information based on cost estimating and accounting records and conjectures of future activities. A budgeted cost unit should be the smallest unit to which a cost can be clearly traced, provided there is a balance between excessive and too little detail, consistent with the cost of preparing the budget. Dimensions are in dollar or hour units for a specific period, such as a quarter or year. We study this important topic using the following outline:

- Types of budgets and relating cost accounts
- Budgets for overhead calculation

4.8.1 Types of Budgets and Relating Cost Accounts

Business budgets exist in great variety. Appropriation, fixed, and variable budgets are common types. An *appropriation budget* may be directed toward proposed expenditures, for example, the purchase of a machine tool, or construction cost of a plant addition. A *fixed budget* may be directed toward a production department with only one level of throughput activity for a future period. This budget may be satisfactory if the department activities, for example, can be predetermined accurately, or "*fixed*." During the operating period of the budget, it may not be adjusted to actual levels.

A budget that is prepared for a number of levels of activity is known as a *variable budget* or flexible budgeting. A variable budget requires greater knowledge of cost behavior. Regardless of the level of productive activity, some costs are almost completely fixed per time period, such as depreciation of a building. Others are constant within certain ranges of activity, such as superintendence. Some change as the activity fluctuates, such as consumption of factory supplies. This type of budgeting is akin to semifixed cost behavior, and it is similar to semifixed break-even cost analysis, a topic covered in Chapter 9. As a future prognostication, variable budgets are intended for cost control of the activities of the unit. If the unit is operating less than budgeted, what are the causes for this behavior? Variable budgets are an advanced operating cost control tool that answers this question.

The cost accounting cycle is framed about the skeleton of the manufacturing process or the physical arrangement or the service for jobs. Because cost accounts are

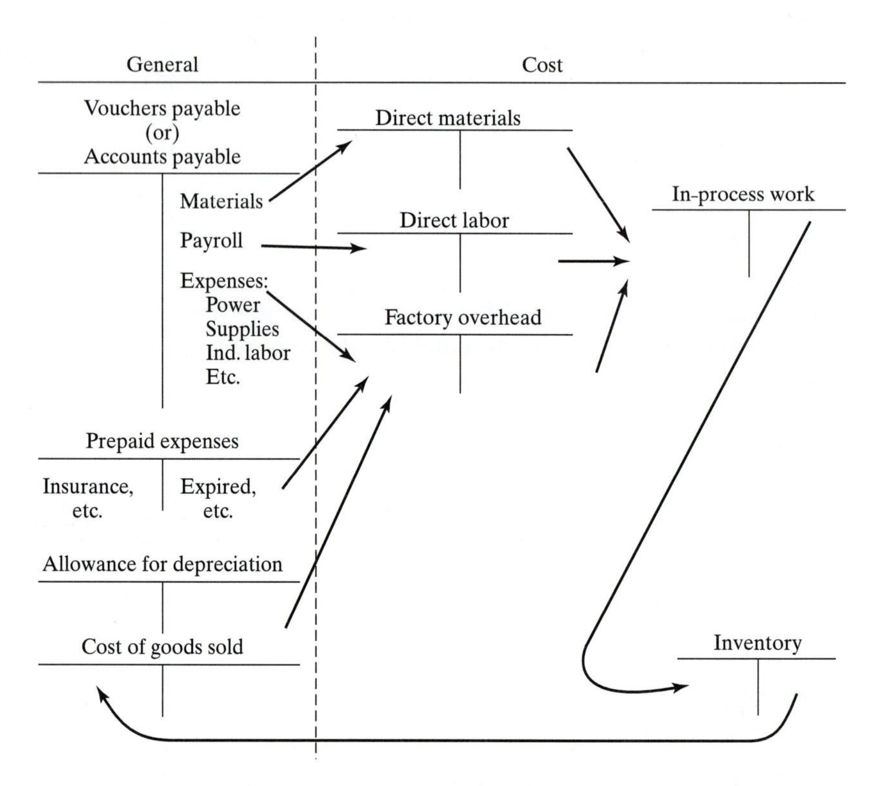

Figure 4.3 Relationship between general accounts and cost accounts.

an expansion of general accounts, cost accounts should, as a basic accounting procedure, be related to general accounts. Figure 4.3 presents the relationship between general accounts and cost accounts. Budgets can duplicate this pattern, but only if the type of budget is for the general plant operation and business planning. Other texts are required for this development.

The larger an organization, the greater its span of accounting records. To illustrate: A materials account controls hundreds of different material items; the payroll account controls departmental labor costs and payroll records for each employee; and the factory overhead account controls indirect labor, supplies, rent, insurance, repairs, and many other factory expenses.

As is shown in Figure 4.3, direct materials, payroll (or direct labor), and expenses are transferred to in-process work, to inventory, and eventually to cost of goods sold. In-process work describes material incompletely processed. This material is usually on the factory floor. An inventory account (as shown in Figure 4.3), if it is finished, has completed processing and is ready for shipment to a customer. This tracing of cost flow aids the management of budgeting for plant operations.

4.8.2 Budgets for Overhead Calculation

Our discussion of budgets will facilitate an understanding of overhead calculation, which is studied shortly. Overhead has been mentioned before, but we now construct budgets in a way that accurately frames the cost accounting cycle described in Figure 4.3, and which coordinates closely with engineering estimating activities.

Budgets can be very large, and in some businesses may encompass many volumes. This book illustrates the methods, but for a smaller circumstance. To set the stage, consider a small plant that produces product with four production centers. Management, sales, and engineering efforts are companion efforts to production. (Chapters 2 and 3 discussed labor and material, and direct labor and direct material were defined; the student may want to review those topics.)

There are three budgets that are assembled as examples: (1) production equipment, (2) labor, and (3) factory. There is a relationship that exists between these three budgets and with direct labor and direct material. The eventual objective is to conduct a cost analysis of future business and to have the costs of producing product recovered in the estimated costs. It is a simple statement, but a vital one to understand at this point: The cost of a product is composed of direct labor to operate production equipment, direct material, and the indirect costs of the business to bring it all together for the market. The purpose of overhead arithmetic is the connection of all direct and indirect costs in a way that accurately reflects the consumption of those future costs in the production of product.

Now consider Table 4.8. Our manufacturing plant has heavy- and light-duty production equipment, which are dramatically different in size and supporting requirements.

The light machining production center has 20 machines and the floor space for the equipment is known. Notice in Table 4.8, hours are listed. This number is the anticipated operating hours for this production center for the upcoming year. Marketing forecasts are helpful, and along with the expected number of standard hours for operation of this production center, the budgeted hours can be evaluated and listed.

A depreciation amount shown was found using group calculation methods. Tooling expenses for the center are determined knowing the expected product design and anticipated sales. These are not simple determinations, but are typical of what is required, and why engineering is necessary for the budget preparation. Notice that depreciation and tooling expenses are directly assigned to the production center.

TABLE 4.8 Production Center Budget for January–December Period

Production Center	Number of Machines	Floor Space, ft^2	Hours	Horsepower Hours	Depreciation Expenses	Tooling
Light machining	20	3,200	49,300	443,000	$52,500	$200,000
Heavy machining	2	3,400	6,800	748,000	95,000	80,000
Assembly	15	1,100	17,000	24,000		10,000
Testing	8	1,700	13,600	68,000	22,500	65,000
		9,400	86,700	1,283,00	$170,000	$355,000

There are only two heavy machines, but their horsepower hours are significantly larger than for light machining. Horsepower hours are the equipment's name plate rating times the number of annual operating hours.

An assembly production center is listed with similar information. However, note that there is no depreciation, as the equipment is fully depreciated. Testing is the fourth production center.

Note that in Table 4.8, which is a simplified budget for production centers, the planned number of machines, floor space, horsepower hours, and other data determined for those cost centers are presented. It is necessary to recognize that the January-December period must be further identified as 49 weeks and a two-shift operation. This budget is for factory equipment used to make product for sale, its space, hours, depreciation, and tooling expenses. The next budget deals with the operation of the equipment.

A second closely related budget is given in Table 4.9. This table connects the direct labor to operate the equipment of the identical production centers. For example, light machining requires 29 direct-labor operators, while 4 operators are required for heavy equipment.

The wage and fringe rates are matched to the same annual period. The gross hourly cost, as typically found by Table 2.6, would be entered for the appropriate production center. Table 4.9 shows the budgeted direct-labor hours and cost. Notice that the gross hourly wage rate is significantly different between the centers, it being the most costly in heavy machining.

The budgeted direct labor hours are 15% greater than machine hours, as shown by Table 4.8. This recognizes that the equipment utilization is 85%, or it will be down 15%, but that the direct labor will be on-job, even though the equipment is not operating.

Efficiency and utilization relate to two different concepts. *Efficiency* deals with labor, which can be more or less than 100%. *Utilization* is found with equipment, which is measured against a budgeted time. In Table 4.9, the utilization is 85%, meaning 15% of downtime, for a 49-week and a two-shift operation. It is usually accepted that utilization cannot be more than 100%.

TABLE 4.9 Budgeted Manpower for Production Centers for January–December Period

Production Center	Number of Direct Labor Workers	Average Direct Labor Wages, $/hr	Average Direct Labor Fringes, $/hr	Gross Hourly Wages, $/hr	Direct Labor Hours	Total Direct Labor Cost Budget
Light machining	29	21.40	6.42	27.82	58,000	$1,613,560
Heavy machining	4	25.75	7.73	33.48	8,000	267,800
Assembly	10	18.65	5.60	24.25	20,000	484,900
Testing	8	20.05	6.02	26.07	16,000	417,040
Total	51				102,000	$2,783,300

Tables 4.8 and 4.9 budgeted the production floor, equipment, and labor and those immediate cost requirements. However, the business is more than the factory, as Table 4.10 shows.

Note in that table that factory, engineering, and general management expenses are planned for the upcoming year. These planned expenses are for the entire plant and are not identified to a production center. These overhead costs cannot be mapped

TABLE 4.10 Factory, Engineering, and Management Budget for January–December Period

Overhead Budget	
Factory	
Space	
Rent	$200,000
Repairs to factory	40,000
Heating	16,000
	$256,000
Utilities	
Electricity	$189,625
Water	50,000
	$239,625
Indirect labor	
Material handlers	$64,000
Inspectors	48,000
Supervisors	90,000
Clerical personnel	27,000
	$229,000
Tooling services	
Repairs and maintenance	$30,000
Perishable supplies	38,000
Nonperishable tooling	355,000
Repairs on tools	28,000
Outside tooling services	18,600
	$469,600
Engineering	
Professional budgeted salary	$105,000
Clerical support labor	67,000
Rent	42,000
Utilities and software	18,000
Depreciation	15,000
	$247,000
Management	
Professional budgeted salary	$280,000
Clerical support labor	125,000
Rent	67,000
Utilities and software	18,000
Depreciation	15,000
	$505,000
Total factory, engineering, and management budget	$1,946,225

to the equipment, because the work, cost, and total effort is given to the business that is involved with engineering, sales, and management. In other words, the cost and effort is "general." The terms "general, administrative, and sales" are sometimes identified with this slate of costs.

Factory expenses are budgeted for space, utilities, indirect labor, and equipment and tooling services. The $469,600 tooling services identified for the factory is different from those for the production center. There are general costs for the factory that are separate from the production centers. Space, utilities, and indirect labor such as material handlers, inspectors, and tool room clerks are necessary for the operation of the total plant, and these costs cannot be slotted to the production centers, which are the direct-labor producing centers.

Engineering and management costs are budgeted and forecast for next year. The space accounts, rent, repairs, and heating are conjecturally stated based on past information and future business. The total overhead budget is the grand sum of the individual items, or $1,946,225.

These three tables are the future budgeted costs of the upcoming year. Notice that these budgets did not identify any particular product that will be engineered and made using the plant resources. Those specifics will be discussed shortly. However, the purpose of these budgets is for overhead calculation.

From an accounting point of view, costs are often identified as to their behavior, either as a function of time or as a function of the object of the manufacturing or service performed. A business's fixed and semifixed costs, if they are for several simultaneous jobs, cannot be spread costwise into one manufacturing job. To do so would create gross inaccuracies. Some costs are *periodic* (time based); others are *object based*, such as direct materials or direct labor; and some are *joint*. For accounting analysis, a new means of looking at joint costs is required. This brings us to the study of overhead.

4.9 OVERHEAD

By definition, overhead is that portion of the cost that cannot be clearly associated with particular operations, products, or projects and must be prorated among all the cost units on some arbitrary basis. Broad details regarding the posting of direct labor cost, time, material, and other indirect costs have been given earlier. What will be discussed here are those overhead aspects that concern engineering. The key to this puzzle is the way in which indirect expenses are allocated, unitized, and charged to individual estimates.

Overhead costs become complicated whenever the costs are joint or have a commonness with different levels of variability. Joint costs, or costs incurred jointly, are depreciation, insurance, property taxes, maintenance, and repairs and are dependent on one another. However, joint costs are handled differently for accounting analysis than for the direct labor and direct material discussed in Chapters 2 and 3. Accounting joint costs are handled entirely by the overhead process.

Direct costs, such as direct labor and direct materials, present little if any allocation problems. Those costs do not exist unless the product is made.

We consider the following topics:

- Importance
- Traditional methods
- Allocation methods
- Productive hour cost
- Activity-based costing

4.9.1 Importance

The underestimating or overestimating of overhead rates is a serious issue in view of the proportion of the total estimated cost. As an illustration, consider two different operations where the labor rates are $21.25 per hour. Machine A, a numerical controlled milling machine, is initially worth $150,000. Machine B, a standard general purpose milling machine, is worth approximately $15,000. By using an average burden rate of 200%, it would be indicated that machines A and B would each cost, on a machine hour basis, $63.75 per hour. However, this is false machine hour costing, because the investment in machine A is 10 times that in machine B. The proper cost base and sensible allocation of overheads to handle discrepancies of this sort are necessary.

In recent decades the ratio of fixed cost to variable cost has risen, and the simple expediency of overhead distribution by means of the single rate is misleading. Traditional cost accounting systems were developed in the early part of the twentieth century for labor-intensive industries. At that time, overhead cost was not a major component of product cost. Direct labor was the major focus of cost containment. It was reasonable to use a single plantwide rate to allocate overhead charges to product. This rate is based on a single volume related cost driver or on substitutes such as direct labor costs or hours.

During that time, machine investment per worker was lower, and it was not uncommon that overhead rates were uniformly distributed over the direct-labor dollar or time base. Indeed, the expression "burden" is sometimes heard, and it means that the unassigned costs of the business were burdened "on the back" of the direct labor employee. The word "overhead" has a historical legend too. In early twentieth-century shops, the boss would be on the second floor, or "over-the-head" of the machine tools (which were driven by steam engines and line shafts with pulley and leather belt attached to the machine tool) and workers, and thus those costs became known as overhead costs.

4.9.2 Traditional Methods

The traditional methods of dealing with overhead consist of assigning the various overhead costs to several departments or divisions or cost centers within the unit, or factory. Sometimes, cost centers combine or separate departments to form homogeneous groups and are a logical point for the accumulation of costs. In making this original distribution no distinction exists between a producing department (heavy machining center) and a management department. This first distribution is then followed by a secondary redistribution.

Overhead is allocated to a designated base. A base for distribution of manufacturing overhead, for example, may be floor area, kilowatt hours, direct labor dollars or hours, machine hours, or number of employees.

The redistribution of costs originally assigned to service cost centers (management, for example) to the production cost centers is termed secondary distribution. We do not worry about these bookkeeping manipulations. This bookkeeping transfer technique of management overhead costs to production departments is for subsequent recovery within operation or product cost.

A secondary distribution in the case of management might be to prorate the total management center costs to production departments based on the number of employees. If the light machining production center has 29 employees while heavy machining has 4, then we could reason that the costs that it receives should be borne on a 29:4 ratio.

The building depreciation, insurance, maintenance, and taxes for non producing centers are often distributed on a floor area basis of the producing centers. Electrical power poses a confusing choice because it may be distributed based on machine weight, machine hours, horsepower hours, or even direct labor hours.

The development of the overhead rate differs from one company to the next. It is not uncommon to find several overhead methods used within one company. A classification is given as follows:

- The basis used to determine and apply overhead, such as direct labor dollars, direct labor hours, machine hours, or material costs
- The scope of the rate, such as the plant, cost center or department, or to a specific machine
- Whether the rate applies to all designs (such as product lines) or to one line of the design or to a unit of product

A historical or *actual burden rate* has the merit of distributing the incurred factory overhead among the jobs. The rate is subject to certain defects, because it is unavailable until the close of the accounting period. Historical rates delay cost calculations until the end of the month and often later and fluctuate because of seasonal and cyclical influences acting on the actual overhead costs and on the actual volume of activity for which overhead cost is spread. Our discussion will disregard historical rates in view of the dependency on predetermined overhead rates. Overhead rates are determined in most situations from information developed from operating budgets, such as those given in Section 4.8.

4.9.3 Allocation Methods

A general procedure for overhead is as follows:

- Collection of overhead charges
- Finding concurrent basis
- Calculation of rate
- Applying the rate

The four steps of overhead analysis are illustrated in Figure 4.4. The horizontal axis is "period," where E is the "now" time of the estimate. Past periods are −1 and −2, and future ones are 1, 2, and 3. The period term is general, typically meaning month, quarter-year, or year. In Figure 4.4(a)–(c), actual amounts are shown as a dash, while

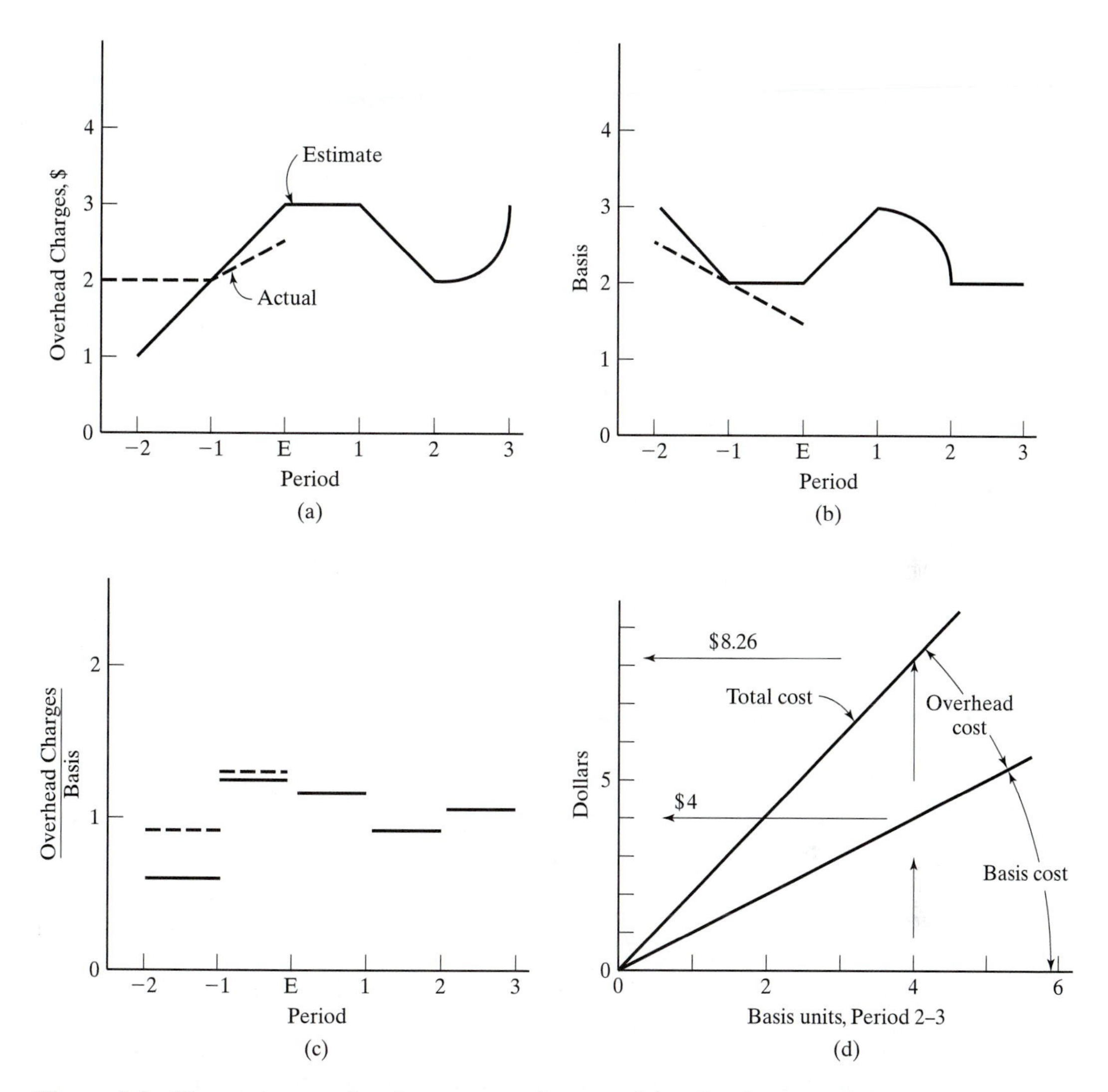

Figure 4.4 Illustrative overhead sequence of events: (a) collecting and plotting overhead charges; (b) collecting and plotting the basis; (c) calculating overhead rate; (d) applying overhead rate.

the future estimate value is shown as a solid line. Of course, we can only have actual values up to the now-time but charges can be forecast and budgeted for future periods. Figure 4.4(a) plots Overhead Charges versus Period.

This gathering is sometimes referred to as "pooling." Note that the behavior of the charges can be linear, flat, or nonlinear. A pooling indicates that despite changes to the charges during any one period, cost behavior can be modeled as a lump sum quantity.

In Figure 4.4(b) the vertical axis is identified as "basis," a term used frequently in accounting to mean simply any future cost parameter. The units can be one of several different quantities. Think of the basis as machine hours, direct labor dollars, or cost. There are other basis factors that can be useful.

In Figure 4.4(c) a definition of overhead is given as

$$\text{Overhead rate} = \frac{\text{overhead charges}}{\text{basis}} \tag{4.7}$$

This equation expresses the fundamental relationship as a rate—i.e., dollars per basis unit—though percentages are also possible. In overhead calculations, the basis is the "denominator."

It is the total area in Figure 4.4(a) divided by total area of Figure 4.4(b) for one period that gives a constant rate value shown in Figure 4.4(c). In period 2–3, the area of the "charges" is 2.13 units, and the area of the "basis" is 2. The rate becomes 1.065 (= 2.13/2). Note that actual charges, basis, and rates do not coincide with the estimated values. This is usually what happens, though an effort is made to have coincidence.

The overhead rate of Figure 4.4(c) is used by engineering. Though accounting procedures will perform the first three parts of Figure 4.4, it is usually engineering that handles the application shown in Figure 4.4(d). The procedure begins with the need for a cost estimate for inside future period 2–3. If for instance, 4 basis units are necessary to do some work, we multiply this estimate by the rate of period 2–3, or 1.065. At 4 estimated basis units, the overhead is 4.260, and when added to the basis units, the point on the line is 8.26 (= 4.26 + 4). If the dimension is dollars, then the total cost is $8.26. Similarly, all other values along the line of (d) are linear.

4.9.4 Methods of Direct Labor Hours and Dollars

The following method of finding overhead as a ratio of the direct labor dollars is one of the oldest methods and remains popular:

$$R_{dl} = \frac{C_o}{C_{dl}} \tag{4.8}$$

where R_{dl} = overhead rate on basis of direct labor, dollars

C_o = overhead charges for a cost center, dollars

C_{dl} = direct labor, dollars

The scope for the numerator and denominator of Equation (4.8) may be the factory or department or cost center. Though it would appear that Equation (4.8) is dimensionless, it is more useful to remember R_{dl} as dollars of overhead charge per dollar of direct labor cost.

Another method uses prime cost as the denominator:

$$R_p = \frac{C_o}{C_p} \tag{4.9}$$

where R_p = overhead rate on the basis of prime cost

C_p = prime cost, dollars

Prime cost is defined as direct materials dollars plus direct labor dollars. This method will work for a few industries that have identical product units, such as cement, petroleum, and foodstuffs. We do not discuss this method further.

It is sometimes assumed that many factory overhead charges are proportional to labor time. This situation occurs if labor is paid on an hourly basis and the rate per hour is substantially the same for all workers. The rate of overhead per direct labor hours is calculated as

$$R_{dl} = \frac{C_o}{H_{dl}} \tag{4.10}$$

where R_{dl} = overhead rate on basis of direct labor hours

H_{dl} = budgeted direct labor hours

The dimensions for Equation (4.10) would be dollars overhead charge per hour of direct labor. Now examine Tables 4.8, through 4.10 and note that C_o = \$2,471,225 (= 1,946,225 + 170,000 + 355,000), which is the grand total of all overhead. The direct labor hours H_{dl} = 102,000 hours. Then R_{dl} = \$24.23 (=2,471,225/102,000), which means that for each direct labor hour expended an overhead cost of \$24.23 will latch on it, so to speak. The scope of the overhead includes production center depreciation and tooling expenses as well as factory, engineering, and management expenses. The direct labor gross hourly wages are given in Table 4.9.

Example: Table 4.11 Finding Estimated Cost Using Direct Labor Hour Overhead Basis

A company receives a request for an estimate, and engineering estimates the hours to manufacture the design. Those numbers are shown below. In addition, a material takeoff of the design indicates a unit cost of \$173.80. There are 210 products to be made and shipped. The estimated cost using the Equation (4.10) approach is shown in Table 4.11.

TABLE 4.11 Method to Find Product Cost Using Labor, Material, and Overhead

Production Center	Estimated Hours	Gross Hourly Wage, \$/hr	Direct Labor Costs	Overhead Cost Rate, R_{dl}, \$/hr	Overhead Costs	Labor + Overhead + Material
Light machining	7.31	27.82	\$203.36	24.23	\$177.12	\$380.49
Heavy machining	471.23	33.48	15,776.78	24.23	11,417.90	27,194.68
Assembly	21.26	24.25	515.56	24.23	515.13	1,030.68
Testing	7.93	26.07	206.74	24.23	192.14	398.88
Subtotal of labor and overhead costs						\$29,004.73
Material cost per unit	\$173.80					
Material cost for 210 units						\$36,498.00
Total product cost						\$65,502.73

This method assumes that engineering and management costs are uniformly spread or allocated against each direct labor hour, but many overhead charges do not relate to direct labor time, such as engineering change orders, or engineering design time, for example. Furthermore, the overhead charges that are directly caused by the machine depreciation are spread over all machines, and light machining will disproportionately carry depreciation of heavy machining. There are other problems of the method that are resolved by the productive hour cost rate method, which is discussed next.

4.9.5 Productive Hour Cost Rate Principles

When the machinery employed in production is not reasonably equal in value, the productive hour rate method is successful. This method is exploited for advanced technology manufacturing and employs the principles of activity-based costing. This method has the fundamental relationship

$$R_{mh} = \frac{C_o}{H_m} \tag{4.11}$$

where R_{mh} = overhead rate on the basis of budgeted hours of machine or equipment, or common grouping of equipment

H_m = productive machine or equipment budgeted time, hours

Equation (4.11) can be calculated for single equipment or machine grouping, as the following example demonstrates. Budgets shown by Tables 4.8, through 4.10 are necessary. Now we consider the physical factors associated with the machine centers, such as number of machines, floor space, machine horsepower hours, depreciation, and tooling expenses. Those data are shown in Table 4.8, which deals with the machine center budget. From Table 4.9 we use the manpower cost requirements to operate those machines, such as number of direct labor employees, wage, fringe rate, budgeted direct labor hours, and the total direct labor cost. From Table 4.10 we use factory overhead costs, engineering and development, and general management charges.

The allocation factor or the basis for assignment is described for the several columns and may be MACRS, floor space, horsepower hours, direct-labor hours, and direct labor cost.

Observe in Table 4.12 the distribution of \$256,000 space costs to each of the four production-centers. The floor space cost to each center is based on its shop area to total production area. For instance, the allocation fraction 256,000/9400 = \$27.23/ft^2 would result in \$87,149 (= 3200 × 27.23) for light machining. The \$256,000 is total overhead for space, and 8300 ft^2 is total area.

Consider the overhead assignment of electrical power. There is budgeted \$239,625 for next year's power costs. Although each machine could have its own power meter, that is very impractical if the factory is electrically connected jointly. From Table 4.9 we find the horsepower hours for each production center. The total horsepower hours are 1,283,000 while light machining records 443,000 horsepower hours. The factor for the assignment of the electrical power is \$0.187 per horsepower hour. Then for light machining the applied overhead for power is \$82,739 (= 0.187 × 443,000).

TABLE 4.12 Productive Hour Cost Assignment to Production Centers

Production Center	Depreciation	Tooling Expenses	Space	Utilities	Indirect Labor	Tooling Services	Engineering	Management	Total Overhead
Light machining	$52,500	$200,000	$87,149	$82,739	$130,216	$162,146	$140,451	$292,763	$1,147,960
Heavy machining	95,000	80,000	92,596	139,703	17,961	273,781	19,373	48,589	767,000
Assembly		10,000	29,957	4,482	44,902	8,784	48,431	87,980	234,530
Testing	22,500	65,000	46,298	12,700	35,922	24,889	38,745	75,667	321,720
	$170,000	$355,000	$256,000	$239,625	$229,000	$469,600	$247,000	$505,000	$2,471,220
Basis for allocation	MACRS	Directly assigned	Shop area	HP Hr	DLHr	HP Hr	DLHr	DL$	

The factory's indirect labor of $229,000 is spread to the production center by the ratio of the production center's direct labor hours to the total. For example, in light machining indirect labor costs are $130,216 (= 229,000 × 58,000/102,000).

Tooling services overhead are assigned to the production centers by the horsepower hour basis. For light machining, the distributed overhead is $162,146 (= 469,600 × 443,000/1,283,000).

Engineering costs are allocated to the production centers by the direct labor hour basis. That amount is $140,451 (= 247,000 × 58,000/102,000). Management costs are allocated to the production centers by the direct labor cost basis, or $292,763 (= 505,000 × 1,613,560/2,783,300). The total of reassigned overhead to the light machining production center is $1,147,963.

In a similar way the other production centers are systematically analyzed for their fair share of overhead expenses. The sum of all overhead expenses is shown to be $2,471,225, which is shown in Table 4.13.

The productive hour rate method builds on the budgeted tables as shown earlier, but the distribution is based on the gross hourly rate and the machine hour, and it is formulated as follows:

$$\text{Productive hour cost rate} = \text{machine hour cost rate} + \text{gross direct labor hourly rate} \quad (4.12)$$

TABLE 4.13 Calculation of Productive Hour Cost Rates

Production Center	Total Overhead	Budgeted Hours	Machine Hour Cost Rate, $/hr		Gross Hourly Wages, $/hr		Productive Hour Cost Rate, $/hr
Light machining	$1,147,963	49,300	23.29	+	27.82	=	51.11
Heavy machining	767,003	6,800	112.79		33.48		146.27
Assembly	234,538	17,000	13.80		24.25		38.04
Testing	321,721	13,600	23.66		26.07		49.72
	$2,471,225	86,700					

The machine hour cost rate is the production center total overhead divided by the budgeted hours. We now move to finding the productive hour cost rate.

Consolidating Overhead Costs. As an example, let's say $23.29/hr (= 1,147,963/49,300) is the machine hour cost rate based on light machining budgeted hours. The gross hourly cost for direct labor is added to each machine center.

Productive hour cost rates are the sum of the gross hourly costs and the machine hour costs. Comparison of this column in Table 4.13 shows that heavy machining costs about four times as much as assembly. Productive hour cost rates afford an accurate method of allocating overhead expenses and, from the engineering point of view, are preferred for estimating manufacturing from specifications and route sheets. The productive hour rate method is best whenever operations are performed by equipment, which is significant in terms of the final cost of the product.

The productive hour rate method is adaptable. For example, if there is no direct labor associated with the operation, then there is no need to add gross hourly costs to the productive hour. Mechanization is an example, because it is possible for equipment to operate unattended. For mechanization, then, the labor hourly costs would be limited, perhaps zero. Consider the case for an operation, such as hand assembly. Suppose that the hand assembly did not use any depreciable equipment. The depreciation costs in this case could be zero, indicating that the operation was mostly labor, as one would expect for a labor intensive operation. The versatility of the productive hour cost method is further illustrated by the situation where there are multiple machines per operator. For this case, a fraction of the gross hourly cost would be prorated among the various equipment.

Look again at the earlier example, Table 4.11, where an estimate was given using the overhead method of direct labor hour. For comparison that example is extended for overhead effects using the method of the productive hour cost rate (Table 4.14). The estimated hours and the material costs are identical, but the results with the PHCR are different.

The earlier example of Table 4.11 found the total product cost as $65,503, but we now realize that the heavy machining component was underrepresented in cost contribution to that analysis. Product cost is significantly higher, as the heavy machining

TABLE 4.14 Finding Estimated Cost Using Productive Hour Cost Overhead Basis

Production Center	Estimated Hours	Productive Hour Cost Rate, $/hr	Product Cost, $/lot
Light machining	7.31	51.11	$373.61
Heavy machining	471.23	146.27	68,926.81
Assembly	21.26	38.04	808.73
Testing	7.93	49.72	394.28
Subtotal	507.73		$70,503.43
Lot material cost, 210 units at $173.80 each			36,498.00
Total product cost			$107,001.44

production center costs are being recognized in the quotation. Submitting a lower than qualified estimate would cause financial losses for the job.

Overhead calculation and application are important for engineering. First, there is the realization that the cost of engineers contributes to the product cost via overhead procedures. Second, the proportion of overhead costs—especially for technological products—can be more significant than the direct costs of labor and material.

Accounting practices have other elaborate schemes for the calculation and application of overhead, and space does not permit their study.

4.9.6 Activity-Based Costing Principles

The emphasis with productive hour cost and activity-based costing is to find variables that demonstrate cause and effect for overhead charges. This means that a cost driver, or a variable of some kind, unequivocally must give overhead costs.

One approach identifies two types of cost drivers: volume related and nonvolume related. Volume-related cost drivers have a direct positive relationship with production quantity. An example would be direct labor and machine hours. If more units are manufactured, it is expected that variable costs of production to make more units would be linear with the number. For instance, one more product unit means that one unit of direct labor, direct material, and electrical power are required.

An example of nonvolume-related cost drivers is the number of different products. The number of different products certainly affects the engineering support requirements, but what is the relationship? If product A requires 5 engineers, do 2 products, A and B, require 10 engineers? The relationship is often murky, but it is important to accurate overhead costing. An illustration of volume and nonvolume cost drivers is shown here.

Volume Cost Drivers	Nonvolume Cost Drivers
Direct labor hours	Input
Machine hours	Number of suppliers
Direct labor costs	Number of engineering changes
Production volume	Number of sales orders
Kilowatt hours	
	Output
Utilities	Number of products
	Inventory levels
	Defect and scrap levels
	Process
	Number of schedule changes
	Amount of rework
	Downtime
	Number of material moves

There are many nonvolume cost drivers that are not given above, but the essence of effective overhead analysis and its recovery through product costing methods requires an intelligent array of methods to do the work.

For a company to implement ABC, the costs of all of the activities involved in making a particular product can be accumulated into a product cost estimate. Engineering determines which activities are required to produce the product and how many units, such as hours or number of occurrences of each activity, are used. The costs of the units for each activity are necessary. In manufacturing, the activities are based on the number of hours per individual part or product, number of hours per batch, or the number of times an activity is used per batch (purchase orders, scheduling actions, design changes, etc.). Products for each batch can be costed in the ABC method by

$$C_{uabc} = \Sigma(h_{pi}C_{hi}) + \Sigma\left(\frac{h_{bi}C_{hi}}{N_b}\right) + \Sigma\left(\frac{M_iC_{oi}}{N_b}\right) + \Sigma C_{oi} \tag{4.13}$$

where C_{uabc} = product cost using the ABC method, dollars per unit

h_{pi} = activity i, hours per part

C_{hi} = activity i, dollars per hour

h_{bi} = activity i, hours per batch

N_b = number of parts in the batch

M_i = number of occurrences of activity i per batch of product

C_{oi} = activity i, dollars per occurrence

Equation (4.13) is not considered to be general and does not apply in all situations. If the ABC system is being used to determine the complete product cost including the indirect labor and overhead, all of the activities used must be accounted for. Some overhead costs occur once for each unit. In this case you need a unit- or volume-based driver like labor hours. This would be the case for equipment costs. The utilities, consumables, and maintenance depend on the hours of operation. However, many overhead costs—for example, inspection—occur once for each batch. The cost driver then becomes the number of inspections, and the cost will be applied to the batch of units that is inspected and then divided over the number of units in the batch. Section 8.4.2 provides an example.

Next, we consider the cost control side of accounting—what happens after the fact and what procedures are available that measure the accounting predictions.

4.10 JOB AND PROCESS ACCOUNTING AND VARIANCE PROCEDURES (OPTIONAL)

Cost accounting procedures are established to provide historical cost information and are known as job order or process cost. The computerized procedures record actual material, labor, and burden charges to control accounts. These activities lead us to study the following:

- Job order recording procedures
- Process recording procedures
- Accounting variances

If many different types of products are made and there are various customers, then a job order accounting procedure is used. In job or lot production, every "run" of product is assigned a production order number. This number is a convenient way to collect the material requisitions and labor job tickets. A typical job cost sheet describes the costing points to which the run refers. A job order may cover the production of one unit or a number of identical units. Examples might be a large seagoing ship or several similar electric generators when the items are complex or costly. The total quantity is divided in smaller production lots, and the job order for the total contract may be supported by a separate job order for each lot. In a job order, the production cycle and the cost cycle are equal in the time allowed for gathering the costs.

Process-cost procedures are used whenever a large volume or a repetition of a highly similar process exists. The completed item is a consequence of a series of processes, each of which produces some change in the material. Process-cost procedures are found in industries such as oil, steel, and chemical that operate 24 hours per day. The production cycle continues without interruption, but the cost cycle is terminated for each accounting period, such as a month, to determine the results of operations. Costs are accumulated on computerized process-cost sheets that show input and output material and labor.

Sometimes these records can be queried to determine standards, both labor or equipment, since the records will have hours or days paid and the quantities for work completed. This historical information can be reused for time and cost estimating for similar and future designs.

Upon completion of the job order and the process accounting period, overheads are applied by using actual information, if available. Thus, average cost-per-unit information is obtained from both procedures, although both procedures are substantially different approaches. Both procedures fundamentally divide the total manufacturing cost by the number of product units produced and compute the unit costs of the individual processes and total them. In both cases, those historical unit values can be forecast into the future to provide a rough glimpse of estimated cost.

Manufacturing estimates are for the immediate future period. Because of this brief lead time, a meaningful comparison of actual to estimated values is sometimes possible.

The term "estimate" is often used synonymously with "standard" for operation work. Accountants are involved with standard costs. Engineering, when referring to standards, thinks in terms of a rigid specification, but experts from other fields have dissimilar viewpoints. A *standard cost* as discussed here provides a dollar amount that is a "should be" amount and is not an immutable natural law. For this discussion we use the term "estimate" meaning standard, although the two are not identical in meaning.

A standard unit cost of a labor operation, part, or product is a predetermined cost computed even before operations are started. In constructing the standard unit cost of an item, it is necessary to study the kind and grade of materials that should be used, how each labor operation ought to be performed, how much time each labor operation should take, and how the indirect services should be best administered. The aim, of course, is to specify the most effective method of making the item and then, through adherence to the specifications in the actual operations, achieve the lowest practical unit cost.

Despite the care in establishing standard or estimated costs, the reported actual costs are likely to deviate from the standard or estimated. These differences are known as *variances* and are expressed as dollar amounts or percentages. They are favorable variances when the actual costs are less than the standard costs and unfavorable when actual costs exceed standard costs. It should not be interpreted that excess of actual cost or of standard cost is nonbeneficial to the firm. Similarly, not all favorable variances represent actual benefits to the company. The terms favorable and unfavorable when applied to variances indicate the direction of the variance from standard cost.

Some businesses use variance analysis to understand inflation, deflation, schedule delays, engineering change orders, or rework, for example. Those variances are the result of a poor or excellent performance by the manufacturer, buyer, or seller. Other causes are possible. A consistent approach allows for reconciliation and explanation of the total cost increase.

The elements used to estimate direct material cost are quantity, shape, and the raw material unit cost. Eventually, actual material costs are assembled and may differ from the estimate because of a quantity variance or raw material cost variance or both. For instance, we estimate 200 units at \$17.38 per unit (= \$3,476 total). In due course, a cost system returns an actual quantity of 210 units at \$17.05 per unit (= \$3,580.50 total).

An unfavorable variance of \$104.50 results in part from an excess of 10 units and in part from a deficit unit cost of \$0.33. The analysis is described further in Figure 4.5. There is a simple way to analyze the difference. The figure starts with actual material cost per unit times actual units. We change one of the factors (cost or quantity) to the estimate. The difference gives a variance due to the factor that was changed first. The remaining difference is the variance for the other factor.

In Figure 4.5(a) the quantity factor is changed first, and for Fig. 4.5 (b) the material cost per unit factor is changed first. Those procedures result in a difference in the quantity or material cost per unit variance, depending on the order of calculation. However, the final net variance is the same regardless of the order of calculation. When one follows the calculations downward, the procedure becomes evident.

The graphical approach shows that an ambiguity exists, and it depends on the order of calculation. The net variances are the same, and a typical though arbitrary resolution is given by equations, as follows:

$$V_m = (N_a - N_e)C_e$$

$$V'_m = (C_a - C_e)N_a \tag{4.14}$$

$$\text{Net material variance} = V_m + V'_m$$

where V_m = variance for material owing to quantity change, dollars

V'_m = variance for material owing to material cost per unit change, dollars

N_a = actual quantity, number

N_e = estimated quantity, number

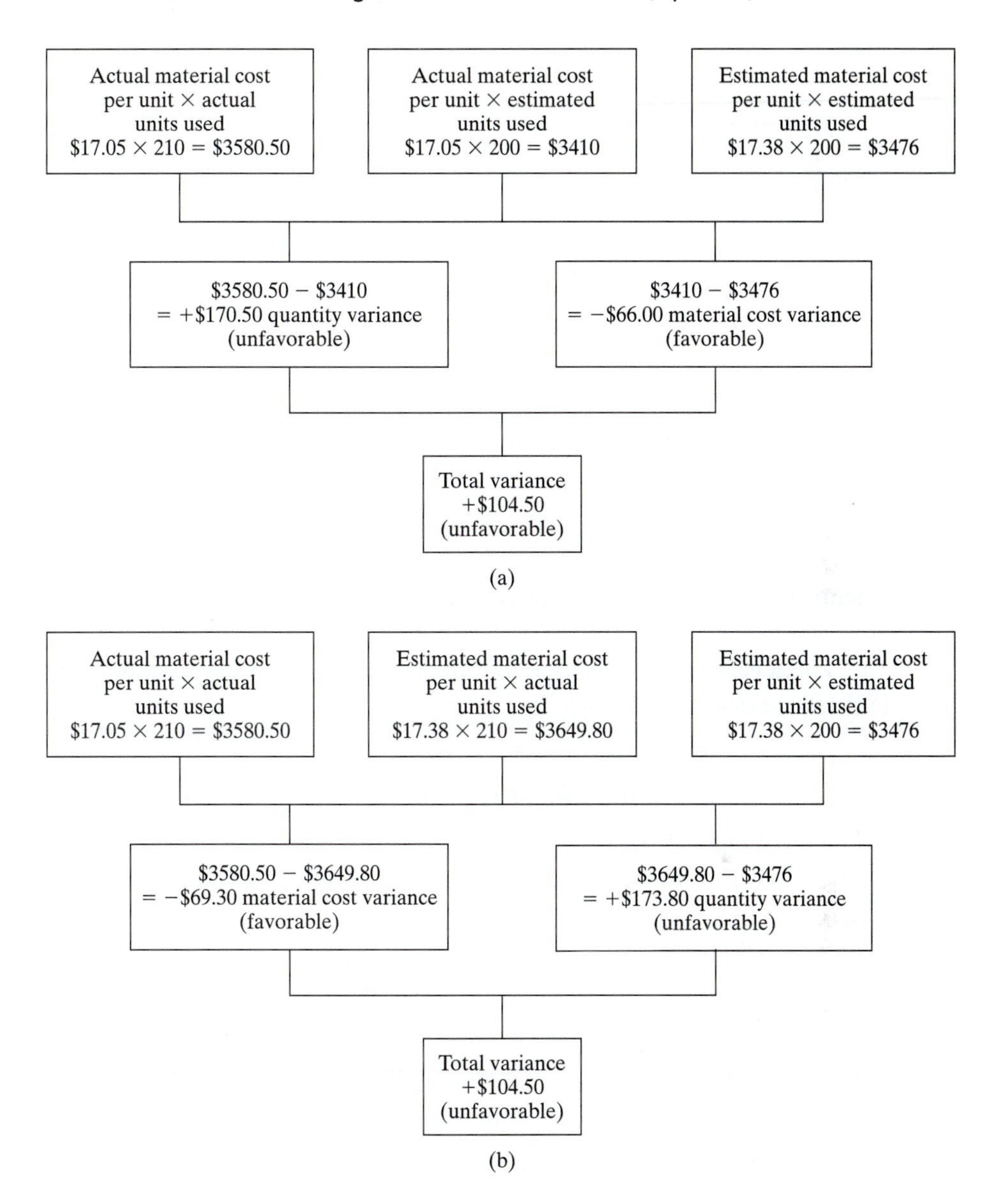

Figure 4.5 Finding variance on (a) quantity first and (b) material cost per unit first.

$$C_e = \text{estimated material cost per unit}$$

$$C_a = \text{actual material cost per unit}$$

The formulations of this equation are shown in Figure 4.6(a). Both favorable and unfavorable variances are possible. By using Equation (4.14) the solution to Figure 4.6 becomes $V_m = +\$173.80$ and $V'_m = -\$69.30$. The net material variance is an unfavorable \$104.50 (= 173.80 − 69.30).

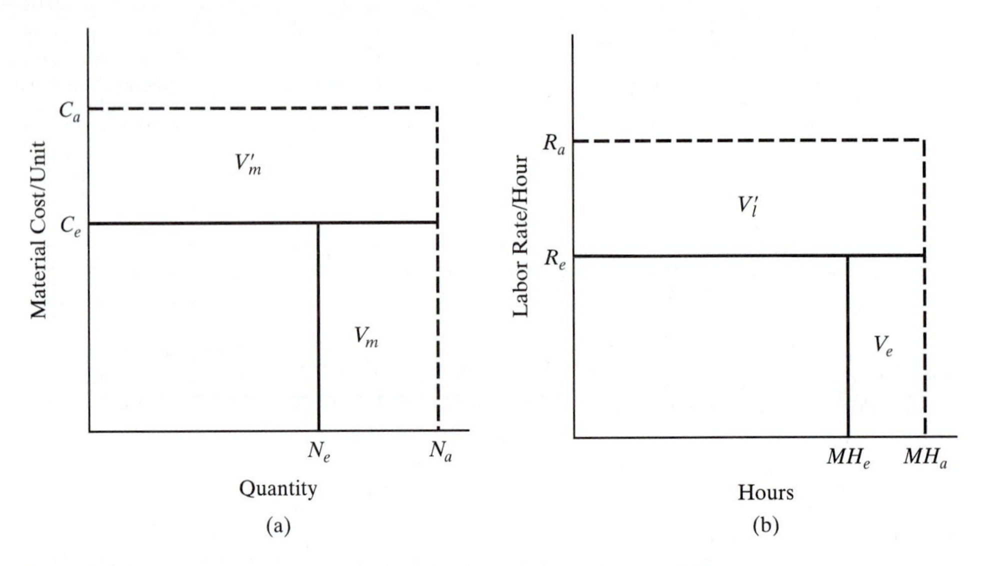

Figure 4.6 (a) Variance for material, and (b) variance for labor.

Labor costs can be analyzed for both labor-hours and the labor dollar rate. Figure 4.6(b) illustrates the approach for Equation (4.15).

$$
\begin{aligned}
V_l &= (MH_a - MH_e)R_e \\
V_l' &= (R_a - R_e)MH_a \\
\text{Net labor variance} &= V_l + V_l'
\end{aligned}
\tag{4.15}
$$

where V_l = variance for labor owing to difference from estimated lot hours, dollars

V_l' = variance for wages owing to difference from estimated hourly rate, dollars

MH_e = estimated total hours for operation

MH_a = actual total hours for operation

R_e = estimated labor wage rate, dollars per hour

R_a = actual labor wage rate, dollars per hour

Example: Labor Variances

Consider the example where the estimated wage rate is \$28.76 per hour and labor hours are 283. The actual wage rate and lot hours were \$27.32 and 325. The unfavorable variance owing to changes in hours is \$1208 [= (325 − 283)28.76], though the favorable rate variance is −\$468 [= (27.32 − 28.76)325]. The unfavorable net labor variance is \$740 (= 1208 − 468). Remember that man-hours are the product of a labor time estimate and the number of lot units.

One purpose of variance is to quantify and control the variances. The cause of the variance will allow management an opportunity for correction. Quantity variance measures how many additional (or fewer) units were required to produce the production run. Production can be held accountable for this variance. Price variance measures how much additional (or less) was paid for the material. Purchasing can be held accountable for this variance.

Remember, cost estimating is done before the product is produced, while cost accounting and variance analysis are performed using actual production data after the work is done. Overhead variance analysis is not often done. Overhead is a collection of costs and is not easy to analyze.

Accounting costs, if purely historical and derived from job-order or process-cost procedures, are usually inadequate as estimates of future costs. The design may have changed or labor or material costs may have increased or decreased. However, for static situations, variance analysis can be helpful from a cost-control point of view. The exact information needed to control job progress varies with the jobs and the people who manage them. The variance analysis in particular is more useful for commodity products, or situations where there is large volume and standard labor rates are employed. However, for fast-moving enterprises with a significant overhead and less labor effort, the variance procedure is not found as frequently.

SUMMARY

Cost accounting is important to the functioning of diverse engineering cost analysis. Accountants provide overhead rates, some historical costs, and budgeting data. Engineering reciprocates with manpower and material estimates for the designs. The engineering estimate in many situations serves as a forecast of a profit-and-loss statement for products. Interestingly, the accountant provides the verification of the P&L after the project has closed out. Thus there is mutual dependence between accounting and engineering.

Engineering is more curious about balance sheets, P&L statements, and overhead rate determination than about the intimate details of the structure of accounts and the bookkeeping maneuvers. We have stressed this interpretative understanding of the accounting fundamentals. The chapter problems reinforce the learning of the important financial documents.

QUESTIONS FOR DISCUSSION

1. Define the following terms:

Accounting equivalence	Depreciation
Accrual	Double-entry bookkeeping
Activity-based costing	Fixed assets
Balance sheet	MACRS
Basis	Net worth
Budgeting	Nonvolume cost drivers
Consistency	Profit

Contra account	Profit and loss
Credit	Retained earnings
Current liabilities	T-accounts
Debit	Variance

2. List 10 kinds of liabilities for a business.
3. Why do expense accounts normally have debit balances and income accounts normally have credit balances?
4. What is the purpose of a trial balance? Into what main groups are trial balance accounts divided?
5. Distinguish between the cash and accrual bases of accounting. When are they suitable?
6. What are the functions of special and general journals?
7. What is the purpose of the budget? How would you define a cost center for an engineering budget? How do you prevent the budget from being meaningless?
8. Why is equitable distribution of cost essential throughout the organization to cost finding, analysis, and prediction?
9. Define overhead. What is the essential and philosophical purpose of overhead?
10. What is the difference between volume and nonvolume cost drivers? How does that affect overhead calculation?
11. Prescribe and contrast several methods for the distribution of overhead cost.
12. How do job order and process cost procedures differ?
13. Under what conditions are variance cost procedures useful?

PROBLEMS

4.1 Evaluate the effects of transactions by constructing a daily balance sheet showing an *asset* side and a *liabilities and equities* side. The business is called John Smith, Machine Shop.

January 1: John Smith starts his business, depositing $10,000 of his own money in a bank account that he has opened in the name of the business and creating capital stock.

January 2: The business borrows $5,000 from a bank, giving a note, thereby increasing the assets and cash and the business incurs a liability to the bank.

January 3: The business buys inventory for $10,000, paying cash.

January 4: The firm sells material for $300 that cost $200.

(*Hint*: Balance sheets follow a formal business style. Create a heading for each day's activities using the name of the business, type of business document, and date. For each transaction there are two entries.)

4.2 For the ledger T-accounts shown on page 171 from Weichman Consulting Company, set up a balance sheet for the end of March. (*Hints*: Note that for each T-account entry there are two dates, amounts, and title of transaction that are cross-identified.)

4.3 Construct a profit-and-loss statement and a balance sheet statement for the month of June by using the ledger T-accounts shown on page 172. The name of the business is Precision Manufacturing. (*Hint*: Determine the profit-and-loss statement first, and notice the location of the gross profit into the Capital T-account. Ignore taxes.)

P4.2

Cash

March	1	Capital	$1000	March	3	Rent	$200
	10	Consulting fee	250		20	Salaries	350
	25	J. A. Wilson on acct.	500				

Customers (accounts receivable)

March	10	J. A.. Wilson	$1200	March	25	Cash on account	$500

Supplies on hand

March	5	Accounts Payable	$360	March	31	Supplies used in March	$110

Equipment

March	4	Notes payable	$3200				

Accounts payable

				March	5	Supplies	$360

Notes payable

				March	4	Equipment	$3200

Weichman Consulting Co. Capital

March	3	Cash	$200	March	1	Cash investment	$1000
	20	Salaries	350		10	Consulting	250
	31	Supplies used	110		10	J. A. Wilson	1200

4.4 Construct a year-end balance sheet for CCM Manufacturing:

Account Title	$
Retained earnings	610,000
Cash	150,000
Outstanding debt	450,000
Raw materials	90,000
Finished goods	50,000
Current liabilities	40,000
Stock ownership	400,000
Fixed assets	1,100,000
In-process materials	110,000

4.5 Using the table at the top of page 172, prepare the profit-and-loss statement and find the retained earnings by using the account balances of Billerbeck Manufacturing Supplies, Inc., for the year ended December 31. BMS is a publicly held corporation.

Account Title	$	Account Title	$
Dividends paid	45,000	Retained earnings, January 1	90,000
Sales	700,000	Rent	70,000
Sales returns	40,000	Salaries	130,000
Inventory, January 1	120,000	Interest earned	2,000
Purchases	270,000	Sales discounts	10,000
Purchase returns	20,000	Interest expense	5,000
Inventory, December 31	160,000	Taxes, all kinds	47,000

P4.3

Equipment

June	1	Balance	$7210				

Accounts payable

June	2	Cash, Meyer Co.	$350	June	1	Balance	$360
					6	Supplies	600
					28	Miscellaneous expenses	40

Notes payable

				June	1	Balance	$3200

Capital

				June	1	Balance	$1790
					30		1570

Income

June	30	To P&L	$2500	June	4	Account, Proprieter A.B. Jones	$2200
					15	Cash, I. N. Smith	300

Lease expenses

June	1	Cash	$200	June	30	To P&L	$200

Miscellaneous office expenses

June	10	Telephone cash	$60	June	30	To P&L	$100
		Electricity	40				

Salaries

June	20	Cash	$350	June	30	To P&L	$350

Supplies expense

June	30	Supplies used	$280	June	30	To P&L	$280

4.6 Evaluate the effects of the following transactions by constructing a balance sheet showing an assets side and a liabilities plus net worth side.

Eastwood Manufacturing is organized with a capital stock of \$250,000 cash for the entire stock.

Bought from Culpepper Co. on credit \$100,000 of manufacturing material.

Borrowed \$80,000 cash from First National Bank.

Paid Culpepper \$30,000 on account.

Returned \$10,000 of defective material to Culpepper.

Loaned \$50,000 to Robert Gondring.

Paid \$20,000 cash for building site.

Erected a building at a cost of \$120,000 cash.

Borrowed \$70,000 from Friendly Insurance, giving a mortgage for collateral.

(*Hints*: Find the T-account balances for these titles: Cash, Accounts receivable, Material, Land and building, Accounts payable, Notes payable, Mortgage, and Capital stock. Each T-account will have a debit and credit column. Determine the account balance, then write the balance sheet.)

4.7 Evaluate the effect of the following transactions by constructing a balance sheet showing an assets side and a liabilities plus net worth side.

(a) Samuel Specthrie establishes SS Manufacturing, paying \$250,000 cash for the entire capital stock.
(b) Paid \$50,000 cash for a building site.
(c) Erected building costing \$200,000, paying \$50,000 cash and issuing a \$150,000 first mortgage for the balance.
(d) Borrowed \$60,000 cash from First National Bank.
(e) Bought furniture, costing \$15,000 on a open account from Wood Furniture Company.
(f) Bought tools, costing \$50,000 from Universal Tool on credit.
(g) Bought computer equipment, costing \$30,000 from Byte Company for cash.
(h) Returned \$15,000 of faulty tools to Universal Tool.
(i) Paid \$25,000 in reduction of bank loan.
(j) Bought U.S. Treasury bonds, costing \$40,000, for cash.

(*Hint*: The cash account $\$k = 250 - 50 - 50 + 60 - 30 - 40 = \$115k$.)

4.8 Patrick Lyell Manufacturing started the year with the following balances:

Account Title	Balance as of January 1, \$
Cash	100,000
Inventory	100,000
Manufacturing equipment	400,000
Accounts payable	50,000
Net worth	550,000

Transactions during the year were limited to the following: paid \$100,000 for labor; purchased \$150,000 worth of materials; noted equipment depreciation of \$50,000, adding to inventory 300,000 units costing \$1 to the manufacturer; sold 300,000 units for \$2 each, cash; and purchased new equipment costing \$200,000.

Make an end-of-year income statement and balance sheet as of December 31. (*Hint*: Neglect income taxes. Accounts payable and inventory at the end of year were the same as at the beginning of the year. For instance, the balance sheet for the cash account, per $1000, is 100 − 100 − 150 + 600 − 200 = $250.)

4.9 Eastwood Engineering Services, a family-owned firm, is chartered to provide consulting services, design, and a limited range of manufacturing services. Its designs have developed into registered patents. Some designs are licensed by subcontractors, the firm choosing not to use them itself, but it does receive income from that business arrangement. These accounts and their balances are summarized below.

Find the income statement and balance sheet for titled "Eastwood Engineering Services, Second Quarter." (*Hint*: These values are $1,000, but ignore that. For each account, first identify if it is an asset, liability, net worth, income, expense, or tax. Each will have a spot in either the income or balance sheet statement.)

Account Title	$	Account Title	$
Cash on hand	3	Receivables from projects	55
Fees received	73	Royalties on patents	40
Bonds owned	12	Equipment book value	58
Sale of design	7	Interest on owned securities	2
Salary expense	28	Office rent	1
Equipment lease	6	Travel	4
Utilities paid	5	Supplies expense	9
Patent assets	11	Accounts payable	30
Capital stock	25	Retained earning balance at start of quarter	27
Taxes, all kinds	12		

4.10 The following data are the assets, liabilities, incomes, expenses, and taxes of Warren Andrews, Design-Build Inc., an employee-owned firm.

Prepare an income and expense statement for the 6-month period ending June 30. Prepare a balance sheet at June 30. (*Hint*: Remember to have the right headings for the documents. It is necessary to prepare the income and expense statement first because the yearly retained earning amount is needed to complete the balance sheet.)

Account Title	$	Account Title	$
Securities owned	200,000	Office fixtures owned	50,000
Accounts payable	70,000	Bank loan	100,000
Interest from owned securities	5,000	Surplus, January 1	75,000
CAD supplies expense	8,500	Traveling expense	7,500
Real estate owned	250,000	Rental income on owned properties	30,000
Interest expense on bank loan	1,500	Automobiles owned	40,000
Telephone expense	2,500	Capital stock	425,000
Taxes (all kinds)	20,000	Cash in bank	65,000
Income from fees	150,000	Staff salary expense	120,000
Receivables from clients	90,000		

4.11 The following data are the assets and liabilities at December 31, and incomes and expenses for the business year for Wilmer Hergenrader, P.E., Consultant. Use this information to prepare an income statement and a balance sheet.

Account Title	$	Account Title	$
Cash in bank	80,000	Furniture and fixtures	30,000
Fees earned	420,000	Rental income on building lease	26,000
Interest received from bonds	2,500	Bonds owned	120,000
Land	60,000	Traveling expense	35,000
Receivables from clients	50,000	Taxes paid	15,000
Staff salaries	270,000	Office expense	10,000
Buildings	180,000	Telephone expense	5,000
Bank loan payable	70,000	Accounts payable	25,000
Capital stock	275,000	Automobiles owned	20,000
Surplus, January 1	65,000	Association dues expenses	2,500
Interest expense on bank loan	4,000	Donations to charity	2,000

4.12 The following data, expressed as $\$10^4$, are the closed ledger accounts showing assets, liabilities, income, and expenses of Blue River Fabricators, Inc., a family owned firm that does not declare dividends. Prepare an income statement and balance sheet for the year.

Account Title	$\$10^4$	Account Title	$\$10^4$
Cash in bank	1560	Office expense	200
Project income	3000	Manufacturing fixtures owned	1000
Interest income, owned securities	100	Accounts payable	1400
Rental income	600	Interest expense on bank loan	50
Salary and wage expense	2400	Receivable from contracts	1800
Bank loan	2000	Equipment owned	800
Capital stock	8300	Materials expense	150
Retained earnings, January 1	1700	Buildings owned	5000
Securities owned	4000	Taxes, all kinds	140

4.13 A heavy-duty truck is purchased for \$35,302. Besides depreciation to zero, costs include \$20,792 for repairs and maintenance; \$16,768 for gas and oil; \$14,472 for garaging and parking; \$10,800 for insurance; and \$10,560 for state and federal taxes. It will cost its owner \$108,694 by the time it has been driven 10 years and 100,000 miles.

Find the depreciation cost per mile. What are the yearly and per mile costs? What are the percentages for the elements of ownership?

4.14 The value and upkeep costs of a Chevrolet 1500 work truck are found to follow the schedule shown in the table below:

End of Year	Drop in Value, %	Drop in Value, $	Upkeep Costs, $
1	28	$4904	$412
2	21	3600	540
3	15	2700	1076
4	11	2000	1372
5	9	1504	1288
6	6	1036	1388
7	4	756	1828
8	3	484	968
9	2	340	1216
10	1	192	308

(a) How many years does it take for a truck to depreciate two-thirds and three-fourths of its value?
(b) If a truck is driven 13,500 miles per year, then what are the operating yearly costs in dollars per mile? Plot those operating costs. When is the advantageous time to trade a truck assuming that the chief criterion is per-mile economy?
(c) Discuss: If the immediate cost of repairing an old truck is less than first-year depreciation on a new one, is the best policy to buy a truck and drive it until it is ready to be junked? (*Hint*: How does "prestige" influence your decision to keep an old truck rather than buying a new one?) Does it influence your decision if the truck is self-owned or owned by the company?

4.15 Manufacturing equipment has a capital cost of $43,000, salvage value of $3000, and an asset life of 10 years. Compute the depreciation expense for the first 3 years under (a) accelerated cost recovery and (b) straight line.

4.16 Business A purchases manufacturing equipment that costs $250,000. Economic life is estimated as 5 years with a salvage value of $15,000. This company chooses the accelerated cost recovery method for depreciation. Business B, in competition with company A, buys the same equipment at identical cost. The management of company B uses straight-line depreciation. Determine the yearly depreciation charges and the end-of-year book value for companies A and B. Discuss the two methods under this competitive situation, and comment on the importance of the depreciation method for cost estimating.

4.17 A printed circuit supplier will buy a flow-soldering machine for a delivered price of $250,000. Life is 6 years or 12,000 hours of production for total solder capacity of 144,000 in.^{2} during the life period. Salvage value is expected to be nil. At the end of its life, the machine will be retired to secondary and emergency service.

Determine the annual depreciation charge and book value for accelerated cost recovery, straight-line, and units of production methods. Plot the "book value" for the three methods. Discuss the merits of book value.

4.18 General Engineering has the following $\$10^5$ balances on its accounts for the year to date, June 30:

Income	$11,000
Assets (other than fixed assets)	64,600
Capital stock	45,000
Expenses (other than depreciation)	5,150
Liabilities	18,000
Equipment (at beginning of period)	3,000
Taxes	1,250

The depreciation for the period is $600. Prepare the income statement and the balance sheet for the period. Comment on the importance of depreciation to the cash flow of the business.

4.19 Using the data below, find the graphical solution for variance on quantity first and material cost first, then find the total variance.

Product Element	Estimate	Actual
Material cost per unit	$25	$40
Number of units	100	200

4.20 The estimates for a material are 650 lb at \$17.15/lb. The record shows an actual usage of 665 lb at \$17.10/lb. Find the variance for material cost, quantity, and total. (*Hint*: Use the protocol "quantity first" followed by "material cost per unit first.")

4.21 Specification AISI 1018 cold-rolled steel material is estimated to cost \$0.0293 per unit for 20,000 units. Actually, material cost \$0.032 per unit and 20,500 units were necessary because of greater scrap and waste. Determine the material, quantity, and total variance. (*Hint*: Use the quantity first followed by the material first protocol.)

CHALLENGE PROBLEMS

4.22 Analyze the following transactions using the approaches shown in Tables 4.2, 4.3, 4.5, and 4.6.

Transactions	Accounts Affected
1. Founded AJAX Manufacturing paying \$300,000 cash for capital stock	Cash Capital stock
2. Bought material from Contractor Supply on credit for the amount \$150,000	Inventory Accounts payable
3. Paid rent for month, \$4000	Rent Cash
4. Sold material to M. Meyers for \$35,000 cash	Cash Sales
5. Paid \$50,000 to Contractor Supply on account	Accounts payable Cash
6. Borrowed \$80,000 cash from First National Bank	Cash Note payable
7. Retired \$60,000 of capital stock	Capital stock Cash
8. Sold material on account to K. Wilson for \$60,000	Accounts receivable Sales
9. Returned \$20,000 material to Contractor Supply	Accounts payable Inventory
10. Paid salaries and wages, \$3,000	Salaries Cash
11. Paid taxes of \$5,000 from cash	Taxes Cash

4.23 A contractor acquires truck-mounted concrete pumping equipment. Owing to the initial and operating costs, the contractor is in doubt as to the preferred method to charge depreciation. The contractor considers three methods: straight-line, MACRS, and units of production.

The facts are as follows: initial cost, \$1,460,000; useful life, 10 years; and a 20% salvage value. Production output starts at 200 yd^3/hr for a billing year of 2000 hr. It is estimated that the pumping rate will decline 10 yd^3/hr each year.

For the data, plot the book value of the asset against time for the methods and select the best choice. Discuss the merits of these allocation schemes. (*Hint:* The best choice is the minimum book value as time proceeds, thus minimizing taxable income.)

4.24 An 18-in. long 1020 steel bar weighs 83 lb. A lot estimate is required for 40 parts. Material cost is estimated as \$0.73/lb. An invoice shows that 43 parts were consumed for \$2960. Setup and cycle were estimated as 0.94 hour and 7.33 minutes per unit. Records show that 6 hours were needed, and the labor rate was \$27.25 instead of the planned \$26.90. Find the material, labor, and net variances for material and labor.

4.25 A lapping machine, which is able to grind surfaces to a flatness of 0.4 μ in., is estimated to require 0.5 hour for setup and 1.721 minute per unit for the cycle. The planned lot quantity is 7500 units and the estimated labor wage is \$22.58 per hour. Actual total man-hours and wage were 225 and \$22.61. Find the dollar variances for hours and hourly rate. Calculate the net labor variance and productivity factor. (*Hint*: Lot hours $= SU + N(H_s/60)$; consider the productivity on the basis of hours.)

4.26 A company is composed of five cost centers. Each month a budget is prepared anticipating the distribution of overhead costs to the centers. Let c_w be costs incurred within the cost center, such as depreciation, supplies and indirect labor, and let c_m be miscellaneous costs. The table shows the initial distribution of all overhead costs to each cost center. Additionally, for each center and month, a schedule of direct labor dollars and hours is matched.

Cost Center	c_w	c_m	Direct Labor, \$	Direct Labor, Hours
Fabrication	\$300,000	\$10,000	\$201,600	16,000
Assembly	80,000	5,000	72,960	6,400
Testing	20,000	40,000	37,440	3,200
Engineering	40,000	90,000		
Administration	20,000	10,000		

(*Hint*: Each direct department such as fabrication, assembly, and testing will have indirect labor costs that are organized for the purpose of the department and are included in c_w. Dimensions for the overhead rate is dollars of overhead cost to direct labor hours.)

Make a distribution of engineering and administration costs to the producing departments based on the proportion of direct labor dollars. Find the total overhead costs for the producing departments. Find the overhead rate for the producing departments based on hours.

4.27 Carbon Fabricators, a supplier of graphite components, has issued the following machine-center schedule of direct labor and overhead costs for next month. Depreciation overhead and tooling costs are assigned to the machine centers at this point.

Machine Center	Area, ft^2	Direct Labor Workers	Direct Labor Hours	Horsepower Hours	Depreciation, \$	Tooling, \$	Direct Labor Budget, \$
Fabrication	25,000	10	1,600	2,200	\$27,000	\$1,000	\$30,165
Assembly	6,000	4	640	325	7,200	500	17,296
Testing	1,900	2	320	650	1,800	4,000	7,744

Calculate the gross hourly rate for the machine-center direct labor. Other plant, engineering, and administration overhead costs are summarized below.

Overhead Summary	Amount	Allocation Basis
Plant space	\$7,000	Area
Indirect supplies	16,000	Direct labor hours
Utilities	4,200	Horsepower hours
Indirect labor	4,000	Direct labor hours
Engineering	13,000	Direct labor hours
Administration	3,000	Direct labor hours

Find the productive hour costs for the machine centers. (*Hint*: Initially, use the allocation basis for assigning the overhead amount to the producing machine centers, then determine the machine hour rates. The sum of the machine hour rate and the gross hourly rate is the productive hour cost.)

4.28 Hercules is a job shop that utilizes a yttrium aluminum garnet (YAG) laser to cut stainless and carbon steel thin-sheets. Laser cutting is cost effective for quantities where the shape geometry of the component is too complicated and expensive for special tooling that would blank the part. Hercules is grouped into two machine centers for this type of work. Some of the future facts about Hercules annual budget follow.

Cost Center	Worked Hours	Overhead, $	Direct Labor, $
YAG laser cutting	5,200	$178,000	$104,000
Surface finishing	1,750	65,000	32,000

The overhead is exclusive of the depreciation of a new laser power unit, which is an unassigned capital cost of $100,000 and requires straight-line depreciation for 5 years to a nil value. Find the productive hour cost rate for the two cost centers.

A firm is seeking a quote from Hercules for a quantity of 481 units. Using the customer's specifications, drawings, and request for quotation, Hercules determines the unit cost of the material as $2.50, and the standards for the cost center work are 2.500 hr/100 units and 1.000 hr/100 units for cutting and finishing. Find the lot and unit cost that Hercules will quote to the customer.

4.29 Columbia Production provides fabrication, assembly, and testing for its customers' designs. It supplies the subproducts within a one-month period after receiving the request for quotation. Indeed, it prepares a monthly budget of future direct labor and overhead costs, and quotes its customers on the results of this analysis, and then delivers the product if it wins the competitive bidding.

The schedule of costs for Columbia's three cost centers is given below.

Cost Center	c_w	c_o	Direct Labor Hours	Direct Labor $
Fabrication	31,000	$10,000	1,600	$30,165
Assembly	8,640	$4,000	640	17,296
Testing	2,500	$2,000	320	7,744

The information is analyzed to give c_w, where these overhead costs are incurred within the cost center, and c_o, where these overhead costs are engineering and administration but have been assigned to the producing center. For the cost centers, determine productive hour cost rates. (*Hint*: The basis for the machine hour cost rate is direct labor hours.)

A product cost model can now be defined as $c_p = c_{dm} + c_{phc}$, where c_{dm} = direct material cost and c_{phc} = direct labor and equipment cost. Let fabrication, assembly, and testing cost center standards be 0.480, 0.200, and 0.064 standard hour per unit. If $c_{dm} = \$1$ per unit, what is the quoted unit and lot cost for 1215 units?

4.30 A three-dimensional variance analysis is possible for direct labor operations, if we consider quantity, labor wage rate, and the standard unit time. Consider the following:

$$V_{dl} = (N_a - N_e)H_eR_e$$
$$V'_{dl} = (H_a - H_e)N_aR_e$$
$$V''_{dl} = (R_a - R_{e)})N_aH_a$$
$$\text{where net labor variance} = V_{dl} + V'_{dl} + V''_{dl}$$

A supplier orders a sputtering operation, which places a gold thickness of 0.000002 in. on a surface. This operation is estimated to require a standard rate of 1.415 hours per unit with a wage rate of $18.67. The buyer orders a quantity of 200. After the shipment is concluded, an actual rate of 1.548 hours per unit is recorded, 210 units are processed, and the labor rate is $17.32.

Find the variances for direct labor owing to quantity, standard performance, and wages. Also find the net direct labor variance.

PRACTICAL APPLICATION

The understanding of financial management of manufacturing is important. Indeed, if the student aspires to leadership in this business, then it is critical that he or she comprehend the important business documents, such as the income statement and the balance sheet. This practical application asks you to find a current annual report of a business and summarize the financial position of the company. Annual reports have three parts: prose, pictures, and tables. It is the tables that state the strengths or weaknesses of a business. Basically, the income statement tells you how the company did this year in comparison to last year, and the balance sheet tells you how strong the finances are by indicating what the company owns and what it owes as of a certain date.

Your instructor will suggest several businesses that are able to respond to your questions about financial management of manufacturing. Many manufacturing businesses are openly held—that is, their financial reports are public knowledge—but finding these annual reports can be challenging. There may be a business department on your campus that may have access to these resources. You can look at the *Wall Street Journal*, which sometimes lists companies that are willing to provide annual reports, or go to the Web sites of companies you might be interested in.

Your assignment is to prepare a written summary of a business, where the information is gleaned from their annual report, and then to give an oral report to your class. Your instructor will amplify these directions.

CASE STUDY: MACHINE SHOP

Dennis Schultz is the engineer-owner of Precision Job Shop, a company that prides itself on its ability to maintain ±0.00005 in. tolerances and surface finishes of 4 μin. The factory layout is shown in Figure C4.1, and the 200 × 150 ft are equally spaced into 10 cost centers.

The machine shop is more advanced than its competitors. Engineering design uses the popular "Pro-E" software and is able to load the computer numerical equipment with the files ready for manufacturing, without any intervening attention. The

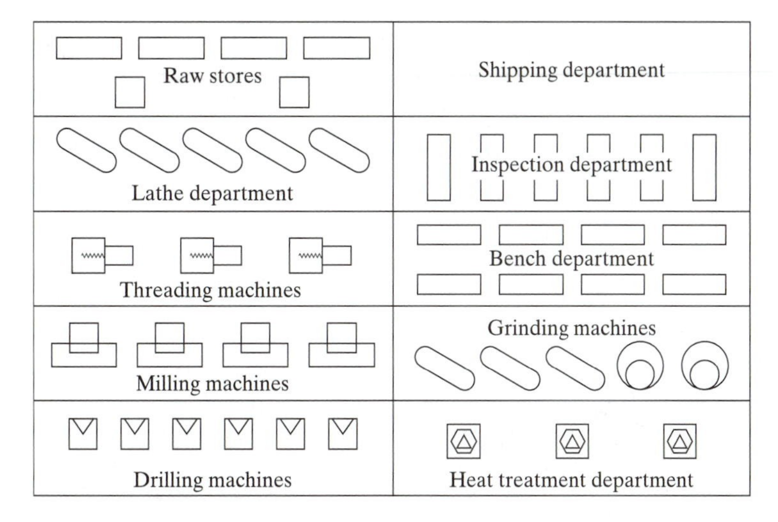

Figure C4.1 Plant layout for Precision Job Shop.

company is connected to customers via business-to-business servers and Internet links, and a customer will couple its designs for quotation directly to Dennis without any in-between handling. Dennis counts as his customers well-known aerospace and gas turbine companies. The shop, if it wins the competitive bidding, is able to deliver product on a just-in-time schedule, often within a month of the receipt of the order.

Dennis is making more estimates in recent years, and the "capture" percentage of estimates-won to estimates-made is also falling. Equally bad, he is able to keep his lower-priced equipment loaded, but the more expensive equipment is operating at lower capacity. Dennis has said he knows that his direct labor costs are competitive, but he is suspicious of his overhead computation, which is a general plantwide rate based on direct labor hours. Thus the spreading of overhead costs is relatively greater to the inexpensive equipment, as they are being loaded more costwise than the more expensive machinery.

With this developing e-to-e business opportunity and the recognition of the overhead distribution problem, PJS accountants have developed an activity-based system, and overhead costs are mapped to the producing cost centers on these new principles. The accounting system is reconfigured to give the overhead budget as shown in Table C4.1.

The plant is organized into seven producing machine centers as shown in Figure C4.1 and Table C4.1. Because of product demand, some of the machine centers are more fully loaded than others, and a partial second shift is used. This is reflected in the budget hours and the direct labor workers, because in some centers the number of workers exceeds the number of machines in that center. The direct labor hours are larger than the machine center hours, because of the machine downtime. The machine may be down for maintenance, but the direct labor worker continues to consume time. The machine center has its undepreciated amount known, and a remaining yearly life is matched to the amount, as seen in Table C4.1.

The plant layout shows three indirect centers: raw stores, inspection, and shipping. These centers along with engineering and administration provide the additional overhead costs that are now mapped according to activity-based costing principles

TABLE C4.1 Budget Forecast for Machine Centers

Machine Center	Monthly Machine Center Hours	Monthly Horsepower Hours	Direct Labor Workers	Gross Hourly Wages, $/Hr	Monthly Direct Labor Hours	Total Annual Undepreciated Dollars	Life Years
1. Lathe	13,328	160,000	8	23.45	15,680	30,000	5
2. Threading	6,664	66,000	4	21.04	7,840	120,000	7
3. Milling	6,664	53,000	4	26.75	7,840	110,000	12
4. Drilling	19,992	100,000	12	20.50	23,520	4,000	4
5. Bench	26,656		16	15.75	31,360	2,000	2
6. Grinding							
Horizontal	3,332	50,000	2	28.52	3,920	50,000	10
Rotary	1,666	13,000	1	24.53	1,960	90,000	10
7. Heat treating	6,664		4	19.20	7,840	15,000	3

TABLE C4.2 Budget Forecast for Indirect Cost Centers, Engineering and Administration, and Plant Overheads

Overhead Costs	Monthly Amount, $	Overhead Basis
Wages, indirect centers	575,000	Direct labor hours
Supplies	265,000	Horsepower hours
Plant supervision	174,000	Direct labor hours
Tooling	1,188,000	Horsepower hours
Utilities	123,970	Horsepower hours
Maintenance	136,480	Machine center hours
Engineering and Administration	700,000	Direct labor dollars
Plant undepreciated amount	1,600,000*	Horsepower hours and 240 months

*Annual amount.

(see Table C4.2). An overhead basis is given, which follows the principle of activity-based costing and is the dominant cost driver that linearly connects the overhead cost to the producing centers.

Help Dennis compute a productive hour rate for his seven producing machine centers. (*Hints*: Each machine center has its book value of depreciation that is an undepreciated amount divided by the remaining years and 12 months. For instance, the lathe center will allocate to its own monthly depreciation an amount of \$500 (= 30,000 × 1/5 × 1/12). The indirect wages \$575,000, for example, are prorated to the lathe machine center by the ratio of the direct labor hours used in the lathe center to the total for the plant and month, or \$90,196 (= 575,000 × 15,680/99,960). Ordinary straight-line methods of depreciation are satisfactory for the plant amount of \$1.6 million. For instance, the plant undepreciated amount is distributed monthly over the remaining 20 years and the horsepower hours simultaneously. The total activity-based overhead is the sum of those overheads that are directly connected to the machine center, plus the overhead accounts that are adjusted and related to the machine center by the basis.)

PSB receives a request-for-quote to machine the Christmas-tree roots of cast high-temperature-material turbine blades, and its computer estimating system establishes the following lot hours for 471 units.

Machine Center	1. Lathe	2. Threading	3. Milling	4. Drilling	6. Grinding	7. Heat treating
Lot hours	26.15	15.23	219.26	15.11	11.79	2.31

Material cost per unit is found to be $113.63. What is the job cost with the new productive hour cost rates?

Chapter 5

Forecasting

Despite all statements to the contrary, "emotional estimating" or "guesstimating" has not disappeared from the cost analysis scene. Nor has its substitute, estimation by formula and mathematical models, been universally nominated as a replacement. Somewhere between those extremes is a preferred course of action. In this chapter we consider the ways to enhance cost analysis by graphical and analytical techniques.

Business forecasting comprises the prediction of prices of material, availability and cost of labor, market demand, and costs of manufacturing. The usual approach to business forecasting involves extrapolation of past data into the future. Most business forecasts are made for the short-run period of up to two years. Medium-term forecasts cover two to five years, and long-term forecasts are for more than five years.

In this chapter we consider basic statistical and indexing methods. Forecasting does not mean estimating. Forecasting is a small but important part of the professional field of engineering cost analysis. Students who have learned these approaches have found that cost advice is sharpened with a clearer understanding of these methods.

Software programs are available in such rich variety that we make no mention of them, except to remind the student that the effective handling of these conveniences starts with a knowledge of fundamentals. Forecasting begins with a graphical interpretation of facts, whenever that is possible. We open with that first principle.

5.1 GRAPHIC ANALYSIS OF FACTS

Descriptive statistics are concerned with the following methods:

- Collecting, organizing, and analyzing data
- Summarizing and presenting visual images
- Drawing conclusions and making decisions

In a more restricted sense, the word "statistic" implies the data or the numerical measures derived from the data (for example, the average). The facts gathered for descriptive statistics and graphical presentations may be either discrete or continuous. They may be the result of a series of observations taken over time or another observable variable.

Raw data communicate little information. A frequency distribution, which compacts the facts into viewable records, offers an improved way to convey information. Such a distribution begins with the collection and organization of observations into a tabular arrangement by intervals. The prices are observable, and with the number of observations the survey is considered a factual representation.

Example: Survey of Product Price

A survey is made of the price for a roll of heavy 100 × 4 ft plastic film.

Price, ($/roll)	Number of Observations	Relative Frequency	Cumulative Frequency
12.35–12.75	1	0.003	0.003
12.75–13.15	6	0.019	0.022
13.15–13.55	33	0.102	0.124
13.55–13.95	51	0.157	0.281
13.95–14.35	121	0.373	0.654
14.35–14.75	50	0.154	0.808
14.75–15.15	44	0.136	0.944
15.15–15.65	13	0.040	0.984
15.65–16.05	5	0.016	1.000
	324	1.000	

The price range of $12.35–$12.75 is called an interval, and its end numbers are called limits. The size of the interval is 0.40 (= 12.75 − 12.35). The first midpoint is 12.55 [= (12.35 + 12.75)/2].

Relative frequency is the number of observations for each interval divided by all observations. Graphical representations of relative-frequency distributions are called *histograms*.

Example: Relative-Frequency and Cumulative-Probability-of-Occurrence Histogram for Survey Data

The plot of the percentage observations against price is called a histogram. If the relative frequencies are consecutively summed, then a cumulative frequency results. That is shown by the dashed line and right-hand scale. If those data are considered representative of the parent population, then the term *cumulative probability of occurrence* can be used.

If the midpoints of various histogram cells are joined by a line and smoothed, the frequency curves will appear as shown in Figure 5.1, which presents several types of frequency curves obtainable from analysis of data. A common title for the y axis would be "percentage of observations" or "count."

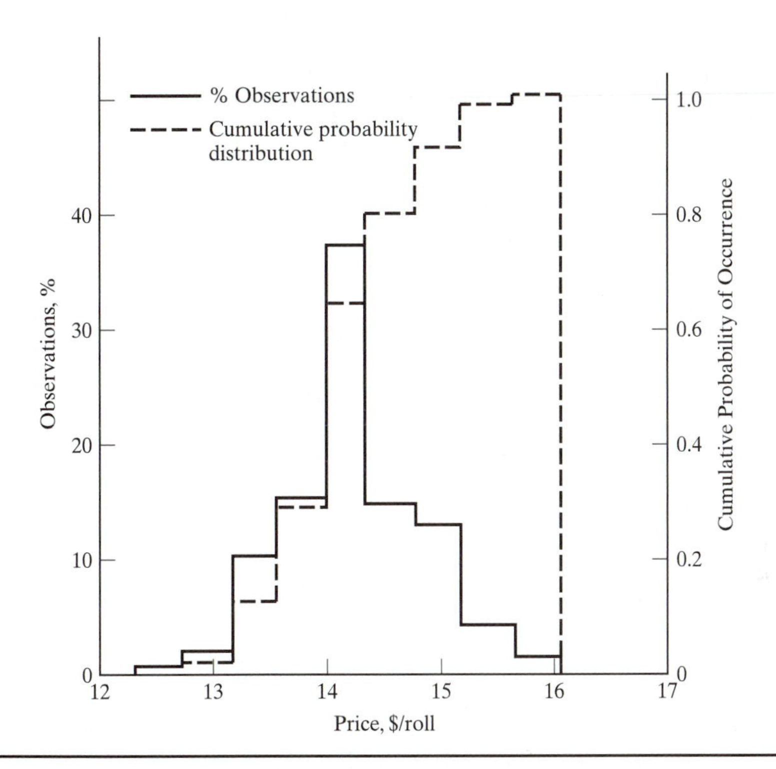

Curve construction calls for a "trained eye." For instance, if a wealth of data exists, then the bimodal or multimodal plot, special distributions that are more rare, may be evident. But without an abundance of facts, graphical conclusions of this sort are seldom found. Mathematical analyses are preferred for this and for other reasons. After the graphical plot is concluded, we calculate a measure of central tendency, such as mean, median, or mode, and a measure of dispersion, such as the standard deviation or range.

An average is a number typical of a set of data. Sometimes called the *arithmetic mean*, it is found as

$$\bar{x} = \frac{x_1 + x_2 + \cdots + x_n}{n} = \frac{\sum_{i=1}^{n} x_i}{n} = \frac{\Sigma x}{n} \tag{5.1}$$

where $\bar{x}$ = sample mean of set of data, number

x_i = i^{th} observation, $i = 1, 2, \ldots, n$

n = sample number

The middle value for an odd number of data, or the mean of the two middle values if the set number is even, is the *median*. It is found by first sorting the data in ascending or descending order of magnitude. The *mode* is the value that occurs with greatest frequency. The mode may not exist or may not be unique. The set, −1, 0, 2, 4, 6, 6, 7, and 8, has the mean 4, median 5, and mode 6. The set, −1, 0, 2, and 4, has no mode. The set, −1, 0, 0, 4, 6, 6, 7, and 8, is bimodal with the modes 0 and 6.

The degree to which numerical data spread about a mean value is called the *dispersion of the data*. Various measures of dispersion are available, the most common

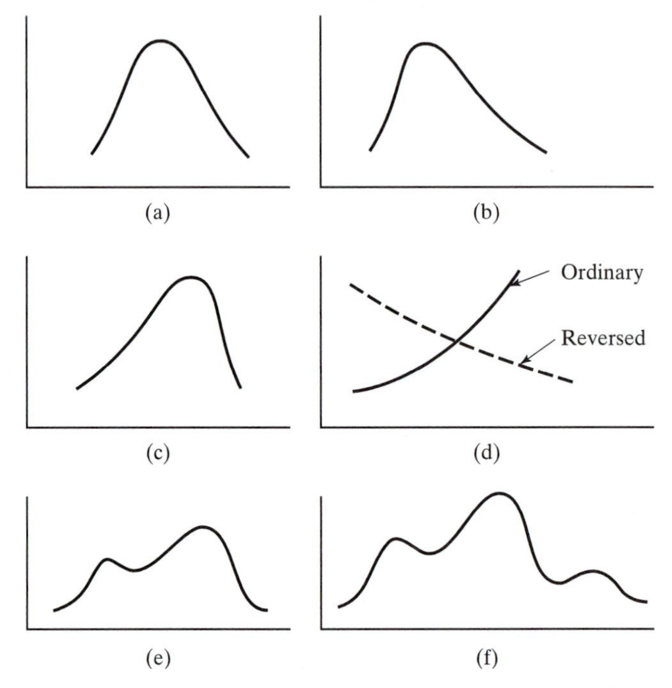

Figure 5.1 Frequency curves obtainable from analysis of data: (a) symmetrical; (b) skewed to the right; (c) skewed to the left; (d) J shaped; (e) bimodal; and (f) multimodal.

being the range and standard deviation. The *range* of a sample of data is the difference between the largest and smallest numbers in a set. The *standard deviation* of a set of n numbers, $x_1, \ldots, x_n$, is denoted by s and is defined by

$$s = \sqrt{\frac{\sum_{1}^{n} (x_i - \bar{x})^2}{n - 1}} \tag{5.2}$$

where s = sample standard deviation, number

The standard deviation is determined relative to the mean of the sample. The *variance* of a set of data is defined as the square of the standard deviation, or s^2. The set $(-1, 0, 2, 4)$ has the range of 5, $s = 2.22$, and $s^2 = 4.92$.

Plotting of facts is an important step in graphical analysis. Despite the ease with which numerical facts can be mathematically and statistically analyzed with computers, plotting of two-variable data, y versus x, for example, is useful to gain a "feeling" from the actual data. Mathematical relationships, void of a visual or real-life experience, lead to misjudgment in estimating. Very frequently you need to see the plot to understand what is going on. Knowledge, skill, and practice derived from direct observation of or participation in the statistical analysis help to develop judgment that is so necessary for engineering cost analysis.

A *graph* is a pictorial presentation of the relationship between variables. Most graphs have two variables, where we presume for manufacturing-cost estimating that x is the independent or controlled variable and y is the dependent variable. Once raw data

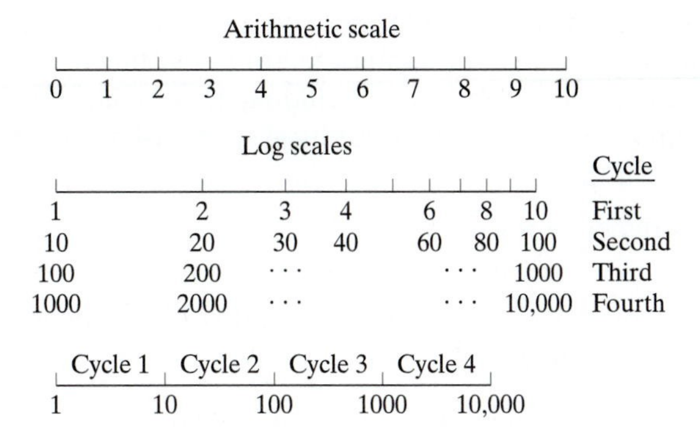

Figure 5.2 Arithmetic and logarithmic axes.

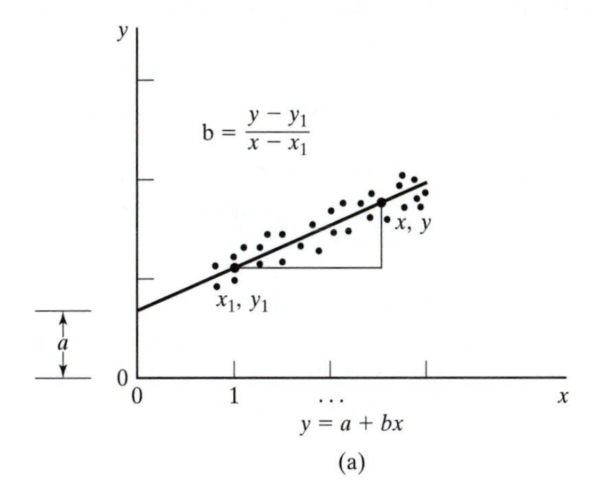

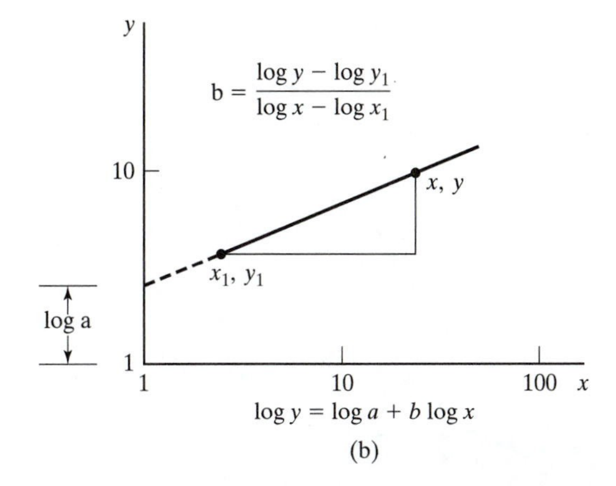

Figure 5.3 Plotting straight lines on (a) arithmetic and (b) logarithmic scales.

are gathered, the next steps select the axes and divisions, locate the points, and draw the straight line (initially). A rule of thumb to follow in plotting by eye is to have half of the points above the line and half below it, excluding those points that lie on the line.

With the line drawn we find its graphical straight-line equation by using

$$y = a + bx \tag{5.3}$$

where y = dependent variable

a = intercept value along the y axis at $x = 0$

b = slope, or the length of the rise divided by the length of the run

x = independent or control variable

Measurements of a and b are determined from the graph, because the intercept and slope can be measured from the line drawn through the points. This step, frequently overlooked in cost analysis, improves the gaining of experience. Both arithmetic and logarithmic axes are popular in cost analysis. They are shown in Figure 5.2.

Trial plots should attempt different pairs of the axes, such as semilog or log-log, in addition to arithmetic axis. That plot that best "straightens out" scattered data is selected. In Figure 5.3(a), the axes are arithmetic, and slope b is calculated from the right triangle. The line is extended to the y axis at $x = 0$, and at this point the intercept a is measured from the graph.

The logarithmic copy of Equation (5.3) is given by Figure 5.3(b). The slope is found by using the points (x, y) and (x_1, y_1) after the line is plotted by using the rule of thumb either manually or with graphical tools.

Example: Plot of Data Showing Cost Variability

Actual cost data are plotted in the figure below with a linear line. Note that the points are left unerased, meaning that all data points are plotted and remain on the graphical plot, which indicates the variability that may exist with time or cost information.

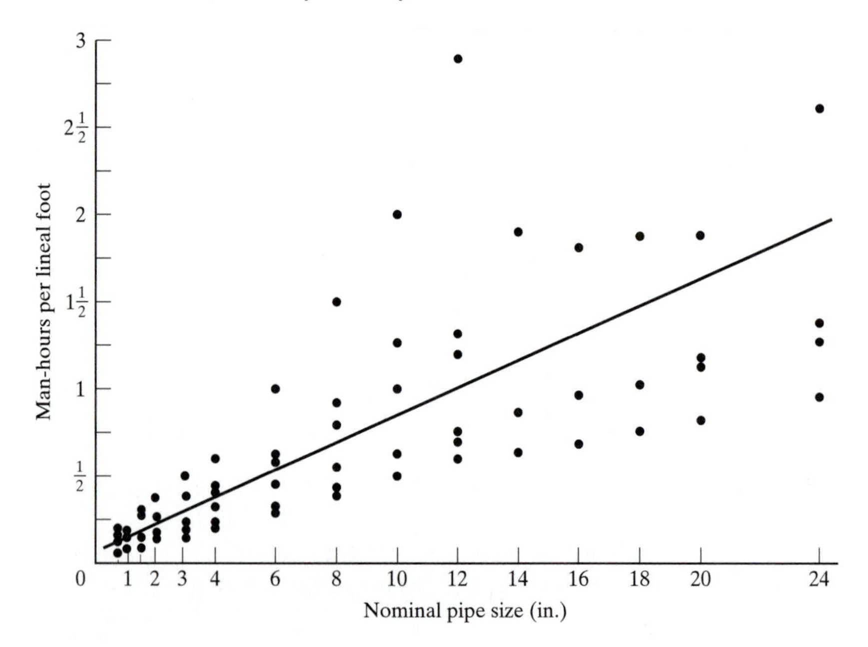

At the nominal 10-in. pipe size, the variability of labor hours is four times from a minimum to maximum, not an unusual event for cost facts. Interpretations of this kind are possible whenever there is a visual opportunity for evaluation of data.

Why graphical plots in the age of information technology? Frequently, a graphical line may show a sharp slope change that is not evident in mathematical analysis. Graphical plotting shows this knee jerk in data. "Seeing" is an important substitute for vicarious "doing." When possible, graphical plots are important to give this feel.

Initially, graphical plots screen and aid the selection of the best model that fits the data as a straight line, and then the method of least-squares analyzes data for their equation. The simple formula $y = a + bx$ is sometimes sufficient for practical day-to-day cost analysis work, but when the line appears nonlinear or is logarithmic straight, curves other than those found by using arithmetic scales are used.

Graphical plots will often be drawn differently by two people, but their mathematical equations should be identical. After the plot and the graphical equations are concluded, attention turns to mathematical analysis.

5.2 LEAST-SQUARES AND REGRESSION (OPTIONAL)

Descriptive statistical methods are usually concerned with a single variable and its frequency variation. But many problems in the engineering cost analysis field involve several variables. Now we explain methods for dealing with data associated with two or three variables. These important tools are as follows:

- Regression and least-squares
- Confidence and prediction limits
- Curvilinear regression
- Correlation

In regression, and on the basis of sample data, we want to find the value of a dependent variable y corresponding to a given value of variable x. This is determined from a least-squares equation that fits the sample data. The resulting curve is called a regression curve of y on x, because y is determined from a corresponding value of x. If the variable x is time, then the data show the values of y at various periods, and the equation is known as a time series. A regression line or a curve y on x or the response function on time is frequently called a trend line and is used for forecasting. Thus regression refers to average relationships between variables.

5.2.1 Least-Squares

The notion of fitting a curve to a set of sufficient points is essentially the problem of finding the parameters of the curve. The best-known method is that of least-squares. Since the equation is to be used for estimating purposes, the equation should be so modeled as to make the errors between the observed value and the regression-line equation value small. An error of estimation, as discussed in this context, means the difference between an observed value and the corresponding fitted curve value for the specific value of x. It will not do to require that the sum of these plus or minus differences be as small as possible. The requirement is that the sum of the absolute values of the errors be as small as possible. However, sums of absolute values are not mathematically convenient. The difficulty is avoided by requiring that the sum of the squares of the errors be minimized. If this procedure is followed, the values of parameters give what is known as the best curve in the sense of least-squares difference.

The principle of least-squares states that if y is a linear function of an independent variable x, then the most probable position of line $y = a + bx$ will exist whenever the sum of squares of deviations of all points (x_i, y_i) from the line is a minimum. Those deviations are measured in the vertical direction of the y axis. The underlying assumption is that x is either free of error (a controlled assignment) or subject to negligible error. The value of y is the observed or measured quantity, subject to errors that have to be minimized by this method of least-squares. The value y is a random-variable value from the y-population values corresponding to a given x. For each value of x_i, we are interested in the corresponding value for y_i.

If a large number of experiments are made, a histogram and eventually a normal curve, or Gaussian curve, or a bell-shaped curve, could be constructed. If natural processes are conducted in the experiment, 68.27% of the data fall within $\pm 1\sigma$ (where σ is defined as the population standard deviation), 95.45% within $\pm 2\sigma$, and 99.73 within $\pm 3\sigma$, as is evident in Figure 5.4 (a). The area under the normal curve is 1. Appendix 1 tabulates the normal distribution.

In Figure 5.4(b) the normal distribution is compared to the Student t distribution. Notice that the t-distribution is wider, and Appendix 2 gives the table of values for this situation. The Student t distribution is wider than the normal distribution because Student t does not presume a value of the standard deviation but instead uses the sample standard deviation s as an estimator of σ, and since this is a random variable in its own right, this gives greater variability of the Student t distribution.

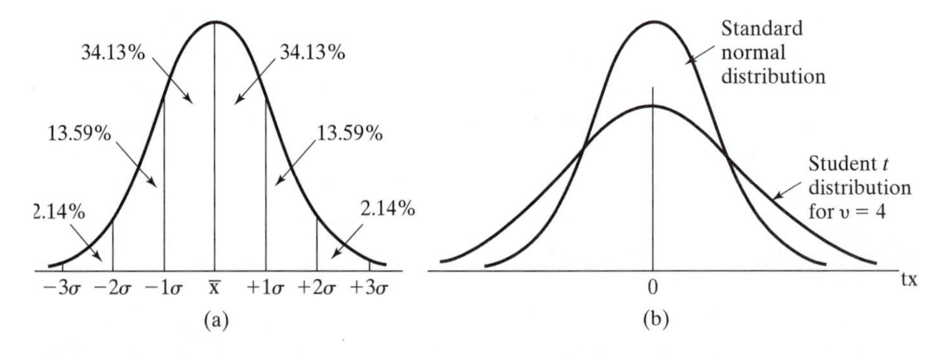

Figure 5.4 Distributions: (a) standard normal distribution showing proportion of area in standard-deviation zones, and (b) normal and Student t distributions compared.

PICTURE LESSON A Meeting between Ford and Edison

Smithsonian Institute.

Henry Ford (1863–1947), one of the greatest American engineers and industrialists, built his first working gasoline engine in 1893. By 1896, he had completed his first horseless carriage, the "Quadricycle," so called because the chassis of the four-horsepower vehicle was a buggy frame mounted on four bicycle wheels. In 1903 Ford was ready to market an automobile and the Ford Motor Company was incorporated with a mere $28,000 put up by ordinary citizens. The company was a success from the beginning.

"I will build a motor car for the great multitude," Ford proclaimed in announcing the birth of the Model T in 1908. In the 19 years of the Model T's existence, he sold 16,500,000, a production total amounting to half the auto output of the world. The motor age arrived owing mostly to Ford's vision of the car as the ordinary man's utility rather than as the rich man's luxury. The Model T was the chief instrument of one of the most rapid changes in the lives of the common people in history, and it effected this change in less than two decades.

The remarkable Model T was made possible by the most advanced production technology yet conceived. After much experimentation by Ford and his engineers, the system was able to deliver parts, subassemblies, and assemblies (themselves built on subsidiary assembly lines) with precise timing to a constantly moving main assembly line, where a complete chassis was turned out every 93 minutes, an enormous improvement over the 728 minutes formerly required. The minute subdivision of labor and the coordination of a multitude of operations produced huge gains in productivity. In 1914 the Ford Motor Company announced the Five Dollar Pay Day (compared to an average pay day of $2.34) and reduced the work day from nine hours to eight, extraordinary news then.

What Ford dreamed of was not merely increased capacity but complete self-sufficiency. The plant he built in River Rouge, Michigan, embodied his idea of an integrated operation encompassing production, assembly, and transportation. To complete the vertical integration of his empire, he purchased a railroad, acquired control of coal mines and acres of timberland, built a sawmill, acquired a fleet of Great Lakes freighters to bring ore from his Lake Superior mines, and even bought a glassworks. It was quite a vertically organized juggernaut of production efficiency. Ford, a complex and not always admired man, and his many legacies are legend.

Thomas A. Edison (1847–1931) was the quintessential American inventor in the era of Yankee ingenuity. He singly or jointly held a world record of 1093 patents, and he created the world's first industrial research laboratory. His career began during the adolescence of the telegraph industry, when virtually the only sources of electricity were primitive batteries giving a low-voltage current. He played a critical role in introducing the modern age of electricity. From his laboratories and workshops emanated an astounding array of inventions—the phonograph, the carbon-button transmitter for the telephone speaker and microphone, the incandescent lamp, a revolutionary generator of unprecedented efficiency, the first commercial electric light and power system, an experimental electric railroad, and key elements of motion-picture apparatus, among other things.

Ford admired Edison, as Ford credited Edison with encouraging his inventive aspirations. The photograph included here shows Ford (second from left) describing the process of flash butt welding. Edison, too, had welding patents and was an early giant of welding technology.

Suppose that our observations consist of pairs of values as (x_i, y_i), which are assumed to give a linear plot as in Figure 5.5. The symbol $\overline{y_i}$ ("y - bar") is the average

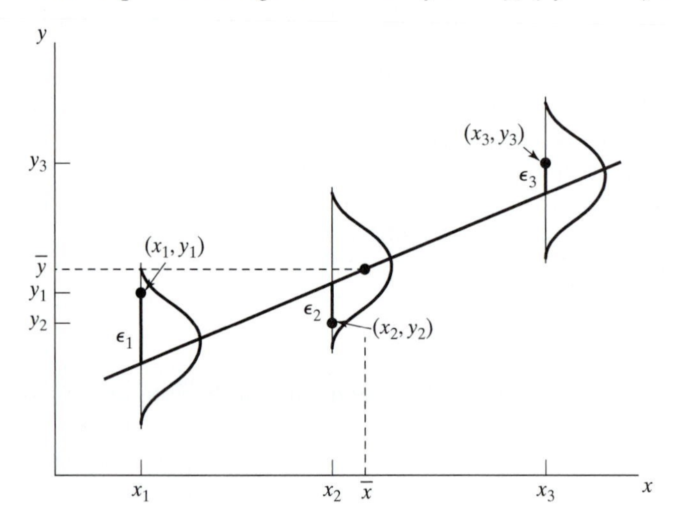

Figure 5.5 Regression line of $y = a + bx$ showing the differences ε.

value resulting from the x_i controlled variable. For instance, if the value of $x_i = 5$ and the experiment were repeated 100 times, $\overline{y_i}$ would be the mean value of 100 observations. Notice that repetitions of the controlled x_i value results in a normal distribution, and a particular y_i value deviates from the best fit line by an amount ε_i.

The bell-shaped curve is positioned with respect to the y-axis value. Our problem is to uncover a and b for what is known as *best fit*. For a general point i on the line, $y_i - (a + bx_i) = 0$, but if an error ε_i exists, then $y_i - (a + bx_i) = \varepsilon_i$.

For n observations, we have n equations of

$$y_i - (a + bx_i) = \varepsilon_i \tag{5.4}$$

where ε = difference between actual observation and regression value

Summing, we can write the sum of squares of these differences as

$$\sum_{i=1}^{n} \varepsilon^2 = \sum_{i=1}^{n} [y_i - (a + bx_i)]^2 \tag{5.5}$$

If a function is to have a minimum value, it is necessary that its partial derivative vanishes there, or

$$\frac{\partial \Sigma \varepsilon^2}{\partial a} = 0, \frac{\partial \Sigma \varepsilon^2}{\partial b} = 0 \tag{5.6}$$

Those equations thus give what is called a normal set of equations (not in the same sense as the "normal" distribution), which are

$$\begin{aligned} \Sigma y_i &= na + b\Sigma x_i \\ \Sigma x_i y_i &= a\Sigma x_i + b\Sigma x_i^2 \end{aligned} \tag{5.7}$$

and simultaneously solving those two normal equations for a and b we have

$$a = \frac{\Sigma x^2 \Sigma y - \Sigma x \Sigma xy}{n\Sigma x - (\Sigma x)^2} \quad \text{and} \tag{5.8}$$

$$b = \frac{n\Sigma xy - \Sigma x \Sigma y}{n\Sigma x^2 - (\Sigma x)^2} \tag{5.9}$$

Those a and b values are substituted into $y = a + bx$. This least-squares equation passes through $(\overline{x}, \overline{y})$, which is the coordinate mean of all observations. The calculations of those coefficients are handled with a computer or calculator when serious numerical problems are given. Transparent computer and calculator calculations obscure the steps that are key in figuring the regression coefficients from a set of data, so we show a spreadsheet example.

Example: Spreadsheet Calculations for Least-Squares Coefficients

We want to find the coefficients a and b regression values for the equation $y = a + bx$. The two left-hand columns x and y are the original data, and with the x^2 and xy columns, we are able to calculate Equations (5.8) and (5.9).

Year, x	Index, y	x^2	xy	y	ε	ε^2
0	87	0	0	84.875	2.125	4.516
1	89	1	89	87.264	1.736	3.013
2	90	4	180	89.654	0.346	0.120
3	92	9	276	92.043	−0.043	0.002
4	93	16	372	94.432	−1.432	2.051
5	99	25	495	96.821	2.179	4.746
6	97	36	582	99.211	−2.211	4.887
7	100	49	700	101.600	−1.600	2.560
8	101	64	808	103.989	−2.989	8.936
9	106	81	954	106.379	−0.379	0.143
10	106	100	1060	108.768	−2.768	7.661
11	109	121	1199	111.157	−2.157	4.653
12	115	144	1380	113.546	1.454	2.113
13	118	169	1534	115.936	2.064	4.261
14	122	196	1708	118.325	3.675	13.506
105	1524	1015	11337			63.168

For $Y = a + bx$, the constants a and b are given by

$$a = \frac{\Sigma x^2 \Sigma y - \Sigma x \Sigma xy}{n\Sigma x - (\Sigma x)^2} = \frac{(1015)(1524) - (105)(11{,}337)}{15(1015) - (105)^2} = 84.875$$

$$b = \frac{n\Sigma xy - \Sigma x \Sigma y}{n\Sigma x^2 - (\Sigma x)^2} = \frac{15(11{,}337) - (105)(1524)}{15(1015) - (105)^2} = 2.389$$

If Y = index and X = year for the index, the equation of the least-squares line is $Y = 84.875 + 2.389X$.

The limitations of the method must be pointed out. The method of least-squares is applicable when the observed values of y_i correspond to assigned (or error-free) values of x_i. The error in y_i (expressed as a variance of y) is assumed to be independent of the level of x. If inferences are to be made about regression, it is also necessary that the values of y_i corresponding to a given x_i be distributed normally, as presented in Figure 5.5 with the mean of the distribution satisfying the regression equation. The variance of the values of y_i for any given value of x must be independent of the magnitude of x.

Though the evidence of this statement can be statistically shown, experience shows that only a small number of the distributions met within cost analysis can be described by normal distributions. Cost data are limited at the zero end. Distributions influenced by business, engineering, and human factors are generally skewed. Despite those drawbacks, the least-squares method is widely used. Imperfection, apparently, does not reduce popularity.

5.2.2 Confidence Limits for Average Values and Prediction Limits for Individual Values

If variations around the universe regression are random, then the least-squares method permits the computation of sampling errors and provides the reliability of the estimate of the dependent variable from the fitted line. Confidence limits for regression values can be found through the extension of simple statistics. The confidence limits for individual regression values and for the straight line are quadratic in form around the sample line of regression. (Notice Figure 5.6.)

If we are finding the prediction limits for single values, then they are wider than the confidence limits. Look at Figure 5.6 again and note the outside quadratic lines for the prediction limits.

The variance of an estimate permits the forming of the confidence limits of the estimate. The approach in this case, similar to the variance of a sample, reckons the deviations from a line instead of a mean. The variance of y, estimated by the regression line, is the sum of squares of deviations divided by the number of degrees of freedom available for calculating the regression line, or

$$s_y^2 = \frac{\Sigma \varepsilon_i^2}{v} \tag{5.10}$$

where s_y^2 = variance around the regression line

We can assume that ε_i is as defined previously by Equation (5.4). Only two bits of information are required to determine the regression line: means $(\bar{x}, \bar{y})$ and either slope b or intercept a. With n as the number of paired observations, v is defined as

$$v = n - 2 \tag{5.11}$$

where v = degrees of freedom, number
n = paired observations, number

The degrees of freedom is $n - 2$ because 2 degrees have been absorbed, so to speak, through the estimation of the intercept and slope. Also,

$$s_y^2 = \frac{\Sigma \varepsilon_i^2}{n - 2} \tag{5.12}$$

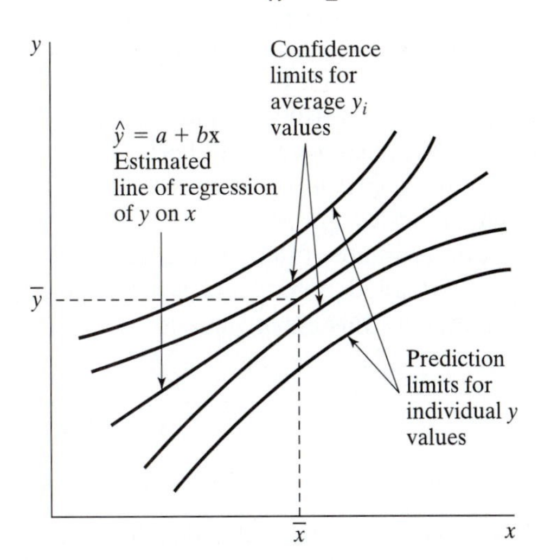

Figure 5.6 Confidence bands for average values and prediction limits for individual values.

Define

$$s_{\bar{y}}^2 = \frac{s_y^2}{n} \tag{5.13}$$

where $s_{\bar{y}}^2$ = variance of the mean value of y, or $\bar{y}$

We can now give the confidence limits for $\bar{y}$. A table of the Student t distribution and of the values of t corresponding to various values of the probability (level of significance) and a given number of degrees of freedom v is found in Appendix 2. The Student t distribution is similar in shape to the normalized standard variable (compare Appendixes 1 and 2), except that the t distribution is flatter—i.e., it has area and probability farther in the tails, both right and left [see Figure 5.4(b)]. With this t value we state that the true value of $\bar{y}$ lies within the interval

$$\bar{y} \pm ts_{\bar{y}} \tag{5.14}$$

where t = value of t distribution for probability α and degrees of freedom v

The probability of being wrong that the true value is outside the interval is equal to the level of significance of the value of t as α. Because the regression line must pass through the mean, an error in the value of $\bar{y}$ leads to a constant error in y for all points on the line. The line is then moved up or down without change in slope.

If limits for an individual y value are desired, then a different approach must be asserted. The equations for the interval of a single value are not the same as for the mean value. An index value, for instance, is a single calculated number and would lead to a future individual value, not a future mean value.

The statement that usually describes the limits for individual values goes like this: If we use the sample line of regression to estimate a particular value for y, then we add to the error of the sample line of the regression some measure of the possible deviation of the individual value from the regression value.

For individual values, a new set of parabolic loci may be viewed as prediction limits. Figure 5.6 presents the prediction loci for individual values as well as the confidence loci for average values. Note that the prediction limits for y get wider as x deviates up or down from its mean, both positively and negatively. This implies that predictions of the dependent variable are subject to the least error when the independent variable is near its mean and are subject to the greatest error when the independent variable is distant from its mean.

If we require an estimate of the confidence limits corresponding to any x_i, we calculate the limits for $\hat{y}$ (called "y-hat" and, strictly speaking, a statistical estimate). The variance of the estimate of this mean value is

$$s_{y_{\bar{i}}}^2 = s_y^2\left[\frac{1}{n} + \frac{(x_i - \bar{x})^2}{\Sigma(x - \bar{x})^2}\right] \tag{5.15}$$

where $s_{y_{\bar{i}}}^2$ = variance of each mean value of y

Notice that a bar is over the sub-subscript i. A new $s_{y_{\bar{i}}}^2$ is computed for each x value. The confidence interval for the mean estimated value of $y_{\bar{i}}$ corresponding to specific x_i is

$$y_{\bar{i}} \pm ts_{y_{\bar{i}}} \tag{5.16}$$

The interior confidence limits of Figure 5.6 correspond to Equation (5.16). For a predetermined level of significance, we are able to predict the limits within which a future mean estimated value of y_i will lie with an appropriate chance of error.

We find the prediction interval of a single estimated value of $\hat{y}_{\bar{i}}$ using the variance of a single value, which has as its variance

$$s_{y_i}^2 = s_y^2\left[1 + \frac{1}{n} + \frac{(x_i - \bar{x})^2}{\Sigma(x - \bar{x})^2}\right] \tag{5.17}$$

where $s_{y_i}^2$ = variance of the individual value of y

This variance is larger than s_y^2 because the variance of the single value is equal to the variance of the mean plus the variance of $\hat{y}$ estimated by the line, or

$$s_{y_i}^2 = s_{y_{\bar{i}}}^2 + s_y^2 \tag{5.18}$$

Each x has its own value of $s_{y_i}^2$. The prediction interval for a single value is greater, or

$$y_i \pm ts_{y_i} \tag{5.19}$$

In Figure 5.6 the outside quadratic lines are computed by using Equation (5.19). The terms confidence interval and prediction interval have different meanings. A confidence interval deals with an expected average Y value. A prediction interval deals with a single Y value. The prediction interval is greater in magnitude than the confidence interval.

Example: Confidence and Prediction Limits

If the anticipated y is expected to give an average value, we are interested in the confidence interval for $x = 15$, a new value that is in the future one year. Contrariwise, if the expected number is an single value, then the interval is called a *prediction interval*. Continuing the example with the "period" and the "index" illustration given above, we find the degrees of freedom $v = n - 2$, or $15 - 2 = 13$, and

$$s_y = \left(\frac{\Sigma\,\varepsilon^2}{n-2}\right)^{1/2} = \left(\frac{63.168}{13}\right)^{1/2} = 2.204$$

$$s_{y_{\bar{i}}} = s_y\left[\frac{1}{n} + \frac{(x_i - \bar{x})^2}{\Sigma(x - \bar{x})^2}\right]^{1/2} = 2.204\left(\frac{1}{15} + \frac{(15-7)^2}{280}\right)^{1/2} = 1.198$$

For a 5% level of significance, and using the t distribution with $v = 13$, we obtain $t = 2.160$. Then for the confidence interval of the mean value of y for $x = 15$, we have $120.71 \pm 2.160(1.198) = (123.301, 118.127)$.

If the y value is expected to a single value, as is most likely with an index projection, the prediction limits are expected to be wider than with the average value, though the expected midpoint number will be the same as 120.71. The prediction interval is then found as follows:

$$s_{y_i} = s_y\left(1 + \frac{1}{n} + \frac{(x_i - \bar{x})^2}{\Sigma(x - \bar{x})^2}\right)^{1/2} = 2.204\left(1 + \frac{1}{15} + \frac{(15-7)^2}{280}\right)^{1/2} = 2.509$$

The prediction interval is $120.71 \pm 2.160(2.509) = (126.133, 115.953)$. Notice that the prediction interval is wider than a confidence interval. Remember that we are expressing the confidence of two different things, expectation or mean value versus an individual value itself.

In a similar manner, the intercept a and slope b can have calculated limits that correspond to a prescribed probability. The variance of the intercept a is a special case of the variance of any mean estimated y_{y_i}. If we substitute $x_i = 0$ in Equation (5.15), then the variance of intercept a is

$$s_a^2 = s_y^2\left[\frac{1}{n} + \frac{\bar{x}^2}{\Sigma(x - \bar{x})^2}\right] \tag{5.20}$$

Its confidence band is given by

$$a \pm ts_a \tag{5.21}$$

Refer to Figure 5.7 (a), which corresponds to Equation (5.21). The variance of slope b is given as

$$s_b^2 = \frac{s_y^2}{\Sigma(x - \bar{x})^2} \tag{5.22}$$

Its confidence band is given by

$$b \pm ts_b \tag{5.23}$$

Refer to Figure 5.7 (b), which corresponds to Equation (5.23).

The confidence limits for the intercept are parallel lines. The confidence band for the slope is represented by a double fan-shaped area with the apex at the mean.

The number of degrees of freedom is $v = n - 2$ for all confidence intervals in this section.

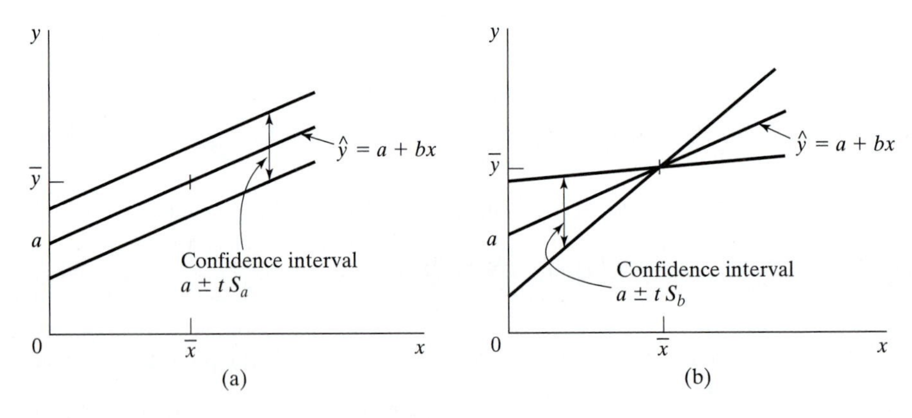

Figure 5.7 (a) Confidence interval for intercept a, and (b) confidence interval for slope b.

Example: Confidence Intervals for Intercept and Slope

The confidence interval for the intercept is as follows:

$$s_a = s_y\left(\frac{1}{n} + \frac{x^2}{\Sigma(x - \bar{x})^2}\right)^{1/2} = 2.204\left(\frac{1}{15} + \frac{49}{280}\right)^{1/2} = 1.083$$

The intercept interval is $84.875 \pm 2.160(1.083) = (87.214, 82.536)$.

The confidence interval for the slope b is found as

$$s_b = \frac{s_y}{[\Sigma(x - \bar{x})^2]^{1/2}} = \frac{2.204}{(280)^{1/2}} = 0.132$$

The slope interval is $2.389 \pm 2.160(0.132) = (2.674, 2.105)$

5.2.3 Curvilinear Regression and Transformation

The world of linearity, largely an imaginary one, is a tidy and manageable assumption about which generalizations can be asserted with boldness. The function $y = a + bx$ is a characterization frequently employed for known nonlinear situations. Many prefabricated tools are known that can be used with those assumptions. But computers and a greater awareness are forcing a reevaluation of linearities.

We plot data in a variety of ways, hoping to "straighten out" an arithmetic curve by means of semilog or log-log plots. For instance, if samples of paired data straighten out on semilog paper, we conclude that the form is exponential or $y = ae^{bx}$ and we would be tempted to apply the least-squares methods that used this form.

A brief listing of nonlinear relationships for engineering cost analysis purposes would give the following:

$$\text{Exponential:} \quad y = ab^x \tag{5.24}$$

$$\text{Power:} \quad y = ax^b \tag{5.25}$$

$$\text{Polynomial:} \quad y = a + b_1x + b_2x^2 + b_3x^3 + \cdots + b_px^p.. \tag{5.26}$$

The exponential and power equations, Equations (5.24) and (5.25), are used frequently. The regression methods of Section 5.2 can be made to work with those curvilinear models. Those equations appear like $y = a + bx$ if transformed using the logarithm. As an example, the exponential function $y = ab^x$ transforms to $\log y = \log a + x \log b$.

Let $y = \log y$, $a = \log a$, $b = \log b$, and $x = x$. This equation gives a straight line when plotted on arithmetic-logarithmic scales. Similar to Equations (5.8) and (5.9), the values of the parameters of Equation (5.24) can be solved by using

$$\log a = \frac{\Sigma x^2 \Sigma \log y - \Sigma x \Sigma y \log y}{n\Sigma x^2 - (\Sigma x)^2} \tag{5.27}$$

$$\log b = \frac{n\Sigma x \log y - \Sigma x \Sigma \log y}{n\Sigma x^2 - (\Sigma x)^2} \tag{5.28}$$

The power function $y = ax^b$ is transformed from a curved line on arithmetic scales to a straight line on log-log scales if we let $y = \log y$, $a = \log a$, and $x = \log x$. Intercept and slope equations can be found by making those substitutions into Equations (5.8) and (5.9).

$$\log a = \frac{\Sigma(\log x)^2 \Sigma \log y - \Sigma \log x \Sigma(\log x \log y)}{n\Sigma(\log x)^2 - (\Sigma \log x)^2} \tag{5.29}$$

$$b = \frac{n\Sigma(\log x \log y) - \Sigma \log x \Sigma \log y}{n\Sigma(\log x)^2 - (\Sigma \log x)^2} \tag{5.30}$$

The power equation models the important "learning" concept, useful for estimating manufacturing products. The practical applications of learning are found in numerous companies. As a product continues to be manufactured, it takes less time or cost to make what has been made earlier. The principle has been applied to an extensive list of products—from airplanes to computers.

Example: Practical Application of the Power Equation for "Learning"

We have five sets of unit number and labor hours to manufacture an engine-generator set, and the unit number x and the labor hours y are collected to construct that design. Data (unit number, hours) are (10, 510), (30, 210), (100, 190), (150, 125), and (300, 71). Thus the 10th and 300th units require 510 and 71 hours, respectively. Notice the decline in hours as the unit numbers increase.

The purpose is to model the unit number and labor hours using the power equation in both logarithm and original units. The intention of this example is to forecast the unit time for the 350th unit.

Use the power equation $y = ax^b$, which is of the form $\log y = \log a + \log x$. Transform the variables with logarithms, or $y = \log y$, $a = \log a$ intercept, $b = b$ slope, $x = \log x$, and $n =$ sample size. The first two columns of the table below are the original data, and the next four columns are the transformed calculations. The spreadsheet gives the results.

Unit, x	Labor hours, y	$x = \log x$	$y = \log y$	$(\log x)^2$	$\log x \log y$
10	510	1.0000	2.7076	1.0000	2.7076
30	210	1.4771	2.3222	2.1819	3.4302
100	190	2.0000	2.2788	4.0000	4.5575
150	125	2.1761	2.0969	4.7354	4.5631
300	71	2.4771	1.8513	6.1361	4.5858
		9.1303	11.2567	18.0534	19.8441

The sums of the four columns are substituted into Equations (5.29) and (5.30), giving the following:

$$\log a = \frac{\Sigma(\log x)^2 \Sigma \log y - \Sigma \log x \Sigma(\log x \log y)}{n\Sigma(\log x)^2 - (\Sigma \log x)^2}$$

$$= \frac{(18.0534)(11.2567) - (9.1303)(19.8441)}{5(18.0534) - (9.1303)^2} = 3.1921$$

$$b = \frac{n\Sigma(\log x \log y) - \Sigma \log x \, \Sigma \log y}{n\Sigma(\log x)^2 - (\Sigma \log x)^2}$$

$$= \frac{5(19.8441) - (9.1303)(11.2567)}{5(18.0534) - (9.1303)^2} = -0.5152$$

The transformed equation is $\log y = 3.19211 - 0.5152 \log x$. Upon substitution of $x = 350$, $\log y = 1.8814$, and when transformed to regular units, the time $y = 76.10$ hours is forecast for the 350th unit.

In original units of the power equation, $y = 1556.38(350)^{-0.5152} = 76.10$ hours, identical to the forecast from the logarithmic equation. We call these approaches "transformed" (in the logarithm units) and "original."

Each of these nonlinear functions, Equations (5.24–5.26) can be statistically evaluated for fitted data such as the following:

- Coefficients of the equations
- Listing of calculated estimates
- Standard error of y, $\Sigma\varepsilon_i^2 = \Sigma(y_i - \hat{y})^2$

Computer software will do these calculations quickly.

Normally, the best-fit line for regression is judged by the smallest value of the standard error. Other computed values provide an intuitive feel for the accuracy of the correlation.

Polynomial regression—that is, a calculation where for any x the mean of the distribution of y is given by $a + b_1x + b_2x^2 + b_3x^3 + \cdots + b_px^p$—is used to obtain approximations whenever the functional form of the regression curve is a mystery. Consider the following quadratic least-square fit of a fuel-cell firing circuit.

Example: Least-Square Fitting of Quadratic Equation for Fuel-Cell Combustion Exciter

Hydrogen fuel cells use an electronic spark exciter to ignite the gas, where y, or dollars for the manufactured unit, is related as $y = a + bx^2$, where x = milliseconds of firing cycle.

x	Observed Value, y_i	Error-free Value, y	Deviation, $y_i - y$
0	1	a	$1 - a$
1	1.4	$a + b$	$1.4 - (a + b)$
2	1.8	$a + 4b$	$1.8 - (a + 4b)$
3	2.2	$a + 9b$	$2.2 - (a + 9b)$

The sum of the squares of the deviations is

$$\Sigma\varepsilon^2 = (1 - a)^2 + (1.4 - a - b)^2 + (1.8 - a - 4b)^2 + (2.2 - a - 9b)^2$$

For a minimum we are required to satisfy

$$\frac{\partial\Sigma\varepsilon^2}{\partial a} = 0, \text{ and } \frac{\partial\Sigma\varepsilon^2}{\partial b} = 0$$

or (1 − a) + (1.4 − a − b) + (1.8 − a − 4b) + (2.2 − a − 9b) = 0
and (1.4 − a − b) + 4(1.8 − a − 4b) + 9(2.2 − a − 9b) = 0

$$4a + 14b = 6.4$$

$$14a + 98b = 28.4$$

Hence, $a = 1.171$ and $b = 0.122$. When substituted back into the general form, the fitted equation becomes $y = 1.171 + 0.122x^2$. A better fit is obtained from an equation of a different form or $y = a + bx + cx^2$.

The previous example is straightforward, but the application of the method of least-squares to nonlinear relations usually requires a good deal of computational effort, made easier with computers/software. In most cases we can "transform or rectify" a nonlinear relation to a straight-line relation. This manipulation simplifies handling of the data and permits a linear graphical presentation that may be more revealing for certain facts. With a rectified straight line, extrapolation is simpler, and the computations of certain other supportive statistics, such as the standard deviation or confidence limits, are simpler.

Now, those previous functions indicated a regression of y on x that was linear in some fashion. Sometimes a clear relationship is not evident, and a general polynomial is selected. A predicting equation of the polynomial form [see Equation (5.26)] requires a set of data consisting of n points (x_i, y_i). We estimate the coefficients a, b_1, b_2, b_p of the pth-degree polynomial by minimizing the partial derivatives of

$$\Sigma\varepsilon^2 = \sum_{i=1}^{n}[y_i - (a + b_1x + b_2x^2 + \cdots + b_px^p)]^2 \tag{5.31}$$

which is the least-squares criterion. This results in $p + 1$ normal equations of the shape:

$$\begin{aligned} \Sigma y &= na + b_1\Sigma x + \cdots + b_p\Sigma x^p \\ \Sigma xy &= a\Sigma x + b_1\Sigma x^2 + \cdots + b_p\Sigma x^{p+1} \\ &\vdots \\ \Sigma x^p y &= a\Sigma x^p + b_1\Sigma x^{p+1} + \cdots + b_p\Sigma x^{2p} \end{aligned} \tag{5.32}$$

where summation superscript and subscript notation has been ignored. We will have $p + 1$ linear equations in $p + 1$ unknowns $a, b_1, \ldots, b_p$.

Example: Polynomial Least-Squares Fit of Fuel-Cell Exciter

The hydrogen fuel cell exciter for the combustion ignition of the gas is closely related to the clearance positioning of the igniter. The data (with x = in., y = cost) are given as follows:

Clearance of igniter, in.	Cost, $
0.01	7.00
0.02	8.40
0.03	9.20
0.04	10.10
0.05	10.30
0.20	26.20

Spreadsheet tabulations give the following:

$$\Sigma x = 0.35 \qquad \Sigma y = 71.2$$

$$\Sigma x^2 = 0.0455 \qquad \Sigma xy = 6.673$$

$$\Sigma x^3 = 0.008225 \qquad \Sigma x^2 y = 11.0225$$

$$\Sigma x^4 = 0.00016979$$

With those values we solve the following system of three normal linear equations,

$$6a + 0.35b_1 + 0.0455b_2 = 71.2$$
$$0.35a + 0.0455b_1 + 0.008225b_2 = 6.673$$
$$0.0455a + 0.008225b_1 + 0.00016979b_2 = 11.0225$$

for which $a = -31.73$, $b_1 = 1660$, and $b_2 = -7025$, and the predicting polynomial equation becomes $y = -31.73 + 1660x - 7025x^2$.

In practice it may be difficult to determine the degree of the polynomial to fit data, but it is always possible to find a polynomial of degree at most $n - 1$ that will pass through each of n points. For practical reasons, we prefer the lowest degree that describes our data. Higher order polynomials will have more peaks and valleys to the equation and curve than lower order, thus inviting unwarranted speculation as to what the process is really doing.

5.2.4 Correlation

The methods of regression show the association of one dependent variable to an independent variable that are considered linearly related. There is a closely related measure, called *correlation*, that tells how well the variables are satisfied by this linear relationship. If the values of the variables satisfy an equation exactly, then the variables are perfectly correlated. With two variables involved, the statistician refers to simple correlation and simple regression. When more than two variables are involved, they are multiple regression and multiple correlation. Only simple correlation is considered in this book.

Figure 5.8 indicates the location of points on an arithmetic coordinate system. If all the points in a scatter diagram appear to lie on or near a line, as in Figure 5.8 (b) or

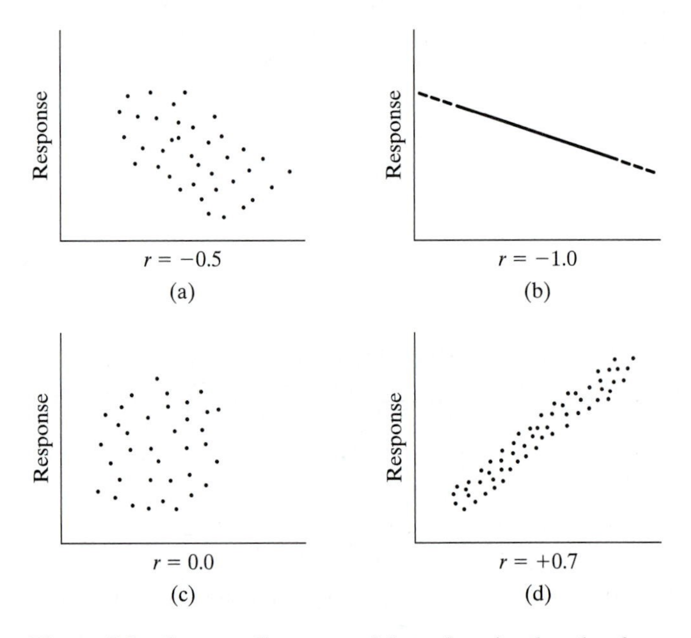

Figure 5.8 Scatter diagrams of data showing levels of correlation coefficients: (a) negative linear correlation; (b) negative linear correlation with points exactly on line; (c) no correlation; and (d) positive linear correlation.

(d), the correlation is presumed to be linear, and a linear equation is appropriate for regression or estimation. If there is no relationship indicated between the variables, as in Figure 5.8 (c), then there is no correlation (i.e., the data are uncorrelated, and any number of straight lines can be drawn through the randomly mixed cloud of observations). In Figure 5.8 (b) the correlation coefficient is negative linear. For Figure 5.8 (d) a positive linear correlation coefficient is suggested.

With a fitted curve from data, it is possible to distinguish between the deviations of the y observations from the regression line and the total variation of the y observations about their mean. A calculated difference between the two variations gives the amount of variation accounted for by regression. The higher this value, the better the fit or correlation. For $y = a + bx$, no correlation exists if $b = 0$, as in Figure 5.8 (c), and the line plots as a horizontal line. Thus x and y are independent. There is no correlation of x on y if x is independent of y.

Advanced statistical mathematics lead to the following definition of correlation:

$$r = \frac{n\sum xy - \sum x \sum y}{\left\{\left[n\sum x^2 - \left(\sum x\right)^2\right]\left[n\sum y^2 - \left(\sum y\right)^2\right]\right\}^{1/2}} \tag{5.33}$$

where r = correlation coefficient, number, $0 \le |r| \le 1$.

Correlation is concerned only with the association between variables, and r must lie in the range $0 \le |r| \le 1$.

Example: Calculating Correlation Coefficient

Review the example on page 196. Recollecting that $n = 15$, $\Sigma xy = 11{,}337$, $\Sigma x = 105$, $\Sigma y = 1524$, $\Sigma x^2 = 1015$, and $\Sigma y^2 = 156{,}500$, then

$$r = \frac{(15)(11{,}337) - (105)(1524)}{\left\{\left[(15)(1015) - (105)^2\right]\left[15(156{,}500) - (1524)^2\right]\right\}^{1/2}} = +0.98$$

which is a high score for correlation. The correlation is positive, meaning that as x increases, so does y.

The magnitude of the correlation coefficient r determines the strength of the relationship, and the sign of r tells us whether the dependent variable tends to increase or decrease with increasing magnitude of the independent variable. For instance, a positive value indicates that the index is increasing positively with years. If two variables are linearly related, r is a useful measure of the strength of the relationship between them. The value of r will be equal to $+1$ or -1, if and only if the points of the scatter lie perfectly on the straight line, which is unlikely in cost analysis.

The interpretation of the correlation coefficient as a measure of the strength of the linear relationship between two variables is a purely mathematical interpretation and is without any cause-and-effect implications. It is possible to have high correlation coefficients between variables that are only bogus. Can you think of situations that give high correlation having no real-life cause and effect?

5.2.5 Multiple Linear Regression

It may happen that the method of least-squares for estimating one variable by a related variable yields poor success, as real systems are rarely that simple. Although the relationship may be linear, frequently there is no single variable sufficiently connected to the dependent variable being estimated to yield good results. Some problems are not simply characterized by $y = a + bx$, but then the addition of more linear variables can be successful in their prediction. The extension to two or more independent variables is natural.

Because linear functions are simple to work with and estimating experience shows that many sets of variables are approximately linearly related, or can be assumed so for a short duration, it is reasonable to estimate the desired variable by means of a linear function of the remaining variables. Problems of multiple regression involve more than two variables but are still treated like those involving two variables.

This approach is not limited to time-trend problems. Time is a catch-all, which takes into account gradual changes due to different factors, both known and suspected.

For three or more variables, a regression plane is a generalization of the regression line for two variables. Multiple linear regression has the form

$$y = a + b_1x_1 + b_2x_2 + \cdots + b_kx_k \tag{5.34}$$

where $x_0 = 1$

a = intercept of kth + 1 plane, constant

$b_1, b_2, \cdots, b_k$ = partial regression coefficients

We do not say that the result so obtained is the best functional relationship. We simply state that, given this assumed function and criterion, and using the least-square criterion, we have chosen the best estimates of the parameters.

This is a plane in $k + 1$ dimensions. The plane intersects the means of the variables, similar to the two-variable case shown in Figure 5.5.

Example: Two-Variable Linear Regression for Disk-Drive Motor

The need for newer high-speed disk-drives to run cooler and quieter and with less vibration and improved reliability is important as RPMs increase. With disk motors approaching 10,000 to 15,000 RPMs, the spindle motor bearings—precision components that are pressed on the shaft of the motor to ensure that the spindle turns smoothly without wobble or vibration—cause noise and heat. Heat created by the spindle motor can eventually cause damage to the hard disk.

A newer technology uses fluid-dynamic bearing motors. Internal air spaces within the bearing are now filled with viscous oil. This oil can withstand temperature changes far greater than traditional ball-bearing drives having air spaces. The technology is not new, as it has been used in gyroscopes and high-accuracy machine tools for 50 years or more. However, the application to disc motors is revolutionary.

Of the many possibilities that increase the cost of a 10,200-RPM fluid-dynamic bearing motor are two variables, viscosity and average time latency (half rotation of the platter to gain full spinning velocity), which contribute to more than 85% of the motor cost.

A two-variable linear equation of the form $y = a + b_1x_1 + b_2x_2$ is able to model the cost of the disk motor.

As before, the coefficients are determined by using the method of least-squares. To illustrate, consider the case of two independent variables with n sets (y, x_1, x_2) of points.

We minimize $\Sigma\varepsilon^2$ as before, which requires that the partial derivatives of $\Sigma\varepsilon^2$ with respect to a, b_1, and b_2 be set equal to 0. Then the normal equations corresponding to the least-squares plane for the y, x_1, and x_2 coordinate systems are

$$\begin{aligned} \Sigma y &= na + b_1\Sigma x_1 + b_2\Sigma x_2 \\ \Sigma x_1 y &= a\Sigma x_1 + b_1\Sigma x_1^2 + b_2\Sigma x_1x_2 \\ \Sigma x_2 y &= a\Sigma x_2 + b_1\Sigma x_1x_2 + b_2\Sigma x_2^2 \end{aligned} \tag{5.35}$$

Subscripts and superscripts for summation notation were dropped for convenience. The solution of this system of three simultaneous equations gives the values of a, b_1, and b_2 for Equation (5.35) and is referred to as y on x_1 and x_2.

If we keep x_2 constant, then the graph of y versus x_1 is a straight line with slope b_1. If we keep x_1 constant, then the graph y versus x_2 is linear with slope b_2. Because y varies partially because of variation in x_1 and partially because of variation in x_2, we call b_1 and b_2 the partial regression coefficients of y on x_1 keeping x_2 constant, and of y on x_2 keeping x_1 constant.

This is a regression plane, but more complicated regression surfaces can be imagined—for example, with four-dimensional or five-dimensional space.

Example: Multiple Linear Regression of Disk-Drive Motor

A campaign is conducted to design and manufacture a 10,200 RPM fluid-dynamic bearing motor disk drive. One of the major engineering tasks to be done is the determination of a model that estimates the cost per 100 units. The motor has 15 design iterations where working prototypes are constructed and tested, with numerous engineering properties measured during actual operation.

Each design package is estimated for cost, and two variables are measured: viscosity and average time delay. Viscosity is measured in units of poises (dyne-seconds per square centimeter) while delay is counted in microseconds. The data of the (y, x_1, x_2)-points (= cost per 100 units, viscosity and time) are not shown, but the sums of calculations are given in the table below.

$n = 15$	$\Sigma x_1^2 = 156{,}872.75$	$\Sigma x_1 y = 180{,}565.04$
$\Sigma y = 1756.30$	$\Sigma x_2 = 32.37$	$\Sigma x_2 y = 3860.886$
$\Sigma x_1 = 1520.10$	$\Sigma x_2^2 = 72.01092$	$\Sigma x_1 x_2 = 3357.3750$

Substituting the values of the table into in Equation (5.35), the normal equations are

$$1756.30 = 15a + 1520.10b_1 + 32.37b_2$$

$$180{,}565.04 = 1520.10a + 156{,}872.75b_1 + 3357.3750b_2$$

$$3860.886 = 32.37a + 3357.3750b_1 + 72.01092b_2$$

for which $a = 29.1181$

$$b_1 = 0.7052$$

$$b_2 = 7.658$$

The multiple linear equation then becomes $y = 29.1181 + 0.7052x_1 + 7.6458x_2$. For the selected design where $x_1 = 95$ poises and $x_2 = 2.07$ microseconds, the cost y is equal to \$111.94 per 100 motors.

Although there are other models available that analyze the powerful tool of linear regression that should be considered, we leave that study to other statistical texts. To conclude our discussion of linear regression, we can make the following assumptions:

- The x_j values are controlled and/or observed without error. Perfection remains a difficult requirement within engineering estimating practices, but it is nominally met.
- The regression of y on x_j is linear.
- The deviations $y - [j|x_j]$ are mutually independent.
- Those deviations have the same variance whatever the value of x_j.
- Those deviations are normally distributed.
- The data are taken from a population about which inferences are to be drawn.

- The model contains all important variables, leaving no extraneous variables that make the relationship of little real value.

Other models are useful for forecasting, and we now consider time series, a class of functions that offers opportunity for regression to determine future values.

5.3 TIME-SERIES MODELS (OPTIONAL)

Knowledge of cost behavior is an important subject in estimating, especially those products having long activity times. The government, for instance, develops forecasts of many economic time series such as the gross domestic product and exports. Most companies develop forecasts of product sales. While the forecasting of engineering-business factors can be handled by regression and indexing methods, there are additional procedures known as *time series models*. Because of the importance of time series analysis, students need a basic appreciation of these procedures. We cover the following fundamentals:

- Consistent data collection
- Types of time-series behavior
- Moving average
- Smoothing

There are many business situations where new information becomes available and may be added to a data set on a revolving basis. Periodic observations of labor and material cost, overhead, and product demand are examples. Those data that have recurrent additions are referred to as time-series data. The data are refreshed with these costs (or prices), and as the process continues, data that are more senior and perhaps not as important are removed from the data set.

Indeed, a time series is a set of data collected at successive points or over periods of time. A sequence of quarterly data on the price of a 7200-RPM computer disk drive by suppliers or a sequence of weekly data on product sales are examples of time series. Usually, the data are collected at equally spaced periods of time, such as day, week, month, or year.

In interpretation of time-series analysis, we are interested in the response of these observations. Is the underlying process constant, variable, trend-cycle, seasonal, or regular? Samples of graphs, where time is the X-axis and a response variable (consider this to be any business factor) is the Y-axis, are shown in Figure 5.9. Looking at part (a), we see that the simplest of the time-series cases is the mean. For the trivial case of a mean that is constant, or nearly so, over the time interval for which the forecast is required, the dependent variable is nonsensitive. In other words, there is no trend, and a future value is the same as it is today, which is the situation in Figure 5.9 (a).

The linear model with a trend, Figure 5.9 (b), is found more widely than the quadratic (c) and exponential models (d). Movements are generally considered to be cyclical if they recur after constant time intervals. A typical example of cyclical movements is the so-called business cycles representing intervals of boom, recession, depression, and business recovery. Seasonal movements refer to identical or nearly

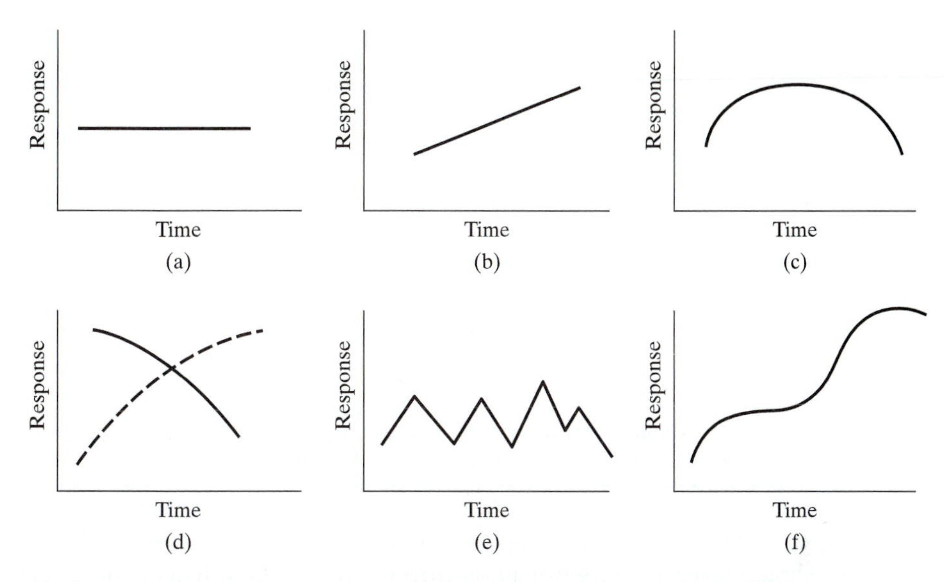

Figure 5.9 Typical time-series models: (a) constant (no trend); (b) linear; (c) quadratic; (d) exponential; (e) cyclic or seasonal; and (f) linear and cyclic.

identical patterns that a time series appears to follow during corresponding months of successive years such as Figure 5.9 (e). Those events may be illustrated by peak summer manufacturing activity and lesser demand preceding the Christmas period. Certain of those effects are sometimes superimposed on other effects, Figure 5.9 (f), where a linear and a cyclic pattern are superimposed.

Cycles may be interpreted in either of two ways: as random-cause deviation from established long-term trends, or as significant fluctuation due to some economic and business effect. In the case where the cycles are interpreted as deviations from a trend, the peaks and troughs are normally referred to as errors from the estimate and are caused by a collection of unknown factors. Again, refer to Figure 5.9 (f). Those deviations that are caused for economic and business reasons are suspicious and need to be studied further.

Example: Time Series of the Cost of Computer Disk Platter Assembly

Every hard disk contains one or more flat disks called *platters* that are used to actually record and hold the data in the drive. These disks are composed of two main materials: (1) a substrate polymer material that forms the bulk of the platter, which gives structure and rigidity, and (2) a magnetic media coating, which actually holds the magnetic impulses that represent the data. Hard disks get their name from the rigidity of the platters used, as compared to floppy disks and other media, which use flexible "platters." The media coating of platters is critical, and the surface of each platter is precision machined and treated to remove any imperfections. The hard disk itself is assembled in a clean room to reduce the chances of dirt getting onto the platters and the assembly.

Cost data for the disk platter assembly has been collected for 10 periods. The most recent period is 10, and the oldest is period 1.

Period, t	Data x_t
1	\$6.020
2	6.050
3	6.070
4	6.094
5	6.155
6	6.232
7	6.471
8	6.397
9	6.517
10	6.539

A reasonable forecast of the platter cost is given by the average price, or \$6.255 each. But are there other values that can be discerned from the information?

If we believe that the computation of an ordinary average at any single point in time should place no more weight on current observations than on those achieved previously, then the forecast for any future observation could be the ordinary average value, or \$6.255. But a major motivation for the moving average places greater reliance on more recent values than on senior values.

In this \$6.255 average price, the denominator is 10, and that suggests that the average is best analyzed by dividing the total number of observations by 10, or the size of the current data set. But what happens when the next data point becomes available? New values are added, while senior values are dropped. Thus a *moving* average, denominated by an arbitrary selected value of N, so it is reasoned, better predicts the future value. With this logic, we state the following: The moving average of N most recent observation, computed at time t, is given by

$$M_a = \frac{x_t + x_{t-1} + \cdots + x_{t-N+1}}{N} \tag{5.36}$$

where M_a = moving average of response variable, price, dollars

x = information data, such as cost, price, yield

t = time period, such as year, month, week, or day

N = selected denominator of group of time period, number

There is a restriction that the number of terms in the numerator be equal to N and that x_t be the latest term added. Another arrangement, of course, is to use for example, the most recent three, four, or five observations and divide the sum by 3, 4, or 5. The model might then be arranged into another form:

$$M_a = M_{t-1} + \frac{x_t - x_{t-N}}{N} \tag{5.37}$$

where for t, $N = 1,2,3,\cdots$, and $t > N$

Example: Moving Average for Platter Assembly

Assume a time series of information x_t is given above for periods 1-10. If a 3-year moving average is desired, then the data would be arranged as shown in the table below. The moving average is the moving 3-year total divided by $N = 3$.

Period, t	3-Year Moving Total, $	M_a, $/unit
1		
2		
3	18.140	6.047
4	18.214	6.071
5	18.319	6.106
6	18.481	6.160
7	18.858	6.286
8	19.100	6.367
9	19.385	6.462
10	19.453	6.484

One of the considerations with moving averages is selection of rate of response. The rate of response is controlled by the choice of N of the observations to be averaged. If N is arbitrarily chosen large, then the estimate is stable. If N is selected small, then fluctuations due to random errors or other legitimate causes can be expected. We are able to take advantage of those properties. If the process is expected to be constant, we may select a large value of N in order to have a stable estimate of the mean. However, if the process is fluctuating, small values of N provide faster indications of response.

For most forecasting problems, some type of moving average is desired that reflects both historical and current trends. A smoothing function is defined as

$$S_t(x) = \alpha x_t + (1 - \alpha)S_{t-1}(x) \tag{5.38}$$

where $S_t(x)$ = smoothed value of the estimated quantity, cost, price, etc.

α = smoothing constant, $0 \leq \alpha \leq 1$.

The α is similar but not exactly equal to the fraction $1/N$ in the moving-average method. Whenever this operation is performed on a sequence of observations, it is called *exponential smoothing*. The new smoothed value is a linear combination of all past observations. Statistically speaking, the expectation of this function is equal to the expectation of the data, which is its average.

When the smoothing constant α is small, the function $S_t(x)$ behaves as if the function provided the average of past data. When the smoothing constant is large, $S_t(x)$ responds rapidly to changes in trend. Though precise statements cannot be made regarding this smoothing, the table below generally describes the effect of a smoothing constant on time-series data.

	Variations in α Values		
Drift in actual data	Small, $\alpha = 0$	Little, $\alpha = 0.5$	Large, $\alpha = 1$
None	None	None	None
Moderate	Very small	Small	Moderate
Large	Small	Moderate	Large

Initial conditions are established for exponential smoothing, typically $S_{t-1} = M_{t-1}$. If there are no past data to average, then smoothing starts with the first observation, and a prediction of the average is required. The prediction may be what the business activity intended to do. Those predictions can also be based on similarity with other business activities that have been observed for some time. If there is confidence in the prediction of initial conditions, then a small value of the smoothing constant, $\alpha \rightarrow 0$, is satisfactory. On the other hand, if there is little confidence in the initial prediction, then it is appropriate to have α as a larger value, $\alpha \rightarrow 1$, so that the initial conditions are quickly discounted. This is the counter to the argument about flexibility of response to a change. If we believe that the real value is like the prediction, then there is little reason to have a change. On the other hand, the contrary viewpoint would have a quick response between the prediction and the real value.

Example: Exponential Smoothing and Forecasting for Future Period

A platter assembly can be analyzed using exponential smoothing. Consider the example with exponential smoothing for $\alpha = 33\%$.

Date, t	Data, \$/unit, x_t	Smoothed Data, \$/Unit $S_t(x) = 0.33x_t + 0.67\,S_{t-1}(x)$
1	6.020	6.020
2	6.050	6.030
3	6.070	6.043
4	6.094	6.060
5	6.155	6.092
6	6.232	6.139
7	6.471	6.249
8	6.397	6.299
9	6.517	6.371
10	6.539	6.427
11		6.438

Period 11 is then forecast using regression or trending and is given as \$6.438 per unit. The original data gave an average of \$6.255 each.

Most forecasting methods assume that historical data are collected from some relatively stable information process. These models develop forecasts where time or period is the independent or predictor variable. But beyond these methods is cost indexes, another special calculation that is used widely in engineering cost analysis. Because of their importance in escalation and preliminary estimates, we give them special attention.

5.4 COST INDEXES

From time to time escalation adjustment for a contract quote or component cost is an important need. Once extrapolation through time-series of cost indexes for future periods is finished, this fine-tuning becomes possible. Thus the student will see how engineering analysis may exploit time-series cost indexes.

Cost indexes have been used for those purposes for a long time.[1] We will study the following aspects of them:

- Applications
- Finding the periodic rate of an index
- Construction of a composite index

A *cost index* is an amalgam of labor, material, and services, and it compares cost or price changes between periods for a fixed quantity of goods or services. The index is merely a dimensionless number for a given period showing the cost at that time relative to a certain base period or *benchmark*. The cost of a similar design from the past to a present or future period can then be forecast without going through detail costing. If discretion is used in choosing or developing the proper index, then a reasonable approximation of cost will occur.

Index numbers are useful in other ways. With time for cost analysis usually scarce, we are tempted to exploit previous designs and costs, which are likely rooted on outdated circumstances. As costs vary with time because of changes in demand, economic conditions, and prices, indexes convert costs applicable at a past date to equivalent costs now or in the future. This may be stated formally as

$$C_c = C_r\left(\frac{I_c}{I_r}\right) \tag{5.39}$$

where C_c = desired cost, present, future, or past, dollars

C_r = reference cost, dollars

I_r = index chosen to correspond to time period of C_r, number

I_c = index chosen to correspond to time period of C_c, number

If a design cost for a previous period is known, then present cost is determined by multiplying the original cost by the ratio of the present index value to the index value applicable when the original cost was obtained.

[1]An Italian, G. R. Carli, devised the index numbers about 1750. Using indexes, he investigated the effects of the discovery of America on the purchasing power of money in Europe.

Example: Forecasting Construction Cost for Factory

New construction of a 700,000-ft^2 factory is planned for a future period. Several years ago a similar factory was constructed for a unit area estimate of \$162.50/ft^2 when the index was 118. The index for the construction period is forecast as 143, and cost per ft^2 will be

$$C_c = 162.50\left(\frac{143}{118}\right) = \$196.93/\text{ft}^2$$

Total construction cost for the factory is 700,000 $\times$ 196.93 = \$138 million. The method is speculative, but for a quick and dirty budget cost, it may be the best we are able to do at the moment.

Though general-purpose indexes are openly published and are widely accepted, their making, alteration, and application are worthy subjects because it may be better for engineering to develop its own index.

Arithmetic development of indexes falls into several types: (1) adding costs and dividing by their number, (2) adding the cost reciprocals and dividing by their number, (3) multiplying the costs and extracting the root indicated by their number, (4) ranking the costs and selecting the median value, (5) selecting the mode cost, and (6) adding actual costs of each year and taking the ratio of those sums.

Though formulas are straightforward, determination of indexes bears little resemblance to formulas because of the variety and complications. The weighted arithmetic method is the most popular. For most cost-analysis situations, a tabular approach is the best technique.

A cost index is a dimensionless number representing the change in cost of material or labor or both over a period of time. Prices, which are the input of an index, must relate to specific material or labor. An index for an aluminum ingot is based on the price of a specific quantity and type of ingot, such as A323 40-lb aluminum ingot. Quantity and quality must remain constant over the periods so that price movements represent a true price change rather than a change in quality or quantity. This is tough for indexes that are charted over many periods.

A cost index expresses a change in price level between two points in time. A cost index for aluminum ingot in the year 2010 is meaningless alone, and an index for material A has no relationship to the index for material B. Nor will the cost indexes for material A in two geographical areas be comparable.

To compute a price index for a single material, a sequence of prices are sampled for two or more epochs of time, for a specific quantity and quality of the material. The prices gathered for the material may be average for the period (month, quarter, half-year, or year) or they may be a single observed value, as found from invoice records for one purchase. There are numerous sampling biases that need to be understood in the gathering of the data but are not discussed here.

Example: Index Finding for Laser-Glass Material

Prices have been collected for a standardized unit of a laser-glass material. Details were initially designated at period 0, which for this example is the origin. Subsequently, other prices are found for periods 2, . . . , 5. Period 5 is considered the present while period 6 is future.

Period	0	1	2	3*	4	5
Price	$43.75	$44.25	$45.00	$46.10	$47.15	$49.25
Index	94.9	96.0	97.6	100.0	102.3	106.8

*Benchmark period.

Index numbers are computed by relating each period to one of the prices selected as the denominator. If period 3 is arbitrarily chosen as the benchmark period, then the period 2 price divided by the period 3 price = 45.00/46.10 = 0.976. When the period 3 index is expressed as 100.0, the period 2 index is 97.6. The index can be expressed on the basis of 1 or 100 without any loss of generality. The benchmark period is defined as that period that serves as the denominator in the index calculation and has an index value of 100.

Movements of indexes from one period to another are expressed as percentage changes rather than changes in index points:

Current index, period 5	106.8
Less previous index, period 4	102.3
Index point change	+4.5
Divide by previous index	102.3
Equals	+0.044 = +4.4%

Thus, between period 4 and 5, the price for the laser glass inflated 4.4%. The average periodic change resulting from these indexes can be found by using

$$r = \left[\left(\frac{I_e}{I_b}\right)^{1/n} - 1\right] \times 100 \tag{5.40}$$

where r = average rate per period, percentage
I_e = index at end of period, number
I_b = index at beginning of period, number
n = number of periods

The time period can vary in length, but monthly to yearly intervals are common. For an index beginning with 94.9 and ending with 106.8 over a 5-year period, the average rate is +2.39%.

If the average index rate is expected to persist, then Equation (5.40) is reformed to give

$$I_e = I_b\left(1 + \frac{r}{100}\right)^n \tag{5.41}$$

Example: Future Cost of Laser-Glass Material Using Indexes and Past Cost

If the average rate of increase for laser-glass material is 2.39%, find a future cost of a quantity of this material that will occur in period 7. First, the index for period 7 is

$I_7 = 94.9(1 + 0.0239)^7 = 112.0$. If cost C_2 = \$370,000 is known from records and for a design that is similar to the composition of the index, use of Equation (5.39) will give for $n = 7$.

$$C_7 = 370{,}000\left(\frac{112.0}{97.6}\right) = \$424{,}600$$

Thus, in period 7, the laser-glass material will cost \$424,600 for a similar situation that was present in period 2.

But engineering indexes are more complex than any single material index. A composite index is often required, say, for adjustment of "quote-or-price-in-effect" type of inflation/purchase contracts. Equally important is the updating of estimates of complicated components, assemblies, and plants.

Example: Finding Composite Index for Laser Product

A product called a 10-cm disk aperture laser amplifier is selected for a composite index. The 10-cm disk amplifier was constructed only during period 0, but cost tracking of selected items has continued. With a design having over 1250 amplifier components and assemblies, individual tracking of all components is very unlikely, and perhaps unnecessary. A snapshot of the amplifier design is consolidated into six materials, which are identified and prices are gathered over 3 years.

Material	Quantity	Quality Specification	0*	1	2	3
Laser glass	3–10 cm disks	Silicate	\$26,117	\$24,027	\$22,345	\$21,228
Steel tubing	18 kg	AISI 1035	1,913	2,008	2,129	2,278
Aluminum extrusion	4 kg	AISI 304	418	426	439	456
Printed circuits	31 boards	Mil Std 713	637	643	656	657
Harness cable	4 braid, 4 m	Mil Std 503	2,103	2,124	2,134	2,305
Glass tubing	12 m × 35 mm I.D.	Tempered	4,317	4,187	4,103	4,185
Total			\$35,505	\$33,415	\$31,806	\$31,119
Index, %			100.0	94.1	89.6	87.6

*Benchmark period.

The material column identifies those items that are significant cost contributors to the laser. Usually, a representative group is selected if the design is complex. If the design is simple, however, all materials may be chosen. The quantity column is proportional to the needs of the design. The quality specification column identifies the technical nature of the material.

Cost finding begins once those three columns are determined. Year 0 is the first period of the cost facts, and the index 100.0 is determined once we divide the total by itself. Cost facts are collected for each subsequent period, and each total is divided by the benchmark total to obtain the index. Thus the indexes are 100.0, 94.1, 89.6, and 87.6. A general decline in prices is suggested by those indexes. Apparently, there is technological improvement that reduces prices.

One may argue that materials, quantities, and qualities are not consistent, especially over any extended period of time. Indeed, if technology is progressive, then reduction in the cost and index is possible. This is counter to the widely held belief that indexes increase because of inflation. A reminder, often forgotten in recent decades, is that deflation is also a possibility.[2]

Index creep results from changes in quality, quantity, and the mix of materials or labor. This phenomenon may be unseen in the collection of long-term information for indexes.

The effects of changes in product mix, quantity, and quality on the index scheme are called *technology creep*. Indexes may become unsuitable for high-technology products and aggressive manufacturing progress. Every so often it may be necessary to reset the benchmark year whenever delicate effects are influencing the index and are not being removed.

There are several kinds of indexes:

- Material
- Labor
- Material and labor
- Geographical influences
- Design
- Quality

Virtually any amalgam of materials, labor, services, products, and projects can be evaluated for an index. An interesting contrast to a price type of index is a quality index. Instead of noting price changes, the purpose of a quality index is to remove price effects and show quality changes between the periods.

In the development of the index, there is a choice in the selection of the information. Wholesale prices or retail prices, wages or type of manufacturing, and proportion of labor to materials are typical alternative choices. Indexes apply to a place and time—that is, period covered or region considered, base year, and the interval between successive indexes, yearly or monthly. Additionally, indexes are varied as to the compiler and sources used for data.

Every industrialized nation regularly collects, analyzes, and divulges indexes. Many government index listings are given in the *Statistical Abstract of the United States*, a yearly publication that includes material, labor, and manufacturing. A yearly publication of the U.S. Department of Labor is the *Indexes of Output per Man-Hour for Selected Industries*.[3] This volume contains updated indexes such as output per man-hour, output per employee, and unit labor requirements for the industries included in the U.S. government's

[2]But the last time general wage rates declined in the United States was in 1931, a depression year with 30% unemployment.

[3]The Bureau of Labor Statistics of the U.S. Department of Labor provides a large listing of statistics that is broadly used in commerce. The most widely known statistic is the Consumer Price Index (CPI). The CPI provides monthly data on changes in the prices paid by urban consumers for a representative basket of goods and services. The Producer Price Index (PPI) measures the average change over time in the selling prices received by domestic producers for their output. For additional information, see www.bls.gov/ppi/home.htm.

PICTURE LESSON The Paris Exhibition of Automobiles

Bardon Automobile.

It was during the early years of automobile invention that many ingenious prototypes were created. There were no boundaries to the possibilities for the exhibitions. Engines, suspensions, fuels, and all sorts of Rube Goldberg adaptations were presented for these showcase extravaganzas.

The Paris Exhibition of 1902 featured several examples of fore-carriages, which were provided with motor and driving wheels, and were fitted under horse carriages doing away with horse traction. It was said that "the idea, while a very good one, showed that the carriage was unable to withstand the heavy strains set up with mechanical traction."

The engraving included here shows the Bardon model. Some of the automobiles used the fringe and openness of the horse-drawn carriage of that time. This model was driven by a spirit motor, with a measured fuel consumption that did not exceed 0.195 pint per ton mile. The Bardon was very successful in the European races of the time.

One automobile demonstrated a petroleum engine driving a dynamo, which then supplied the wheel motors with the necessary current to speeds up to 28 miles on the level.

Steam cars were not overlooked. All the essentials of the steam engine were constructed and fitted under the carriage. The steam engine developed 12 horsepower using superheated steam for a car seating eight persons. The boiler was carried in the rear of the car and when lamp petroleum was used, the consumption was 0.7 pint per mile. In large traction engines, the boiler was in the front and heating was by coke fuel.

productivity-measurement program. Each index represents only the change in output per man-hour for the designated industry or combination of industries. The indexes of output per man-hour are computed by dividing an output index by an index of aggregate labor hours. For an industry, the index measures changes in the relationship among output, employment, and labor-hours. The Bureau of Labor Statistics publishes monthly producer prices and price indexes and covers over 4000 product groupings.

Subjective evaluation or human intervention may produce a more refined forecast, particularly necessary when some event outside the usual run of economic activity inevitably has an economic consequence. In forecasting there are always dangers ahead. It is to this topic that we now turn.

5.5 CALCULATION, INTERPRETATION, AND UNINTENDED CONSEQUENCES

Most forecasting systems assume that historical facts are derived from some relatively stable historical process. Those data form the base for forward-looking images of the future. We have studied some well-established models for snooping ahead, but there are pitfalls such as the following that we now discuss:

- Randomness
- Causal effects
- Murphy's laws
- Individual reckoning

The mathematics presented in this chapter are digestible, but when we consider the prospects of many variables and equations, the opportunity for a tidy solution seems daunting. Computers, with spreadsheet and statistical software, make short work of massive computations, generating linear and nonlinear regression equations. For it was in the 1960s that the computer became common in engineering cost analysis.

How do we filter the mass of engineering and business information that is thrown at us daily? How can we understand what is important and separate the good information from the bogus? Inquisitive methodologies within the field of statistics need to be understood more broadly than developed so far.

The computer has without a doubt created many benefits for cost analysis, but unfortunately it has also created traps. Perhaps, the greatest danger is encountered at the outset when the source and type of data are being selected. Despite the excellence of computation, final results are dependent on the reliability of data used, the interpretation of the computations, and judgment as to reasonableness of the conclusion.

Have you noticed how pundits comment on one observed random outcome and interpret it as significant?[4] Even so a commonly stated belief is that most engineers cannot accurately understand probabilities and randomness. How does 95% differ from a 90% level of confidence, despite the perception that bigger is more reliable though having a wider zone of the interval? Human judgment rarely includes the experience of probabilities in everyday commerce and life.

[4]Read the business pages of any newspaper or magazine to observe this phenomenon.

The attempt to evaluate cause and effect is continually hampered by random occurrences and our responses to them. Although engineers gain much predictive information from their impressions, most matters of fact depend upon reasoning about causes and effects, even though engineers do not typically experience causal relations in the arena of engineering cost analysis.

What are causal relations? In order for x to be the cause of y, x and y must exist adjacent to each other in space and time, x must precede y, and x and y must invariably exist together. There is nothing more to the idea of causality than this; in particular, there is no magic and secret force that *causes* possesses and that it transfers to the *effect*.

Still, all judgments about causes and their effects are based upon experience. The engineer is able to grasp the experience vicariously, if you will, on the basis of analysis of past data, on the visualization of the graphs, and on the realization of the importance of correlation and the partial regression coefficients. There is a transfer of experience from that which cannot be personally felt to that which is objectively understood.

Sometimes data for a variable will include one or more values that appear unusually large or small and out of place when compared with the other data values. There is an old theory[5]—the problem of rejecting data due to suspected errors, those that have a large difference from the sample mean—that is available but is not discussed here. These values are called *outliers* and often are arbitrarily included or excluded in the data set. Strictly a mathematical rule, the criterion is unable to provide an interpretation of bad points and to distinguish if rejected points arise from spurious information or if the point(s) is in accord with likely random values. It is more important than the rule, however, that the engineer take steps to study outliers and carefully verify their inclusion status.

It was "Murphy" who first observed that if anything can possibly go wrong, it will go wrong.[6] Another sage said that if there is a possibility of several things going wrong, the one that will cause the most damage will be the one to go wrong, and yet another offered this corollary: If there is a worse time for something to go wrong, it will happen then. These "laws" are the reality of unintended consequences.

In forecasting the prices of mutual fund shares in the stock market, short time periods are volatile with much noise, while longer-period conjecture is stable. In the social sciences, forecasting is known as econometric modeling, which is vulnerable to the same predilections as stock prediction. We believe the opposite is true for engineering cost analysis, as short period estimates are steady, but longer-term ones are less so.

One of the untested beliefs of the authors is that in forecasting, if you intend to look ahead for the next year, the historical process should backcast no more than one year. Furthermore, if the plan is to consider a longer midrange period of three years ahead, then the database should be of equal length. In other words, use a backcasting history that is approximately equal to the forecasting period. Use short-term data for short-term forecasts, and long-term data for distant predictions.

[5]Chauvinet's rule.

[6]It was named after Captain Edward A. Murphy in 1949, an engineer working on an Air Force project designed to see how much sudden deceleration a person can stand in a crash. One day, after finding that a transducer was wired wrong, he said, "If there is any way to do it wrong, he'll find it." The contractor's project manager kept a list of "laws" and added this one, which he called Murphy's Law. Actually, what he did was take an old law that had been around for years in a more basic form and give it a name.

Equations of 150 variables or more about the general economy with all the accompanying statistical measures of reliability are known. But in engineering cost analysis, the usual number of variables is much less. Indeed, the fewer the variables that have cause-and-effect strength, the better the nature of the equation. Some say that five or so variables is a maximum when dealing with engineering cost analysis.[7]

While in this chapter we have stressed quantitative practice and statistical approaches to forecasting, it is naive to suggest that experience and judgment are not significant to success of the forecast. An engineer may decide to adjust a forecast that was made by traditional methods to take account of other unique conditions; she may, for example, decide that suppliers will alter their pricing patterns because of special circumstances

Any practical forecasting system needs to accommodate human judgment. A engineer may decide that the circumstances of the moment are unique and that a forecast produced by the usual statistical methods should be modified to consider special current circumstances. Material escalation, for example, is influenced by complicating factors such as political events, world currency fluctuation, and regional instability. These mathematical models may require the intervention of experienced opinion.

While some steps of the cost estimating process are mechanistic and conducive to computer automation, estimating is a predictive process for which judgment and experience add value. Effective cost estimating requires an understanding of the work being planned. Sanity testing, subjective evaluation, and producing a more refined forecast are learned skills. When using any method for forecasting, one must exercise a performance measure to assess the quality of the method, but ultimately the individual's sanity check is the most important.

SUMMARY

As is now evident, forecasting is centered around the analysis of costs and when coupled with estimating, takes imperfect information and adds the ingredients of judgment and knowledge about engineering designs to provide the setting for future costs.

The applications of graphical analysis, single and multiple linear regression, correlation, indexes, moving averages, and time series are the principal statistical techniques used in forecasting for the future. In forecasting information, it should be remembered that the processes mimic the situation in the future. Experience and engineering judgment are crucial contributions to the success of this effort.

QUESTIONS FOR DISCUSSION

1. Define the following terms:

Backcasting	Prediction limits
Best curve	Regression
Confidence limits	Relative frequency

[7]In a large undertaking for the U.S. Air Force, the Rand Corporation of Santa Monica, California, studied the cost of aircraft frames, and in the construction of many statistical models, they provided predictive equations usually having three or four variables, finding that number to be sufficient.

Correlation
Histogram
Indexes
Intercept
Least-squares criterion
Mean
Moving average
Multiple linear regression
Power function
Slope
Smoothing constant
Standard deviation
t table
Technology creep
Time series
Transformed units
Trend line

2. "Statistics never lie, yet liars use statistics" is a common statement. Discuss.

3. Why are graphical plots preferred initially over mathematical analysis of data?

4. Is cost estimating more concerned with empirical evidence or with theoretical data? Illustrate both.

5. Discuss what regression analysis is. What are its underlying assumptions?

6. What is minimized in a least-squares approach?

7. What is meant by correlation analysis? What does $r = 0$ or 1 imply?

8. Distinguish between correlation and causation. Can you have causation without correlation?

9. What is the purpose of a moving average? How does smoothing relate to a moving average?

10. If an engineer is confident of past data, then would the smoothing constant be large or small?

11. What are the differences between cycles and trend cycles?

12. Define a cost index.

13. Why would one use a cost index? What are the qualifications of a manufacturing index?

14. Three moving-average cost trend lines are assumed for a steel-laced fibrous material used for a hydrogen fuel cell employed in automotive technology and are shown as Figure Q14. Write a paragraph that evaluates the pros and cons of the proposed moving average for the three cases.

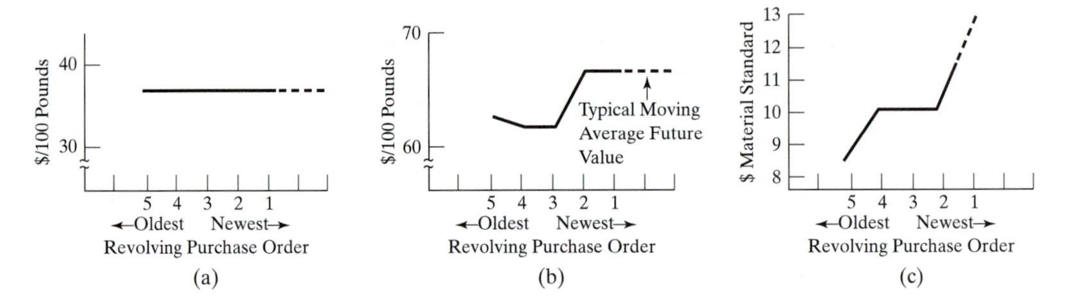

Figure Q14

PROBLEMS

5.1 A labor survey is conducted for area job shops, and the following wage groupings and their frequency of occurrence are found:

Sketch the frequency distribution, relative-frequency curve, and cumulative-frequency curve. (*Hints*: Notice that the wage-interval end numbers are open. Use

Wage, \$/hr	Number
<12	5
12 < 15	11
15 < 18	28
18 < 21	36
21 < 24	52
24 < 27	15
27 -	13

engineering computation paper for the graphs. A sketch is freehand and rendered with a pencil and straightedge.)

5.2 Time studies are collected for a handling element, and frequency of the time for this task are gathered into intervals and are as follows:

Time, minutes	Number of Observations
<0.04	2
0.04 < 0.08	12
0.08 < 0.12	14
0.12 < 0.16	15
0.16 < 0.20	14
0.20 < 0.24	13
0.24 < 0.28	12
0.28 -	6

Plot the frequency-distribution table, relative-frequency curve, and cumulative-frequency curve. Find the mean value of the time. What is the modal value? (*Hint*: Use engineering paper or a spreadsheet for the plotting.)

5.3 Find the mean, median, mode, range, variance, and standard deviation for one of the following sets, as selected by your instructor:

(a) $3, 5, 2, 6, 5, 9, 5, 1, 7, 6$
(b) $41.6, 38.7, 40.3, 39.5, 38.9$
(c) $2, -1, 0, 4, 6, 6, 8, 3, 2$

5.4 A regional market survey has determined the economic worth of the fabrication industry over a 10-year period.

Year	1	2	3	4	5	6	7	8	9	10
$\$10^6$	1467	1216	1360	1400	1518	1678	1818	2160	2290	2460

(a) Graph the linear line of the data on engineering computation paper by using the "one-half" rule. Then determine the graphical equation $y = a + bx$ of the line. Forecast the value in year 11 by using the curve and equation.
(b) Find the least-squares equation and forecast the value for year 11. How do the two equations found in (a) and (b) compare?

5.5 An index for the wage rate of assemblers has followed the following pattern:

Year	0	1	2	3	4	5	6	7	8	9
Index	100.0	104.4	107.8	112.3	117.2	122.5	132.8	145.6	159.3	171.6

(a) Plot a straight line of the data on arithmetic coordinates by using the "one-half" rule. Find the graphical equation $y = a + bx$ of the line. Forecast the value in year 10 by using the line and equation.
(b) Visually determine the year for which there is an apparent change in slope, and construct two new straight lines for the data, finding their graphical equations. Also forecast the value in year 10. Does there appear to be a variation in the data to warrant the dropping of earlier data?
(c) Find the regression equation for the best set of data.

5.6 A model of the form $y = ax^b$ is to be fitted to (x, y) data: $(1, 100)$, $(10, 10)$, and $(100, 0.1)$. Find the regression equations in transformed and original units. What are the values for $x = 150$? (*Hint*: "Transformed" implies logarithm conversion of the data. For this simple set of data, use your calculator. Notice that the data are in powers of 10, so use logarithms to the base 10 rather than the Naperian system for any calculation.)

5.7 Aerospace companies estimate the cost of the airplanes, major assemblies, and vendor materials by an important technique known as "learning." This function, which was discovered as an empirical truth in the 1930s, states that as production continues, the time or cost to produce one more unit declines at a constant rate between doubled units. The function that models this relationship is $T = KN^s$, where T = time or cost of a specific unit, and N = unit number. The intercept K and slope s are empirical quantities determined by regression calculations. Assume that a contractor determines data $(y = T, x = N)$ such as $(450, 15)$, $(325, 30)$, and $(200, 45)$. Determine the regression equation in original and transformed units. Find the estimate for the 60th unit by both methods.

5.8 The life of a lathe cutting tool is determined by standard tests upon a rotating bar stock, and two principal variables are cutting velocity in feet per minute (V) and tool wear, which is measured against a predetermined limit, and once the wear limit is reached, the minutes T are noted for a specific V. The famous tool life equation, as developed by Frederick W. Taylor, is $VT^n = K$, where n is the slope and K the intercept, which are statistically determined using regression methods. Assume that a simple set of information is $(V, T) =$ (1000 fpm, 1 min), (100, 10) and (10, 100). Notice as velocity reduces, tool life increases. Find the regression equation that models tool life. What velocity will give a tool life of 5 min?

5.9 (a) Find t_α if $\alpha = 0.10$ and $v = 10$ degrees of freedom.
(b) Find $t_\alpha = 0.05$ if $n = 20$. (*Hint*: Assume that the t_α is necessary for regression of two variables.)
(c) Compare t_α and Z (from Appendixes 1 and 2) for $\alpha = 0.05$ and $n = 120$.
(d) If $t_\alpha = 2.8$, then find α and n for $v = 16$.

5.10 (a) Find the mean and individual dependent value for the example on page 196 for year 16.
(b) Determine the variance of the mean value y and the individual y for $x = 16$.
(c) Determine the confidence and prediction limit for parts (a) and (b) for 95%.
(d) Repeat the above for $x = 17$.

5.11 An index for gross hourly labor wage is given for 6 years:

Year	0	1	2	3	4	5	6
Index number	100.0	106.0	111.1	117.2	121.3	125.3	128.0

(a) Graph the data as a linear line by using the "one-half rule."
(b) Compute linear trend values, find a least-squares line fitting the data, and construct its graph. Compare this to the "eyeball" fitted line.

(c) Predict the labor index for year 7 and compare it with the true value, 132.6. What is the range for the individual year 7 index by using a 95% prediction interval? (*Hint*: Use the most appropriate technology for the arithmetic.)

(d) The actual gross hourly wage rate for year 6 is \$27.15. What is the expected range for year 7. Use the following equation:

$$\text{Expected wage range} = \frac{27.15}{\text{year 6 expected index}} \times \text{year 7 range indexes}$$

5.12 An important product is cost sampled for significant components and over the years a product index is faithfully calculated. The following is the table of indexes.

Year	Product Index, y
0	95.1
1	97.7
2	98.4
3	100.0
4	101.1
5	102.2
6	103.5
7	104.9
8	106.6
9	109.7

(a) Determine the values of parameters a and b of the linear regression equation using the least-squares method.

(b) Find the 90% prediction interval for the year 10 value. Estimate the 95% confidence interval for the slope.

(c) The product cost for year 9 is \$17,500 per unit. Find the expected cost and the 90% upper and lower values of the cost for year 10.

(d) Determine the coefficient of correlation, r. What can you say about the strength of the prediction of the cost range?

5.13 Manufacturing time studies are conducted on a spot welding operation. Three operations are summarized. (*Hint*: Your instructor will assign a subset. Consider the application of spreadsheet software for the work.) Plot the raw data for a subset as a linear line on engineering computation paper. Find the regression equation and overplot the visual equation. How good was your personal eyeball plot?

Time Study	Number of Spot Welds	Spot Time, Min	Load Part, L + W + H, in.	Load Time, Min	Unload Part, L + W + H, in.	Unload Part Time, Min
1	8	0.28	29	0.08	37	0.10
2	3	0.13	46	0.10	46	0.17
3	9	0.34	101	0.27	106	0.28
4	14	0.87	60	0.21	60	0.32
5	36	2.06	53	0.19	53	0.51

(a) Find the regression equation for number of spots x versus spot time Y. Determine the 95% confidence limits for 15 spots.

(b) Find the regression equation for load part x versus load time Y. Determine the 95% confidence limits for $L + W + H = 50$ in.

(c) Find the regression equation for unload part x versus unload part time Y. Determine the 95% confidence limits for $L + W + H = 100$ in.

5.14 Product cost learning has been found to imitate the function $T_u = KN^s$ where T_u = unit time for the Nth unit, K = labor hours estimate for unit 1, and s = slope of the improvement rate. Transform this into a log relationship, and determine the log regression line. Also determine the equation in original units. Estimate the labor hours at $N = 50$ units in both log and original units.

(a)

N	Labor Hours, T
5	155
8	143
13	137
17	97
25	75

(b)

N	Labor Hours, T
7	210
13	160
21	142
26	128
31	121

5.15 Historical data of direct labor for assembly of a 12-kW four-cylinder diesel generator are plotted on arithmetic graph paper (see Figure P5.15). The engineer/designer has aggressively designed features into these generator packages to ease their manufacturing, thus allowing the reduction of direct labor.

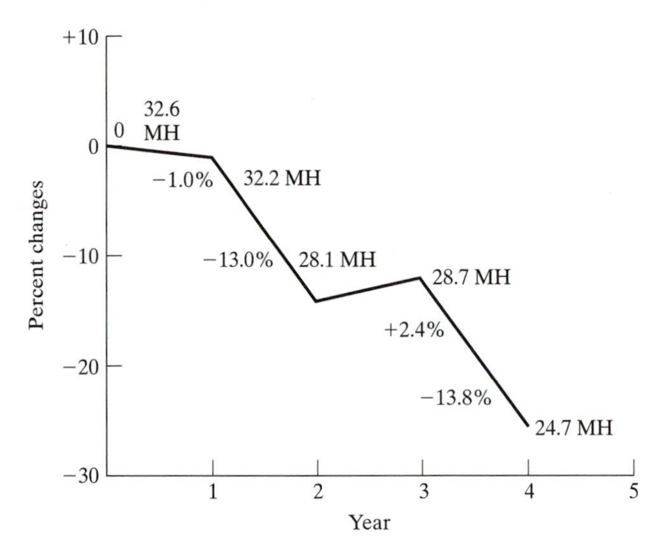

Figure P5.15 Evidence of labor learning for the 12-kW four cylinder diesel generator set.

(a) Replot these data on semilog paper. (*Hint*: Have the y-axis as the logarithm. The x-axis requires a zero location, and thus requires arithmetic scales.)

(b) Find the semilog regression equation. Use the model $T = Ks^x$, where T = labor hours, x = year, K = intercept, and s = slope

(c) What is the expected value for year 5 from your log plot and equation?

5.16 Technical catalogs are useful in many ways. For example, a catalog provides cost and horsepower information for enclosed capacitor motors. They are fan-cooled, 1725-RPM, having a 5/8-in. keyway shaft, when purchased in lot quantities. The listing gives costs on 1/4, 1/3, 3/4, and 1 hp of $67.50, $75.00, $88.50, and $105.00. Determine the cost equation by using the following:

(a) Semilog equation, $y = ab^x$
(b) Power equation, $y = ax^b$
(c) Which equation, using the test of $\Sigma(y - \hat{y})^2$, is best? (*Hint*: Use the original equation for the test.) What is the forecast cost of $1\frac{1}{4}$-hp motor?

5.17 A company that has been converting manual production methods to automatic equipment and robotic methods over the last seven years is interested in evaluating its effort. Is there correlation between the number of employees and shipment value? What is your assessment?

Contract Award, $\$10^6$	Number of Workforce Employees
573	1573
606	1550
648	1530
720	1550
765	1540
798	1550
848	1560

5.18 Analyze the following data by using multiple linear regression: The data are two dimensional, where weight and length are the independent variables. The response-dependent variable is cost/1000 units.

	Length		
Weight	1.50	1.75	2.00
33.3	2100	2390	2450
42.3	2950	3090	3120
52.6	3460	3660	4215
60.5	4540	4770	5070

5.19 Production of computer hard drive disks is in large volumes. For instance, one design and manufacturer produces about 4 million hard drives per quarter. Especially sensitive to the production rate is the automatic die casting and machining of the A383-aluminum alloy for the base of the disk drive. Higher production rates cause many problems, including blow holes, misplaced flashing, unstable dimensions, and warping. The following array is the scrap percent for the automatic die casting and machining line of the hard drive base.

Production, x (10^3 units per period)	1000	2000	3000	3500	4000	4500	5000
Scrap, y (% of production rate)	5.2	6.5	6.8	8.1	10.2	10.3	13.0

(a) Determine an exponential and a second degree polynomial equation.
(b) Which model gives the smallest error sum of squares? Which model is preferred for estimating future scrap losses?

5.20 Material costs are collected from historical purchase orders, and the information is arranged by period and quantity. The quantity represents the volume of units purchased during the period that is specified. The dependent variable is listed as dollars per unit. (*Hint*: During period −3, or three periods ago, and for the quantity of 2 units, for example, the cost per unit is $8.00.)

Cost per Unit

Period	Quantity 1	2	3
−3	10.00	8.00	7.00
−2	10.70	8.60	7.50
−1	11.10	8.95	7.85
0	11.25	9.00	7.90

(a) Use multiple linear regression to find the unit-cost equation y, where y = cost/unit, x_1 = period and x_2 = quantity. Use computer spreadsheet software to find the equation.
(b) If the quantity is 2 units, forecast the unit cost for the next future period 1.
(c) Discuss the implications of inflation and quantity purchasing upon costs. (*Hint*: Notice that a column has increasing cost as you go from the older −3 to the 0 period, which is due to inflation. A row has declining unit costs as the quantity of the order increases.)

5.21 A company uses an extensive amount of polymer for a coating operation. This material is a major part of the product cost. Determine a three-period total and moving average for the information given below. Also, using an exponential smoothing function, calculate the smoothed data for $\alpha = 0.33$. What is the expected cost for period 11? (*Hint*: Use trending approaches to find the forecast value for period 11 for the moving average and exponential smoothing method.)

Period	Unit Price	Period	Unit Price
0	$60.0	6	$65.2
1	63.2	7	65.8
2	64.2	8	66.4
3	64.3	9	67.5
4	64.4	10	70.0
5	65.0		

5.22 Commodity prices of important engineering raw materials are volatile, but their response, though difficult to forecast sometimes, becomes clearer if plotted.

(a) Plot the following raw data of a 40-lb aluminum ingot.
(b) By using a smoothing function with $\alpha = 0.1$ determine smoothed data and overplot the raw information. Discuss business cycles with the familiar function that the information resembles.

Period	Value	Period	Value
1	$7.50	11	$4.00
2	8.00	12	2.50
3	9.50	13	1.80
4	9.70	14	0.80
5	10.00	15	0.20
6	9.90	16	0.10
7	9.80	17	0.30
8	9.00	18	0.90
9	7.60	19	2.00
10	6.00	20	4.00

5.23 The purchase price of the 304-stainless steel alloy cover and damper component of a computer hard drive, which is an adhesive-joined sandwich of two flat 0.020 and 0.024 in. punch and formed components, is given as shown in the table.

Years Ago	10	9	8	7	6	5	4	3	2	1	Now
Cost (Cents/Unit)	23.2	24.1	26.3	25.7	26.8	27.2	28.0	27.8	28.0	28.5	28.3

The company utilizes off-shore manufacturing and employs forensic engineering of its own and competitor components in what is called "competitive analysis." Important to this analysis is the cost of its own and its competitor component. Because of standardization, many external design features are similar; for example, mounting hole location and sizes and thickness between the firm's and competitor's component.

Analyze these data using a time-series model. Make a table for a 3-year moving total and average price. (*Hint*: Make assumptions for the first two years of the moving average.) Plot the moving average price versus time and describe the movement. Discuss the practical factors that influence the price of components that have several million units per quarter and are outsourced to the Far East for manufacture. What is your recommendation for N? Can you discover any indications from the data?

5.24 A composite manufacturing labor-hour index is based on 1987 = 100 and includes 1969, 41.9; 1979, 48.0; 1989, 107.6; and 1999, 205.8. Convert the index to a 2009 basis = 100. (*Hint*: Assume the indexes are linear between years. An index given as 100 is also 1.00. This problem demonstrates that indexes can be adjusted to other years, if the composition and value of the index are acceptable. This is a subjective evaluation of the quality of the index.)

5.25 A firm contracts and buys its computer hard drives from East Asia and uses several independent suppliers having an extensive and qualified manufacturing base. The firm contracts for 40 million units per year. Engineering design and manufacturing liaison are conducted from the United States. Preliminary cost estimating is guided by a product index developed for computer hard drives and is given for two periods as follows:

Year	Index
2	1200
5	3108

A 80.43-GB 7200-RPM hard drive costs $127.45 for 100 units to this firm in year 3. What is the preliminary cost of a similar but slightly better performing drive, which resembles the contents of the product cost index, in year 6? (*Hint*: It is first necessary to find the average rate of increase.)

CHALLENGE PROBLEMS

5.26 A vessel is fabricated for a chemical plant and is a part of a project estimate. A pressure vessel is a cylindrical shell capped by two elliptical heads. The cost-design database begins with a standard vessel, which is fabricated in low carbon steel to resist an internal pressure of 50 psi having average nozzles, manways, supports, and size. If a new design is different than the standard, exceptions are considered using indexes.

Job shops that specialize in fabricating chemical plant pressure vessels have developed rules for accommodating off-standard designs, and a model that adjusts for special features for a nonstandard material, pressure, and size is given as

$$C_v = C_b \times \Pi f_i \times I$$

where C_v = cost of special pressure vessel, dollars

C_b = cost of standard pressure vessel at time of estimate, dollars

f_i = i factors that adjust from standard vessel to special design, number

I = inflation factor adjusting to future period of fabrication, number

This vessel differs from the standard in size (8 ft dia, height, 15 ft); shell material (stainless steel specification 316 plate); and operating pressure (100 psi).

The carbon steel material costs = \$240,000 for a standard design; factor for stainless steel material = 3.67; factor for 100 psi pressure = 1.06; and fabrication is 3 years from the time of the estimate and has a 5% material escalation per year. Determine the cost for a pressure vessel.

5.27 Raw data of industrial unemployment for machinist journeymen and apprentices are gathered for two separate years, as shown in the table.

Age Range	March, First Year	March, Second Year	Age Range	March, First Year	March, Second Year
14–15	20	26	34–35	38	47
16–17	87	93	36–37	53	42
18–19	636	709	38–39	36	85
20–21	206	191	40–44	89	30
22–23	202	50	45–49	101	97
24–25	81	229	50–54	86	107
26–27	15	37	55–59	111	67
28–29	13	29	60–64	117	173
30–31	19	73	65–69	144	180
32–33	25	83	70+	101	102

(a) Construct the frequency histogram, relative-frequency curve, and cumulative-frequency curve for both years. Describe their appearance. (*Hint*: Spreadsheet charting will be helpful.)

(b) Construct the frequency histogram using intervals 14–19, 20–24, 25–34, 35–44, 45–54, 55–64, and 65+. Describe their appearance. Comment about nonuniform-sized classes and any deliberate or unintentional statistical distortion. How would you present the data?

5.28 A study of past records of installed insulation cost (including labor, material, and equipment to install) for central steam-electric plant equipment (including turbine, boiler, pipes) revealed the data shown in the table.

Equipment Cost, $\$10^6$	Insulation Cost, $\$10^5$	Equipment Cost, $\$10^6$	Insulation Cost, $\$10^5$
5.5	3.5	14.1	9.2
10.7	5.5	14.8	9.3
34.1	28	15.5	14.1
2.0	1.4	15.3	13.8
6.0	6.4	21.3	15.0
1.5	2.1	34.0	15.8
8.1	7.2	24.1	9.8
10.1	6.4	26.0	15.8

(a) Plot a chart of this information on arithmetic coordinates and on log-log coordinates. Which is more suitable for cost-analysis purposes? Major equipment costs have been estimated as $3 million for a new project. What is the estimate for the installed insulation cost?

(b) Find the arithmetic regression equation and correlation. What is the insulation cost value for $3 million of equipment?

(c) Find the logarithmic equation and the value of insulation cost for $3 million of installed equipment.

5.29 Requirements for a firm's material are consolidated and purchase requests are issued for a quantity each period. The table below is a summary of the recent purchased cost history of a single material where −3 represents three periods ago and is the oldest record. Period 0 is the "now" time of a purchase request. The units of purchase are of small lots, 1, 2, and 3 units. For example, in period −2 (or two periods ago), we purchased 3 units, and the price was $7.50 per unit. The next future period is the integer 1. There is no information for values not shown.

	Quantity		
Period	1	2	3
−3	10.00		
−2			7.50
−1		8.95	
0			7.90

(a) Use multiple linear regression and find the equation where x_1 = period, x_2 = quantity, and y = material cost per unit.

(b) Forecast the unit cost for period 1 and a quantity of 2.

5.30 A manufacturer uses extensive amounts of A383-aluminum alloy ingot for its principal activity of die casting computer drive disk bases. Purchased material constitutes the major cost for its bids. Past data of a standardized amount of this aluminum ingot are as follows:

Period	Unit Price	Period	Unit Price
0	$60.00	11	$72.80
1	63.20	12	73.90
2	64.20	13	74.80
3	64.30	14	74.90
4	64.40	15	75.00
5	65.00	16	75.40
6	65.20	17	75.80
7	65.80	18	76.20
8	66.40	19	77.20
9	67.50	20	80.00
10	70.00		

(a) Determine a four-period total and moving average. Plot the unit price.
(b) Use an exponential smoothing function and determine the smoothed data for $\alpha = 0.25$. Find the price for period 25.

5.31 A U.S. firm will choose between two suppliers of a 1.3 megapixel optical lens system for a digital camera. Selection is based on low price for identical design and quality specifications. One supplier is based in the United States, while the second is located in France. While the final assembly site is in the United States, differences in shipping cost between the suppliers are insignificant and disregarded. Material cost is equal between the suppliers. Preference is based on labor value-added cost and analyses depend on production rates, efficiency indexes and indirect labor rates.

The direct-labor standard cost (labor only) is $99.65 per 100 units and is lower in France. In the United States the standard is $153.53 per 100 units. The efficiency index for production is 1.3 French man-hour = 1.0 U.S. man-hour. Indirect cost percentages are 120% France = 75% U.S. Find the unit cost of the optical lens system for the two countries. Which of the two suppliers should this firm buy from?

(*Hint*: Conversion of French euros to U.S. dollars is not required, as the cost of labor in France is in U.S. dollars. Use the model of direct-labor value-added cost = (direct cost of labor × efficiency + indirect cost of labor).)

5.32 Intellectual work may be estimated by indexing. Consider the following: An engineering software firm contracts work from major companies. For each job the software firm employs teams that consist of programmers (sometimes offshore employees), engineer developers, and scientists. Many software jobs are ongoing at the same time, and a typical team is analyzed as to their proportion of effort and salary. For instance, programmers constitute 70%

Skill	Index Weight, %	Period		
		0	1	2
Programmer	70	100.00	103.02	112.25
		32.50	33.48	36.48
Engineer Developer	20	100.00	105.85	117.38
		41.20	43.61	48.36
Scientist	10	100.00	104.00	113.18
		39.75	41.34	44.99
Total weight	100			
Weighted average index		100.00	103.68	113.37

of typical job effort, and the team's proportion percentage, hourly salary, and index are given. In period 1, for example, a programmer is paid $33.48 per hour, and the corresponding index is 103.02. (*Hint*: Each skill index relates to the benchmark period, or period 0.)

(a) Forecast the period 3 index for programmers only. (*Hint*: First find the average index for two years.) Repeat for the team. A job that cost $250,000 in period 2 is similar in effort to a potential job for period 3. Forecast its cost.
(b) Discuss: While the rates relate to the amount paid the skilled professional for 1 hour, the indexes do not consider the efficiency with which that hour is utilized in period 3.

PRACTICAL APPLICATION

Consumer information on prices has been collected in the United States since the 1880s by the Bureau of Labor Statistics. While there have been changes in the specifications and quality of the list of items below, the table gives an indication of the "people-economics" of U.S. society.

Item	1930	Recent Year
Gallon of milk	$0.50	$2.80
Loaf of bread	$0.08	$1.79
New auto	$530	$24,050
Gallon of gas	$0.10	$1.82
New home	$6,796	$188,080
Average income	$1,443	$30,836
Dow Jones Industrial Average	77.90	9,750

If we overlook the matter that specifications and quality of the items have changed dramatically, what ratios of the performance of income to food, transportation, and home can you construct that identify the factors of this "quality of life index"? What are the item values for this year? Are we better off economically today than in either 1930s or the 2000s? Discuss.

(*Hint*: A "hedonistic index" isolates the "improvement of quality" as independent of the cost of the item. For instance, a chemical engineering index measures the quality of a type of chemical plant. They remove the cost basis and the index considers only the quality improvements in the design of the plant over the years. Also, from the table above, while a gallon of gas has gone from $0.10 to $1.82, the gas is not the same. Aren't cars better today than in the 1930s? The same is true for the other commodities in the list.)

CASE STUDY: FORECASTING PRODUCTION QUANTITY AND BUDGET REQUIREMENTS FOR A GAS ENGINE

Motorcycles are realizing greater performance and popularity. An engine has the following specifications: 270 kg dry-weight bike having an engine bore × stroke of 100 × 72 mm, compression ratio of 11.3:1, displacement of 1130 cm^3, and an electronic sequential port-fuel system giving 47 mpg highway and 37 city. This revolutionary engine is seeing the increasing production demands shown in Table C5.1, which gives the historical monthly sales

Table C5.1 History of 1130 cm^3-Displacement Motorcycle Engine Sales

	Engine Sales, Units		
Month	3 years ago	2 years ago	Last year
January	4600	4500	6000
February	6200	7800	9100
March	7800	11100	12100
April	9900	12500	14500
May	12400	15400	17200
June	11800	13200	16100
July	10200	12200	14200
August	9600	11900	13900
September	7500	9300	11100
October	5100	8200	9600
November	3900	5100	6200
December	6000	9600	13300
Total	95,000	120,800	143,300

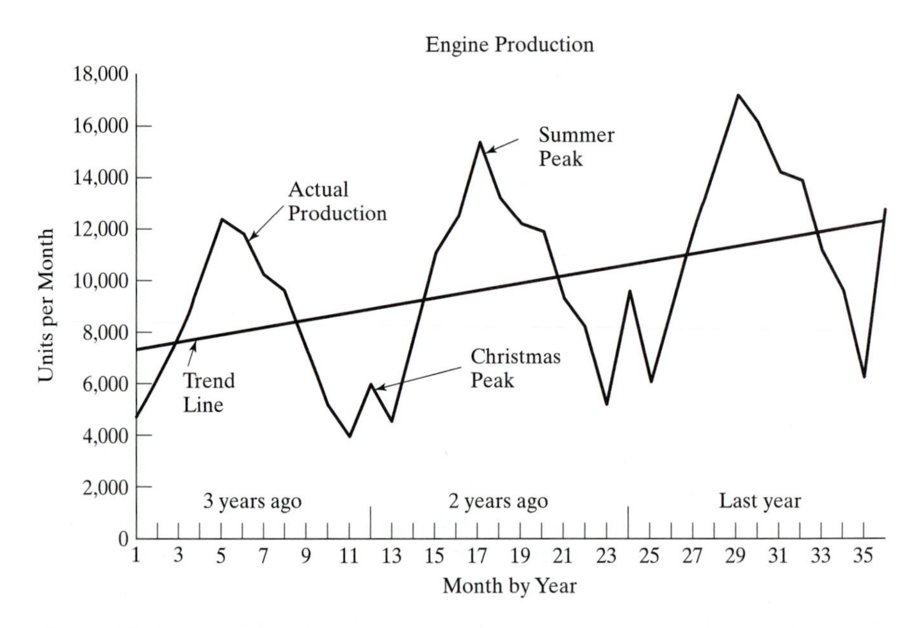

Figure C5.1 Actual and trend line for month-by-year sales quantities of gasoline engine.

of the engine for the past three years. The engine is purchased from an offshore supplier, and cost per unit is \$1380.79.

Figure C5.1 graphs the monthly rise and fall of product sales about an increasing linear demand. The high point for annual sales occurs during the summer months and to a lesser magnitude during December for the Christmas demand. The least-squares line for those data is monthly sales $= 7344 + 142.2X$, where X is the first month January 3 years ago. For instance, January 3 years ago has the trend value of 7486 units, while December of last year has 12,463 units.

The line does not give a satisfactory trend, as other market conditions are influencing the production requirement. Notice the deviations from the trend line and little reliability can be given to this projection for future production requirements.

Evaluate trial and error methods to find the first-quarter production for the upcoming year. Budget cash flow predictions for the first quarter for next year are necessary too. But first contrive a time-series model to forecast the quantity that will be necessary to meet production and sales demand. The objective is to forecast future production quantities and budget requirements for "next year, first quarter." The process used to forecast production quantities is heuristic, or trial and error using rules of thumb.

(*Hint*: Try developing quarterly quantities and see if their deviations from trend line values are meaningful. The deviations would be from the projected trend line. Spreadsheets can be utilized to do "what if" analysis.)

Chapter 6

Estimating Methods

Even though engineering designs differ greatly, estimating methods are remarkably similar, as we will see in this chapter, which introduces general estimating methods. These procedures are discussed and their advantages and shortcomings are noted. Methods range from experience and judgment to mathematics. Students need to grasp the importance of general methods for any engineering cost analysis or estimating.

6.1 ESTIMATING FOR THE ENTERPRISE

Estimates are called many things. Much lip service is given to a particular description of the estimate. The noun *estimate* will have an adjective modifier, such as preliminary, order-of-magnitude, contractor, owner, final, definitive—and the list is seemingly endless.

The nature of the estimate is summarized in the following way:

- Timeliness
- Purpose
- Accuracy
- Effort
- Accountability

A design is without specific form and shape in the early stages of its evolution. For instance, engineering may have progressed through problem definition, concepts, engineering models, and evaluation with the final design step remaining. The preliminary estimate is requested during the initial evaluation. With a lack of facts and specific information, engineering is asked to provide this first estimate. By using various methods, rules of thumb, and simple calculations, a quick and relatively inexpensive estimate is provided. Obviously, the accuracy of the estimate depends on the amount and quality of information and the time available to prepare the estimate.

The preliminary estimate may cause the firm to take some sort of action. An estimate, such as an operation cost, product price, or project bid can have serious financial implications. More frequent, however, is the case where the preliminary estimate screens designs and aids in the formulation of a budget. The preliminary estimate is used by engineering and management to commit or stall additional design effort, to appropriate requests for capital equipment, or to cull out uneconomic designs at an early point. Although decisions based on the preliminary estimate may not lead to legal obligations or authorization for capital spending, mistakes can be costly by eliminating potentially profitable designs. Jargon used in practice include conceptual, battery limit, schematic, green field, and imply preliminary estimate. Their purpose is to choose and eliminate unsound proposals without extensive engineering cost.

A preliminary estimate is an estimate made in the formative stages of design. Overlooked in this interpretation are the accuracy, type of design, nature of the organization, dollar amount, and the purpose for the estimate. A precondition of accuracy for preliminary estimates cannot be imposed, because special designs or objectives create a unique set of requirements. An estimate involving, say, pennies is no less a challenge than one involving millions of dollars, because many of the same methods are used at both ends of the dollar scale.

If early estimates lead to a continuation rather than a dismissal decision, then additional methods are required. Attention turns to methods that are more thorough in preparation and more accurate, as well as more costly in design and estimating. At this time the designer has filled out the details, and engineering constructs an estimate on enlarged quantities of verified information. In some cases the detail estimate is a reestimate of the preliminary one, because only limited updating needs to be done. Naturally, the effort for these estimates varies.

Detail estimating methods are more quantitative. Arbitrary and excessive judgmental factors are suppressed. While fudge factors are never eliminated, emphasis for the detail estimate shifts to comprehensiveness. Whether the detail estimating model is computer software, recapitulation columnar sheet, computational code, or whatever, the intent is the same: We follow formal rules in the detail estimate to find cost.

If, as often happens, we identify estimates on the basis of *purpose*, we find that they are made for the verification of a supplier's bid, or for a budget, or for project funding, etc. Estimates are used as evidence for price preparation or are advisory. The most common purpose for the estimate is to supply the best and most factual information to the pricing process. There is no better information for pricing than the estimate.

Some firms require that the tolerance of the estimate be ±5%, or ±10%, or +25% and −10%. They reason that *accuracy* is crucial to business practice. At the other extreme, some companies assign their estimates accuracy limits to a ROM (rough order of magnitude). Other businesses think of the estimate as a not-to-exceed maximum cost. Some managers view the estimate as a target to gauge whether the job is good or bad.

The work hours devoted to estimating the operation, product, or project design naturally vary with circumstances. The case for increased accuracy from a detail estimate is often made. Some practitioners claim that detail estimates are within 5% about a future actual value. Whether the particular value is 5% or 50% is not significant now. However, we assert that methods generally more accurate have an increase in cost of

PICTURE LESSON The 90-inch Gun Lathe

Turning, drilling, boring, and milling equipment were some of the first machine tools that were built during the Industrial Revolution. The engraving shown here is of a 90-in. gun lathe intended for turning steel ingots into ordnance. Ingots, those mammoth castings from the steel mills, would weigh between 60 to 200 tons. They were positioned by cranes between the centers of the huge lathe for fabrication into barrels for artillery application.

Lathes are designated by the diameter (90 in. for this lathe) of bar stock that is rotated between the driving chuck and the tailstock and the length of the bar stock, which for this lathe is 60 ft. The machine tools of the late nineteenth century did not have covered gearing, so all of the principal mechanisms were visible.

The cutting tools were generally made of high carbon steel that would be hammer forged while the bar stock was red hot. (High carbon steel is also used for piano wire, which is known for hardness and high tensile strength.) Tool wear and replacement of the cutting bit was frequent for gun turning applications.

preparation. Thus, data are purified, design has increased detail, and in actual estimating situations management stipulates that the estimate be within an interval about the future actual or standard value.

An estimate is an attempt to forecast the *actual* cost. It is an important principle to us that the accuracy of the estimate, when compared to actual cost, will be such that in half of the cases it will exceed the actual cost, and in the other half fall short of it. Contrariwise, some would argue that there needs to be a factor of safety in estimates, as there is in design, for example. The factor of safety would increase the costs to ensure that none are understated. We categorically rule against "factor of safety" estimates.

Because engineering tasks vary from the simple to the complex, and with time and resources always limited, it is important to have knowledge about a number of estimating methods. Universal methods are found in many types of applications and for the sake of versatility, students need to learn these methods

6.2 UNIVERSAL METHODS

There are practices that are so common that their description, let alone any thorough report of them, seems unnecessary. Yet a book with our title would be negligent if discussion were omitted. These universal methods can be listed as follows:

- Opinion
- Conference
- Comparison
- Unit

Perhaps these nonanalytical methods are common because of the ease of understanding.

6.2.1 Opinion

Personal opinion is inescapable in cost analysis and estimating. In the business of estimating, it is sometimes called "guesstimating." While it is easy to be critical of opinion methods for engineers who work with analytical tools, in the absence of information and with a shortage of time, there may be no other way to evaluate designs but to use opinion. The key to effective opinion analysis and estimating is "humanware."

An individual is selected for a job because of his or her expert experience, common sense, and knowledge about the designs. The mettle of the domain expert is tested in judging the economic worth of a design. People respected as "truth tellers" are sustained in this activity. Others are not.

Time, cost, or quantities about minor or major items are estimated with this inner experience. That the individual be objective in attempting to measure all future factors that affect the out-of-pocket cost is understood. Opinion estimating is also done collectively in conferences.

6.2.2 Conference

The conference method is a nonquantitative technique of estimation and provides a single-value estimate. The procedure, although having many rituals, involves representatives from various functions or skills who confer with engineering in a roundtable fashion. Various kinds of cost are estimated, ranging from engineering-work items to total cost for minor projects. Sometimes, labor and material are isolated and estimated, with overhead and profit being added later.

The conference method is managed in several ways, depending on the available knowledge. A conference moderator provides questions such as, "What is the labor and material cost for this design?" Various gimmicks sharpen judgment. A "hidden-card gambit" has each of the committee experts reveal a personal and written value to a question. This could provide a consensus. If agreement is not initially reached, discussion and persuasion are permitted as influencing factors. Sometimes this is called estimate-talk-estimate (E-T-E). The hidden-card idea prevents a brainstorming session, which generally gives optimistic estimates.

A ranking scheme along the "good-better-best" approach can be applied in a cost sense. We rank two or more designs by giving their single-cost value. Ranking seems to help, provided that the number of choices is not large. Sometimes this scheme is called *ordinal ranking*.

The lack of analysis and a trail of verifiable facts leading from the estimate to the governing situation are major drawbacks to the conference method. Although the procedure lacks rigor and there is little faith in its accuracy, these factors seldom deter its use.

6.2.3 Comparison

The comparison method attaches a formal logic to estimating. If we are confronted with an unsolvable or excessively difficult design and estimating problem, we designate it as design a and find another design for which an estimate can be found. The existing design is called b. This design might arise from a clever manipulation of the original design or an alteration of the technical constraints on a. In addition, by branching to b we gain information, as various facts may already be known about b. Indeed, the estimate may be in final form, or portions may exist and there need be only a minor restructuring of information to allow comparison.

The alternative design problem b must be selected to bound the original problem a in the following way:

$$C_a(D_a) \leq C_b(D_b) \tag{6.1}$$

where $C_{a,b}$ = value of the estimate for design a or b, dollars

$D_{a,b}$ = design a or b

Further, D_b must approach D_a as nearly as possible. We adopt the dollar value of our estimate C_a as something under C_b. The sense of the inequality in Equation (6.1) is for a conservative policy of estimating. It may be a management policy to estimate cost slightly higher at first, and once the detail estimate is completed with D_a thoroughly explored, we comfortably confirm that $C_a(D_a)$ is less than the original comparison estimate.

An additional lower bound is possible. Assume a similar circumstance for a known or nearly known design c, and a logic can be expanded to have

$$C_c(D_c) \leq C_a(D_a) \leq C_b(D_b) \tag{6.2}$$

Assume that designs b and c generally satisfy the technical requirements (but not the business estimate) as nearly as possible. The inescapable judgment requires that the unknown design be squeezed between two designs that are above and below the unknown cost. Comparison principles apply to any complexity of design, but consider a simple example.

Example: New Design for Ball-Bearing Product

The rolling speed of a ball bearing must be decreased relative to the rotational speed of the shaft. Among other factors the maximum operating speed of a ball bearing depends on its size and the rolling speed of the balls within the raceways. A preliminary design is to add an intermediate ring with raceways, a new patented technology, on its inner and outer peripheries for light radial loads. This version will have one less circular roll of balls as compared to a similar performance models. This would cut the relative rolling speed to approximately half that of the balls in an unmodified bearing. Figure 6.1 (a) indicates a possible arrangement design A. The adaptation of an additional outer roll of balls achieves many of the similar features required of design A and is known as design B. Velocities are reduced, and other technical advantages are achieved.

The design requirements for B satisfy most of A. The cost of B can be determined as the outer roll of balls in many ways is similar to single-roll ball-bearing technology. The technology for conventional double roll ball bearings [Figure 6.1(c)] has known costs and is our design C. With costs known for designs B and C, comparison becomes possible.

A key question must now be answered: Is design (a) functionally fixed between (b) and (c)? If this premise is accepted, then the cost of design A is between these two cost numbers.

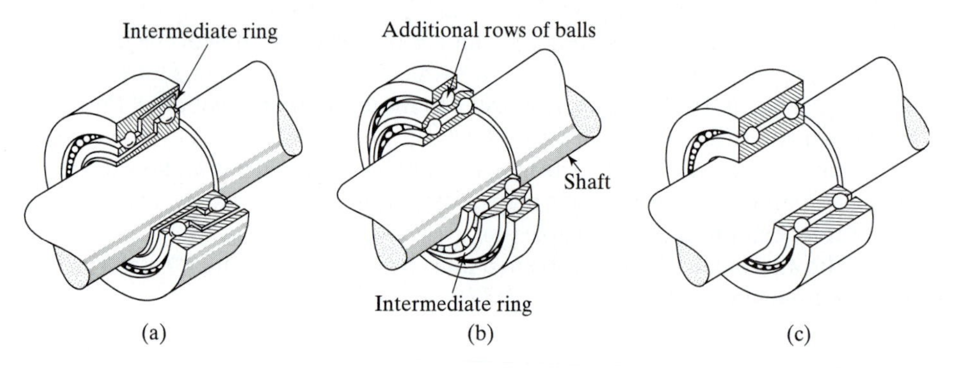

Figure 6.1 Comparison method: (a) unknown design cost, (b) higher-cost design, and (c) lower-cost design.

Precautions for this method need to understand that up-to-date costs, processing variables, similar production quantities, and spoilage rates are factors that call for insight in picking a specific value for design A between the B and C range. Other commonsense hints are helpful. The tighter the tolerance, the better the comparison.

Typically, engineering will avoid estimating from a zero base of design/cost information, and a system of *adds and deducts* begins with a selection of the boundary estimates. The lower or upper bound is the reference to either add or deduct costs for elements that are known.

The comparison logic is simple and useful. Indeed, we believe that all estimating is basically a comparison of the unknown to the fathomed. For the estimating process does not start without some knowledge of design and costs. The comparison method is sometimes called *similarity* or *analogy*.

6.2.4 Unit

The previous methods are qualitative and depend very little on facts. The unit method, on the other hand, uses historical and quantitative evidence and leads to a *cost driver* easily understood. The unit method is the most popular of all estimating methods. Other titles describe the same thing—average, order of magnitude, lump sum, function, parameter, module estimating—which involve various refinements. Extension of this method leads to the factor estimating method, discussed later.

Examples of unit estimates are found in many engineering activities—for example, machining cost per inch of length, aluminum casting cost per pound of weight, and chemical plant construction cost per barrel of oil capacity. Perhaps, the most common unit cost estimate for buildings is the cost-per-square-foot. Area is perceived to have a powerful effect on costs, and thus its popularity.

Though typically vague in those contexts, the strongest assumption necessary is that the design to be estimated is like the composition of the parameter used to determine the estimate. Notice that the estimate is "per" something. The unit estimate is defined as the mean, where the divisor is the principal cost driver, or

$$C_a = \frac{\Sigma C_i}{\Sigma n_i} \tag{6.3}$$

where C_a = average cost per unit of design, dollars

C_i = value of design i, dollars

n_i = design i unit (lb, in, ft, ft^3, count, etc.)

The unit cost is nothing more than the *average cost*. An average, like Equation (5.1), where $\bar{x}$ is found from a set of data is one simple way to find the statistic.

Example: Machining Steel Bar Stock on Lathe

Metal machining cost is often related to the linear length of the bar stock that is turned on a lathe. By using actual job tickets, the total cost for several jobs for a lathe operation are known. If we divide this total cost by the number of inches turned, we have a unit estimate. For instance, three designs of a steel bar stock are similar in all respects except that the length of turning various. Historical observations of total material and labor cost and length for turning a bar are as follows:

Design	C_i, \$	Turning length, in.
1	\$20	2
2	30	3
3	60	4
Total	\$110	9

The turning cost per inch of length is C_a = \$12.22 (= 110/9). The estimated cost for a different length of machining of bar stock is found by using its turned length and multiplying by \$12.22.

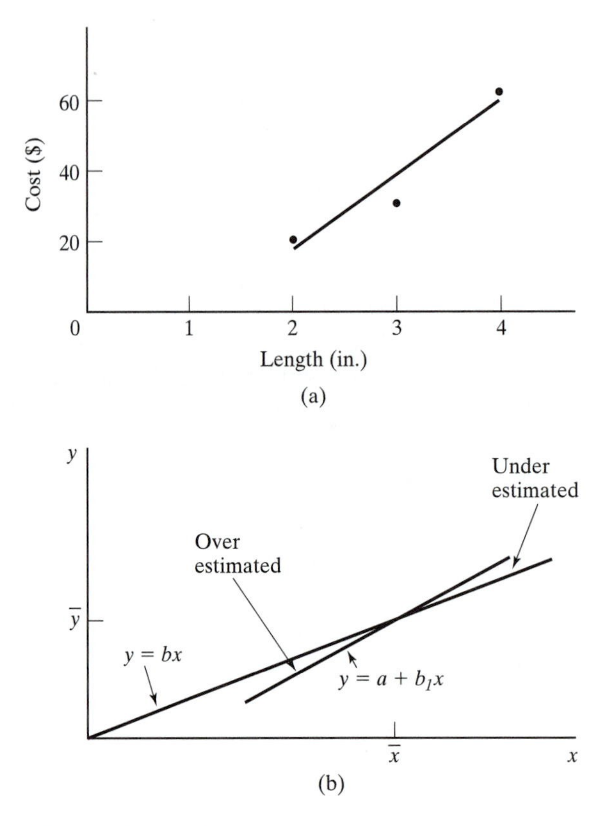

Figure 6.2 (a) Plot of three machining observations, and (b) unit model compared to $y = a + bx$.

The data are plotted on Figure 6.2(a). Effectively, the unit method is nothing more than the general slope b, where $a = 0$, or $y = 0 + bx$. The unit estimating method is improved when the function $y = a + bx$ is statistically fitted with the data. When using $C_a = 12.22$, the unit method either overestimates or underestimates the design parameter of length when compared to $y = a + b_1x$. Notice Figure 6.2(b), which shows the two simple methods compared. The only location of linear inch where there is no estimating error between the two methods is at $\bar{x}$ and $\bar{y}$ of the data, or average length = 3 in. and C_a = \$36.67.

Although the reliability of \$12.22/in. is improved by employing more observations, the unit method suffers from a more fundamental fault. If the data are plotted, and values of $y = a + bx$ are found by the regression methods of Chapter 5, then Cost = −23.33 + 20.00 (in.). Note that for these data the value of the intercept a is a negative fixed cost, an unlikely situation. Whenever $a < 0$ is found in estimating equations, we have a signal for faulty data, or the equation model is suspect. Notice Figure 6.2(a), where the three observations are plotted with the line. In Figure 6.2(b) the plot of the unit method when compared to the regression model results in an over- and underestimate error, except at the mean of the x and y data.

Further the unit method fails to apply the principle of economy of scale. For example, bar stock with a 2-in. machining-cut length costs \$24.44, while a 4-in. cut length bar stock costs \$48.88. At least a linear regression improves on the estimated value as \$16.67 and \$56.67, which is not simple doubling.

The unit method is used extensively. While an improvement over other qualitative methods, it is refined by other methods of estimating, which we now consider.

6.3 OPERATION METHODS

Engineering activities change material from one state to a second state. This requires that operations be planned and completed in precise ways. These technical operations call for a broad understanding, and experience is necessary in their conclusion. The fields of manufacturing and construction have developed these skills in extraordinary ways. These enterprises are the major contributors to the economic welfare of people.

Operators perform operations on materials with the help of equipment and tools. In other cases, equipment performs operations without labor. Two major methods are used to estimate operations:

- Cost- and time-estimating relationships
- Performance time data

These methods are analytical and are the foundation for estimating engineered products and projects.

6.3.1 Cost- and Time-Estimating Relationships

Cost-estimating relationships (CERs) and time-estimating relationships (TERs) are mathematical models or graphs that estimate cost or time. Simply, CERs and TERs are formulas that characterize the cost of an operation, product, or project as a function of one or more independent variables. These principles were discussed in Chapter 5, but forecasting deals with time-serial and minor analytical problems. Indeed, CERs and TERs are formulated to estimate significant end items. Rules of thumb and the unit method are not recognized as CERs. Also these models can be based on the physics or engineering of the process. All production processes are capable of CER and TER development.

Example: Parameters for Cost-Estimating Relationships

The word *parameter* is often used in the CER context. Usually it refers to the coefficients found in an equation—e.g., $C_a = -23.30 + 20.0$ in.—but it has a different meaning for the CER. After the development of a CER, and in the estimating phase, we substitute a new design value, what is loosely called the "parameter." We then calculate a cost. For example, the parameter of a bar stock with a 5 in. machined length, which we substitute in the equation, gives a cost of \$76.70. Sometimes CERs are called *parametric equations*.

These approaches are not new. We use output design variables—physical or performance parameters such as lb or tolerance—to predict cost, since the preliminary design parameters are available early. Sometimes those parameters are called *cost drivers*. CER estimates are considered preliminary.

Example: Arc Welding Power Costs

Welding is an example where the physics of the process can result in performance equations. Such parameters as arc travel speed, welding current and voltage, metal yield, wire speed rate, and flux consumption rate give models for welding processes. For the arc-welding process cost of power, we have

$$C_{pc} = \frac{(I \times V \times C_{ur}/1000) \times 12 \text{ in./ft}}{S \times 60 \text{ min / hr} \times M}$$

where C_{pc} = power cost, dollars per foot
I = welding current, amperes
V = welding voltage
C_{ur} = utility power cost, dollars per kWhr
S = travel speed of arc welding process, in. per minute
M = machine efficiency, percentage

With substitution of typical factors for a 200 amp, 20 volt arc welding process for a \$0.10 per kWhr, 10 in./min travel speed, and machine efficiency of 95% we have

$$C_{pc} = \frac{(200 \times 20 \times 0.10/1000) \times 12}{10 \times 60 \times 0.95} = \$0.01 \text{ / ft.}$$

As in all functional estimating models there must be a logical relationship of the variable to cost, statistical significance of the variables' contribution, and independence of the variables in the explanation of cost. This is sometimes referred to as *causality*, where the design factor contributes significantly to the cost.

CERs are developed by using a variety of steps. We suggest this approach: obtain actual costs, interview experts who have knowledge of cost and design, find cost-time drivers, plot data roughly and understand anomalies, replot and conclude regression analysis, review for accuracy and communication, publish, and distribute the CERs to engineers and cost analysts. Those steps constitute the basis for much of cost analysis, of course.

CERs are used in aerospace and for high-level engineering and marketing planning. In these environments, hardware and software costs are estimated based on weight, lines of code, functions to be performed, quantities of test hardware, heritage, complexity, etc. These cost predictors are used with confidence limits, as discussed in Chapter 5. Parametric estimating then uses aggregate costs as functions of high-level product characteristics or parameters. Such equations are helpful when detailed technical specifications are unavailable. They are useful for long-range planning as the following example illustrates.

Example: Famous CER

CERs have existed since the mid-twentieth century, and a famous series dealt with aircraft physical and performance characteristics. The equations were derived from historical data on 29 post–World War II military aircraft, which were constructed of aluminum. The cost data were obtained directly from ten airframe contractors. The research appropriately describes treatment of the data, fitting of the regression equations, and the selection of the preferred CER. The compilers indicate the limitations that should be observed in their use and provide a method of measuring the uncertainty in the predicted airframe cost. It should particularly be noted that the uncertainty is greatest when the model is used for applications outside of the original data—a good warning. For example, if the airframe material is titanium, then the aluminum sample is unsuitable. Furthermore, if the speed is greater than Mach 2.2, the application is suspect. The CERs were developed for a variety of cost applications, such as the following:

$$F = 0.001244A^{1.160}S^{1.371}Q_D^{1.281}$$

where F = flight test operations cost, 1970 base year dollars

A = weight without engines, fuel and fluids, pounds

S = maximum airspeed, nautical miles per second

Q_D = number of flight test airframes

coefficient of correlation = 0.97

coefficient of variation = 34%

If enumerated information is an option, then engineers will exploit performance time data methods, a topic studied next. CERs and TERs are useful in the development of performance time data.

6.3.2 Performance Time Data Algorithm

Time study, man-hour reports, and work sampling are some methods of work measurement that were introduced in Chapter 2. But time data in this raw form are unsuitable for estimating. Frequently, those data include bad methods, inefficient conditions, or untrained operators. Furthermore, those measurements may be restricted to a narrow range of the equipment's capability. Worst of all, if only time measurements from time studies are used, then they are unavailable for estimating new designs which have not started the production cycle. These deficiencies are overcome by developing performance time data (PTD) from the measurements, and a process that is thus available during the design period.

PTD is also known by names such as standard time data or estimating data. Indeed, performance time data are a catalog of manufacturing tasks and times that are used for estimating new designs, which have no actual history. A production plant, for instance, that has 30 production families—turning, milling, CNC, etc.—will have performance time data for each family. Parenthetically, performance time data are arranged in a systematic order and are used over and over. The advantages over direct observation methods are the lower cost of making estimates and greater consistency and improved accuracy in estimating new work. Description of the production methods is provided in advance of the need for the estimating data. More people are able to use performance time data, especially if it is simplified in presentation.

Performance time data are ordinarily determined from any of the methods of measuring work. In manufacturing, time study is the preferred source. But the starting place for raw data can spring from job tickets, historical data, equipment manufacturer's recommendation, similar processes, and even guesstimates. In recent decades with the decline of the time study practice, other means become necessary.

Starting with the measurements, engineering creates new information that is widely useful for estimating in manufacturing. The algorithm to determine performance time data is as follows:

1. Collect several or more measurements of the manufacturing operation.
2. Classify the elements of the measurement into common groups.
3. Regress the element time against one or more independent and causal variables, and find an equation of the element, such as a time-estimating relationship (TER).
4. Determine if each element TER is a constant number or variable according to conditional rules.
5. If the element TER is a constant number, find a typical time that represents the element.
6. If the element TER is variable, then convert the TER into a time table or leave as equation.
7. Collect all constant and variable elements for the operation, and print performance time data.
8. If the accuracy is acceptable, then use the information for estimating designs immediately after the design is drafted.

The performance data are distributed in either of two commensurable forms: equations or tables. Tables are preferred, despite the naturalness of computers with equations. A greater number of people are more at ease with printed tables.

Although the number of basic studies used for time data is a statistical question, accuracy is a function of the number of observations. A dozen or so time studies of a single manufacturing family are necessary before time data can be constructed.

Engineers use regression analysis to extend those raw time study data into a more digestible form. However, it is not the original time measurements that engineering ultimately desires. Rather, it is a set of engineering performance data, or standard time data, or more briefly standard data, that engineering uses for estimating.

Performance time data are now defined as time values arranged in a concise form from which a manufacturing time can be found that estimates the cost of an operation performed under usual conditions. The key word in this definition is "usual." It is unwise to devise performance time data without first knowing conditions, specifications, methods, and procedures. If those precautions are heeded, then the time values are based on the local conditions in which the standard is to apply.

Assuming that the basic measurements—for example, time studies—were rightly determined (see Chapter 2), one of the first steps is to determine which of the several elements of the measurements are constant or variable. Estimating data for a manufacturing process are composed of constant and variable elements. Because accuracy is necessary for estimating, variable elements provide sensitivity to the major drivers of an operation. Constant elements, while necessary for the absolute amount of the estimate, are nonresponsive to operation drivers but are required each and every time the operation is run. The dependent variable time, or $\hat{y}$ is related to an independent variable by a graph, a table, or regression formula.

A constant element is permitted because of convenience and assumed to have less than an arbitrary percentage P_1 slope between the limits of work scope, x_{min} to x_{max}. Very little work is ever constant, but to get products estimated efficiently we overlook some variability. If the time slope, say a positive 20%, from the minimum to the maximum is found, then the analyst classifies this element as a constant if the P_1 limit value is chosen to be 100%. In this case a constant value is found by averaging the values of the process element.

If the slope of the element is greater than 100%, then the element is classified as conditionally variable. A particular element may be one of many elements that define a set of performance time data for some production operation. Moreover, if the element is P_2 or less of the performance time of the total expected mean operation time, it could be classified as a constant irrespective of its elemental variability. If the element is greater than P_2, then it is classified as a variable element.

A TER element is judged "constant or variable" on the outcome of two tests. The technician will measure time study conditions that include extremes, minimum x_{min} to maximum x_{max}, of the work design. The two tests are as follows:

- *Test 1.* An element is *conditionally* variable if

$$\frac{\hat{y}_{max} - \hat{y}_{min}}{\hat{y}_{min}} \times 100 \geq P_1\% \tag{6.4}$$

where $\hat{y}_{max}$ = maximum dependent value related to time driver x_{max}

$\hat{y}_{min}$ = minimum dependent value related to time driver x_{min}

P_1 = percentage, arbitrarily fixed

The $\hat{y}$ values are preferred over actual raw y data because the least-squares time value minimizes observational errors. If Test 1 $< P_1$, then the element is constant; otherwise the second rule comes into play.

- *Test 2.* An element is variable if

$$\frac{\hat{y}_{ave}}{\hat{y}_t} \times 100 \geq P_2\% \tag{6.5}$$

where $\hat{y}_{ave}$ = average dependent value for x_{ave}

$\hat{y}_t$ = total dependent value for $\bar{x}$ of operation elements

P_2 = percentage, arbitrarily fixed

Test 1 is concerned with inner element variability, and Test 2 evaluates the effect of the element average to the total average for the operation. An element could be variable on the basis of rule 1, but it might have so little effect on the average time of the operation that it would be more practical to consider it constant and overlook small variability.

Both rules are subject to discretion by selection of P_1 and P_2 variability and a plant will have experience in the selection of P_1 and P_2.

Example: Constant or Variable Element?

Five different time measurements are conducted for a manual handling task where product weight and time varies for each measurement. Weight and time for the task are as follows (wt, min): (3, 0.20), (1, 0.15), (5, 0.40), (4, 0.30) and (7, 0.50). Including other tasks, the average total operation time for several tasks = 1 min, or y_t = 1 min. Is the task a variable or constant element? Let P_1 and P_2 = 100% and 10%.

The first step is to organize the data into a table, with weight recognized as the independent variable:

Weight, lb	1	3	4	5	7
Time, min	0.15	0.20	0.30	0.40	0.50

The next step is to find the regression equation, or $\hat{y} = 0.06 + 0.0625x$ where x = product weight, lb.

You can then check for Tests 1 and 2. This requires that $x_{min} = 1$ and $x_{max} = 7$ be substituted into the regression equation, and that Equation (6.4) is solved for $(y_{max} - y_{min})/y_{min} \times 100 = (0.4975 - 0.1225)/0.1225 = 306\%$, which is greater than P_1, so the element is conditionally variable.

The value y_{ave} is found substituting $x = 1$ and $x = 7$, and dividing by 2, which gives 0.1225 and 0.4975 for an average of 0.31 min. Next find the test value for Equation (6.5), and $y_{ave}/y_t = 31\%$, which is greater than P_2. Finally, the task is variable rather than constant.

After the TER time data elements are judged constant or variable, the engineer organizes the information into tables. Constant elements are combined, and an average time value is calculated. One or several elements will be found constant for the operation. An abbreviated description of the elements is written, together with the best value, and is prepared as printed or typed or computer entry. Obviously, constant values are easy to use, but they are insensitive in estimating cost. A trade-off between ease of estimating and sensitivity is necessary.

Each variable element is converted to a table or left as TER. What is adequate and accurate for one variable element could be too detailed for the next. The simplest method of tabulating the variable element is to divide the time driver into classes (i.e., small, medium, large) and then select a time value for each class. This can be improved by subdividing the $\hat{y}$ into several or many unequal cells, or equal percentage jumps, and finding the x value corresponding to each specific $\hat{y}$. Construction of a performance time table is by trial and error. Ease of understanding, speed in labor estimating, compression of tables into the fewest number, and accuracy are desirable features.

Example: Developing Performance Time Data for Spot Welding Process

Spot welding is a form of resistance welding in which two sheets of metal are held between copper electrodes. A welding cycle is started when electrodes contact the metal under pressure before electric current is applied, then a squeeze time under pressure with current, and finally a pressure dwell without current.

Five time studies have been summarized and tabulated below. Time is normal (i.e., observed time study elemental time that has been rated, see Chapter 2).

The time in units of minutes is associated with their causal variables. Among the many choices of an independent variable, girth, or length plus width plus height

$(= L + W + H)$, and the number of spots are selected for handling and welding. Five elements are identified. Girth for handling the major part is identified as x_1, which is the load-major-part time and includes moving part from skid to the electrodes.

In spot welding, a second and third part, if necessary, are assembled by C-clamping and welded to the first or major part. The secondary part and third part time includes moving the part from skid and positioning on major part with clamps and fixtures, and the variables are identified as x_2 and x_3.

Spot-welding time includes position the clamped and assembled part for spot weld, spot weld, and occasional cleaning of the electrodes. The number of spots is specified as x_4.

The unload time starts after the last spot and includes removing fixtures and clamps and moving the welded part onto the skid. The size for the welded assembly is labeled as x_5.

Table 6.1 is a listing of time studies with time drivers, $x_1, \ldots, x_5$ identified for each of the five elements. For element 1, the time driver x_1 is the size of the major part girth. Then x_2, x_3, and x_5 are the values of the girth $L + W + H$, which are identified as the causal variables for elements 1, 2, 3, and 5. The number of spots is identified as the causal variable for element 4, or x_4.

There are five regression equations, which are normal time without allowances, at this point. For example, y_1 is the load time for the major part size, and y_2, y_3 and y_5 are the times for handling of the major part, second and third part, and unloading of the spot-welded assembly. Those original time values within the time studies are not shown.

Instead, the elements have their minimum and maximum x's and y's shown, along with the minimum and maximum ranges. The values x_{min} and x_{max} are determined from the time study, and corresponding y_{min} and y_{max} values are found by using the equations (see Table 6.2).

Once the data have been classified into the extremes for the operation, we move on to finding whether the work task is a constant or variable type. Remember the constant

TABLE 6.1 Element Regression Equations from Job Measurements

Element	Parameter	Regression Equation
1. Load major part	Major part girth, x_1	$0.0139 + 0.0027x_1$
2. Load second part	Second part girth, x_2	$-0.1282 + 0.0216x_2$
3. Load third part, if necessary	Third part girth, x_3	$0.0642 + 0.0133x_3$
4. Spot weld assembly	Number of spots, x_4	$-0.1156 + 0.0608x_4$
5. Unload	Final assembly girth, x_5	$0.1907 + 0.0014x_5$

TABLE 6.2 Ranges of Independent and Dependent Variables for the Five Elements

Element Work Task	x_{min}	x_{ave}	x_{max}	$\hat{y}_{min}$	$\hat{y}_{ave}$	$\hat{y}_{max}$
1	29	65	101	0.092	0.189	0.287
2	8	18	28	0.045	0.261	0.477
3	5	24.5	44	0.131	0.390	0.649
4	3	19.5	36	0.067	1.070	2.073
5	37	71.5	106	0.243	0.291	0.339
					$\hat{y}_t = 2.201$	

element is important to identify because it reduces the difficulty of estimating, and variable elements give the sensitivity to the estimate.

The information of Table 6.2 is evaluated according to Equations (6.4) and (6.5), and work tasks 1, 2, and 5 are classified as constant, while work tasks 3 and 4 are variable. For this selection, $P_1 = 100\%$ and $P_2 = 10\%$ are plant practice (see Table 6.3).

Once the selection is complete, the next step consolidates the information into simple tables that permit wide usage by interested people. The equations in Table 6.1 represented normal minutes, and need to be increased for allowances, such as personal, fatigue and delay effects (see Chapter 2) where the values are increased by 15% using Equation (2.1).

The constant elements are gathered together and are summarized in Table 6.4.

The 0.872 min is posted on the performance time estimating sheet (Table 6.5) and all details that were used to measure and calculate the numbers are stored in the *back up* development. The procedures to develop the PTD are retained. Only a summary is finally displayed.

Variable elements 3 and 4 provide the sensitivity to estimating the operation. They could be left as regression equations, but they can be converted to tabular form. In Table 6.5 the two work tasks are tabulated into several steps between the limits of the time driver. Pairs of the time driver and the dependent time could more or less be shown, but the choice is a practical one. A setup time, expressed in hours, is also given.

Table 6.5 is the final summary of the performance time data that are used to estimate any spot welding operation. It is printed and distributed or made available on computer links to engineering.

If a new-to-be estimated design gives a third part girth and number of spots that falls between two listed values, the practical rule of thumb is to use the higher value for the estimate. Once the performance time data are prepared and distributed, it is used in a variety of ways, as the following example illustrates.

TABLE 6.3 Selection of Elements into Either Variable or Constant Type

Work Task	Test 1		Test 2	
	$\frac{\hat{y}_{max} - \hat{y}_{min}}{\hat{y}_{min}} \times 100\ (\%)$	Type	$\frac{\hat{y}_{ave}}{\hat{y}_t} \times 100(\%)$	Type
1	212	Variable	9	Constant
2	960	Variable	12	Constant
3	396	Variable	18	Variable
4	2995	Variable	49	Variable
5	40	Constant	—	—

TABLE 6.4 Summary of the Consolidation of Constant Elements

Constant Elements	Normal Average Time (min)
1. Load major part	0.189
2. Load second part	0.261
5. Unload welded assembly	0.291
Total	0.741
Standard time with allowances	0.872

TABLE 6.5 Illustrative Performance Time Database for Estimating Spot Welding Operations

Spot Weld Estimating Data

Setup	1.2 hr
Operation work elements	
Load major part, load secondary part and clamp	0.87 min
Load third part and clamp, if required	

L + W + H	Time (min)	L + W + H	Time (min)
5.0	0.16	24.6	0.45
9.4	0.22	37.5	0.65
15.6	0.32	56.2	0.94
19.7	0.38		

Spot weld

Number Spots	Time (min)	Number of Spots	Time (min)	Number of Spots	Time (min)
3	0.08	9	0.50	21	1.35
5	0.22	12	0.71	28	1.85
7	0.36	16	1.00	39	2.62

Example: Estimating a Spot Welded Assembly

An operation of spot welding three separate parts is to be estimated for lot hours. A manufacturing engineer will study the design and determine the number of spot welds that the design engineer is requiring, which for this example is 26 spots for a formed, cold-rolled steel chassis. Additionally, the estimate database requires that girth be determined from the design, which corresponds to the time driver for the estimating data. The girth for the third part in the assembly is found to have $L + W + H = 27.2$ in. The operation process sheet indicates that 175 units will be spot welded in a batch production mode. The estimate requires that setup hours, cycle minutes, and lot hours be found.

The approach is to use Table 6.5 and to follow the rules for using tabular estimating data. While the equations could also be used, for familiarity and clearness, many companies depend on tables because of their simplicity. "Lookup" rules are adopted to increase time-to-estimate speed, consistency, and accuracy in supplying the estimate. The lookup rule of thumb applies to the use of tables, and says that if a driving variable falls between two entry choices, use the higher entry value, as that is more conservative, which gives a greater time for the element (see Table 6.6).

The time for the handling and clamping of the third part, having 27.2 in. $= L + W + H$ in. of girth, falls between two time values in Table 6.5 (24.6 and 37.5). The higher time value is selected by using the lookup rule of thumb. Similarly, the number of spots required for this design falls between two entries, and the higher value is used. For a lot of 175 units, it is necessary to have a setup of 1.2 hours and then to begin the cycle production. The time per unit is 3.37 min, and total lot hours $= 1.2 + 175(3.37/60) = 11.03$ hours. These estimates are before the actual production. As will be seen in Chapter 7, cost extensions are a simple matter, and performance time data are important in this planning.

TABLE 6.6 Estimating Production Operation Using Estimating Data of Table 6.5 and Rules for Tabular Data

Table Description	Rule	Time
Setup	Constant value	1.2 hr
Load major part, load secondary part and clamp	Constant value	0.87 min
Load third part and clamp	Variable element, using next higher value of 37.5 = $L + W + H$	0.65 min
Spot weld	Variable element, using next higher time for 28 spots	1.85 min
	Total cycle time	3.37 min
	Hours/100 units	5.617
	Units/hr	17.8
	Lot hours = 1.2 + 175(3.37/60)	11.03

The illustrative PTD development for spot welding is typical for all processes in manufacturing. If the plant has CNC, drilling, chemical milling, etc., each of these process families will have a PTD development similar to the spot welding. Performance time data are in advance of the product being produced in a manufacturing operation. Thus the engineering requirements for a new design can be estimated even before the product enters production.

6.4 PRODUCT METHODS

Simply defined, a product is something produced. Operations make changes to materials within a for-profit enterprise, and when linked to one or more components, become a product. The complexities of product design and manufacture are more than can be covered in this section, and the student will need continuing study of later chapters. Product methods include special tools such as the following:

- Learning
- Operations planning
- Bill of material
- Methods of profit and loss, balance sheet, etc. for the product

Only the renown learning CER is discussed now. We find it more instructive to leave other techniques to later chapters.

6.4.1 Learning

It is recognized that manufacturing repetition with similar products reduces the time or cost required for the product. This phenomenon can be modeled with ordinary estimating techniques. The improved performance is called *learning*. Knowing how much the time or cost can be lowered, and at what point learning is applied in estimating procedures, are reasons for studying learning.

The first applications of learning were in airplane manufacture, which found that the average number of man-hours spent in building an airplane declined at a constant rate over a wide range of production.[1] Though the original airplane application stressed only time, practice has extended the concept to cost.

The learning model rests on the following assumptions:

- As production continues, the time or cost required to complete a unit of manufacturing is less each time.
- Unit time or cost decreases at a decreasing rate.
- Reduction in unit time or cost follows a specific estimating model, such as $y = ax^b$.

To state the underlying hypothesis, the man-hours or cost necessary to complete a unit of manufacturing will decrease by a constant percentage each time the manufacturing quantity is doubled. Assume that a rate of improvement is 20% between doubled quantities. This establishes an 80% learning rate, which means that the man-hours to build the second unit will be the product of 0.80 times that required for the first. The fourth unit (doubling 2) will require 0.80 times the man-hours for the second, the eighth unit (doubling 4) will require 0.80 times the fourth, and so forth. The rate of improvement (20% in this case) is constant with regard to doubled manufacturing quantities, but the absolute reduction between amounts is less. Note Figure 6.3, which generalizes for the doubling concept, where the known information is an 80% learning experience and actual value, called *a*.

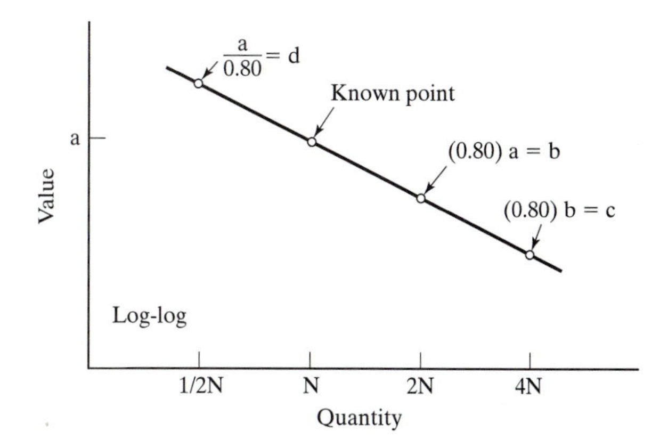

Figure 6.3 Graphical construction of learning with historical value *a* and 80% learning.

[1] T. P. Wright, "Factors Affecting the Cost of Airplanes," *Journal of Aeronautical Sciences* 3, no. 4 (February 1936).

The notion of constant reduction of time or cost between doubled quantities is defined by a *unit* formula such as the following:

$$T_u = KN^s \tag{6.6}$$

where T_u = product man-hours or cost per unit required to manufacture the Nth unit

N = unit number

K = constant, or estimate, for unit 1 in cost or time dimension compatible with T_u

s = slope or function of the improvement rate, decimal

Learning slope s is negative because time or cost decreases with increasing units of manufacturing. T_u plots as a curved line on arithmetic coordinates, or as a straight line on logarithmic coordinates.

Notice Figure 6.4 for T_u. Notice that (a) is Cartesian grid with a "zero" value, and the lines are nonlinear on the plot. In (b) the principal line is T_u, which is linear, and the other two lines, T_c and T_a, are derived from T_u. In (b) the line for T_a curves over T_u slightly until about 20 units, at which point they T_a and T_u are parallel. A logarithmic graph will not have a zero value on either the abscissa or ordinate. In practice both types of graphs are found, and one needs to be aware of the differences.

Reviewing algebra, we transform both sides of Equation (6.6) with a logarithm, thus obtaining

$$\log T_u = \log K + s \log N \tag{6.7}$$

which is of the form $y = a + bx$, the equation of a straight line. Recalling the ideas of transforming the variables, a topic discussed in Chapter 5, then $y = \log T_u$, $a = \log K$,

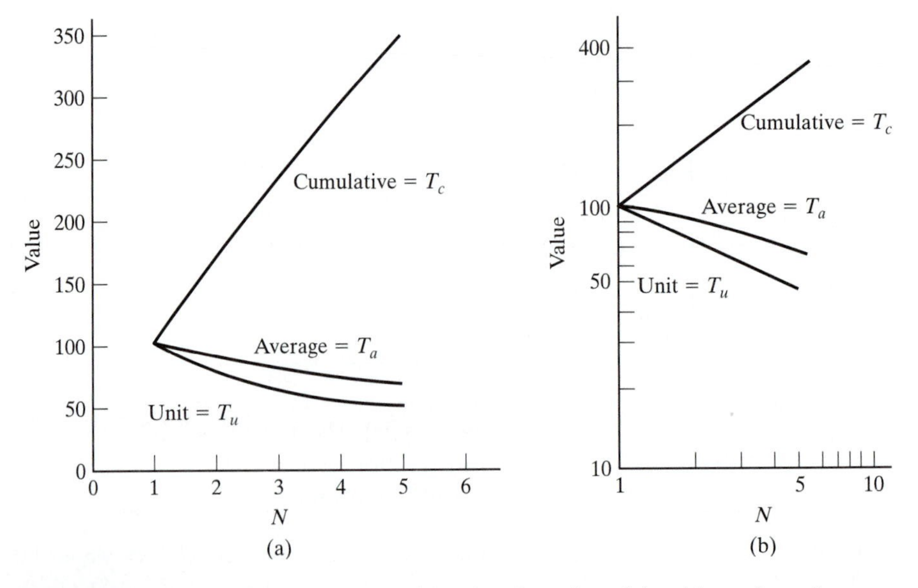

Figure 6.4 Illustrative plots of the learning function: (a) arithmetic and (b) logarithmic.

$b = s$, and $x = \log N$. From this point, we are able to statistically determine the parameters of the equation, K and s. The student is encouraged to review page 202.

To understand the presentation of learning on logarithmic graph paper, first compare the characteristics of arithmetic and logarithmic graph paper. On arithmetic graph paper, equal numerical differences are represented by equal distances. For example, the linear distance between 1 and 3 will be the same as between 8 and 10. On logarithmic graph paper, the linear distance between any two quantities is dependent on the ratio of those two quantities. Two pairs of quantities having the same ratio will be equally spaced along the same axis. For example, the distance from 2 to 4 will be the same as from 30 to 60 or from 1000 to 2000.

Learning is usually plotted on double logarithmic (log-log) paper, meaning that both the abscissa and the ordinate use a logarithmic scale. The exponential function $T_u = KN^s$ plots in a straight line on log-log paper. This function can be plotted from either two points or one point and the slope (e.g., unit number 1 and the percentage improvement). Also, by using log-log paper, the values for a large quantity of units can be presented on one graph, and values can be read relatively easily using the graph. Arithmetic graph paper, on the other hand, requires many values to sketch in the function. Refer to Figure 6.4(a).

Define Φ as the decimal ratio of time or cost per unit required for doubled manufacturing, and then

$$\log \varphi = s \log 2 \tag{6.8}$$

$$\text{and } s = \frac{\log \varphi}{\log 2} \tag{6.9}$$

This relationship is tabulated in Table 6.7 for typical learning rates.

There is a need to extend the unit formulation Equation (6.6) to other models. The cumulative time or cost that a product will require is found using

$$T_c = T_1 + T_2 + \cdots + T_N = \sum_{u=1}^{N} T_u \tag{6.10}$$

where T_c = cumulative total cost or time from unit 1 to N

Look at Figure 6.4, where T_c is shown. Additionally, the average time or cost, T_a, for a unit N can be found by consecutively dividing the cumulative value by the quantity and is found as

$$T_a = \frac{\sum_{u=1}^{N} T_u}{N} = \frac{T_c}{N} \tag{6.11}$$

where T_a = average unit time or cost from unit 1 to N

TABLE 6.7 Decimal Learning Slopes for Ratio Φ

Φ	1.0 (no learning)	0.95	0.90	0.85	0.80	0.75	0.70	0.65
Exponent, s	0	−0.074	−0.152	−0.234	−0.322	−0.415	−0.515	−0.621

TABLE 6.8 Illustrative Factor Table for Learning with ϕ = 80%

N	Learning Table ϕ = 80%		
	T_u	T_c	T_a
1	1.0000	1.0000	1.0000
2	0.8000	1.8000	0.9000
3	0.7021	2.5021	0.8340
4	0.6400	3.1421	0.7855
5	0.5956	3.7377	0.7475

Notice Figure 6.4, where T_a is positioned above the unit line. Observe that T_u, T_c, and T_a join together at $N = 1$. An approximation to Equation (6.11) is given by the following:

$$T_a \cong \frac{1}{(1 + s)} KN^s \tag{6.12}$$

where the approximation improves after 20 or so units. The T_u and T_a lines become parallel at that point. Prior to $N < 20$, T_a is joined to T_u at $N = 1$ and it remains above T_u.

In addition to calculations using the formulas or with graphs, tables are frequently employed because of their simplicity. For instance, an abbreviated table for $\phi = 80\%$ and $N = 1, \ldots, 5$, would be found using Equations (6.6, 6.10, and 6.11) as shown in Table 6.8.

The values of T_u are first found and then T_c and T_a are determined. Appendix 1 gives typical tables for 80% and 90%.

Example: Learning for Welding Assembly

An engineering prototype identified as "welding assembly" results in 1826 actual direct labor hours at the conclusion of the development project and $N = 1$. Factual learning experiences have statistically verified a learning rate of 80% from other similar product development. Estimate the direct labor for unit, cumulative, and average hours for units 2, 3, and 4.

N	Direct Labor Hours for Welding Assembly		
	Unit Hours	Cumulative Hours	Average Hours
1	1826	1826	1826
2	1461	3287	1643
3	1282	4569	1528
4	1169	6247	1434

The 1826 hours are multiplied by the factors corresponding to $\phi = 80\%$, N, T_u, T_c, and T_a. Similarly, it is possible to find the values using equations or graphs. These values, say for $N = 2$, 3, and 4, become estimates and are useful in various ways.

If only two actual product experiences are known, then it becomes possible to find the slope using

$$s = \frac{\log T_i - \log T_j}{\log N_i - \log N_j} \tag{6.13}$$

where s = slope of learning function, decimal

T = time or cost, units of time or dollars

N = unit number, $N_j > N_i$

Example: Audit of Two Production Units

A company audits two production units (the 20th and 40th) and finds that 700 and 635 hours are consumed, respectively. Now at the 79th unit, they want to estimate the time for the 80th unit. Then

$$s = \frac{\log 700 - \log 635}{\log 20 - \log 40}$$

$$= \frac{2.8451 - 2.8028}{1.3010 - 1.6021} = -0.1406$$

$$\text{and } \log \phi = s \log 2$$

$$= (-0.1406)(0.3010) = 9.9577 - 10$$

Taking the antilog of both sides gives $\phi = 0.907$. The percentage learning ratio is 90.7%. By using the information for $N = 20^{\text{th}}$ unit, we obtain

$$700 = K(20)^{-0.1406}$$

$$\log 700 = \log K - 0.1406 \log 20$$

$$\log K = 2.8451 + 0.1406(1.3010) = 3.0279$$

$$\text{and antilog } K = 1066 \text{ hours.}$$

The learning theory function is $T_u = 1066N^{-0.1406}$. The unit time for any unit can be calculated directly, and for the 80th unit, $T_{80} = 1066(80)^{-0.1406} = 576$ hours.

An identical answer could be found using the doubling notion, or for $T_{80} = T_{40} \times 0.907 = 576$ hours. For this example, only two data points were available. If three or more historical points had been available, regression methods to find the slope s would be used.

Learning theory has been successfully applied to gas-fired power plants, nuclear power plants, ships, airplane and electronic product developments, large scale products and many other projects. Its best application arena is in the estimating of large-scale projects. Small jobs, with little or no engineering design improvement and follow-on and planning, are unsuitable.

How does learning happen? Even if the actual value for the first several units increase as compared to the estimates, management needs to insist that the learning philosophy prevail. This is so, as there is a perception that management persuasion assists the learning phenomena. The process of learning does not happen without engineering effort and manufacturing planning. There is no serendipity in the learning process, and *wishing so* does not make it happen. The engineering enterprise needs to understand that learning will happen if there are engineering time/cost reduction activities that make it happen.

The phenomena of learning results from the consequences of better planning, improved productivity, and competition, and these effects are driven by engineering design and manufacturing management. We cannot separate these driving factors for any day-to-day estimating. Learning is best practiced as an overall preliminary method.

Practitioners will have general knowledge of slopes s and learning rates ϕ. In some cases, this information is openly published and available. In the estimating process, we estimate the first few units, and then the remaining units are calculated using the learning percentage. The slope and learning-rate factors are transferred from actual situations. The first unit K is identified by any of several estimating techniques.

Furthermore, the slope percentage ϕ is crucial to the sanity of estimating with the learning equations. The slope percentage is important—i.e., while 75% learning is desirable as compared to 90%, it may be impossible or uneconomic. Projects that are routine may be unsuitable, especially if engineering design and manufacturing management are not consciously driven to "improve."

6.5 PROJECT METHODS

Project methods used for estimating are different than those used for operations and products. But certainly estimates about operation and product costs can be information to project cost. Methods that are studied here include the following:

- Power law and sizing CERs
- Cost estimating relationships
- Factor

A project design, whether equipment, plant, capital tooling for a product, or a prototype of some kind, is a single end item. A product has quantity involved, but not so for a project. The design is custom, and there will be only one manufactured or constructed. Project estimating is for singular goods rather than plural goods. Usually, the dollar amount is considered "capital," or large, rather than an expense or minor.

6.5.1 Power-Law-and-Sizing CERs

The *power-law-and-sizing CER* is used for estimating process and plant equipment. This CER models designs that vary in size but are similar in type. The unknown costs of a 2000-gallon kettle can be estimated from data for a 1000-gallon kettle, provided that they are of similar design. We would not expect the 2000-gallon kettle to be twice as costly as the smaller one. The *law of economy of scale* ensures this. In general, costs do not rise in strict proportion to size, and it is this principle that is the basis for the CER.

The power-law-and-sizing CER is given as

$$C = C_r\left(\frac{Q_c}{Q_r}\right)^m \tag{6.14}$$

where C = cost for new design size Q_c, dollars

C_r = known cost for reference design size Q_r, dollars

Q_c = design size of equipment expressed in engineering units
Q_r = reference design size expressed in consistent engineering units
m = correlating exponent, $0 < m \leq 1$

If $m = 1$, then we have a strictly linear relationship and deny the law of economy of scale. For chemical processing equipment, m is frequently near 0.6, and for this reason the model is sometimes called the *sixth-tenths model*. The units on Q are required to be consistent, as it enters only as a ratio.

Furthermore, the model considers changes in cost due to inflation or deflation and effects independent of size, or

$$C = C_r\left(\frac{Q_c}{Q_r}\right)^m \frac{I_c}{I_r} + C_i \tag{6.15}$$

where I_r = inflation index and time coupled to equipment, dimensionless number
I_c = inflation index and associated with new equipment, dimensionless number
C_i = cost items independent of design, dollars

Example: Diesel-Electric Generator Plant

Six years ago an 80-kW naturally aspirated diesel electric set and its operating plant cost \$16 million. The engineering staff is considering a 120-kW unit of the same general design. The value of $m = 0.6$, and the price index for this class of equipment 6 years ago was 187 and now is 194. Differing from the previous design is a precompressor, which when isolated and estimated separately costs \$1,800,000. The future cost, adjusted for size of the design and for inflation, is

$$C = 16{,}000{,}000\left(\frac{120}{80}\right)^{0.6}\left(\frac{194}{187}\right) + 1{,}800{,}000 = \$23 \text{ million}$$

The measurement of m is important to the success of this model, and statistical methods given in Chapter 5 are useful in finding those values. If the statistical analysis assumes constant dollars, then the index ratio I_c/I_r is used for increases or decreases for inflation or deflation effects. A variety of indexes are available, and the index can be matched to the equipment type generally.[2] The consumer price index would be unsatisfactory.

The model usually does not cover those situations where the estimated design Q_c is greater or less than Q_r by a factor of 10. Same values of m are published, and a few exponents for equipment cost versus capacity are given by Table 6.9.

The value of m is important in several ways. If $m > 1$, then we deny the economy-of-scale rule, as shown with the centrifugal compressor. In fact, when $m > 1$, we have

[2]The Department of Labor Producer Price Index (PPI) has over 5000 items that are watched and equipment is charted. Additionally, there are professional societies and consulting groups that maintain indexes for a variety of equipment, materials, labor, and services.

TABLE 6.9 Illustrative Values of the Exponent m for the Power-Law-and-Sizing CER

Equipment	Size Range	Exponent, m
Blower, centrifugal (with motor)	1–3 hp	0.16
Blower, centrifugal (with motor)	$7\frac{1}{2}$–350 hp	0.96
Compressor, centrifugal (motor drive, air service)	20–70 hp drive	1.22
Compressor, reciprocating (motor drive, air service)	5–300 hp drive	0.90
Dryer drum (including auxiliaries, atmospheric)	20–60 ft^2	0.36

diseconomy of scale. If $m = 0$, then we can double the size without affecting cost, an unlikely event.

An equation expressing average cost C/Q_c can be found, or

$$C\frac{Q_r}{Q_c} = C_r\frac{Q_r}{Q_c}\left(\frac{Q_c}{Q_r}\right)^m = C_r\left(\frac{Q_c}{Q_r}\right)^{-1}\left(\frac{Q_c}{Q_r}\right)^m \tag{6.16}$$

$$\frac{C}{Q_c} = \frac{C_r}{Q_r}\left(\frac{Q_c}{Q_r}\right)^{m-1}$$

Total cost varies as the mth power of capacity in Equation (6.14), but average cost C/Q_c will vary as the $(m - l)$st power of the capacity ratio.

6.5.2 Other CERs (Optional)

CERs are popular because of their association to historical records and the ease of their statistical calculation. Sometimes parameter values that are tried and true exist, making CERs effective. Relationships such as the following in addition to those mentioned, exist:

$$C = KQ^m \tag{6.17}$$

where K = empirical constant for plant or equipment, dollars

Q = capacity expressed as a design dimension

m = correlating coefficent, number

However, difficulties exist in using Equation (6.17) because it is necessary to know the capital cost of a similar plant or equipment. Nor is the scale factor m constant for all sizes of the design. Generally, scaling up or scaling down by more than a factor of 5 from a known experience Q should be avoided because of dubious reliability.

The ineffectiveness of Equation (6.17) can be overcome somewhat by separating the capital costs into fixed and variable cost components. The variable cost components are subject to the economy-of-scale rule, while the fixed-cost elements are not. Another relationship is expressed as

$$C = C_v\left(\frac{Q_c}{Q_r}\right)^m + C_f \tag{6.18}$$

where C_v = variable element of capital cost, dollars
C_f = fixed element of capital cost, dollars

A multivariable CER is also possible. For instance,

$$C = KQ^m N^s \tag{6.19}$$

where the symbols have been previously defined. Coefficients are determined by regression methods.

Equation (6.19) deals with the cost principles of *economy of scale* and *economy of quantity*. These are two important principles, and this relationship attempts to tie them together.

It is possible to have multivariable CERs with more than two variables. Indeed, these equations can have many variables, but the knowledge gained from the statistical parameters becomes murky. In engineering, we desire a more hands-on understanding of the variables, where the connection to the variable gives a cause-and-effect relationship that is apparent. Another specialty field is *econometrics*, which is the association of social economics and statistics. Econometrics pays attention to gross national income, expense, and political programs, and measures those effects. These equations can have many variables and the source of the raw data could be national census and federal and state governmental information.

While CERs are equations, they also can be plotted as a graph, which is an effective way to describe cost behavior. All the data are shown, together with the statistically fitted line, whenever a graph is plotted. The original data remain on the graph, since erasing the points destroys the sense of variation. Visual plots give a gut feeling for the data. Plots are limited to two variables.

Example: Plot of Pressure Vessel Cost

Figure 6.5 presents an example of the plot of the cost of pressure vessels used in refining plants. The x axis is gallons per minute multiplied by feet head. The line is nonlinear on logarithms, and the standard deviation, shown as s = 8.78%, is used instead of the

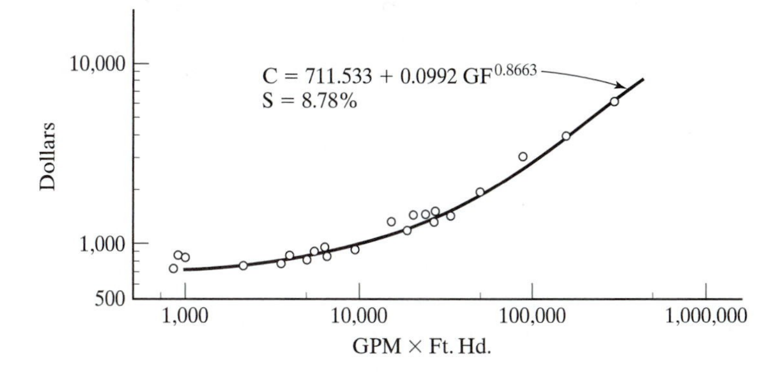

Figure 6.5 Illustrative CER plot and equation derived by using physical variables. The standard deviation is a percentage rather than absolute value. (GPM = gallon per minute × feet head.)

absolute value of the standard deviation. A percentage standard deviation is constant for all values of the variables. The absolute standard deviation varies over the range of the variables and it is a minimum at the means, $\bar{x}$ and $\bar{y}$.

6.5.3 Factor Method

The *factor method* is an important one for plant and industrial construction estimates. Other terms, such as *ratio* and *percentage,* describe the same thing. To estimate a plant with all of the materials, systems, and minutiae using detail approaches is too time consuming and, perhaps, inaccurate. The factor method eliminates these drawn-out details. Essentially, it determines the estimate by summing the product of several quantities, or

$$C = (C_e + \sum_i f_i C_e)(f_I + 1) \tag{6.20}$$

where C = cost of design, dollars

C_e = cost of selected major equipment or basic items, dollars

f_i = factors for estimating major items, building, structural steel, instrumentation, etc., that correlate to major equipment, number

f_I = factor for estimating indirect expenses, such as engineering, overhead, etc., number

$i = 1, \ldots, n$, number

The factors f_i are uncovered by several techniques. An owner's statistical analysis of the firm's operations, industry at large, business associations, contractors, consultants, news journals, or state and federal governments are sources for these factors.

A natural simplification of Equation (6.20) leads to the preliminary unit estimating model $C = fC_e$, where one factor is used to find the composite cost. There are variants, such as $C = \Sigma_i f_i D$, where D is the design parameter (for example, area of construction of building, or capital investment) and f_i is the factor in dimensional units compatible to the design.

The unit-cost estimating method discussed earlier is limited to a single factor for calculating overall costs. The factor method achieves improved accuracy by adopting separate factors for different cost items. For example, the cost of an office building can be estimated by multiplying the area by an appropriate unit estimate such as the dollars-per-square-foot factor. As an improvement, individual cost-per-unit-area figures can be used for heating, lighting, painting, and the like, and their value C can be summed for the separate factors and designs.

In chemical plant construction, the basic item (or items) of the process is identified using the flowchart. The flowchart and the specification sheet are primary input data to the engineer. This *basic item(s)* constitutes a major aspect of a building, such as the structural shell, or tons of concrete for a highway, or process equipment in a chemical plant. The first step is to determine the cost of the basic item. The next step is to find the cost relationship of other components as a percentage, ratio, total cost, or factor of the basic item.

For a building, the tons of structural steel may be selected as the basic item, and factors would correlate brick, concrete, masonry, steel reinforcing, and so forth. For a chemical process, the basic item could be high-pressure high-temperature reactors, and equipment erection, piping and direct materials, insulation, instrumentation, and engineering are factors correlated to process reactor cost. These factors are found by using the statistical methods of Chapter 5.

In some instances there are costs that are independent of the variation in the cost of the basic item. Those independent components (for example, roads, railroad siding, and site development for a new chemical plant are not related to the equipment) must be estimated separately by other methods.

There are common-sense considerations in using the factor method. For chemical plant construction, variations are due to the size of the basic equipment selected, materials of fabrication/construction, operating pressures, temperatures, technology (such as fluids processing, fluids-solids processing, or solids processing), location of plant site, and timing of construction.

Factor behavior is described by Figure 6.6(a). The basic item cost is the x-axis entry. A vertical intersection with the line from the x-axis factor value leads to a factor, as read from the y axis. If the x-entry value is doubled, then we assume that the factor is not doubled, thus ensuring the principle of economy of scale. This is apparent because the factor line in part (a) shows a declining nonlinear curve. Practically speaking, if the physical size of the basic item becomes larger and therefore more costly, the factor

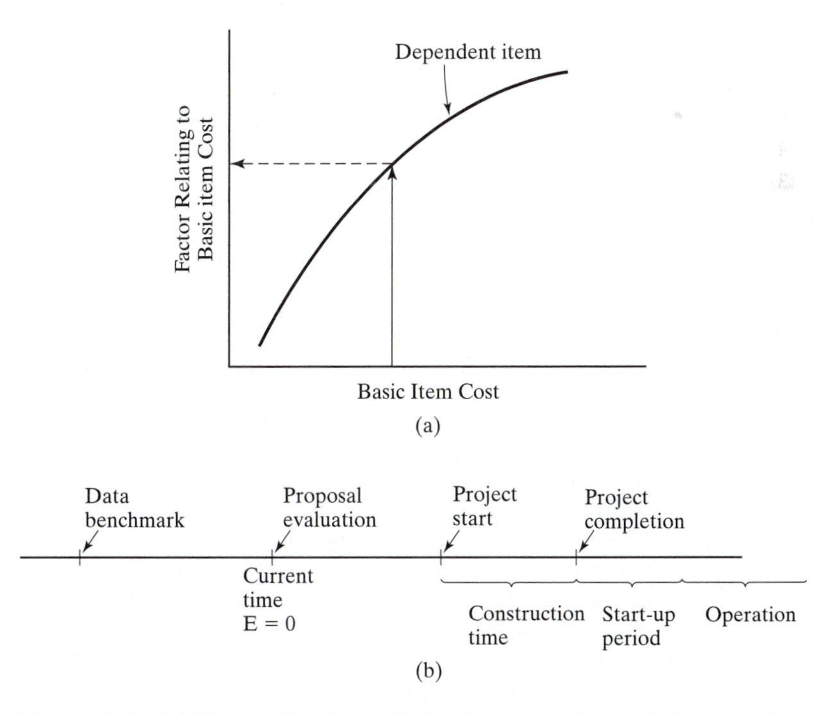

Figure 6.6 (a) Illustrative factor behavior curve for basic item, and (b) time line for project cost evaluation.

relating, say, the engineering design cost is relatively smaller. If the basic item is constructed with more expensive materials, such as stainless steel or glass-lined materials instead of plain carbon steel, then the factors become nonproportional to the item being estimated.

Factor data are typically organized to an index base of 100. An owner or industrywide or government index with a specified benchmark year is used, and all data are referenced to this benchmark year. Indexes (discussed in Section 5.4) are important in the factor method.

We revise Equation (5.39) to show

$$C_r = C_c\left(\frac{I_r}{I_c}\right) \tag{6.21}$$

where C_r = cost at benchmark time, dollars

C_c = cost of equipment at current time, dollars

I_r = index of 100 for benchmark year, number

I_c = index for equipment, plant, etc. at current time, number

The estimate of the basic items is in *current-time dollars*. Thus the cost of important equipment, say high-temperature, high-pressure reactors, is for the now time, $E = 0$, as shown by the time line in Figure 6.6(b). In doing factor estimating an owner maintains indexes allowing it to *backcast* the C_e value to the benchmark year of the cost data. Once the basic item cost is backcast to the period of the factor curve, the factors are read off the curve, or determined from the information. At this point the indexed base-item cost value is multiplied by f_i. Next, the benchmark cost is indexed forward to that point in time when the cost is expected as out-of-pocket, project start, completion, or when progress payment periods are stipulated by the contract between the contractor and the owner [see Figure 6.6(b)].

Example: Estimating a Plant Project

Plant construction cost can be estimated by the factor method. A hydrobromination flowchart, illustrated in Figure 6.7, describes a microprocess plant, and we want to estimate the construction cost of this plant. A hydrobromination process is one step in the sequence of making a liquid soap formulation.

The first step in the factor method estimates the important basic item costs. For the hydrobromination plant, the rising film and ozonation reactors are chosen as the important basic items from the flowchart. Because this process is high pressure and high temperature, we expect the piping, supports, instrumentation, foundation, and so on to correlate with these critical reactors. The basic item costs recognize the distinction between carbon steel and stainless steel reactors, and the factor reflects the distinction.

Figure 6.8 shows the factors of engineering costs and fees, erection, and direct material costs for the construction of the hydrobromination plant. The original data points for the curves are shown along with the lines. All data are left as points on the graphs, an important feature in cost analysis. The entry *x-axis* variable is process equipment costs. Notice that the scales are logarithmic and the lines are nonlinear.

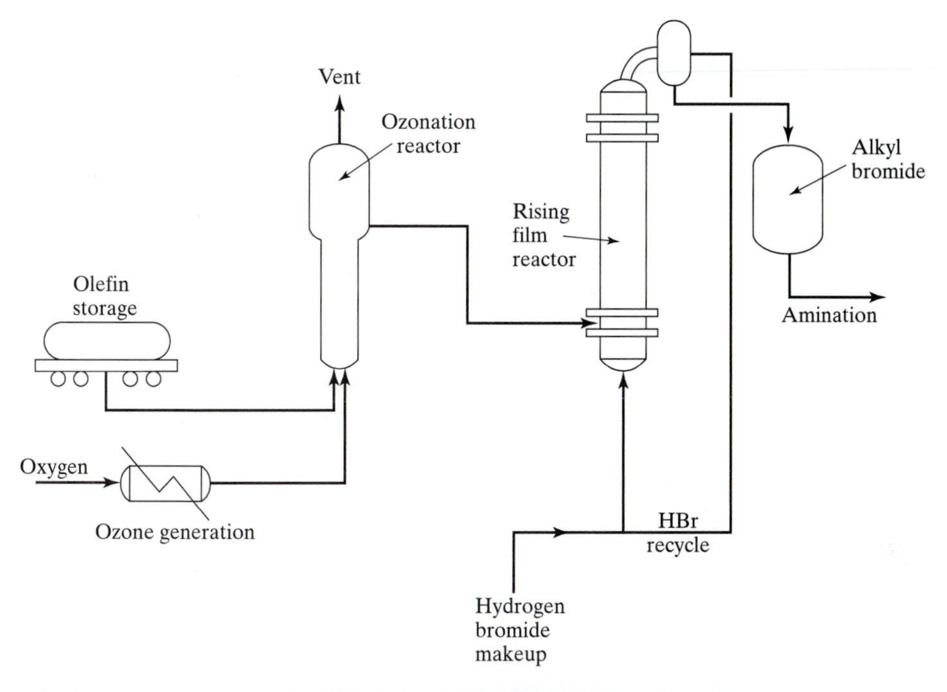

Figure 6.7 Illustrative flowchart for hydrobromination process.

The two reactors from Figure 6.7 are estimated for current time costs, which correspond to the time of the estimate at $E = 0$, as shown by Figure 6.6 (b). The reactor costs could be estimated internally by the plant engineering staff, or suppliers can be asked for bids. In either case, once the basic items are estimated, we have the following:

Major Process Design Equipment	Current Cost, $
Rising film reactor	$2,200,000
Ozonation reactor	700,000
Total	C_e = $2,900,000

The $2,900,000 is the current time cost, and it is backcast to $2,541,630 using the index for this equipment, and the year of the factor data. The factor data are organized at the index number of 100 while the index is 114.1 for the current time. Thus, using Equation (6.21), we determine that $2,541,630 is the cost at the time of the factor data and is found from 2,900,000 (100/114.1).

With this adjusted major component cost as the entry, we find the factors that correlate to the equipment. Figure 6.8 yields engineering costs and fees, 1.1; erection, 1.7; and direct materials, 4.1.

In Table 6.10 the factors become a multiplier for the benchmark cost, or $4,320,770 (= 1.7 × 2,541,630) for erection. This cost is then inflated to the project start

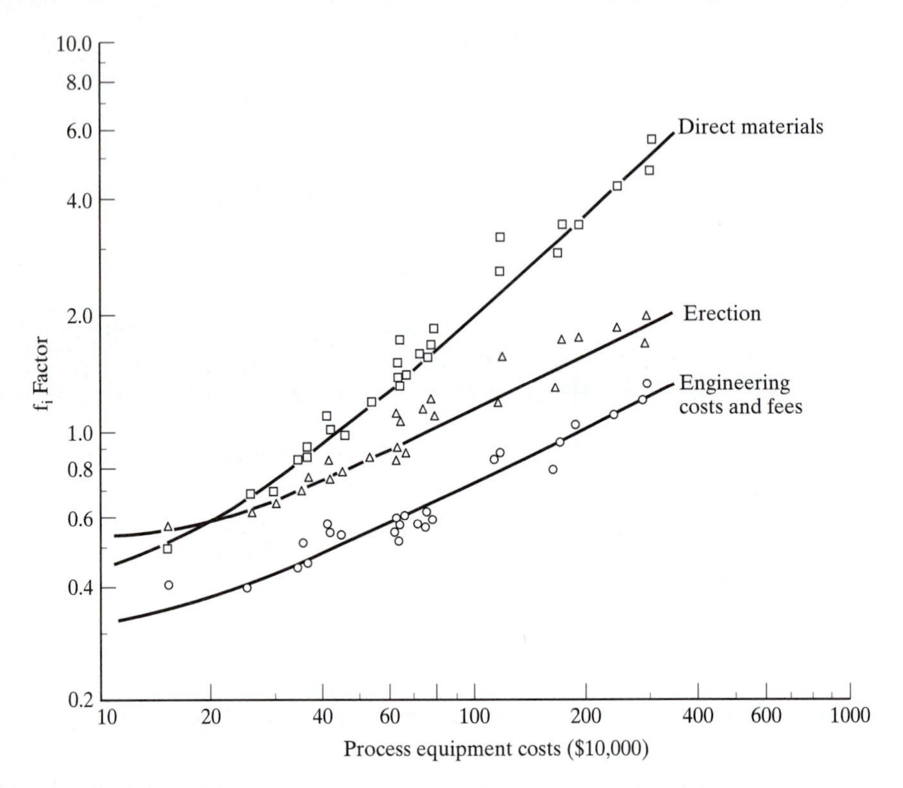

Figure 6.8 Illustrative factor chart based on process equipment costs for benchmark year.

point or the progress payment time stipulated by the contract clauses. This would, for instance, be 4,320,770 × (189/100) = \$8,166,250.

Factor data for chemical process plants are often regionally located in a construction area where significant activity is expected. Chemical plants are frequently benchmarked to the Gulf Coast of the United States. Construction at other sites requires regional indexing to adjust for differences in labor and material cost between the plant site and the benchmark area.

The data in Figure 6.8 are Gulf Coast information, although our hydrobromination plant is constructed in the Midwest where direct materials are 8% more expensive, erection costs are 13.2% higher, and engineering costs and fees are identical. Furthermore, those regional indexes account for the mix of the components. The regional factors are applied to current values. The Gulf Coast cost becomes 8,166,250 × 1.132 = \$9,244,200, adjusting for geographical distinctions of cost.

Finally, the owner's capital investment, \$34,630,210, is provided as the raw construction estimate. A factor for indirect costs, contingency, and profit is applied later to convert raw capital investment into the full cost for construction.

The example deals with plant construction and equipment estimating, but the factor method is used more broadly to estimate labor, materials, utilities, and indirect costs as a multiple of some other estimated or known quantity. Operating costs as a percentage of plant investment or as a percentage of the product selling price are popular.

TABLE 6.10 Spreadsheet of the Factor Method for the Flowchart of the Hydrobromination Plant

Item	Current-Time Cost	Current Index	Major Item Benchmark Cost	f_i	Benchmark Cost	Project Start Index	Gulf Coast Project Start Cost	Midwest Regional Index	Project Start Plant Site Cost
Equipment Major process item	$ 2,900,000	114.1	$2,541,630	1	$ 2,541,630	118.0	$ 2,999,120	—	$ 2,999,120
Erection				1.7	4,320,770	189.0	8,166,250	1.132	9,244,200
Direct materials				1.7	10,420,680	127.0	13,234,260	1.08	14,293,000
Engineering costs and fees				1.1	2,790,000	145.0	4,053,890	1.0	4,053,890
Building site development									1,250,000
Site development									
Process building									2,100,000
Railroad spur									40,000
Utilities									650,000
Total									$34,630,210

PICTURE LESSON Milling Machine

The machine tools of the Industrial Revolution were sometimes manufactured to be product specific. The vertical milling machine shown here was dedicated to machining locomotive cylinder valve faces. Such machine tools were versatile, and could be mobilized for other work as well.

The kinematic motions of the machine tools were pretty much defined by the time of this engraving, which was c. 1890, and the motions of the components, table, headstock and column have not changed very much since that time. The table, for instance, has longitudinal and traverse motions, and already there was automatic feed for drilling and boring. By the early 1900s, almost all of the movements between the headstock, table, ways, end stocks, and the like were widely used in machine tools. Other improvements, for the most part, have been incremental since that time.

Arguably, four prominent milestones of machine tool technology were reached during the twentieth century:

1. Significant progress on machine tools saw the alternating-current electric motor replace the steam engine and line shafts. These motors became integral with the machine tool, and increased equipment utilization substantially.
2. Cutting tool materials progressed from high carbon steel to carbide, where fine powders of tungsten carbide and cobalt were pressed to a shape and hardened by sintering in a hydrogen atmosphere. These materials significantly improved cutting efficiency.

3. Metrology, the science of measurement, improved by orders of magnitude. For instance, linear measurements of 1/64th and 1/128th of an inch were the best a machinist was able to machine. Today, electronic and optical techniques routinely provide ± 0.00005 in. tolerance.
4. The adaptation of automatic motors, called servomechanisms, allowed numeric and computer control of the relative position of the part to the tool. Other improvements were to follow, and nowadays computer numerical control (CNC) is in widespread application.

6.6 ADVANCED APPROACHES (OPTIONAL)

Frequently, an engineer will use a variety of methods to estimate the design. Many methods are rare and are not sufficiently general to include a discussion here. The "spreadsheet" is a software routine that can accommodate a scope of these methods, but the spreadsheet is a means of calculation. We encourage the spreadsheet as a convenience to calculate many problems, but by no means is the spreadsheet a general method for estimating.

Advanced methods that are most popular include the following:

- Expected value
- Range
- Percentile
- Monte Carlo simulation

As usually prepared, the estimate represents an "average" concept. Nor does the estimate reveal anything about the probability of these expected values. It uses information that is called certain, or *deterministic*. Probability concepts are commonly used in estimating. We now examine these methods.

6.6.1 Expected Value

Although a detailed description of a random variable is given by its probability distribution, that information can be concisely summarized by its expected value, or mean. Viewing probabilities as a long-run frequency, the expected value is the long-run average value of the random variable. For this discussion we assume that the engineer is able to give a *probability point estimate* to each element of uncertainty as represented by the economics of the design. This assignment associates nonnegative numerical weights with possible events. For example, if an event is certain, then its associated probability equals 1. The cost of a 2×4 in. stud is \$8.79, and its assigned probability is 1, which implies a probability of 0 to any other stud cost. For the most part, we prefer to deal with the simplest case: certainty. When we say "material cost is \$8.79," we imply, but leave unstated, that the probability is 1, a *certain event*. Any other material cost has the probability of 0. Despite this assertion of certainty, it seldom exists.

Sometimes two events A and B are *mutually exclusive*. For instance, if A happens then B cannot, and event A *and* B is impossible. The probability of the event "either A or B" equals the sum of the probability for each of the events. The addition law for mutually exclusive events is given as $\Pr[A \text{ or } B] = \Pr[A] + \Pr[B]$.

Probabilities are a numerical judgment of the likelihood of future events. The techniques for deriving probabilities include the following:

- Analysis of historical data to give a relative-frequency interpretation, such as the histogram
- Convenient approximations like the normal distribution function
- Expert introspection, or what is called opinion probability

The first two choices are known as *objective probability*. Seldom is there disagreement about how values are found using objective probability methods.

Opinion probability calls for judgmental expertise and a pinch of luck. Better success is ensured when past data are analyzed. On the other hand, data may be unavailable, and it should be remembered that while data are past, probabilities should be indicators of the future. Sometimes both past data and a reshuffling of probabilities are undertaken. This discrimination is not new to professional cost-analysis practice.

The category involving *risk* is appropriate whenever it is possible to estimate the likelihood of occurrence for each condition of the design. These estimates forecast the true likelihood that the predicted event will occur. Formally, the method incorporates the effect of risk on potential outcomes by using a weighted average. Each outcome of an alternative is multiplied by the probability that the outcome will occur. This sum of products for each alternative is called an *expected value*. For the discrete case, the mathematics are as follows:

$$C(i) = \sum_{j}^{n} p_j x_{ij} \tag{6.22}$$

where C = expected value of the estimate for alternative i, dollars

p_j = probability that x takes on value x_j, $0 \leq p_j(x_j) \leq 1$

x_{ij} = design alternative

The p_j represents the independent probabilities that their associative x_{ij} will occur with $\sum_{j=1}^{n} p_j = 1$. The expected-value method exposes the *degree of risk* when reporting information in the estimating process.

Example: Wireless Telecommunication and Real-Time Computer Product

An electronics engineering firm is evaluating a portable television that has special design features of wireless telecommunication and real-time computer functionality. Market research indicates a substantial market available for a small lightweight set if priced at $850 retail. This implies that the set will have to be sold to wholesalers for approximately $650. Several questions need to be answered before the decision can be made to enter the market. Three important ones are: What will be the first year's sales volume in units? How much will the unit cost to produce? What will be the profit? To answer those questions, marketing furnishes a probability estimate of the first year's sales in units.

Annual Sales Volume	Probability of Event Occurring
150,000	0.2
200,000	0.2
250,000*	0.6

*Most frequent case.

Engineering estimates the total cost per unit:

Cost per Unit, $	Probability of Event Occurring
450*	0.7
500	0.2
550	0.1

*Most frequent case.

The example can be divided into two categories:

- Risk not apparent
- Risk apparent

Assume that marketing and estimating use the most frequent case number from their studies and do not report any risk. For "risk not apparent" profit is calculated as

$$\text{Profit} = (650 - \text{cost}) \times \text{volume}$$
$$\text{Most frequent cost} = \$450$$
$$\text{Most frequent volume} = 250{,}000 \text{ units}$$
$$\text{Profit} = (650 - 450)250{,}000 = \$50 \text{ million}$$

Conversely, assume that the organization encourages a policy of reporting risk in estimates. We use opinion probabilities to have three values for cost and three for volume. There are nine possibilities, as shown by Table 6.11.

The total profit is overstated when comparing the case of *risk not apparent* or $50 million to the expected profit of $39.6 million for the case of *risk apparent*.

TABLE 6.11 Calculation of Joint Probability for Cost and Volume of Product

(650–Cost) × Volume	Joint Probability of Volume and Cost	Expected Profit
(650 − 450)150,000 = 30,000,000	0.14	$4,200,000
(650 − 450)200,000 = 40,000,000	0.14	5,600,000
(650 − 450)250,000 = 50,000,000	0.42	21,000,00
(650 − 500)150,000 = 22,500,000	0.04	900,000
(650 − 500)200,000 = 30,000,000	0.04	1,200,000
(650 − 500)250,000 = 37,500,000	0.12	4,500,000
(650 − 550)150,000 = 1,500,000	0.02	300,000
(650 − 550)200,000 = 2,000,000	0.02	400,000
(650 − 550)250,000 = 2,500,000	0.06	1,500,000
Total	1.00	Expected profit = $39,600,000

Another approach is possible. Both marketing and engineering could have computed an expected value. For instance, expected cost = $470(= 450 × 0.7 + 500 × 0.2 + 550 × 0.1) and expected volume = 220,000 units(= 150,000 × 0.2 + 200,000 × 0.2 + 250,000 × 0.6). The expected profit = $39.6 million(= 220,000(650 − 470)), which is the same as in the table.

It is important to point out that there is a distribution or histogram that can be found from these data. Technical people relate to curves, distributions, and so on, and individual cost values are sometimes incomprehensible by themselves, particularly when dealing with dollar values that have little or no relevance to personal experience.

The calculations for this example require that the cost per unit be *independent* of production volume, which is generally not the situation. Another difficulty in applying the expected-value method is the inability of experts to guess opinion probabilities. Seldom are there personal experiences that allow for a development of this ability.

6.6.2 Range

Knowing the weaknesses in information and techniques, we recognize that there are probable errors in the estimate and its procedure. The recognition that cost is a random variable opens up the topic of *range estimating*. A *random variable*, in statistical parlance, is a numerically valued function for the outcomes of a sample of data. Finding the mean or standard deviation—for example, using equations with sample information—gives a random variable. This notion introduces an important probability improvement to single-valued estimating.

The range method involves making three estimates for each major cost element. We bracket a *most frequent* or *modal* estimate value for each cost element. This forms the principle for range estimating.

The following procedure is based on the program evaluation and review technique (PERT) and involves making a most likely cost estimate, an optimistic estimate (lowest cost), and a pessimistic estimate (highest cost).

Those estimates are assumed to correspond to the unimodal *beta distribution* in Figure 6.9 (a). The beta distribution has rich properties that are useful for cost estimating.

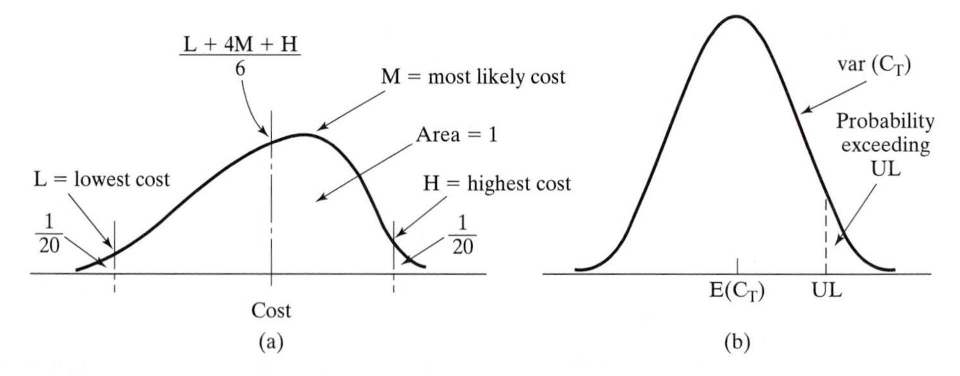

Figure 6.9 (a) Location of estimates for individual-based beta distribution, and (b) normal approximation of sum of individual beta-based element probabilities.

For example, only positive costs are allowed. Remember in the normal or t-distribution, negative costs are theoretically allowable, as the range of the variable can be negative. This beta distribution is skewed left. Symmetric and skewed-right distributions are also possible. The total area under the probability distribution is 1. Although the mathematics of the beta probability density are complicated, the expected values lead to simplified equations.

Almost any method in making the three individual estimates for a cost element is acceptable, from preliminary to detail. For discussion of this range method, we assume that the estimates are already made. A mean cost and a variance cost for the single cost element i are calculated as

$$E(C_i) = \frac{L + 4M + H}{6} \tag{6.23}$$

$$\text{var}(C_i) = \left(\frac{H - L}{6}\right)^2 \tag{6.24}$$

where $E(C_i)$ = expected cost of element i, $i = 1, 2, \cdots, n$, dollars

L = lowest cost, or best-case estimate of cost element i, dollars

M = modal value, or most likely estimate of cost element i, dollars

H = highest cost, or worst-case estimate of cost element i, dollars

$\text{var}(C_i)$ = variance of cost element i, $i = 1, 2, \cdots, n$, dollars^2

If a dozen or more elements are estimated this way, and the elements are assumed to be independent of each other and are added, then the new distribution of the total cost is approximately normal. This follows from the *central limit theorem:* The mean of the sum is the sum of the individual means, and the variance is the sum of the variances. The distribution of the sum is normal, despite the reasoning that says that the individual elements were beta-shaped. A normal-shaped distribution is shown by Figure 6.9(b), which results when several or more individual beta-described elements are added.

The element estimates given by Equation (6.23) are summed and they form the popular normal distribution shown by Figure 6.9 (b). The project total cost is in turn found using the following relationship:

$$E(C_T) = E(C_1) + E(C_i) + \ldots + E(C_n) \tag{6.25}$$

$$\text{var}(C_T) = \text{var}(C_1) + \text{var}(C_i) + \ldots + \text{var}(C_n) \tag{6.26}$$

where $E(C_T)$ = expected total cost of independent elements i, dollars

$\text{var}(C_T)$ = variance of total cost of independent elements i, dollars^2

It is seen that the total expected cost and the variance of the total cost are merely sums of component expected costs and variances, respectively. Also, the distribution of total expected cost is nearly normal regardless of the shape of the distribution of the estimated cost elements.

Various statements regarding the interval for a future estimate value can be made with a normal distribution. For example, what is the probability that a certain cost, called the "upper limit," will exceed the expected cost, or $\Pr[UL \text{ cost} > E(C_T)]$?

What are the probability boundaries for a cost? These are important questions. With the normal distribution assumed for C_T, it is possible to examine this distribution and answer those questions. This probability of exceeding an upper limit is found by using the following relationship:

$$Z = \frac{UL - E(C_T)}{[\text{var}(C_T)]^{1/2}} \tag{6.27}$$

where Z = value of the standard normal distribution (see Appendix 1)

UL = upper limit of cost, arbitrarily selected, dollars

The square root of $\text{var}(C_T)$ is the *standard deviation*, and it has the dimension dollars. The value of Z is used to enter Appendix 1, and the probability of exceeding the upper limit cost is found by using $(0.5 - \Pr(Z))$. The normal distribution is symmetrically centered, and the area under the curve is 1, which is symmetrically divided into two parts with 50% above and 50% below the value of $Z = 0$. Refer to Appendix 1 and observe the characteristics of the table. Manipulation of the normal distribution is a straightforward calculation.

Example: Calculation of Range Estimate for Laser Amplifier

A large laser amplifier for the generation of fission power is to be estimated. The laser amplifier has many elements that are estimated by the range method, and they are independent of one another. In other words, the flash-lamp cost is unrelated to the cost outcomes of a data grid. Three estimates—*L*, *M*, and *H*—are made for each cost element, with 10^3 units, although they are not shown in the table. Table 6.12 shows an example of this calculation.

The expected cost sum of the individual cost elements is $10,602, and the variance representing probable error is 22,301 dollars^2 (or dollars square). The total sum and variance are assumed to be normally distributed, even though the individual cost distributions were something else.

TABLE 6.12 Calculation of the Laser Expected Cost and Variance Using the Range-Estimating Method

Laser Cost Element	Lowest Cost, L	Most Likely Cost, M	Highest Cost, H	Expected Cost, $E(C_i)$	Variance, $\text{var}(C_i)$, $\$^2$
1. Flash lamp	$ 370	$ 390	$ 430	$ 393.33	$ 100.00
2. Data grid	910	940	1,030	950.00	400.00
3. Computer	200	210	270	218.33	136.11
4. Optical isolator	170	180	190	180.00	11.11
5. Power supply	260	290	350	295.00	225.00
6. Switching	171	172	176	172.50	0.69
7. Capacitor	875	925	975	925.00	277.78
⋮	⋮	⋮	⋮	⋮	⋮
n. Frame	$ 2,000	$ 2,100	$ 2,600	$ 2,166.67	$ 10,000
				$E(C_T)$ = $10,602	$\text{var}(C_T)$ = $22,301

The application of Equation (6.27) allows the finding of risk for a project. The location of UL is greater than $E(C_T)$, which is what is usually anticipated. For example,

UL can be the bid value or ceiling price for the project, and management wants to know the chance of exceeding the bid price with actual cost.

Example: Finding the Amount of Risk if an Upper Level Cost for Laser Amplifier Is Exceeded

Management is interested in knowing the risk of an actual cost exceeding $10,850. We now use Equation (6.27):

$$Z = \frac{10{,}850 - 10{,}602}{(22{,}301)^{1/2}} = 1.66$$

$$\text{then } P(\text{upper limit cost} > 10{,}850) = 0.05 = 5\%$$

The value of $Z = 1.66$ is read in Appendix 1, and the corresponding *upper-tail probability* of 5% area gives the chance of exceeding $10,850. Thus the range method is able to say something about the risk of exceeding a particular value, an important assertion by the cost estimate. In other words, there is a 5% chance that actual cost will exceed the $10,602,000 estimate and be above $10,850,000.

According to PERT practices, the optimistic *L* and pessimistic *H* costs would be wrong only once in 20 times, if the activity were to be performed repeatedly under the same conditions. The most likely cost, *M*, is the cost that would occur most often if the activity were repeated many times, or the cost that would be given most often if qualified people were asked to estimate its value. If *M* is ever compared to the actual cost, we should find that with a number of estimates, *M* falls above and below the actual cost equally.

Project work is estimated only once, so those precise requirements are never met. We suggest that these costs be viewed as *best, most likely*, or *worst* cost. Furthermore, it is unlikely that a distribution of major actual cost elements can ever be known. Finally, if costs are internal to a firm, they are seldom independent of each other, although for a project estimate of various subcontractors' costs, the assumption of independence is more valid.

There are advantages to this method. We are able to identify elements of cost that have great uncertainty. For instance, in the earlier example, the "frame" cost element has the highest variance. Management is then able to gather more information, or more bids, or improved engineering, and so forth, to reduce the variation.

The expected-value method discussed earlier required selection of opinion probabilities, which we believe is a dubious practice. Estimating opinion probabilities is puzzling, and experts often give widely differing values for the same situation. However, the range method requires three values of cost; and if commonsense methods of selection are employed, then the range method is acceptable as a preliminary technique.

A less widely used probability technique of the genre of range estimating is known as percentile estimating, a topic we consider next.

6.6.3 Percentile

Estimates reflecting uncertainty may be specified by three values representing the 10th, 50th and 90th percentiles of an unstated probability distribution. The best value is the 50th percentile, however the 10th and 90th estimates allow the examination for cost contributions to high or low cost. The 10th percentile cost is the best-case scenario and represents a 1 in 10 chance that the cost will be lower. The 90th percentile cost is the worst-case scenario and represents a 1 in 10 chance the cost will be greater.

Percentile methods are concerned with frequency information and location. The percentile is a point below which a stated percentage of observations lie. The percentile will be a value in the same units as the observations themselves. For example, an engineer might find that 34 out of 40 units involve assembly times below 46 minutes. The percentage of items requiring such times is $100(34/40) = 85\%$, and 46 minutes is the 85th percentile. If that engineer finds that 62 minutes is the 95th percentile, then 95% of the 40 units, 38 of them, have assembly times falling below 62 minutes.

For several cost items, the selection of which ones and how they combine is considered an independent opportunity, much like the beta distribution approach given previously. A low cost can combine with a midrange cost, and then with a high cost. After estimating the three values for each cost item, the 10th and 90th percentiles are expressed as differences from the 50th value (or midrange). The next step is to square the differences and sum.

Example: Percentile Method

Three cost items estimated at the 10th, 50th, and 90th percentiles are given here.

Cost Item	Percentile			Difference	
	10th	50th	90th	(50 − 10)	(90 − 50)
1	$25	$33	$44	$8	$11
2	9	13	15	4	2
3	3	4	7	1	3

After estimating, the costs are expressed as differences with the 50th percentile, and square and sum. The square root of the sum is the contributor of uncertainty to the 10th and 90th estimates as follows:

	$(50 - 10)^2$	Midvalue	$(90 - 50)^2$
	$64	$33	$121
	16	13	4
	1	4	9
Total	81	50	134
Square root	$9		$11.58

Total estimate at 10th percentile = 50 − 9 = $41
Total estimate at 50th percentile = $50
Total estimate at 90th percentile = 50 + 11.58 = $61.58

Sensitivity analysis can be applied to the percentile method in a simple way, as follows:

Cost Item	Contribution to Low Uncertainty	Contribution to Total Cost	Contribution to High Uncertainty
1	79%(64/81 × 100)	66%(33/50 × 100)	90.3%(121/134 × 100)
2	19.8%(16/81 × 100)	26%(13/50 × 100)	3%(4/34 × 100)
3	1.2%(1/81 × 100)	8%(4/50 × 100)	6.7%(9/134 × 100)

The sensitivity analysis shows that cost item 1 is likely to give a greater cost variance, and thus is subject to watching or redesign of the product and process.

6.6.4 Monte Carlo Simulation

Models are descriptions of systems. In the domain of physical sciences, models are based on theoretical laws and principles. The modeling of engineering cost systems may be almost as difficult as modeling a physical system, because few fundamental laws are available, many procedural elements are untractable, and policy inputs are hard to quantify. Surveys indicate that simulation is a tool that is sufficiently robust to handle engineering cost systems. Monte Carlo is an advanced business simulation tool and business decisions under uncertainty have been modeled this way since the 1960s.

Simulation is divided into the following classes:

- Real versus abstract
- Machine versus man-machine
- Deterministic versus probabilistic

Testing of a prototype airplane or a pilot plant is *real* simulation, while mathematical and logical statements employing a synthetic model are *abstract* simulation.

Machine simulation attempts to program all eventualities for computer running. On the other hand, *man-machine* simulation allows program interdiction by the human being at strategic points where the skill and experience of the human are considered superior to the computer. This division of labor allows the computer to do what it can do well, and encourages the ability of people to interpret qualitatively rather than quantitatively.

While engineering business is a probabilistic undertaking, these situations are usually characterized by constant-value models. In a *deterministic* situation, we assume that the *constants* substituted for the business parameters are an ideal approximation for that design. In a *probabilistic* situation, however, it is necessary to introduce random events such that the operating parameters of the modeled system are affected. Sometimes probabilistic models are titled with the interesting term *Monte Carlo*.

Although simulation uses mathematics, it is not mathematics per se. In simulation you *run* problems, not *solve* them as you do in mathematics. The intent is the collection of pertinent data from the experiment as one runs and watches the outcome of many simulation trials. On the one hand, the actual experiment of a design provides realism. On the other hand, orthodox mathematical solutions to business and engineering problems remains an abstraction. Albert Einstein offered this observation on the topic: "So far as the laws of mathematics refer to reality, they are not certain. And so far as they are certain, they do not refer to reality."[3] Figure 6.10 illustrates the contrast of simulation to closed-form single-value solutions.

A simulation model is formulated to look like the actual engineering system. Other chapters give many examples of systems. Simulation procedures, then, use the common estimating elements of labor, material, overhead, etc., which are coupled with

[3]Spoken by the theoretical physicist and Nobel laureate in a lecture to the Prussian Academy in 1921.

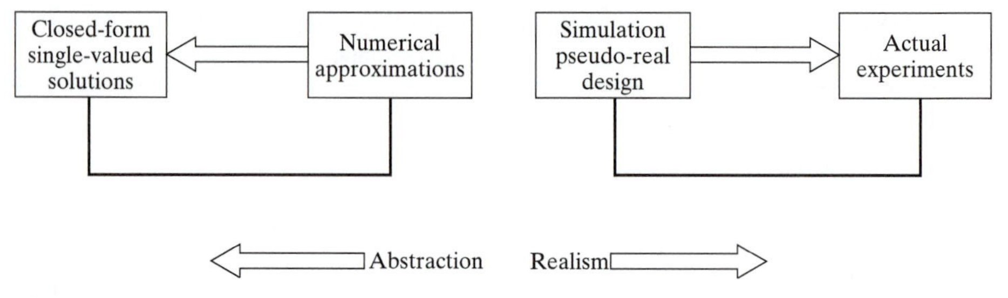

Figure 6.10 Comparison of abstraction and realism for the solution of engineering-business problems.

the governing business condition. This is the pattern for estimating, which is shown by Figure 6.11.

Example: Simulation of Direct Labor and Material

Direct cost is expressed as $A = x + y$, where x and y are probability distributions. The x and y can represent direct labor and direct material, for instance. Furthermore, we have field data for the distribution of $f(x)$, given as follows:

Cost of x, $\$10^6$	Relative Frequency	Monte Carlo Numbers
1.1	0.05	0.00–0.05
1.2	0.10	0.06–0.15
1.3	0.15	0.16–0.30
1.4	0.20	0.31–0.50
1.5	0.20	0.51–0.70
1.6	0.15	0.71–0.85
1.7	0.10	0.86–0.95
1.8	0.05	0.96–1.00

Monte Carlo numbers are found similarly to the development of the cumulative frequency interval for empirical distributions, as shown in Section 5.1. Thus \$1.1 million has a Monte Carlo interval of 0.00 to 0.05. In simulation we assume that random numbers are the probability of the functional cost item. If a random number of 0.18 is drawn from a hat containing a well-mixed collection of 100 separately numbered decimals or from a random-number table, both having the number range between 0 and 1, we say that 0.18 corresponds to \$1.3 million after entry in the Monte Carlo table, because it falls within that range. Tables of random numbers are found in other texts. We provide a small sample when required for a problem.

The collection of field data and determining of Monte Carlo numbers is one approach. A second method uses analytical distributions fitted with empirical or trusted coefficients. There are many distributions, both continuous and discrete, and the student is referred to a text on probability and statistics for more information. In this brief summary, we do the following:

A *probability density function* for the continuous random variable Y is defined as

$$f(y) \geq 0$$

$$\int_{-\infty}^{+\infty} f(y)\, dy = 1 \tag{6.28}$$

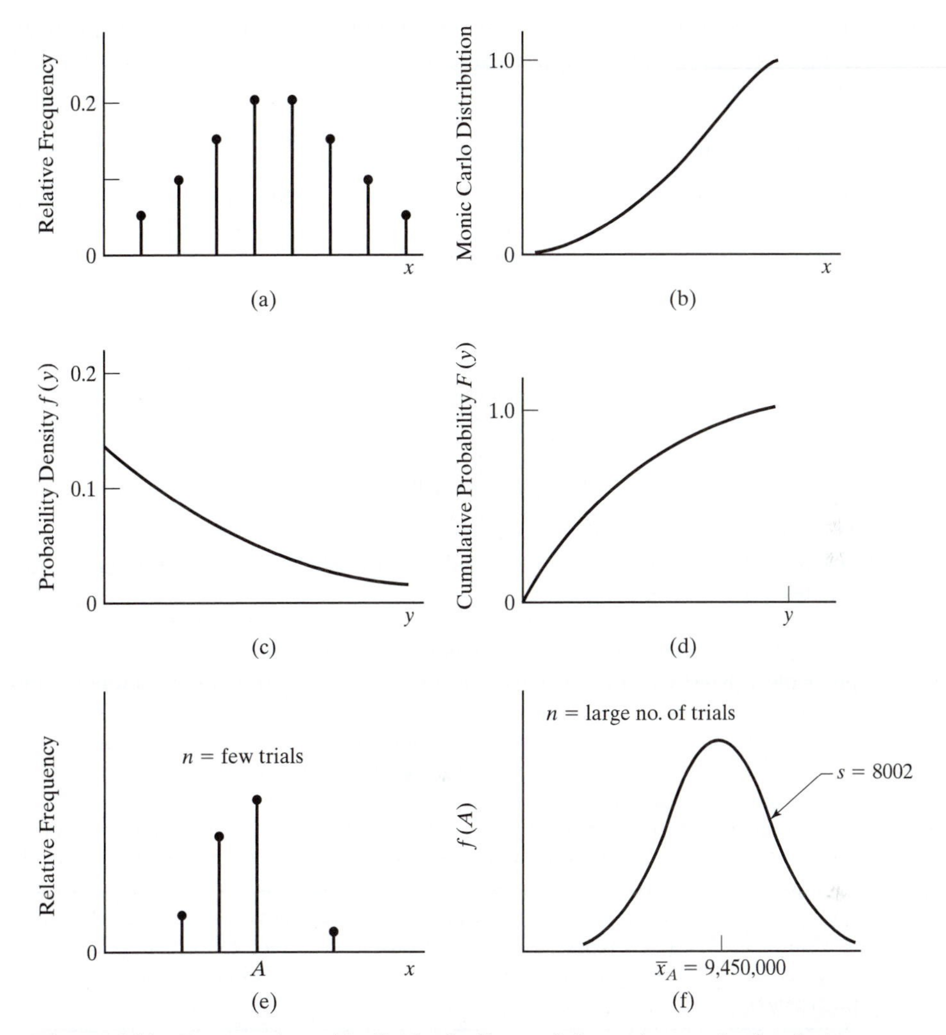

Figure 6.11 Progression of individual effects of the estimate via simulation: (a) discrete histogram obtained from field studies or opinion; (b) Monte Carlo or cumulative distribution of frequency distribution; (c) probability density idealized to an economic factor; (d) cumulative probability distribution of density; (e) histogram of several simulation trials; and (f) distribution of simulation after large number of trials.

$$P(a \leq y \leq b) = \int_a^b f(y)\,dy$$

As an immediate consequence, the *cumulative distribution F(x)* of the random variable is given by

$$F(y) = P(Y \leq y) = \int_{-\infty}^{y} f(y)\,dy \tag{6.29}$$

In Monte Carlo simulation we have various random numbers, 0 to 1, to substitute for $F(x)$ and then we solve for the upper value of integration, y. The student will recognize this as the inverse method of calculus.

Example: Exponential Function Monte Carlo Approach

We continue the illustration of the direct cost A as a sum of x and y. For example, we may assume that the exponential function represents a type of cost. Now assume that $f(y)$ is given by an exponential density with a functional form of

$$f(y) = \frac{1}{8}e^{-y/8}, \; y \geq 0$$

To find the cumulative probability distribution, we use

$$F(y) = \int_0^{C_y} f(y)\,dy = \int_0^{C_y} \frac{1}{8}e^{-y/8}dy = 1 - e^{-C_y/8}$$

$$\text{then } C_y = -8\ln[1 - F(y)]$$

When random numbers, 0 to 1, are individually selected and substituted, we find the Monte Carlo value of C_y.

Our example for the direct cost A requires that the random variable x be added to the random variable y. If a random number is found from a random-number table or computer file, it is set equal to the cumulative probability distribution. A random number of 0.73 is drawn for design x, and a corresponding value is \$1,600,000. In a similar way, we set a random number of 0.18 to the functional model for y, and then $C_y = -8\ln(1 - 0.18) = 1.587$ or \$1,587,000. Our solution is cost A $= 1{,}600{,}000 + 1{,}587{,}000 =$ \$3,187,000, which is accepted as a single trial of our cost A.

This procedure is repeated many times, and finally a distribution of direct cost A is found. The shape of this distribution varies from a jerky histogram to a smooth normal distribution. This depends on the number of trials [see Figure 6.11 (e) and (f)]. For the examples, the expected mean is \$9,450,000 while the standard deviation is 8002, which is shown as Figure 6.11 (f). For a typical engineering-business system, the distribution is tested for its cost-estimate average, standard deviation, confidence interval, and other statistical properties.

In using simulation to estimate engineering-business cost, it is usually necessary that input relationships be independent of each other. Typically we apply the *central limit theory* to say that the concluding distribution is a normal distribution.

There are numerous offspring of statistical-probabilistic and numerical methods that are not discussed in this chapter. Decision trees, Bayesian, neural nets, artificial intelligence, fuzzy logic, etc., are interesting, yet space limits our choices to the more popular methods. For additional information on these approaches, the student is directed to standard references.

6.6.5 Single-Value or Probability-Distribution Comparisons?

Suppose that the statistical distributions of two designs, A and B, are to be compared and that minimum cost is the criterion for selection. A hypothetical question is asked: Which design is the preferred choice if all else is considered equal with the exception of cost?

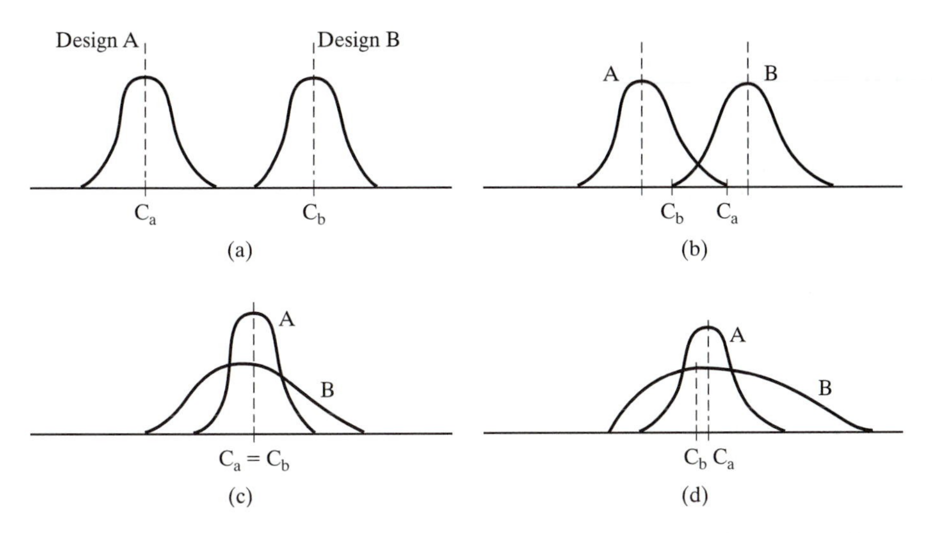

Figure 6.12 Cost analysis of designs A and B shown as probability distributions.

If single-value estimates are adopted, and it is necessary to compare between competing choices, say A or B, it is straightforward to select one based on the criterion of minimum cost. But if probability cost distributions are available for designs A and B, the choice is not always clear-cut. Refer to Figure 6.12 for this discussion.

For Figure 6.12 (a) all probable costs of A are lower than those for B and there is no difficulty in choosing A. Notice that the *expected costs* are identified by C_a and C_b.

The situation for Figure 6.12 (b) has a probability that the actual cost of A is higher than that for B, or $C_a > C_b$ even though the mean cost of A is less than the mean cost of design B. Given that there is a small chance that $C_a > C_b$, we may comfortably select A. But as the amount of overlap between distribution A and B increases, the expected-cost estimate may not be given a clear mandate.

For Figure 6.12 (c) the average cost estimates are equal, although their distributions are not. Certainly, the cost distribution of B is greater and there is a chance of having a lower cost than A. On the other hand, A is less variable. The appraisal of minimum cost and risk will serve as the guide here. How would you make the choice?

For Figure 6.12 (d) the expected-cost estimate of B is lower but less certain than that of A. If only the expected-value estimate is used in this case, then we are likely to choose the more desirable alternative B.

Which is preferred: single-value or probability-distribution estimating? The answer is not always obvious. We see that statistical/probability science may not make the choice of either design A or B any more certain. The engineer and manager will need to rest their choice on experience, even in some of these cases.

SUMMARY

Estimates are assembled on available information. Hypothetically, if there is no information, then there can be no estimate. Conversely, an estimate is unnecessary if all actual costs are available. When all actual costs are available, the cost analysis depends on accounting procedures.

The engineer operates within the limits of incomplete information. In this in-between region we separate estimating methods into preliminary and detail. Preliminary methods are less numeric than detail methods. Accuracy of the estimate is improved by attention to detail, but offsetting this are the speed and cost of preparation that favor preliminary methods. The methods are preliminary because they correspond to a design that is not well formulated. Frequently, a preliminary estimate leads to the management decision of further consideration for the design.

Before detail methods of estimating are attempted, information that has been deliberately collected, structured, and verified to suit the method of estimating must be available. Most engineering-work estimates are shaped by using detail methods. Large-scale projects are less dependent on detail, and they lean more on information specially developed by the owner or contractor.

In the next chapter the methods of detail estimating are examined further. Our study ranges from simple cost factors to encyclopedias of performance time data and extensive tables of material unit costs to estimate manufacturing work.

Here we return to the question: How do we select the right method for a particular application? Regrettably, no textbook can supply an infallible set of rules. You will have to rely on study and experience.

QUESTIONS FOR DISCUSSION

1. Define the following terms:

Benchmark costs	Learning
Central limit theorem	Modal cost
CERs and TERs	Monte Carlo simulation
Conference method	Opinion probability
Cost driver	Parameter
Cumulative learning	Percentile
Detail estimating	Preliminary estimating
Economy of quantity	Range method
Economy of scale	Sixth-tenths model
Hidden card	Unit-quantity method

2. Discuss the timing of preliminary estimates. Is the timing a precise point in a well-ordered organization?
3. Use the conference method to estimate the following: (a) the price for a clean-air car using a hydrogen fuel-cell electric wheel drive in 2010; (b) the ticket price for a 5000-mile one-way air trip; (c) the cost of a year's college education in the year 2020; (d) the time to machine a round 4 in. stainless steel barstock that is 20 in. long for a 0.15 in. depth of cut.
4. What advantages can you cite for the conference method? Disadvantages?
5. Assume that design A can be redesigned into designs B and C with known costs. What safeguards can you suggest to ensure that A is properly estimated?
6. Give the steps for the development of data for the factor method.

7. Point up the human frailties in determining opinion probabilities. Would you think that one would underestimate or overestimate those point probabilities?
8. Describe the kinds of applications suitable for the factor method. What kinds are suitable for the unit-quantity method?
9. What is meant by the law of economy of scale? Apply it to the factor method. Does the unit method follow this law?
10. Aluminum 12-oz containers are ubiquitous in American life. Billions are produced annually. The four major components of their cost are (1) advertising and marketing, (2) distribution cost from bottling and container plant to store, (3) contents of container, and (4) container cost. Rank these four components from cheapest cost-per-filled-and-delivered-container unit to most expensive. Write down your rankings and give reasons.

PROBLEMS

6.1 An inventor has fashioned a new mailbox design with an interesting feature. After leaving the mail, the mail carrier closes the hinged front-cover of a street-unit mail box, and a spring-loaded plastic flag pops up at the back of the steel box, signaling the owner that she has received mail. The inventor surveys popular catalogs and finds the following description and cost data:

Size, in.	Weight	Primary Material	Cost	Best Feature
10×3	1 lb	Plastic	\$14.16	Mail slot
$5 \times 2 \times 11$	2 lb, 11 oz	Steel	20.40	Liberty Bell emblem
$13 \times 7 \times 3$	3 lb, 6 oz	Steel	31.20	Hold magazine, wall mount
$6 \times 2 \times 10$	2 lb, 11 oz	Steel	35.20	Letter size
$14 \times 7 \times 4$	4 lb, 4 oz	Steel	47.20	Holds magazines, wall mount and black wrinkle paint finish
10×3	1 lb, 8 oz	Brass	47.20	Mail slot
$14 \times 7 \times 4$	4 lb, 12 oz	Steel, aluminum	55.20	Holds magazines, wall mount
$6 \times 2 \times 10$	4 lb, 8 oz	Some forged iron	64.80	Letter size
$14 \times 6 \times 4$	4 lb, 15 oz	Aluminum	71.92	Holds papers, magazines, wall mount

(a) In using comparison estimating, where do you place the inventor's design in the table?
(b) Construct a rough unit estimate from the data to help in the evaluation. (*Hint*: Do not expect too much from an analysis of this sort, but it may give insight.)
(c) How much "forgiveness" is necessary for a survey of this type? How would you sharpen the approach?

6.2 Five time studies are taken of a "handling" element. The technician suspects that weight is the major driver. Weight and average time for the element are as follows: (lb, min), (6, 2.1), (13, 3.7), (21, 3.9), (23, 4.2) and (32, 5.2). The average total operation time = 10 min, or $y_t = 10$ min. Is "handling" a variable or constant element? If it is a constant element, then find its value. (*Hint*: Find the regression equation for the data, and use that to find the dependent values for the two tests.)

(a) Let P_1, and $P_2 = 100\%$ and 25%.
(b) Repeat for P_1 and $P_2 = 100\%$ and 50%.

6.3 An element "operator moves casting and place into fixture" has been time studied and information recorded. Three possible independent variables are casting weight, girth, and fixture locating points and data are given in the table below. (*Hint*: Use the correlation coefficient and Test 1 for evaluation of the three variables.)

Determine three linear one-variable regression equations where each variable is a time driver. Discuss criteria for selection of the best time driver among the three candidates for the preparation of productivity data. Which of the three equations is the "best line?" For your best line, provide a table of 25% increasing steps in time. (*Hint*: One metric that shows the strength of relationships is the correlation coefficient r, as discussed in Chapter 5. To convert a regression equation, start with $x_{\min}$, and then increase the x independent variable and solve for y. Show the time dimension to two decimals.)

Time Study	Time, min	Weight, lb	Girth, L + W + H, in.	Fixture Location Points, no.
1	3.4	15	7	3 (3 buttons)
2	1.4	5	6	4 (nest)
3	2.8	12	7	3 (2 edges, 1 pin)
4	2.2	10	3	4 (nest)
5	4.8	20	11	1 (pin only)
6	4.2	17	9	2 (2 edges)

6.4 Find the estimated lot hours and cost, cost per unit, cost per 100 units, and pieces per hour for a spot welding operation. A spot-weld design has a primary, secondary, and a third part that forms a chassis. The productive hour cost is $35. The lot quantity is 175. Use Table 6.5. (*Hint*: The metric hr/100 units depends only on the cycle time, while lot hours includes the setup hours and cycle minutes for the lot quantity.)

(a) The third part girth is 14 in. A total of 22 spot welds are necessary. (*Hint*: Follow the rules of thumb in applying Table 6.5. The dimension "girth" is the sum of the maximum dimensions in the x-y-z coordinate axis). What purposes do these rules serve?

(b) Repeat for another design where third-component girth is 20 in. and the number of spots is 29.

(c) Repeat for another design having two parts and the number of spots is 37. There is no third part.

6.5 (a) Find the estimate for unit 6 when the value for unit 3 is 7961 hours at 75% learning.

(b) Find the unit time for the fourth and sixth units when unit 1 has a learning rate of $\phi = 87\%$ and 18,000 hours. (*Hint*: Use the notion of doubling to find the answer for unit 4. Use Equation (6.6) to find the answer for the sixth unit.)

(c) Find the intercept value K when the tenth unit is 6054 hours with 95% learning. What is the expected unit time for the eleventh unit?

(d) If unit 1 is $2 million, then find the dollars for units 2 and 4 with learning rates of 92% and 100%. What observation can you make for 100% learning?

(e) Units 1–50 have accumulated 10,000 hours. Calculate the time for the twenty-fifth unit if the least-square slope for the learning rate is 92%. What is the time for the fiftieth unit? (*Hint*: Use an approximation.)

(f) If the cost at unit 10 is $100,000, then find the unit, cumulative, and average cost at unit 11 for a learning rate of 92%.

(g) A contractor has collected actual data for units 3 and 4 as 7729 and 7225 hours. Find the unit, cumulative, and average hours for the fifth and sixth units.

6.6 An audit of learning performance revealed that at the 18th product unit the direct product cost was \$1,810,000 and at the 24th product unit it was \$1,690,000. What is the estimated direct product cost for the 27th unit?

6.7 Four projects are designed and built in serial order, and project management believes that the historical data conform to the learning model.

Project No.	1	2	3	4
Unit Cost, $\$10^3$	7815	7424	7205	7053

Roughly plot the unit, cumulative, and average values on both arithmetic and logarithm coordinates. Find the future unit, cumulative, and average values for the fifth and sixth units using the graphical plots. (*Hint*: Each project is separated by a start and a conclusion from its neighboring project, and there is no intermingling of cost values. These unit values are determined after the conclusion of the project. It is the unit value that is considered *actual*, and the cumulative and average values are derived from the actual number.)

6.8 A project has an actual first unit cost of \$72.8 million. Four additional and similar units are to be financed by bank loans and constructed in sequential fashion. The bankers are anxious that a learning model be adopted for budgeting and financial management. Based on other related experiences, the bankers insist that costs follow the 93.7% learning model. Determine a table of unit, cumulative, and average costs for the next four projects.

6.9 There are two competing engineering designs, called A and B, for processing the same product. Product cost estimates are found for unit 1 for the rival engineering designs. The learning rates are chosen based on comparison to earlier experiences for engineering process systems A and B.

	Engineering	
Estimate	Design A	Design B
Initial unit cost	\$600,000	\$710,000
Learning rate	95%	85%
Final no. of units	3	3

If the product from either engineering system design are equal in function and intrinsic value, which engineering design is the preferred economic choice? (*Hint*: Assume that the manufacturing tooling costs for A and B are identical, though the product costs are not.)

6.10 An 80-kW diesel generator set, naturally aspirated, cost \$16 million 8 years ago. A similar design, but 140 kW, is planned. The economy of scale exponent $m = 0.6$. At the time of the original information the index $I = 187$, and now the index is 207. A precompressor is estimated independently at \$1.9 million.

(a) Find the estimated equipment cost.
(b) Repeat for $m = 0.7$.

6.11 A cost estimate is requested for a 34-cm-dia disk laser amplifier equipment. A previous design of 22 cm dia was estimated for a cost of \$16,459,000. The annual Producer Price Index = 7.5% for equipment, and will be used for a 3-year escalation. A value of $m = 0.7$ is adopted. Find the cost of the new equipment. (*Hint*: Consider the PPI to have a compounding effect on cost.)

6.12 The cost of natural-gas fired turbine equipment is suspected to follow the power law and sizing CER. Information on turbine cost and specification is gathered as follows:

kW	Cost, $\$10^6$
3,500	18
5,000	25
7,000	32
8,000	35

A similar design, except that it is larger at 8750 kW, is to be estimated. Determine the slope m and estimate the cost of the new unit. (*Hint*: Use $\log C = \log C_r + m \log {}^{Q}\!/_{Q_r}$ and let $Q_r = 3{,}500$ kW. Plot the relationship on logarithmic graph paper. Have Q/Q_r as the x-axis and cost as the y-axis. You can graphically find the slope by using the rise/run triangle.)

6.13 An electric-motor manufacturer sells totally enclosed capacitor motors, fan cooled, 1725-RPM, $^5/_8$-in.-O.D. keyway shaft, and advertises its costs for $^1/_4$, $^1/_3$, $^3/_4$ and 1 hp as \$90, \$100, \$118 and \$140 in a suppliers catalog. A 3450-RPM, 1-hp motor costs \$134. Determine m for the 1725-RPM motor series, and estimate the cost for a $1^1/_2$-hp motor, which is not advertised in the catalog. Should the 3450-RPM motor be included in the sample to estimate the 1725-RPM motor?

6.14 An equipment bid for a rising-film and ozonation reactor is received from a job shop for \$1,750,000. The current index for this class of equipment is 121.5. The project cost for a soap-formulation chemical plant is required.

(a) Find the benchmark cost for the equipment. What are the factors for engineering, erection, and direct materials? (*Hint*: Use the graph Figure 6.8 for the information.)

(b) An overall index of 123 is forecast for the costs during the construction period. The regional factor for the plant site is 7% more than for the Gulf Coast. Items independent of the factor-estimating process are found to cost \$2,000,000. What is the total factor cost for the chemical plant?

6.15 (a) A designer estimates a project to cost \$5 million. This design contractor believes that a bid of \$5.6 million has a probability of 0.3 of being low and become the winning bidder. Another bid of \$5.3 million has a probability of 0.8 of being low and being selected. Which bid gives the best expected profit? (*Hint*: Notice that as the bid becomes higher, its chance of being selected is lower.)

(b) A salesperson makes 15 calls without a sale and 5 calls with an average sale value of \$200. What is the expected sale value per call?

(c) A engineering reseller accepts old equipment as trade-in for new models and sells the traded-in equipment through a secondary outlet. Analysis shows that the markup is \$2500 on 70% of the equipment and \$4000 on the rest. What is the expected markup?

(d) An insurance company charges \$20 for an additional \$50 increment of insurance (from \$100 to \$50 deductible). What is their assessment of the risk for the increment of insurance?

(e) A company sells two different designs of one item. A study discloses that 65% of its customers buy the cheapest design for \$75. The remaining 35% pay \$110 for the expensive model. What is the expected purchase price?

6.16 Revisit the joint-probability example given in the chapter. Find the *expected profit* for the following changes in the probabilities:

Annual Sales Volume	Probability of Sales Volume	Cost per Unit	Probability of Cost per Unit
15,000	0.1	$450	0.6
20,000	0.3	500	0.3
25,000	0.6	550	0.1

6.17 A newly graduated engineer buys a new car for $26,274. Already she is planning for its replacement at some point in the future. Her estimating model for the decision, she reasons, is

$$\text{Replace car} \leq 26{,}274 - \Sigma\text{depreciation of existing car} - \Sigma\text{maintenance of existing car.}$$

Once the model becomes negative, the point is reached when the replacement is made. Assume that the price of her next car, whenever she buys it, will be equal to that of her first car. Other driving costs are the same regardless of whether she drives a new car or not. The depreciation of cars is well known, and she tabulates that value in a table. A major maintenance cost is $4000. Her *opinion probability* for a major maintenance cost is given as follows:

Life	Cumulative Probability of Major Maintenance	Cumulative Decline in Depreciation
1	0.1	0.24
2	0.2	0.49
3	0.4	0.64
4	0.7	0.75
5	1.0	0.83
6	0.1	0.89
7	0.2	0.93
8	0.4	0.96
9	0.7	0.98
10	1.0	0.99

(a) When should she replace her car?
(b) How would you improve the replacement model?

6.18 (a) Find the expected total mean cost and variance of the cost items in the table below.
(b) What is the probability that the future project mean cost will exceed $325 million?

Cost Item	Optimistic Cost, $\$10^6$	Most Likely Cost, $\$10^6$	Pessimistic Cost, $\$10^6$
Direct labor	79	95	95
Direct material	60	66	67
Indirect expenses	93	93	96
Fixed expenses	69	76	82

6.19 A five-element project is estimated as follows:

Cost Item	Optimistic Cost, $\$10^6$	Most Likely Cost, $\$10^6$	Pessimistic Cost, $\$10^6$
1	4.0	4.5	6.0
2	10.0	12.0	16.0
3	1.0	1.0	1.5
4	4.0	8.0	12.0
5	2.0	2.5	4.0

(a) Determine the elemental mean costs, total cost, and elemental and total variance.
(b) The project will be priced to a customer for \$26 million. Find the probability that cost will exceed \$26 million. What is the probability that the cost will be less than \$26 million.

6.20 Use the percentile method to estimate the following:

	Percentile Cost		
Cost Item	10th	50th	90th
1	\$4.0	\$4.5	\$6.0
2	10.0	12.0	16.0
3	1.0	1.0	1.5
4	4.0	8.0	12.0
5	2.0	2.5	4.0

Find the 10th, 50th, and 90th percentile costs. Also calculate the contributions of the cost items to low uncertainty, total cost, and high uncertainty.

6.21 Direct cost is equal to the sum of direct labor and direct material. Let direct labor be distributed exponentially with the mean of 3. Direct material is distributed continuously with $f(y) = 2/9\ y$, where $0 \leq y \leq 3$. Find the mean of direct cost. (*Hint*: For convenience use the 10 random-number sets from Problem 6.22.)

6.22 Refer to the illustration given in Section 6.6.4 and determine a new value of the Direct Cost A using the following random numbers:

Random Number, Direct Labor	Random Number, Direct Material
0.24	0.64
0.82	0.98
0.83	0.25
0.18	0.94
0.66	0.03
0.76	0.23
0.07	0.96
0.62	0.80
0.61	0.64
0.96	0.99

Sketch a histogram of the random variable Direct Cost A. Find the mean and standard deviation of direct cost, assuming that each of the ten independent random pairs is a trial.

(*Hint*: Because of the small number of trials, it is necessary to make a wide-interval selection of direct cost to allow more than one value in each histogram interval. Perhaps select only 4 or 5 intervals, and have the lower and upper intervals as open ended. The vertical axis of the histogram is count.)

6.23 A simulation model is defined as $C = x/y$, where cost x is given by

$$f(x) = \begin{cases} \dfrac{1}{b - a}, & 2 \leq x \leq 8 \\ 0, & \text{elsewhere} \end{cases}$$

and the cost variable y is given by the frequency:

Cost of y, $	Occurrence
1	0.15
2	0.25
3	0.40
4	0.15
5	0.05

Find the mean C after five simulation trials. (*Hint*: Use the first five random-number sets of Problem 6.22 for $f(x)$.)

CHALLENGE PROBLEMS

6.24 Consider the circumstances of an Internet Application Service Provider (ASP), compilers of information for Web-based estimating software. This firm is evaluating silicon integrated-circuit wafer and device production. In particular, the throughput of wafers per hour has been collected from three suppliers for their equipment, and information is 33.0, 27.2, and 34.7 8-in. wafers per hour. The compiler is wondering on how the work element should be developed and displayed for the Web page. Should the element be listed as an average rate per hour, or should it be referenced to a table where additional information of equipment cost, wafer yield, and equipment utilization are added? (*Hint*: The dilemma is that simplicity gives ease of use by the engineer, or the data can be referenced and annotated to equipment performance, yield, and cost and thus open the choice to the engineer, and possibly cause errors during the product estimating stage.)

Find the process average. If the ASP policy for an element to be variable is 20%, how should the element be considered? Discuss the consequences to the ASP compiler and the user of the database for either choice.

6.25 Develop performance data for the sawing of 8 in. wafers in an electronic fabrication operation. Four tasks are given for the equipment. The time study elements are in minute units and are considered "normal time."

Work Element	Measurement Study				
	1	2	3	4	5
1. Pickup wafer, min	0.023	0.023	0.035	0.041	0.051
Variable: Robot move distance, in., x_1	8	8	13	16	17
2. Table automatically positions for sawing, min	0.031	0.035	0.029	0.030	0.032
3. Machine moves integrated chip to carrier, min	0.024	0.021	0.031	0.030	0.030
4. Saw chip to length, min	0.093	0.126	0.103	0.210	0.193
Variable: Saw length in wafer, in., x_4	0.25	0.25	0.25	1.00	0.875

Develop the performance data in a tabular format with steps of 25% for variable time and present the data ready for application. Elements 1 and 4 are variable. Allowances are 15%.

A performance table may have fixed time steps, which is useful for linear lines. An efficient way of doing this is given by

$$\hat{y}_{i+1} = \hat{y}_i(1 + P) \quad \text{and}$$

$$x_{x+1} = \frac{\hat{y}_{i+1} - a}{b}$$

where x_{x+1} = independent variable, recursively determined
y_i = minimum starting dependent value, time
P = percentage for step, abritrarily fixed
a, b = regression coefficients for intercept and slope, number

(*Hint*: To convert from normal time to standard time increase normal time by the allowance multiplier 100%/(100%–PF&D%).

6.26 The learning phenomenon has been confirmed with the construction of similar natural-gas-fired central-steam power plants in the western plateau region of the United States. In this actual case, construction involved different project design companies and general contractors for each power plant. Despite this mix, the learning technology transferred between the projects.

Actual data are recorded as (N = unit no., T = $\$10^6$, and year no.) or (1, 98.75, 1), (2, 89.86, 2), and (3, 82.85, 3). The cost data are adjusted for escalation.

An owner is interested in exploiting this feature for future plant estimates. Find the unit, cumulative, and average costs for the future construction of units 4, 5, and 6. The future index values are (year, index no.), or (4, 1.2), (5, 1.3), and (6, 1.5).

(*Hint*: The actual cost data of years 1, 2, and 3 are de-escalated to the benchmark period 1. Unit 4 will be built in year 4, and so on. The two-point determination of s is not appropriate. It is necessary to find the slope s and the intercept K of these data using the regression models of Chapter 5 as there are three sets of values. Regression equations are given as

$$\log a = \frac{\Sigma(\log x)^2 \Sigma \log y - \Sigma \log x \, \Sigma(\log x \log y)}{n\Sigma(\log x)^2 - (\Sigma \log x)^2}$$

$$b = \frac{n\Sigma(\log x \log y) - \Sigma \log x \Sigma \log y}{n\Sigma(\log x)^2 - (\Sigma \log x)^2}$$

Spreadsheet analysis will be useful for this regression solution to find intercept K and slope s.)

6.27 The "average learning model" is defined differently than the "unit model," which is the one given in the chapter, and is given by the following:

$$T'_a = KN^s$$

$$T'_c = KN^{s+1}$$

$$T'_u = KN^{s+1} - K(N - 1)^{s+1}$$

where T'_a = average time or cost per unit
T'_c = cumulative total time or cost per unit from 1 to N
T'_u = unit time or cost given that T'_a is defined as the linear line
K = intercept at $N = 1$
N = unit number, 1, . . . , N

(*Hint*: It is pointed out that the average learning model is the original practice, but later years saw the unit model become more prominent. The unit model is also known as the Boeing model.)

(a) Find the average value for the 20th unit when the 10th unit has an average of 100 hours for 68% learning.
(b) The average cost for the 100th unit is $10,000. Find the average cost for the 50th and 200th unit for a learning rate of 87%.
(c) If 400 average hours are required for the third unit and 350 hours for the sixth unit, find s, T'_a, T'_c, and T'_u for the tenth unit.
(d) If 800 average man-hours are required for the tenth unit and 750 hours for the 20th unit, find the percentage learning ratio, average time, cumulative time, and unit time for the 5th and 40th units.
(e) If the unit time at unit 100 is 100, find the unit and cumulative and average times at unit 101 for a learning rate of 92%.

6.28 Engineering equations allow estimating applications. Such is the case for pressure vessels for chemical process plants. Early estimating of vessels use expressions for the shell weight, given by

$$t = \frac{PD_m}{2SE} + C_a$$

where t = wall thickness of vessel, inch.
P = pressure, psi
D_m = average of inside and outside vessel diameter, inch.
E = joint efficiency, dimensionless decimal
S = allowable maximum working stress, psi
C_a = corrosion allowances for additional wall thickness, inch.

$$\text{Vessel cost} = K\rho CtL$$

where ρ = material density, lb/in.^{3}
K = material cost, dollars per pound
C = mean circumferance of shell, inch.
L = vessel length, inch.

The following information is determined from engineering flowchart calculations: $C_a = 1/16$ in., $S = 10{,}000$ psi, $D_m = 220$ in., $E = 1$ for a welded structure, $P = 25$ psi, $\rho = 0.265$ lb/in.^{3}, $K = \$2.45$/lb for AISI 6120 chromium-vanadium steel that is rolled to the vessel diameter, and $L = 18$ ft. For this type of field welding, a waste allowance of 20% is anticipated.

Find the direct material cost of the shell and ends for this vessel. (*Hint*: Calculations for the thickness of the minimum shell will likely give a nonstandard material thickness, and it is necessary to increase upward to the next higher standard thickness. The top and bottom ends will likely be elliptical in shape, but as an approximation use the spherical equation for the area of a sphere, or πD_m^2. This material cost is excluding welding fabrication and erection labor.)

PRACTICAL APPLICATION

Use the conference method to estimate the cost of an engineering task or product. Selection of the assignment and forming of the team will be guided by your instructor. The general description of the task should be familiar to the team, but the exact cost value is a mystery.

Candidates for the task, for instance, are the cost of a local but important manufacturing operation that is important to the economy of the region. Then check the value of the design by making formal inquiry with experts. Write a report detailing the process.

(*Hint*: As important as determining the cost of the engineering task is to the conference method, it is significant to grasp the advantages and foibles of the procedure. The report should evaluate the process. *Evaluation* is where the learning takes place.)

CASE STUDY: INDUSTRIAL PROCESS PLANT PROJECT COST

An industrial process plant is to be constructed. Major basic items are separately quoted by fabrication contractors in current-time cost as $2,720,000 during an index period of 136 as compared to a benchmark index of 100. The estimating graph, which is shown as Figure C6.1, is based on the Gulf Coast region. It is anticipated that the plant will be built in five years, and the index for that period's cash flow is forecast as 172. The plant site is in the northeastern region of the United States, and construction for that area is expected to cost more than the base area by 8.3%. Plant site development is independent of the basic items and is expected to cost $8.2 million. Find the plant cost. Which element is the most expensive? (*Hint*: The estimating graph is constructed for an index of 100. The independent costs are valued at current time and are estimated in the region of the plant site.)

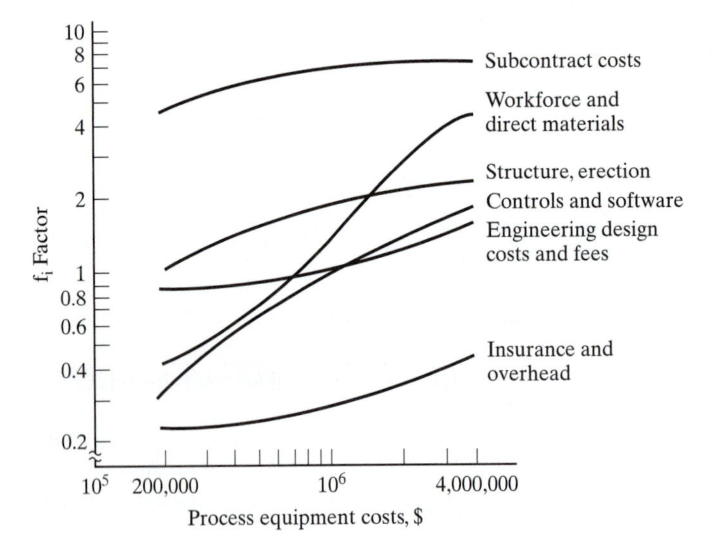

Figure C6.1

Chapter 7

Operation Estimating

A common approach to operation estimating begins with breaking down an operation into its essential elements. A design for this analysis is available, and the engineer uses cost equations and engineering performance data to model the cost details.

In manufacturing, the design is a part print and a part processing plan—for example, in chemical industries a flowchart and layout with engineering calculations and in engineering construction a set of drawings and specifications. Those designs are used for purposes such as estimating, planning, scheduling, methods improvement, and production or construction. The designs communicate the description and sequence of operations and are the hub on which individually and collectively many decisions, both great and small, are made.

Outmoded practices of operation estimating take the total cost of operating the plant for a time and divide by a total production quantity. Other practices determine a labor and material cost for an operation from a test run or prototype. Historical records of like operations are used. Guesstimating, as it is called, is wrong for many reasons. Those methods are unsuitable for predictive cost estimates.

In contemporary practice the engineer begins by dividing the design task into large portions of labor and material. A progressive pattern of fine details is determined until the description of labor and materials is very broad. At this point, dollar extensions of labor and material are made to reflect the cost of the design.

Thousands of labor and material operations exist. Regrettably, our choice of explanation extends only to a very few. Trade books, handbooks, company sources of data, Websites and so on must be consulted for facts for any real-life estimating. Some practical sources of information are listed in the references at the end of this book.

Analysis of operation cost is conducted by a variety of professionally trained people. The effort may include a variety of individuals—from designer to manufacturing engineer to production planner to engineering manager, for instance.

7.1 BACKGROUND

Before an operation estimate is started, some authority must initiate this activity; that is, a request for quote (RFQ), an e-commerce inquiry,[1] work order, production planner, foreman's request, work-simplification savings, employee suggestion, or sales request. Those requests are the commissioning order to begin the work of estimating. The order package will include the design, period of activity, quantity, quality, and any other requirements.

Operation estimating follows many paths. In all these approaches, *microanalysis* is preferred, which is a facility for subdividing a manufacturing operation into physical and economic elements. Because detail estimates are relatively costly to prepare, the risk of achieving success must be offset by the cost of preparation. If risk were not minimal, then the estimates for the operations would not be started. This risk is measured by a *capture percentage*, which we define as (estimates won)/(estimates made) $\times$ 100. For job shops the capture percentage may be low, and quotations are prepared by using preliminary procedures. If the capture percentage is high, or the cost of estimating cost is not proportionately higher than the benefits from success, then detail estimating is mandatory. Though the labor and material may differ between designs, techniques for estimating are based on the same principles and practices.

Manufacturing operations are necessary to produce a change in worth. The fusion of people, equipment, and tools is the primary component. The economic value of material is altered and the measure of the change is called *cost*—a term that in this instance implies a consumption of labor, materials, and tools to increase the value of some object. In some cases, automatic equipment produces units of output and may not require labor. Those operations, however, consume materials and utilities and constitute the operation cost estimate.

Furthermore, manufacturing operations require tools, fixtures, or test equipment. Without such tools the operation is not possible. Those tools, which we refer to as nonrecurring initial fixed costs, or simply "tooling," are estimated along with direct labor and material.

Operation evaluations are limited as to the time horizon. The immediate future period rather than an extended time period is intended. The nature of labor is for a period of time such as "units of labor time per piece" or "units of labor per month or year."

The purpose of operation estimates is to establish the cost for components and assemblies of a product, to initiate the means of cost reduction, to provide a standard for production and control, and to compare different design ideas. Estimates may verify operation quotations submitted by vendors and help determine the economic method, process, or material for manufacturing a product. Whenever an operation is material or labor intensive, the methods described in this chapter are primary.

Figure 7.1 is a description of the manufacturing elements estimated for an operation. The term *prime cost* is accounting terminology, meaning direct labor and material cost.

[1]Web companies that provide facilitating software, or what is called "enterprise software" for this activity are given in the references.

PICTURE LESSON Turning

Kennametal (Bill Kennedy).

The desirable properties for any tool material include the ability to resist softening at very high temperature, low coefficient of friction, good abrasive resisting qualities, and sufficient toughness to resist fracture.

The photograph shows a titanium-nitride-coated carbide metal cutting insert—about the size of a dime and in the shape of a diamond—performing a rough turning operation on a bar stock. The insert is in intermetal contact with the rotating steel bar stock. Notice the curling chip in contact with the tool. The diamond insert has eight corners; after a corner is used up, the insert is "indexed" to another corner, which prevents the surface from becoming excessively rough. The tool body is used over and over again, but the cutting tip, also called an insert, is replaceable once it wears out, as a consequence of the tremendous friction between the tool tip and the workpiece. The insert becomes waste after all eight corners are no longer functionally removing material.

Notice the small groove on the insert. A groove is formed in the green compact prior to its sintering and hardening during powder-metallurgy production. The purpose of the groove is to reverse the direction of the chip, the material that is removed from the stock. A continuous chip—think of a string unwinding off a string ball—will happen under certain conditions. The continuous chip is humanly dangerous and causes removal problems. The reversal creates stresses in the chip, and it fractures as a result. Small chips are seen in the air.

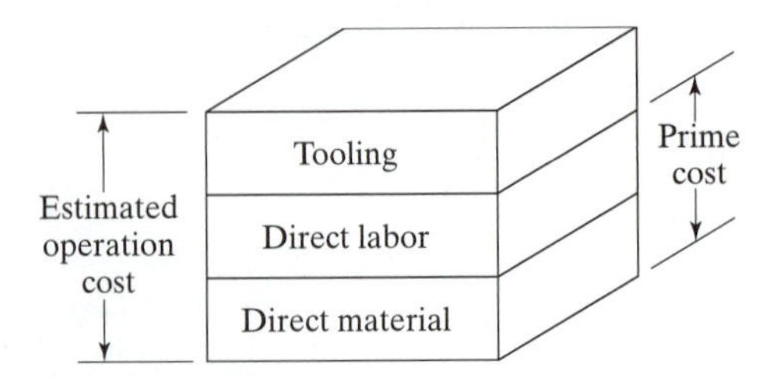

Figure 7.1 Descriptive layer chart of cost elements included in an operation estimate.

7.2 MANUFACTURING: WHAT IT IS, WHAT IT DOES, AND WHAT IT IS NOT

At this point we broadly explain the thinking about "industry." The keys to this narration discuss the plant and its location, output, and the relative value of the output. In Table 7.1 we identify four industries: (1) construction, (2) agriculture, (3) information and service, and (4) manufacturing. The features that delimit these industries are remarkably diverse, and those differences cause substantial variation in the practices of engineering, economic analysis, marketing, and entrepreneurship.

These businesses are profit motivated, pay taxes, and are not generally subsidized by governments that would lessen the competitive drive of self-interest. The players for these industries include owners and shareholders, engineers, employees, and customers.

The *construction industry* is portable. Think of construction plant and equipment such as graders, backhoes, etc., that are mobilized at a plant site. Materials are brought to the plant site and on completion of the work, the output stands fast while the plant moves away. Construction output includes immobile structures, such as airports, buildings, canals, dams, factories, homes, municipal treatment plants, pipelines, power plants, roads, structures, and tunnels, to name a few.

Construction plants vary in size and complexity, ranging from the plant to build the Hoover Dam to a worker, wheelbarrow, pick, and shovel. Construction plants are themselves the output of manufacturing plants, and they construct the manufacturing plants. The student may want to answer the chicken-egg riddle: "Which came first?"

The construction plant produces a single and unique end product. The product is stylized in design and varied in method of erection, and it differs in location. For instance, some homes are basically similar, but the units are site adapted.

TABLE 7.1 The Big Picture About "Industry"

Industry	Plant and Equipment	Location of Plant Site	Output of Industry	Relative Value of Output, Customer's View
Construction	Portable	Variable	Immobile and unique	Expensive
Agriculture	Mobile	Land fixed	Moves off the plant site and is numerous	Inexpensive
Information and service	Transitional	Virtually anywhere	Temporary	Low priced
Manufacturing	Fixed	Fixed	Moves off the plant site and is numerous	Cheap to expensive

The construction industry intersects all fields of human endeavor, and this diversity is enjoined in its projects and economic impact. Total construction employs about 5% of the national workforce, which is gauged at 8 million workers. It is a $1.2 trillion industry in the United States.

The *agricultural industry* is similar to a manufacturing plant except that the equipment moves on and off the land plant site. Tractors and harvesters are typical equipment. Plant location of farms and ranches is fixed.

In the United States agriculture accounts for 2% of the workforce, down from a high of 50% at the time of the Industrial Revolution of the 1850s period. Remember that we are producing more food, but with fewer people.

The *information and service industry* is doing either symbolic-analytic work or providing services. It is difficult to neatly categorize so great a spectrum of product that emanates from the information and service industry. But by the midpoint of the twenty-first century a majority of the workforce will be knowledge-work, which is about the manipulation and analysis of symbols and the provision of related services. Such effort is not permanent; it is a transitional process, and changes are continuous.

For instance, software companies have little fixed hardware, equipment, and land but vast intellectual output. It is not necessary that the information and service plant abide near the source of raw materials, as is necessary for construction, agriculture, or manufacturing. The information and service industry can locate virtually anywhere. Territorial locations are unimportant—wired or wireless communications sees to this.

After making the first copy of knowledge-replete products like music, Web pages and computer operating systems, the marginal cost of every other copy is virtually nil, but its value to the user grows. This explains why a seemingly insane strategy, such as giving away your basic product, has become a strategy in the information/service economy. You can download thousands of software products for almost nothing, while the supplier collects revenue from other sources, such as selling upgrades, support, or advertising. America Online gave away as much product as possible, and the more it gave, the more demand for its product grew. These are not new ideas—Gillette gave away shaver-holders to sell the blades. The output of the information and service industry is temporal, sometimes lasting only days or appearing again in a newly molded intellectual form.

Cost evaluation of the information-service industry is different too. The principle of diminishing returns applies to the ever-shrinking proportion of value-added activity, as that activity increases. Consider the agriculture grain harvest: If you continue to add more and more fertilizer to the same acre, the output declines. Similarly in the steel business: As the larger mills got larger still in terms of capital investment, their profit per unit of investment diminished. But in the information industry, there are increasing returns as volume increases rather than diminishing returns leading to a low-price perception by customers, theoretically. The peculiar economics of digital businesses drives them, almost inevitably, to massive scale. The information network systems can be difficult and costly to build; but once built, they can be expanded almost at will, since the cost of replicating digits is minuscule. Moreover, the information businesses of today often become more valuable to their customers the bigger they get. Adolescents want AOL so they can chat with their friends; adults use Microsoft Word software because that is what their associates use. As a result, traditional economic thinking gets turned on its head. Instead of decreasing returns to scale, which some textbooks argue keep companies from getting too big, like the steel integrated companies of the 1950s, the information economy is characterized by increasing returns to scale.

With the Internet, a company can sell stuff it does not make from stores it does not own, and ship it in trucks driven by people working for someone else. The *transactions cost*, the cost involved in bring a product to market in an Internet economy are so reduced, it reduces the price to the consumer.

We define manufacturing, simply, as "making stuff."[2] The m*anufacturing industry* includes a great variety of factories and job shops. The plant is fixed, as the materials come to the plant, are processed, and products move off the plant site. Mistakenly, manufacturing plants are thought to be only mass production. This is not so, even though there are automobile plants that mass produce and beverage container plants where production of the popular thin-wall aluminum 12-oz. soda and beer container, for example, exceeds many million cans per day worldwide. But manufacturing plants are better termed "lot or batch" plants, as quantity ranges from one to many units. The average product-lot quantity is less than 100 units.

In the United States manufacturing is employing about 16% of the population. By the midpoint of the twenty-first century, only 5 to 10% of the workforce will be doing routine manufacturing. Probably, agriculture has shrunk to a lower level than manufacturing ever will. There are only so many calories that you can eat. This does not mean that we are in a postindustrial or postagricultural society as output is at an all-time high and efficiency is currently at a near maximum.

Manufacturing goods will not shrink in terms of people's buying. Productivity in manufacturing continues to be strong—we are simply exporting more. In addition, much of design is U.S.-based, while manufacturing, especially labor intensive work, is done outside the borders of the United States.

Furthermore, the industry of manufacturing can be separated into continuous or discrete manufacturing. In continuous manufacturing,[3] the process is dictated by chemical reactions, which is the domain of the chemical engineer. Usually special process equipment is connected by pipes and conveyors to perform the operations.

Discrete manufacturing of durable goods—anything from cars to turbines—can be broadly classified as mass production or moderate or job lot production. In *mass production*, sales volume is established and production rates are independent of single orders. In *moderate production*, parts are produced in large quantities and, perhaps, irregularly over the year. Output is more dependent on single sales orders. *Job lot industries* are more flexible, and their production is closely connected to individual sales orders. Lot quantities may range from 1 to 500, for example.

In discrete manufacturing, individual operations are conducted at a *machine*, *process*, or *bench*. Those workstations involve direct labor. A machine is capable of metalworking: examples would be a computer-controlled lathe, an automatic milling machine, or a drill press. A process tends to be chemical or fusion in character, such as

[2]A broader perspective of manufacturing than is discussed here can be found at www.SME.Org.

[3]The American Institute of Chemical Engineers provides an abundance of information about chemical engineering at www.aiche.org.

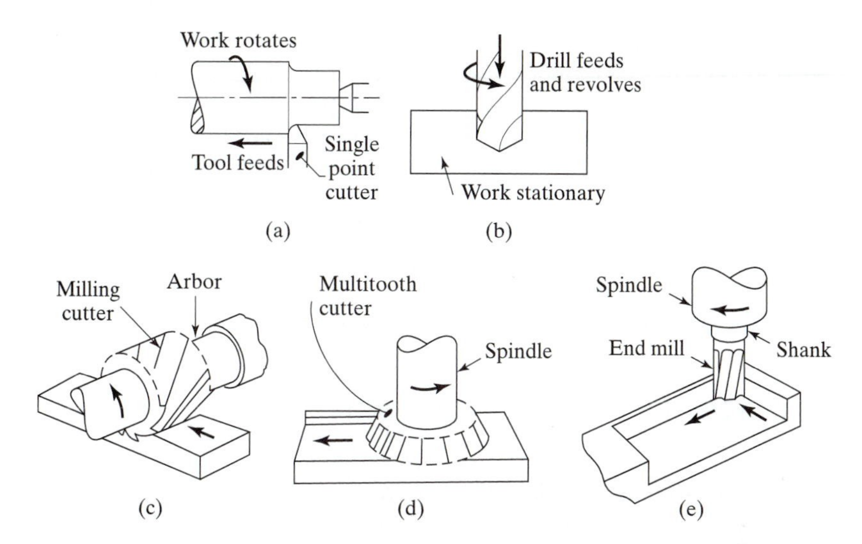

Figure 7.2 Simple metal-working operations: (a) turning, (b) drilling, (c) plain or slab milling, (d) face milling, and (e) end milling.

spray painting, silver plating, welding, or casting. Bench work is an assembler performing joining, fastening, and assembly at a bench or conveyor, which is customarily labor intensive. There are other classifications we could discuss, and the notion of "batch" or "flow line" is important, especially from a cost-estimating viewpoint. The batch and flow line distinctions are discussed later.

Figure 7.2 illustrates turning, milling, and drilling metal-working operations. We will study those operations more closely. Unfortunately, there are thousands of discrete manufacturing operations, and space limits our study to those three well-known processes.

In *discrete product manufacturing* there are several classes of processes: production of raw fabrication materials from earth materials, foundry and casting, machining, grinding, welding, joining, pressworking, forming, electronic fabrication, thread and gear working, finishing, and assembly, to name only some.

Machining produces a part shape by material removal, and forming produces a part by deforming the material and shaping it in a die or mold. Think of car fenders, which are a result of the forming process and in assembly parts are joined together. The knowledge about manufacturing is so vast that other texts are suggested for further study.[4]

In discrete manufacturing, we find the processes related to setup and cycle. *Setup time* includes work to prepare the machine, process, or bench to produce parts or run pieces. Starting with the workstation in a neutral condition, setup includes operator punching in or out, paperwork, obtaining tools, positioning unworked materials nearby, adjusting, and inspecting. After the parts are cycled, setup includes tearing down, returning tooling, and cleaning up of the workstation to a neutral condition ready for the

[4]Phillip F. Ostwald and Jairo Muñoz, *Manufacturing Processes and Systems*, 9th ed. (New York: Wiley 1997).

next job. The setup does not include time to make parts because that is included in the cycle time. Estimating setup time is done for job shops and moderate quantity production. In mass production, setup costs are of less unit importance, although its absolute value remains unchanged for large quantities. Setup is handled as an overhead charge for mass production, but it is estimated for each lot of production quantity.

Cycle time or *run time* is the work needed to complete one unit after the setup is complete. Unlike setup time, cycle time is the repetitive portion of the production quantity.

With those definitions behind us, two options offer different understandings of operation analysis—either classical or contemporary.

7.3 CLASSIC OPERATIONS ANALYSIS (OPTION 1)

Frederick W. Taylor, the great American engineer, is largely credited with advancing the engineering of manufacturing operations. We do a disservice to the student if we fail to study those ideas, even though they are a century old. For Taylor left a legacy that is beyond our discussion here, but his impact is large and his ideas remain a classic for study in the twenty-first century.[5]

This study is concerned with metal cutting, an important manufacturing operation. The operation cycle consists of load work (LW), advance tool (AT), machine, retract tool (RT), and unload work (UW). Those elements are shown in Figure 7.3 (a) and are labeled "one work cycle." Tool maintenance is required after a number of parts are machined, and includes removing the tool and replacing or regrinding the tool point and reinserting the tool to have it ready for metal cutting. Figure 7.3 (b) illustrates setup, a common part of production that is necessary before the cycle is able to start.

A model used in machining estimating describes the cost of a single-point-tool rough-turning operation. It involves Taylor's tool-life formula, which identifies the life of a tool and indicates when the tool should be replaced or maintained and optimizes performance of the metal-cutting system. In this model, operation unit cost is a function of handling, machining, tool changing, and tool cost. An important operating metric for the unit cost of machining is "cutting speed," "machining velocity," or "surface feet per minute." These terms refer to the peripheral velocity of a point on a rotating cylinder.

Handling time is the minutes to load and unload the work piece from the machine. For a particular work material and operation, handling time is constant. With a

[5]The student will find it worthwhile to study Taylor's book, *On the Art of Cutting Metals*, which was published by the American Society of Mechanical Engineers in 1906. Taylor's work led to the development of metal cutting and his tool life formula, design of tools, applications of nomographs (which were made into slide rules for manufacturing operations), and scientific management. His discovery of high-speed steel material (along with others including Henry L. Gantt, who became famous for the Gantt charts) is still in use today. The Midvale Steel Corporation, Taylor's employer for a period of time, prohibited him from working on high-speed steel, their thinking being that it would compete with their other products. Taylor accepted a new position at Bethlehem Steel Company, where he successfully concluded the development of HSS. His promotion of "scientific" management and other advancements are milestone events.

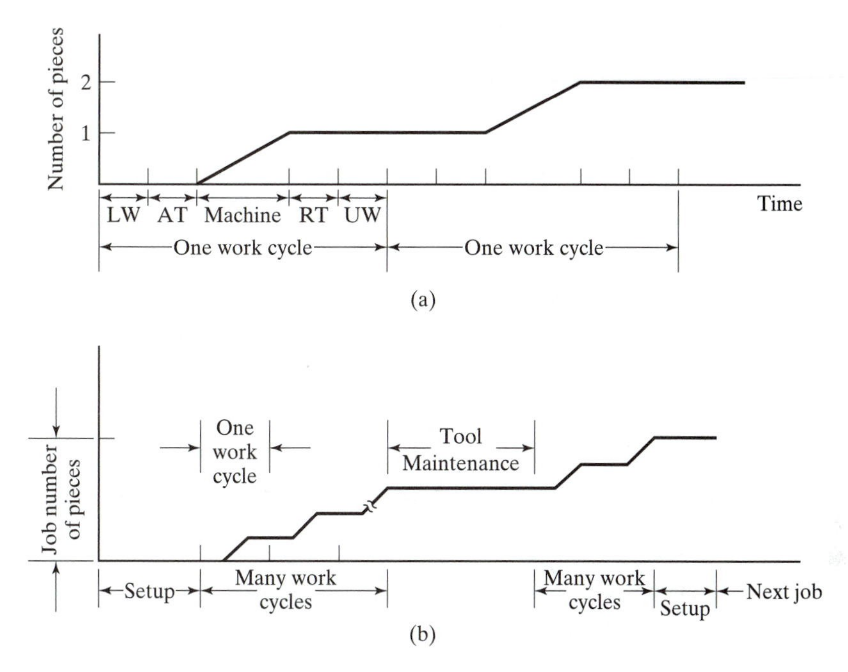

Figure 7.3 An overview of manufacturing operations: (a) cycle and production of pieces, and (b) setup and cycle for job number of pieces.

broader assumption, handling can also include advancing and retracting the tool and dimensional inspection of the part, particularly if those elements are constant. Handling cost is unrelated to cutting speed and is a constant for a specified design and machine. Figure 7.4 (a) is an example of the handling element plotted against cutting speed. Handling cost is found as follows:

$$\text{handling cost} = C_o t_h \tag{7.1}$$

where C_o = direct labor wage, dollars per minute

t_h = handing time, minutes

Machining is the focus of much attention. Machining time t_m is the minutes that the cutting tool is actually in the feed mode or cutting and removing chips. We define

$$t_m = \frac{L}{fN} = \frac{L\pi D}{12Vf} \tag{7.2}$$

$$\text{machining cost} = C_o t_m \tag{7.3}$$

where t_m = machining time, minutes

L = length of cut for metal cutting, inches

D = diameter of bar stock or cutter, inches

V = cutting speed, feet per minute (fpm)

f = feed rate, inches per revolution (ipr)

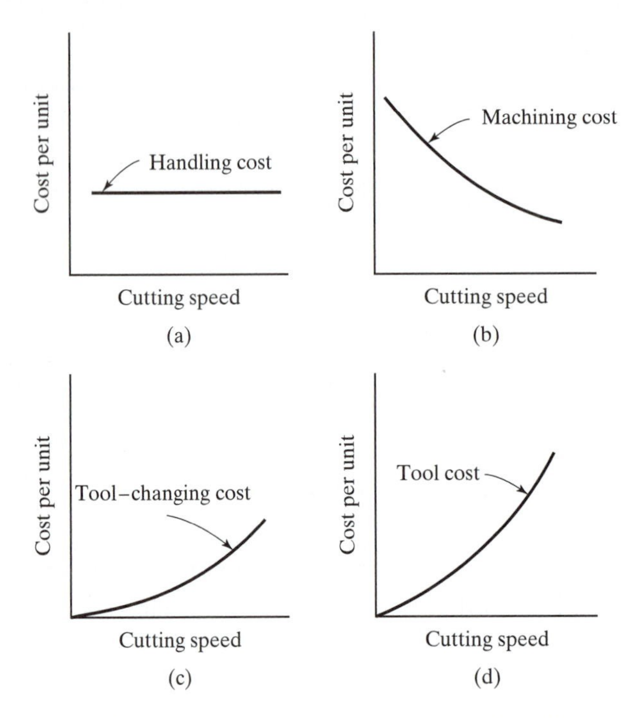

Figure 7.4 Graphical model of costs: (a) handling, (b) machining, (c) tool changing, and (d) tool cost.

$$N = \text{rotary cutting speed} = \frac{12V}{\pi D}, \text{RPM} \tag{7.4}$$

$$t_m = Lf_{dt} \tag{7.5}$$

where f_{dt} = drilling or tapping machining rates, minutes per inch

Each work and tool material will have unique turning and milling cutting speeds and feeds as determined by testing or experience. Those values will differ for rough or finish machining and for tool materials as well as equipment and the method of holding the work piece. A roughing pass removes more stock when compared with a finish pass, but it does not satisfy dimensional and surface finish specifications. A finishing pass will hold closer tolerances and provide a smoother appearance.

Our simple machining explanation deals with turning, milling, drilling, and tapping. For turning and milling, the material removal rate involves Equations (7.2) and (7.17). The material removal rate is minutes per inch for drilling and tapping. As was evident during Taylor's time, machine tools had a kinematic relationship between gears, feed screws, and traversing rods. As the bar stock or cutter rotates, it is geared to a long screw and rod shaft and a time relationship such that rotation, or RPM, and traversing motion or inch per revolution are known. Machine tools are able to provide a consistent relationship.

Table 7.2 is a very small database for machining. Two tool materials are high-speed steel (HSS) and tungsten carbide. Of the million or more specifications work materials, stainless steel, medium carbon steel, and cast iron are selected for

TABLE 7.2 Illustrative Machining Performance Values for Five Metal-Cutting Operations of Three Work Materials

Operation	1. Turn, Face or Cutoff (fpm, ipr)			
Tool Material	High-Speed Steel		Tungsten Carbide	
Machining Pass	Rough	Finish	Rough	Finish
Bar Stock Material				
Stainless Steel	150, 0.015	160, 0.007	350, 0.015	350, 0.007
Medium Carbon Steel	190, 0.015	125, 0.007	325, 0.020	400, 0.007
Gray Cast Iron	145, 0.015	185, 0.007	500, 0.020	675, 0.010

Operation	2. Milling (fpm, $i_t$pr)		3. Slotting (1-in. End-Mill), (fpm, $i_t$pr)	
Tool Material	High-Speed Steel		High-Speed Steel	
Machining Pass	Rough	Finish	Rough	Finish
Work Material				
Stainless Steel	140, 0.006	210, 0.005	85, 0.002	95, 0.0015
Medium Carbon Steel	170, 0.008	225, 0.006	85, 0.0025	95, 0.002
Gray Cast Iron	200, 0.012	250, 0.010	85, 0.004	95, 0.003

Operation	4. Drill (min/in.)		
Work Material	Stainless Steel	Medium Carbon Steel	Gray Cast Iron
Drill diameter, in., HSS			
1/4	0.55	0.20	0.20
5/16	0.61	0.23	0.23
3/8	0.65	0.25	0.25

Operation	5. Tap (min/in.)		
Work Material	Stainless Steel	Medium Carbon Steel	Gray Cast Iron
Threads/in.			
32	0.33	0.18	0.29
20	0.30	0.15	0.22
16	0.33	0.19	0.21
10	0.48	0.32	0.25

illustration. The HSS value for rough turning of stainless steel is 150 fpm and 0.015 ipr. Taylor and his associates are credited with an early development of high speed steel. Incidentally, the term "high speed" was in comparison to the operating performance of other high carbon steels which were used at that time for machining, and, in relative terms, this new revolutionary cutting material was much higher than any current materials. However, when compared with tungsten carbide and other contemporary tool materials, HSS is not high speed anymore.[6]

For a given diameter of a work material, and as the circular cutting velocity, fpm, increases, the unit machining time and cost decreases, as shown in Figure 7.4 (b).

The length of cut depends on the geometry to be machined. In lathe turning, the length, which is being machined, is at least equal to the length dimension given by the component design and is usually longer because of additional stock for roughing or finishing. In a lathe-facing element, the length of cut depends on the diameter, and the length is equal to the distance from the center of the bar stock circle to the outside

[6]An extensive listing of contemporary cutting tool materials can be found at www.Kennametal.com.

diameter (or the stock OD). Thus the facing length is at least equal to one-half of the diameter.

The diameter, D, may be either the work piece diameter, as found in lathe turning and facing, or the cutter diameter, as found in milling work. When a lathe-turning operation is visualized, the largest unmachined bar stock dimension is the diameter for Equation (7.2). For milling and drilling, the diameter is the cutter dimension. If a 6-inch milling cutter is used, the diameter is 6 inches, and it is this dimension that determines the time for milling in Equation (7.17).

Tool life is a measure of the length of time a tool will cut satisfactorily. Most studies of tool life are based on the famous Taylor's tool life cutting speed equation. We define

$$VT^n = K \tag{7.6}$$

where T = average tool life, minutes per cutting edge

n, K = empirical constants resulting from field studies and statistical analysis, $0 < n \leq 1, K > 0$

Cutting tools become dull as they continue to machine. Once dull, they are replaced by new tools or are removed, reground, and reinserted in the tool holder. Usually the tungsten carbide material tools are powder-metallurgy compacts such that a square insert will have eight points that are useful for turning work, and after a corner is worn, the dime-sized tool point is "indexed" to another corner that is sharp. Dull tools or consumed inserts are eventually thrown away. Dull cutting tools increase horsepower, heat, and friction, and roughen up a surface. Scrap is caused by dull tools. Empirical studies relate tool life to cutting velocity for a specified tool and work piece material. Look at Figure 7.5 for the steps to construct a tool life equation.

Tool wear is evident on various locations of a tool point. The wear is close to the end of the tool where the heavy frictional forces and high intermetal temperatures are caused between the tool and the work material during cutting.

Tool life tests are conducted under standardized conditions, if the results are to be comparable. A fixed-shape tool is used under constant conditions of work material,

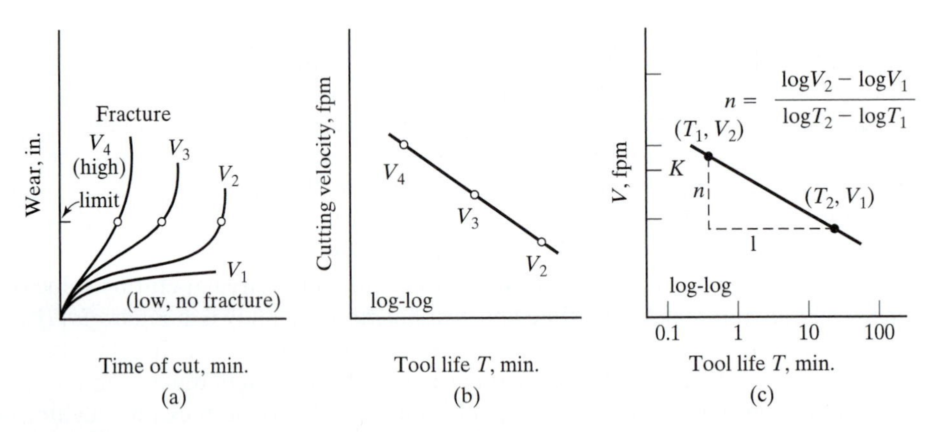

Figure 7.5 Procedure to find tool life: (a) tool life curves for constant turning velocities; (b) replotting wear limits versus cutting velocities; and (c) Taylor tool life curve.

depth of cut, feed, etc. The cutting velocity is held constant, and every so often the linear distance of wear on the flank face of the tool is measured, and the increasing amounts or distance of wear are tabled against the time of cutting, as measured in minutes. Note the four velocities shown in Figure 7.5 (a). Some velocities may not result in wear that is noticeable—for example, turning aluminum bar stock with tungsten carbide. A typical test requires a specified length of wear, say 0.030 in. Note that V_1 does not have much appreciable wear and does not reach the 0.030 in. limit. As $V_2 < V_3 < V_4$, the time of reaching the wear limit reduces as the velocity increases. If the V_4 cutting continues, it is possible that tool fracture and a safety condition may result.

The composite of these times to reach the wear limit are shown as Figure 7.5 (b), which is plotted on log-log coordinates and has a linear line drawn through the sampling of data. In Figure 7.5 (c) the value of n and K are found by graphical or statistical regression methods. The intercept K is found when $T = 1$, and the slope is found using two points and calculating n. Table 7.3 is a very small listing of Taylor's empirical coefficients, n and K. Tool changing times and costs are given.

Example: Taylor's Tool Life Model

Consider the example of an operation on 430F stainless steel, where from Table 7.3 the tool life equation is found as $VT^{0.16} = 400$. We can find either V or T given the other variable. If $V = 200$ fpm, we expect 76 minutes of average tool life. If T equals 50 minutes, we expect V to be 215 fpm.

TABLE 7.3 Taylor's Tool Life Parameters, and Tool Changing and Indexing Times and Costs

Tool Material	High-Speed Steel		Tungsten Carbide	
Work Material	*K*	*n*	*K*	*n*
Stainless steel	170	0.08	400	0.16
Medium carbon steel	190	0.11	450	0.20
Gray cast iron	225	0.12	3000	0.43
Tool Changing Times and Costs				
Time to index a turning type of carbide tool point			2 min	
Time to set a high-speed tool in lathe tool holder			4 min	
Remove drill, regrind and replace in chuck			3 min	
Cost per tool cutting corner for turning, carbide			\$3	
Cost for high-speed steel tool point			\$5	
Cost of drill			\$3	

The third cost is the tool-changing cost per operation. Define it as

$$\text{perishable tool-changing cost} = C_o t_c \left(\frac{t_m}{T}\right) \tag{7.7}$$

where t_c = tool changing time, minutes

The tool-changing time t_c is the time to remove a worn-out tool, replace or index the tool, reset it for dimension and tolerance, and adjust for cutting. The ratio of t_m/T is a percentage of component cutting time to average tool life, and thus prorates the cost over the tool life. Time depends on whether the tool to be changed is a disposable insert or a regrindable tool for which the whole tool must be removed and a new one reset. In lathe turning and milling, there is the option of an indexable or regrindable tool. A drill is only reground. Some data shown for the tool-changing times and costs are shown in Table 7.3. We see the increasing tool-changing cost as cutting speed increases, as shown in Figure 7.4 (c).

As the tool continues to wear, it is eventually thrown away and a new one replaces it. The cost of perishable tooling and its relationship to cutting velocity is found as

$$\text{perishable tool cost per operation} = C_{pt}\left(\frac{t_m}{T}\right) \tag{7.8}$$

where C_{pt} = perishable tool cost, dollars

Tool cost C_{pt} depends on the tool being a disposable tungsten carbide insert or a regrindable tool for turning. For insert tooling, tool cost is a function of the insert price and the number of cutting edges per insert. For regrindable tooling, the tool cost is a function of original price, total number of cutting edges in the life of the tool, and the cost to grind each edge. As the cutting speed increases, the cost for the tool increases, as shown in Figure 7.4 (d).

The total cost of the operation is composed of handling, machining, tool changing, and tool wear. Again, refer to Figure 7.4 for their behavior. Machining cost is observed to decrease with increasing cutting speed and tool and tool changing costs increase. Handling costs are independent of cutting speed. Thus we can say that unit cost C_u is given as

$$C_u = \sum\left[C_o t_h + \frac{t_m}{T}(C_{pt} + C_o t_c) + C_o t_m\right] \tag{7.9}$$

where C_u = unit cost using Taylor's theory, dollars per unit

On substitution of t_m and T by Equations (7.2) and (7.6), and after taking the derivative of this equation with respect to velocity and equating the derivative to zero, the minimum cost is found once we have V_{min}. The following equation gives the optimum velocity for minimum cost directly:

$$V_{\min} = \frac{K}{\left[\left(\frac{1}{n} - 1\right)\left(\frac{C_o t_c + C_{p_t}}{C_o}\right)\right]^n} \tag{7.10}$$

This equation gives the velocity to find minimum unit cost for a rough turning operation.

The above development does not recognize revenues that are produced by the machining. Had we found the marginal cost and marginal revenue, similar to Section 9.3.3, the intersection of marginal revenues and marginal costs provides a higher value of velocity than that demonstrated by Equation (7.10). (Figure 9.9 will later illustrate intersection of the marginal revenue and cost lines.) This intersection is the profit maximum point. Consequently, Equation (7.10) identifies the velocity for minimum cost, which is not the same operating velocity for maximum profit.

If the costs in the basic model are not considered, then the model gives the time to produce a work piece, and we develop

$$T_u = t_h + t_m + t_c\left(\frac{t_m}{T}\right) \tag{7.11}$$

where T_u = minutes per unit

T_u is the production rate, and it may be converted to units per hour or hours per 100 units.

Occasionally, to avoid bottleneck situations, we need to accelerate production at cutting speeds greater than that recommended for minimum cost. In those expedited operations, we assume the perishable tool cost to be negligible, or $C_{pt} = 0$. A similar development as above gives the cutting speed that corresponds to maximum production rate as

$$V_{\max} = \frac{K}{\left[\left(\frac{1}{n} - 1\right)t_c\right]^n} \tag{7.12}$$

The tool life that corresponds to maximum production rate is given by

$$T_{\min} = \left(\frac{1}{n} - 1\right)t_c \tag{7.13}$$

This tool life $T_{\min}$ is less than the value given by Equation (7.6) found earlier, because the cost of the perishable tool is ignored in the operation cost.

Example: Taylor Tool Life Criterion Cost Optimization

Consider an operation of rough turning 430F stainless steel. The work material is 1.750 in. OD bar stock. The cutting length is 16.53 in. The turning operation will use a tungsten carbide, insertable, and indexable eight-corner tool point that costs \$3 per corner. After the tool point has worn and gives diminished performance, the time to reset the tool is 2 min for a new corner. Handling of the part is 0.16 min, and the operator wage is \$15.20/hr.

The Taylor tool life equation is $VT^{0.16} = 400$ for the tool and work material. Feed of the rough turning element is 0.015 ipr for a depth of cut of 0.15 in.

If the four terms of the cost equation are plotted with several velocity V values as the x variable, we have Figure 7.6. The sum of the four costs gives total cost. The bottom point of the total cost curve gives the operating optimum at 200 fpm for a minimum unit cost of \$0.80. Similarly, by using Equation (7.10) we have for the example information

$$V_{\min} = \frac{400}{\left[\left(\frac{1}{0.16} - 1\right)\left(\frac{0.25 \times 23}{0.25}\right)\right]^{0.16}} = 201 \text{ fpm}$$

Substituting $V_{\min} = 201$ fpm in the four cost Equations (7.1, 7.3, 7.7, and 7.8), the Taylor unit cost is \$0.81. The maximum tool life is found as $T = (400/210)^{1/0.16} = 74$ min. The cycle time using Equation (7.11) for handling, machining, and tool changing is $T_u = 0.16 + 2.51 + 0.033 = 2.70$ min and the production rate is 22.2 units per hour, or 4.500 hr/100 units.

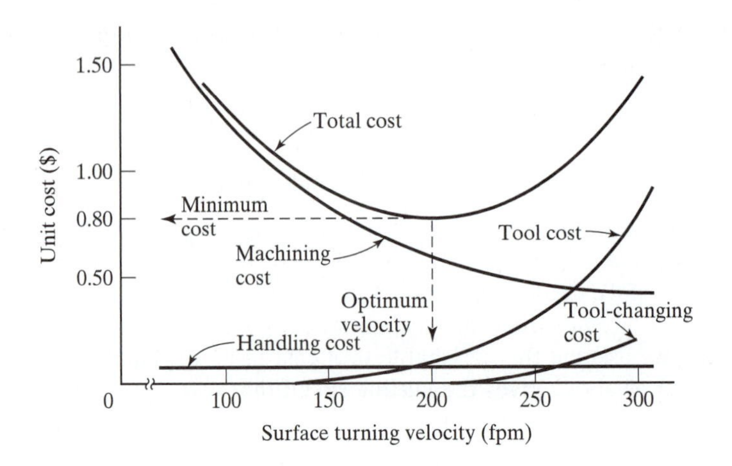

Figure 7.6 Classical optimization of rough turning operation using Taylor's theory. Optimum velocity and minimum unit cost are found as 200 fpm and \$0.80.

The rotary cutting speed N for $V_{min} = 12V/\pi D = 12(201)/\pi(1.75) = 439$ RPM. The maximum cutting speed is $V_{max} = 275$ fpm if we ignore the cost of the cutting tool. Under these circumstances, tool life diminishes to $T_{min} = 10.5$ min.

The Taylor approach, while historically important, does not tell us about contemporary methods. Furthermore, this approach is limited to a narrow range of machining, and most methods of manufacturing are not suitable for this analysis. In addition, contemporary methods must be capable of dealing with the variety of manufacturing operations.

7.4 CONTEMPORARY OPERATIONS ANALYSIS (OPTION 2)

The steps for contemporary operations analysis include the following:[7]

- Factual time and cost estimating catalogues
- Product design
- Process knowledge
- Operations sheet preparation
- Time and cost finding

The sequence is somewhat important, though steps can be out of order, and an experienced engineer understands the intricacy of these maneuvers.

Engineering drawings, marketing quantity, specifications of the work piece material, equipment and processes, and engineering performance data sets are required before detail estimating can begin. Preparation of the operations sheet occurs simultaneously with direct labor estimating. Once the operations are listed, the engineer refers to information, which coincides with the machine, process, or bench identified

[7]Computer-based operations analysis does not really alter basic understanding. The "information revolution" of the late twentieth century simply routinized what was done all along.

by the operations sheet for doing the work. Each operation is detailed into elements that correspond to the time data. (The development of performance time data is discussed in Section 6.3.2.) Those elements are presented by software, and the form is standardized by the company, or marginal jottings or scratch pad calculations may be followed. Computer spreadsheets are used also.

The production batch or lot quantity plays an important role in selection of the production equipment and sequence. For the operation and workstation, and after visualizing the elements of an operation, the engineer selects times from the engineering performance data for setup and cycle.

Data Warehouse Readiness. Cost analysis depends on cataloged information. Data warehouses are simply computer databases, manuals, and handbooks that contain extensive lists of practical time elements matched with a method description. The data encyclopedia is ready to use whenever the operation is planned. Although simple in theory, it is more difficult to achieve. For the most part, manufacturing companies do not devote resources to measure and validate data warehouses for accurate and timely analysis and estimating. Alas, guessing will not do, though it is widely practiced.

Setup and cycle elements are found by time study or job tickets, or from first-hand experience. The raw information is reworked into engineering performance data. The student will want to study Chapters 2, 3, and 6 for reminders.

Setup and cycle operation times are necessary for manufacturing estimating. Table 7.4 is a small sample of setup values. In keeping with our manufacturing

TABLE 7.4 Illustrative Manufacturing Setup-Time Elements

Setup Description	Hours
Punch in and out, study drawing	0.2
Turning Equipment	
First machining tool	1.3
Each additional machining tool	0.3
Collet	0.2
Chuck, fixture	0.1
Milling Equipment	
Vise	1.1
Angle plate	1.4
Shoulder-cut milling fixture	1.5
Slot-cut milling collet	1.6
Tight tolerance $< \pm 0.005$ in.	0.5
Drilling Equipment	
Jig or fixture	0.1
Vise	0.05
NC turret or spindle drill press	
First turret or spindle	0.75
Additional turrets or spindles	0.07

equipment choices, the table shows setup times for turning, milling, and drilling. Typically, each equipment, process or bench will have unique setup values. Although setup values, may appear similar from company to company because of equipment, there can be variations. Technology, machine tool design, operator training and skill, and lean manufacturing affect setup times.

Those setup work elements are briefly described by one-liners—"Punch in and out, study drawing," for example. People familiar with the equipment and plant understand its meaning. In one shop, the understanding implies "going to a foreman's desk and making computer entry with a job code and person number, and on conclusion of the job, returning to the desk to log out." Or the meaning may be that the operator does these elements by cell phone or other wireless notification. The method clearly influences the time. In the context of the plant for which the setup data are pertinent, these nuances are understood.

Setup time data are enumerated in hours because this is customary in the United States. Setup times, which include allowances, are posted once for each occurrence on the operations sheet. Setup element times are additive; that is, time for a work holder and for a tool are added, those values being read from engineering performance data and posted on the operations sheet.

The cycle is the repetitive part of the operation. While setup is a one-time only event for each operation, the cycle elements are repeated for each operation up to the lot quantity N. Too, cycle time elements are dependent on the equipment, plant situation, operator training, skill, and motivation. Table 7.5 illustrates cycle elements for turning, milling, and drilling equipment. Typically, real-life data warehouse sets are much larger.

Not all the elements of Tables 7.4 and 7.5 are used each time an estimate is required. They are picked selectively from the tables. The engineer pictures a virtual time study for the operation. For instance, a job setup will require punching in and out, and so on. Further, a lathe operation will need "starting and stopping" the machine each time the cycle is started, and so on. The analyst picks those elements that are needed realistically for the job.

In the event that the job requires elements that are not shown, the engineer can obtain information from experienced people in the shop, or make comparisons to known information. Lacking an occasional element time is not as serious as complete absence of information.

Product Design. A design called a pinion is shown in Figure 7.7. Production of an annual quantity of 1000 units is planned with five lot requirements of 200 units each. Our firm is responding to a request for quotation. The pinion is a part of a larger product requirement.

In addition to the drawing or computer file of the drawing, the firm will have other intelligence about the design. A bill of material may be at hand, or previous experience on similar products is a possibility. Design information may be missing, inaccurate, or even misleading and cross talk with the buyer or designer is required. If the estimate is to end in a quotation, a file of background information will accompany the print.

TABLE 7.5 Illustrative Manufacturing Cycle-Time Elements for Turning, Milling, and Drilling Equipment

Machine and Element Description	Minutes
Turning, Milling, Drilling Equipment	
Start and stop machine	0.08
Change speed of spindle	0.04
Air clean part or fixture	0.06
Brush chips	0.14
Inspect dimension with micrometer	0.30
Turning Equipment	
End turret advance, return, and index	0.08
Cross slide advance, return, and index	0.09
Advance stock through feed tube	
Length (in.) < 6	0.18
$6 \le$ in. length < 14	0.23
$14 \le$ in. length < 35	0.37
Place, remove oil guard	0.19
Milling Equipment	
Pick up part, move, and place; remove and lay aside	
lb < 5	0.15
$5 \le$ lb < 10	0.18
$10 \le$ lb < 15 lb	0.25
Open and close vise, operator use mallet	0.24
Clamp casting to fixture mounted on angle plate	0.76
Open and close vise, air cylinder	0.06
Numerically-Controlled Turret-Drilling Equipment	
Pick up part, move, and place; remove and lay aside	
lb < 5	0.13
$5 \le$ lb < 10	0.16
$10 \le$ lb < 15 lb	0.22
Clamp part	
Vise, $\frac{1}{4}$ turn	0.05
Air cylinder	0.05
C-clamp	0.26
Thumb screw	0.06
Remove and replace bushing in jig	0.07
Machine Operation	
Change tool, each	0.06
Raise tool, position table to new location, lower tool ready for drilling, per hole position	0.06
Index turret, each	0.03

Does not include machining information such as cutting velocity and feed.

Process Knowledge. Before operations planning can start, one must have knowledge of the company's equipment and processes, capabilities of suppliers, job descriptions of workers and their skills, plant layout, schedule of work and future plant loading, general and special tooling and its availability, tool room ability, standard and special shop materials, and so on. This information might be cataloged in

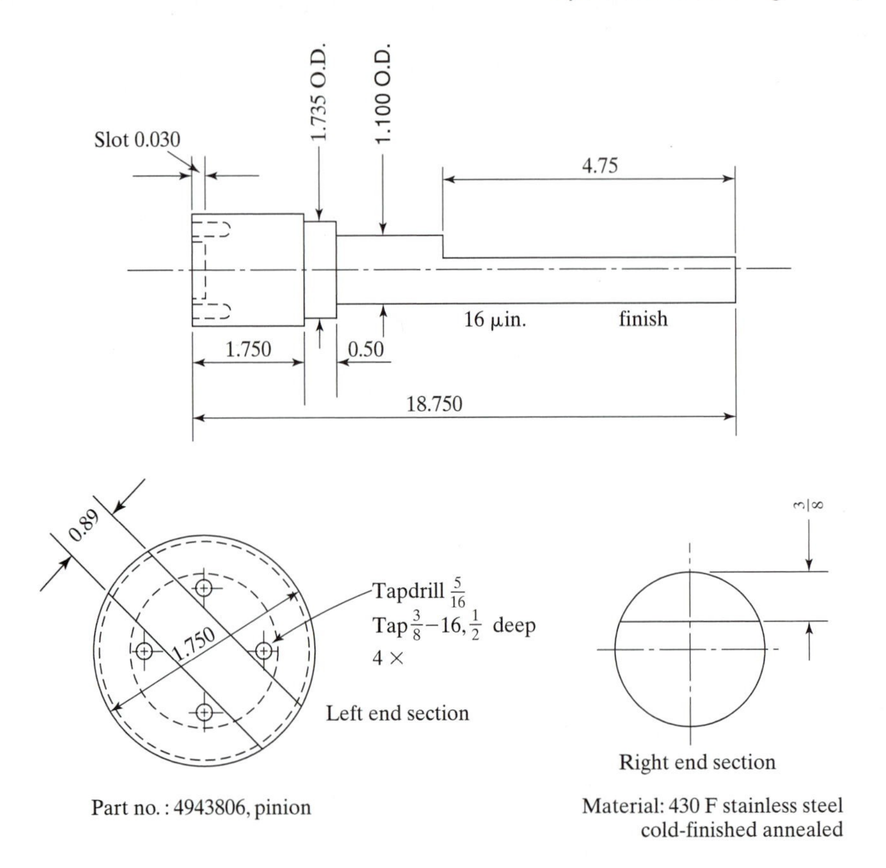

Figure 7.7 Sketch of metal component (used for example, part no. 4943806).

some systematic way, or it might be "mental experience." Company handbooks or computer summaries are the usual way.

Operations Sheet Preparation. The operations sheet—also called a route sheet, instruction sheet, traveler, or planner—is fundamental to manufacturing planning. Next to the product description, or the engineering print, the operations sheet is probably the most important document in manufacturing. There are many styles and each plant will have its own version. Table 7.6 provides a sample of an operations sheet.

The purpose of the operations sheet is to identify the machine, process, or bench that is necessary for converting raw material into product, to provide a description of the operations and tools, and to indicate the time required for the operation. The order of the operations is special too, as this sequence indicates the various steps in the manufacturing conversion. The shop will use these instructions in making the part.

The process description is a step-by-step set of instructions, which can be brief, or detailed. An *operation* is a step in the process sequence and is defined as all work done at a workstation. A job to drill and tap a hole at one drill press would be one operation. If one drill press is used, then there must be two spindles or turret stations for the drill

TABLE 7.6 Operations Process Plan for Part No. 4943806

Part no.	4943806		**Material**		430 F stainless steel			
Part name	Pinion		**Size**		1.750 ± 0.003			
Lot quantity	200		**Length**		12 ft bars			
			Material unit cost					$21.34
Workstation	**Op. no.**	**Description of Operation**	**Setup (hr)**	**Cycle (hr/100 units)**	**Lot (hr)**	**PHC**	**Lot cost**	**Unit cost**
Turret lathe	10	Face 0.015 Turn rough 1.45 Turn rough 1.15 Finish turn 1.10 Turn 1.735 Cutoff to 18.75	3.2	10.600	24.4	39.16	$955	$4.78
Vertical mill	20	End mill 0.89 slot with $^3/_4$ HSS end mill and collet	1.8	8.067	17.9	90.98	$1631	$8.16
Horizontal mill	30	Slab mill 4.75 × 3/8 (nesting vise)	1.3	1.467	2.9	90.98	$267	$1.34
N.C. turret drill press	40	Drill 4-5/8 holes, tap 3/8-16 (collet)	1.12	5.217	10.4	39.16	$409	$2.04
			Total PHC lot cost				$3262	
			Total PHC unit cost					$16.32
			Material, labor, and equipment unit cost					$37.66

and the tap. If for some reason, however, it is necessary to drill the hole on one drill press, and then transfer the work to another drill press for tapping the hole, that would be two operations.

A *process* is one or more operations that are performed on the product from the time it leaves an inventory location until it returns to finished goods inventory. The raw material is released from inventory, and eventually the processed part reaches the shipping dock, or some intermediate point where it is ready for sale and shipment, or for additional work and integration.

The sequence of operations in the process depends on many factors. Indeed, some of these factors will be significant for one plant but are minor for another plant. It depends on the plant producing the product.

Correct sequences will insist on drilling the hole before tapping, or the milling of a pocket before an assembly is welded to the pocket. Avoiding tumbling of the part in a vibratory tumbler for burr removal, if there are external threads, is a foolish and costly blunder. Protection of the part after grinding will prevent the parts from banging on each other, marring the finish. Operations planning requires this practical thinking.

Preparation of the operations sheet involves the following steps:

- Interpretation of the engineering drawing and specifications
- Initial material requirement
- Selection of equipment and workstation

- Selection of tools, special or general for the workstations
- Partitioning the steps into distinct operations
- Detailing the description of each operation
- Data entry of information

For part no. 4943806, raw material is 1.750 in. OD 430F stainless steel and stock is purchased in 12-ft lengths. This raw material is called bar stock.

Note that in Figure 7.7 the final length dimension is 18.750 in. But the pieces have to be separated from the bar stock. Cutting off is a roughing machining element, and it is necessary to smooth the next adjoining surface, which is called "facing." The actual length and the number of bars needed for the turning operation is given by the following information:

Design length	18.750 in.
Facing length	0.015 in.
Cutoff length	0.187 in.
Total length	18.952 in.
Units per 12 ft bar	7
Bars per 200 lot quantity	29

Physical properties for stainless steel grade, 430F, are different than for other plain carbon steels, and the weight per bar foot is found from tables to be 8.178 lb/ft. If the cost per pound for this material is quoted from a steel supplier as $1.50/lb, the costs and material yield are found as follows:

Cost per bar	$8.178 \times 12 \times 1.50 =$ $147.20
Lot cost for material	$147.20 \times 29 =$ $4,268.92
Cost per unit	$4268.92/200 =$ $21.34
Material yield efficiency	$200 \times 18.75/(29 \times 144) = 90\%$

With the material grade selection, amount, and cost per unit concluded, the next thing is to choose a logical plan of workstations and equipment that will start with the bar stock and conclude with the part. This is handled in several ways. Most likely, the process planner will have some hands-on experience with the production facilities in order to write or computer-enter the following choices:

Workstation	Operation Number	Initial Description
Turret lathe	10	Turn and cutoff
Vertical mill	20	End mill slot
Horizontal mill	30	Slab mill flat
Numerical controlled drill press	40	Drill and tap holes

There are computer programs that purport to artificially achieve the intelligence of the process planner's work. However this work is executed, a process plan will emerge into something like Table 7.6. The table identifies workstation, operation number, description, setup time, cycle time in hours per 100 units, and lot hours together with other technical and cost estimating information. Furthermore, the elements are selected for each operation that correspond to data warehouse elements. Knowing the manufacturing required to machine the metal pinion, the engineer posts machine and workstation specifications, setup, handling and machining elements. Calculations for stock removal, dimensions, feeds, and speeds may be involved. Cycle time elements are included for start and stop, machining, inspection, and machine adjustment. This information is collected for operation 10 in Table 7.7.

Now we digress to explain the details of performing the estimating for each operation separately. This could be handled by computer data systems, but for the sake of learning, we show it as if it is done by pad and pencil.

TABLE 7.7 Worksheet for Estimating Setup and Cycle for Operation 10, Table 7.6.

A. Setup Elements	Hr
Punch in and out, study drawing	0.2
Collet for bar stock gripping	0.2
First facing tool	1.3
Additional 5 tools for all cuts	1.5
Setup total, operation 10	3.2

B. Handling and Other Equipment Time Elements	Min
Start and stop machine	0.08
Advance stock through feed tube, 18.952 in.	0.37
Place and remove oil guard	0.19
Speed changes, assume 4 × 0.04	0.16
End turret advance, return and index, 5 × 0.08	0.40
Cross slide advance, return	0.09
Inspect part with micrometer, irregular 1/5 × 0.30	0.06
Subtotal of handling and equipment elements	1.35

C. Calculation of Machining Times

Element	Dimension	Depth of Cut	Length of Cut, L_d	Safety Stock, L_s	Length L,	Diameter, D	Velocity, V, fpm	feed, f, (ipr)	Minutes, t_m
1. Face		0.015	0.875	1/32	0.906	1.750	350	0.007	0.17
2. Rough turn	1.45	0.15	16.5	1/32	16.53	1.750	350	0.015	1.44
3. Rough turn	1.15	0.15	16.5	1/32	16.53	1.45	350	0.015	1.19
4. Finish turn	1.10	0.025	16.5	1/32	16.53	1.15	350	0.007	2.03
5. Turn	1.735	0.0075	0.5	1/32	0.53	1.75	350	0.007	0.10
6. Cutoff	1.75	0.187 W	0.875	1/32	0.906	1.75	350	0.015	0.08
Subtotal of machining times									5.01
Total cycle time for handling and machining, min									6.36

D. Entry Values for Operation Sheet, Operation 10	
Setup hr	3.2
Hr/100 units	10.600

7.4.1 Pinion Operation 10

Earlier we described that a process planner had selected a preliminary process plan where a turret lathe is selected for a raw material of $1\frac{3}{4}$ in. bar stock. Now we continue with the detail explanation of that operation, and the three operations that follow. The student will want to study Figure 7.8, which is a diagram of the machining elements for operation 10.

Consider operation 10, which uses a turret lathe and a turning metal-cutting machine. Tools are positioned in readiness on the turrets for consecutive machining cutting elements. Usually there are two turrets, a six-or-more stations-type turret,

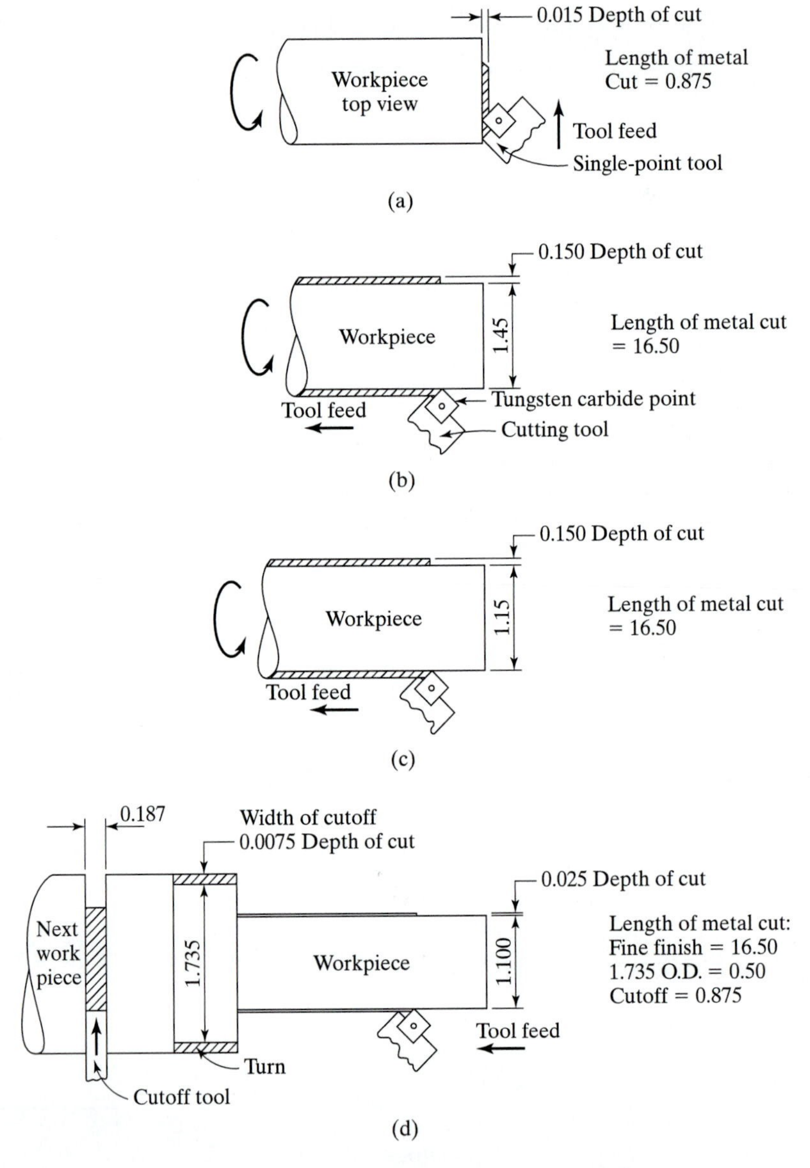

Figure 7.8 Diagram for operation 10, part no. 4943806.

mounted on the lathe-bed ways and a four-position square-type mounted on a carriage. Ten or more different cutting tools can be preset and mounted on those turrets.

Each lathe-machining element, for example, will use a single point tool. These turret tools are preset in the setup and remove stock to achieve intermediate or final dimensions. There are six tools and six machining passes, or face, rough turn the diameter from $1\frac{3}{4}$ in. to 1.45 in., rough turn from 1.45 in. to 1.15 in. OD, finish turn-from 1.15in. to 1.100 in. OD, finish turn from $1\frac{3}{4}$ in. to 1.735 in. OD, and finally cutoff of the bar stock. These machining elements are shown as steps in Figure 7.8. Operation of the turrets to advance and retract the tool from the work piece depends solely on machining elements. Similarly, setup will depend on the number of tools, tolerance, and so on, which are dependent on machining elements. These are interrelated elements.

Figure 7.8 is a description of the consecutive machining elements. A facing cut is made in Figure 7.8 (a). Can you relate the machining passes to Figure 7.8?

Once the outline of the cycle is understood, we proceed to the next step of finding the setup for operation 10. The selection of the time elements is given by Table 7.7 (A). Concurrently, the planner will use Table 7.4 and make selections to match the "visualized" time study. Punching in and out, collet adjusting the first facing tool, and the next five tools are itemized along with the time from Table 7.4 for the setup total of 3.2 hr.

Finding the handling and equipment time for operation 10 is next. Notice Table 7.7(B), where the estimating is shown. Knowing the number of different cuts, each cut will require a separate tool and an "advance, return, and index" turret handling time for either the hex or square turret. Handling and other machine operation elements are now selected from the data warehouse of Table 7.5 and listed in Table 7.7(B). One element, inspection of the part with a micrometer, and the element time of 0.30 min is done on a 1/5 occurrence. This knowledge might come from cross talk communication with the supervisor. The subtotal of handling and equipment elements is found as 1.35 min.

Machining times t_m are calculated next for operation 10, and they are shown in Table 7.7(C). Equation 7.2 is used and substitutions of L, D, V and f are made. Those particular values are given in Table 7.7(C) and found using, Figure 7.7 and Table 7.2. Because those computations are frequent, computer calculations or electronic calculators are programmed to do this work quickly. Observe that values of t_m are listed in Table 7.7 (c). Note that the diameter is successively reduced as a result of rough and finish turning passes. Those differing values of diameter theoretically require different values of the RPM or N, according to Equation (7.4). Each of those RPMs calls for a speed gear adjustment, which also requires equipment time. If the savings in time is small, an average RPM will be used and no speed adjustment made. But for long cuts the savings in machining time is worth the time to change the RPM. That is the plan we use in our examples.

The machining times are calculated for the work of operation 10, and the sum is found as 5.01 min in Table 7.7(C). Total cycle time of 6.36 min and setup of 3.2 hr are finally shown. The cycle time of 6.63 min/unit changes to 10.600 hr/100 units. These are the entry values for operation 10, as shown in Table 7.6.

7.4.2 Length of Cut Calculation

Consider a diversion that deals with the length of cut, L, which varies with the type of operation, depth of cut and geometry of the tool and work piece. More generally, the length of cut L is found as follows:

$$L = L_s + L_a + L_d + L_{ot} \qquad (7.14)$$

where L = length of machining cut at feed velocity f, inches

L_s = safety length, inches

L_a = approach length because of cutter or workpiece geometry, inches

L_d = design cutting length, inches

L_{ot} = overtravel length because of cutter or work piece geometry, inches

The total length L is used to find the time t_m or cost for metal cutting. *Feed* or *ipr* is much less than the rapid traverse velocity, which may be from 100 to 1000 in./min. During this "rapid traverse velocity" movement, the major machine elements, such as the headstock, table, carriage, column, and base, are moving relative to one another, but there is no cutting. Cutting is going on during the feed f period. The rapid traverse movement is done quickly, in order to position the machine tool structural elements and the cutting tool adjacent to the work piece ready to cut at the much slower rate.

A safety length, L_s, is necessary for any stock variation in length, for if the cutter is in rapid traverse velocity and the trips, which cancel the fast mode of the machine table or cutter movement, are set too short, it is possible for the cutter to smack into the work piece, causing safety and damage problems. It is called *safety stock* because it implies that the metal-cutting mode is enabled and the machine is cutting air, so to speak, and not chips. Safety stock may vary from $1/64 \leq L_s \leq \frac{1}{2}$ in. In turning machining, as shown in Table 7.7, a safety stock of 1/32 in. is added to the drawing length.

Approach length, L_a, depends on cutter-work piece geometry, especially for milling and drilling. Approach length for turning is negligible.

For milling, approach depends on the diameter of the cutter, D_c, depth of cut t, and the type of milling, whether it is a vertical spindle [see Fig. 7.2 (d) and (e) with a shell or end milling cutter, or a horizontal arbor-mounted slab or plain cutter as shown in Fig. 7.2 (c)].

In Figure 7.9 (a) the view is from the top, and you are seeing the diameter D_c of the end mill. The operation is to end mill a width that requires two passes, as the machined slot width is greater than the diameter of the end mill. The circle, representing the D_c diameter of the cutter, is rotating, and the beginning station is 1. The part moves from station 1 to 2 to 3 to 4 and back to station 1 where it is in a ready position for the next part. The feed f is shown with an arrow indicating direction. As the width of the machined dimension is greater than the diameter D_c of the $\frac{3}{4}$ in. end mill, it is necessary to move the cutter over a distance such that the full width is machined. Finally, a rough milling pass will always require an approach, but an overtravel may not be required, and the cutter center line would stop at the edge of the part line. If the milling cutter is performing a finishing pass, it is customary to move the cutter of the work piece to avoid a dragging tool mark at the end of the work piece. Remember that the cutter at station 1 needs to move $\frac{1}{2} D_c$ to have full width cutting of the end mill.

Figure 7.9 (b) illustrates a slab, side, saw, or plain milling cutter operation. For a given cutter diameter, as the depth of cut increases, the approach length L_a also increases. In this case the arbor is horizontal, relative to the part on the table.

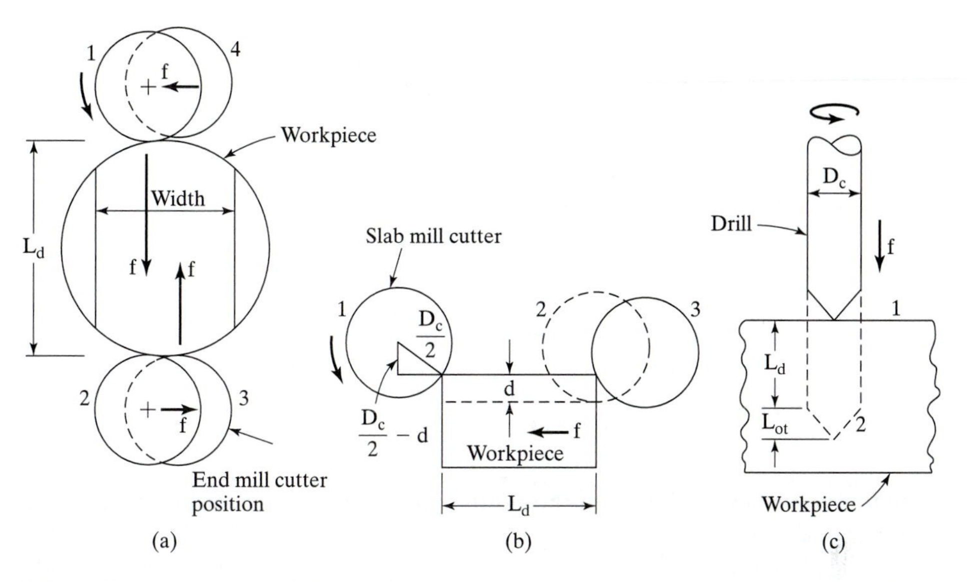

Figure 7.9 Finding additions for length of cut: (a) top view of end-milling path, (b) side view of slab milling, and (c) drilling.

$$L_a = \sqrt{\left(\frac{D_c}{2}\right)^2 - \left(\frac{D_c}{2} - d\right)^2} = \sqrt{d(D_c - d)} \tag{7.15}$$

where D_c = milling cutter diameter, inches

d = depth of cut, inches

For drilling, the drill point adds to the length of cut, and the trigonometric relationship gives the distance accounting for the standard 118° drill point and drill point overtravel as

$$L_{ot} = \frac{D_c}{2 \tan 59} = 0.3D_c \tag{7.16}$$

where D_c = drill diameter, inches

Other work materials, such as stainless steel, employ a drill point angle of 135° giving $L_{ot} = 0.2D_c$. Notice Figure 7.9 (c) for drill point overtravel.

L_d is the design length of cut that is found from the engineering drawing. Typically, it is necessary to increase this number for cutter geometry, depth of cut, and nature of machining. Practitioners knowing these additions use rule of thumb amounts to increase L_d without resorting to the above three equations.

7.4.3 Pinion Operation 20

Consider Figure 7.10, a diagram for operation 20 of part no. 4943806. The pinion is held vertical in a collet, and with vertical-milling machine equipment, a $^3/_4$ in. 4-tooth cutter end mill is selected. An end mill cutter is similar to a drill except that the bottom is flat

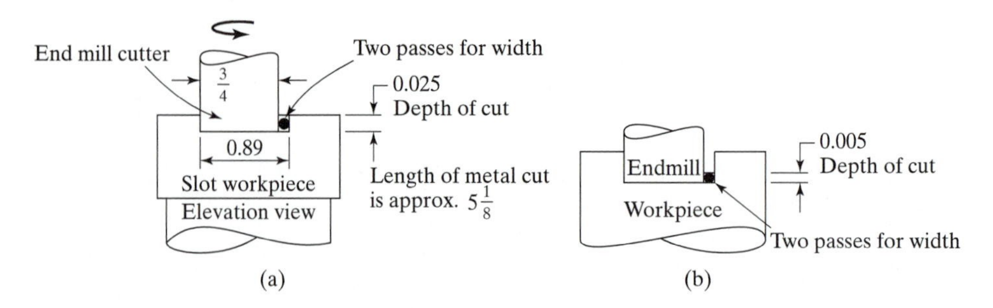

Figure 7.10 Diagram for operation 20, part no. 4943806: (a) Rough end milling of the slot requiring two passes for width, and (b) finish end milling of the slot requiring two passes for width.

and has four cutting flutes on the bottom and side. Two passes are necessary for the width, and a rough cut will remove 0.025 in. depth and the finishing cut will remove 0.005 in. depth with a high-speed steel tool. The width of the slot is slightly wider than the end mill diameter, and the length of cut is affected by this decision.

Worksheets for operation 20 are given as Table 7.8. Setup, handling, and other equipment handling times are given. The length of cut is 6.81 in., as shown by Figure 7.9(a). In this calculation, L is about 50% of total part length.

The formula to calculate machining time is modified for milling operations. A milling cutter is multitooth, and Equation (7.2) is converted to

$$t_m = \frac{L\pi D_c}{12 V n_t f_t} \tag{7.17}$$

where n_t = teeth on multitooth cutter, number

f_t = feed per revolution-tooth, i $_t$pr

D_c = milling cutter diameter, in.

Calculations for the end milling operation are shown in Table 7.8(D). Entry values to the operations sheet in Table 7.6 are setup 1.8 hr and cycle time 8.067 hr/100 units.

7.4.4 Pinion Operation 30

The diagram for operation 30 is shown as Figure 7.11. A horizontal-type milling operation is selected, thus allowing a slab or plain milling cutter. A one-pass machining cut is made because of the rigidity of the horizontal arbor. The milling cutter is holed, allowing the arbor and greater productivity than with end mills. This flat on the metal pinion is relative to the slot cut in operation 20, and a special nesting vise is designed and constructed to ensure dimensional compliance. Because of the greater rigidity of this tool, along with the horizontal arbor milling machine, only one pass is necessary to reach finish dimension.

In operation 30, a 4-in. plain milling cutter, high-speed steel, with a 6-in. face and eight teeth will slab mill the 4.75-in. dimension given on the metal pinion. Figure 7.7 shows a left-end view of the basic features of the machining requirement.

Entries for the worksheet of operation 30, Table 7.9, are made by using information from Tables 7.2, 7.3, and 7.5. Once the worksheet is concluded, a setup of 1.3 hr and a cycle time of 1.467 hr/100 units is transferred to Table 7.6, operation 30.

PICTURE LESSON Milling Cutter

Kennametal (Bill Kennedy).

Milling is a versatile machining operation. The milling cutter is multitooth, unlike turning or boring cutters. A milling cutter will have several inserts and is mounted on an arbor or the end of the shaft, as shown in the photograph.

The insert material is assigned to the tungsten carbide group. Furthermore, this group can be modified by the recent technology of physical vapor deposition (PVD) where the substrate tungsten carbide is enhanced with special surface properties. The PVD coating produces a sharp insert edge to retain full strength allowing milling, grooving, and cutoff applications.

During milling, the cutter is rotating while the stock is in motion or traversing under the cutter. In a turning operation, it is the bar stock that is rotating with small (but important) feeding or traversing motions and the cutting tip is fixed. Even though these are kinematic inversions, there are engineering similarities between the two. Turning is principally working on rounded materials while milling is working on boxlike structures.

TABLE 7.8 Worksheet for Estimating Setup and Cycle for Operation 20, Table 7.6

A. Setup Elements

	Hours
Punch in and out, study drawing	0.2
Slot-cut milling collet	1.6
Setup total, operation 20	1.8

B. Handling and Other Equipment Time Elements

	Minutes
Start and stop machine	0.08
Air clean part	0.06
$10 \leq$ lb $<$ 15 lb load into collet fixture	0.25
Open and close air vise	0.06
Change speeds and feeds, 2 $\times$	0.08
Subtotal of handling and other equipment	0.53

C. Calculation of Machining Times

Element	Length of Cut, L_d	Milling Cutter	V, Velocity, fpm	f_t, feed (i $_t$pr)	t_m, min
Rough mill slot	6.81	0.75 in. end mill, 4 flute	85	0.002	1.97
Finish mill slot	6.81	0.75 in. end mill, 4 flute	95	0.0015	2.34
Subtotal of machining times					4.31
Total cycle time for handling and machining, min					4.84

D. Entry Values for Operation Sheet, Operation 20

Setup hours	1.8
Hours per 100 units	8.067

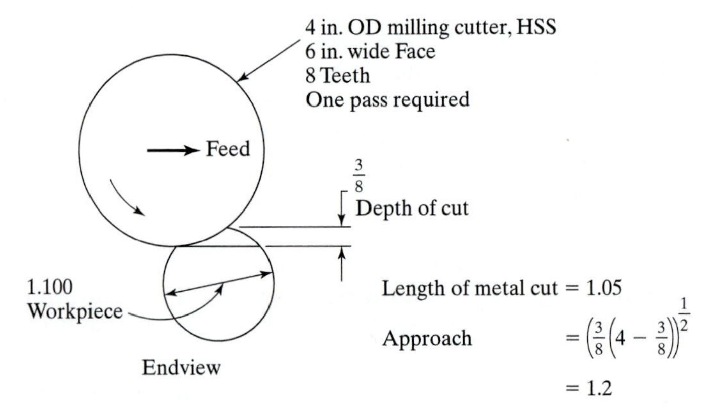

Figure 7.11 Diagram for operation 30, part no. 4943806.

TABLE 7.9 Worksheet for Estimating Setup and Cycle for Operation 30, Table 7.6

A. Setup Elements

	Hours
Punch in and out, study drawing	0.2
Special vise	1.1
Setup total, operation 30	1.3

B. Handling and Other Equipment Time Elements

	Minutes
Start and stop machine	0.08
Air clean part	0.06
$10 \leq$ lb $<$ 15 lb load into collet fixture	0.25
Open and close air vise	0.06
Subtotal of handling and other equipment	0.45

C. Calculation of Machining Times

Element	Length of Cut, L_d	Milling Cutter	V, Velocity, fpm	f_t, feed ($i_t$pr)	t_m, min
Mill flat on end	3.45	4 in. OD, 8 tooth	210	0.005	0.43
Subtotal of machining times					0.43
Total cycle time for handling and machining, min					0.88

D. Entry Values for Operation Sheet, Operation 30

Setup hr	1.3
Hr/100 units	1.467

7.4.5 Pinion Operation 40

Operation 40 is drilling and tapping and the diagram is given by Figure 7.12. The process planner selects a numerically controlled turret-type machine tool. This machine has a 6-sided turret (or more) mounted above the table, and the turret descends for the drilling and

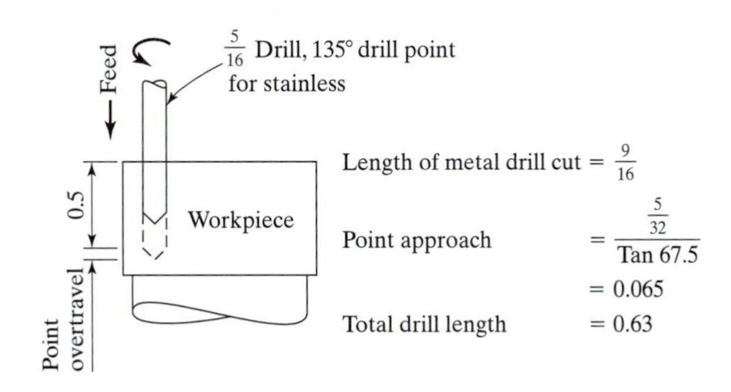

Figure 7.12 Diagram for operation 40, part no. 4943806.

tapping. One side of the turret will mount one tool. The table moves from hole point to hole point while the tool is above the work material. At the conclusion of one size of hole diameter, the turret will "index" or ratchet such that another face of the turret presents a new tool, and each hole of that specification is drilled, tapped, countersinked, counterbored, etc.

The worksheet for operation 40 is shown in Table 7.10. In operation 40, four 5/16-in. holes are drilled and tapped. Drilling will have an overtravel distance required for the length of the 135° conical drill point. Stainless steel uses a different conical point than other materials; the usual included point angle is 118°. A drill length is greater than the required tap length by a thread or so of length. Tapping usually does not require additional distances and only the tap length is necessary. The equation for finding the machining time is given by Equation (7.5), and information for the rate is given by Table 7.2. After finishing the worksheet and finding the setup of 1.12 hr and a cycle of 5.217 hr/100 units, these values are transferred to Table 7.6, operation 40.

The procedures for operation estimating are important for the student to understand. Even though this section has concentrated on simple machining, the general

TABLE 7.10 Worksheet for Estimating Setup and Cycle for Operation 40, Table 7.6

A. Setup Elements

	Hours
Punch in and out, study drawing	0.2
First turret station for drill	0.75
Second turret station for tap	0.07
Collet for holding	0.1
Setup total, operation 40	1.12

B. Handling and Other Equipment Time Elements

	Minutes
Start and stop machine	0.08
Air clean part	0.06
$10 \leq$ lb $<$ 15 lb load part into collet fixture and unload	0.22
Open and close air collet clamping	0.05
Index turret, 2 $\times$ for drilling and tapping	0.06
Raise tool, move to new locations, 8 $\times$	0.48
Subtotal of handling and other equipment	0.95

C. Calculation of Machining Times

Tool	Length of Cut, L_d	f_{dt}	Lf_{dt}	Number of Holes	t_m, min
Drill, 5/16	0.63	0.61	0.38	4	1.52
Tap, 3/8 - 16	0.5	0.33	0.17	4	0.66
Subtotal of machining times					2.18
Total cycle time for handling and machining, min					3.13

D. Entry Values for Operation Sheet, Operation 40

Setup hr	1.12
Hr/100 units	5.217

principles are appropriate for all manufacturing processes. From engineered time and cost performance of production, product design, knowledge of the processes, preparation of the operations sheet to the time and cost calculation, these steps are universally suitable for manufacturing.

Another important part of production are tools—both perishable and permanent. The permanent tools, while they are not ultimately permanent, are designated as such in relation to the perishable tools. Permanent tools are costly and are critical for product and operations. We now turn to this study.

7.5 TOOL COST

Tools are necessary to adapt general-purpose production machinery to a specific operation or product design. The engineer evaluates tool designs for their cost at the moment of the operation sheet preparation. Tool cost is a part of the operation cost and is important.

Officially, we term those tooling devices as *nonrecurring initial fixed costs*—that is, one-time-only costs despite the number of lots or continuous production quantity requirements. Those costs are unlike setup costs, which occur for each batch. The product cannot be manufactured without those tools, and thus they are front end or initial costs. Those designed tools are classified as fixed and capital costs rather than variable costs and may be a depreciable asset subject to federal taxation laws.

Additionally, such tools are classed as permanent tools or "hard" tools, as contrasted to perishable or "soft" tools, such as lathe-turning tools, milling cutters, and drill bits. The tools for Taylor's theory are perishable tooling. Now we say that perishable tools incur an indirect operational expense and are an overhead charge. Perishable tools are not typically estimated, nor are they specially designed, and for the most part, they are ordered from catalogs.

Examples of nonrecurring initial fixed tools are fixtures, jigs, molds, dies, electronic testing and quality control gauges, special-handling devices, and many other types. Tooling for assembly and postoperation inspection could include gauges, holding devices, universal jigs, and specially designed and constructed fixtures. Inspection requirements include tools for the receiving department, in process inspection, postfabrication inspection requirements, and postassembly requirements. Environmental needs where engineering specifications call for peculiar ambient, weather, or shock tests cannot be overlooked. In terms of test equipment and inspection devices, it too is weighed on equal footing with ordinary tooling. Special processing tools may be required for general-purpose or special-purpose equipment. The engineer needs to know what is available for production and what is special that may require design or purchase. These tools require engineering design and are costly.

The tools may be designed and built by the firm building the product, but other arrangements are possible. There is a business sector, which is called a "tool and die" (but that does not describe the activity broadly enough) that provides design and construction; they sell these tools to OEMs, who employ the tools in the manufacture of the product.

We are confronted with dollar signs in every operation in manufacturing. In this vein cost analysis can determine, for example, whether temporary tooling would suffice even though funds are provided for more expensive permanent tooling. Those initial costs are estimated before tool construction. Indeed, an experienced engineer is able to compare an existing part with a tool and cost to a new part design without a tool design and to provide an estimate value. This comparison method is quick but not necessarily accurate.

Many jigs, fixtures, gauges, and dies have common details that permit simplification in methods of estimating. An approach for tool estimating is to use engineering performance data or standard time data. Illustrative estimating data for manufacturing fixtures and drill jigs is given in Table 7.11. The data are described for a variety of base plates, support blocks, minor assemblies, and locators. The time is dimensioned in hours and is specifically for the manufacturing of fixtures and jigs using hardened steel components, which are precision machined and assembled by a journeyman tool and die maker.

The dimension for operation cycle time is in decimal minutes. But tool estimating data are different; though only one tool is to be built, the data need not have a fine degree of resolution. For example, a block to be machined, ground, drilled, reamed, and pinned in place on a larger block is too much detail to be estimated in minutes. Instead, the engineer visualizes the overall character of the tool and selects a time denominated in hours from the provided numbers. This simple practice is preferred because several engineers can use the same data, and accuracy and consistency are improved.

Figure 7.13 is an isometric design of a simple jig for the drilling of one hole. There is a hinged drill plate that lifts upward for the placement and removal of a part that is resting and located against pins, which are accurately positioned. The bushing locates the drill in reference to the part.

Table 7.12 is an estimate for a drill jig, where the part is to have one accurately located hole. The analysis shows that various elements are selected from the time data in Table 7.11, and eventually the total hours are multiplied by the productive hour cost rate of \$75 per hour. Material cost is determined by finding the weight of the tool and then multiplying by the \$25 per pound rate.

Cost of tooling is significant. Indeed, there are occasions when a new product will have greater tool cost than any other single element of the development cost. The manipulation of tooling cost is achieved by one of two ways: (1) apportioning to the product by *amortization*, or (2) by general overhead application. Amortization is an estimating calculation, meaning the division of the tool cost by the quantity of the operational lot or market total. The following relationship is used:

$$C_{ot} = \frac{C_{nif}}{N} \tag{7.18}$$

where C_{ot} = operation cost for permanent tooling, dollars per unit

C_{nif} = nonrecurring initial permanent tooling cost, dollars

N = lot, model, or year quantity, number

Selecting quantity N presents a trade-off problem. If this quantity is too small, then amortization will be overstated, causing greater cost and increased price, perhaps resulting in fewer sales of the product. If the quantity is too large, then other errors may result or the tooling could be made obsolete owing to unplanned design changes required by the part and then sunk tooling costs are created. Furthermore, the cost of the tooling may not be completely amortized, owing to actual production falling short of expectations. Engineering should be able to defend decisions when amortizing the

TABLE 7.11 Illustrative Estimating Data for Work Holders, Milling Fixtures, and Drill Jigs

1. Flat Base Plates

Type	Dimension			Hours
	Thickness	Width	Length	
Drill fixtures				
Small	$^3/_4$–1	3	3	5
Medium	$^3/_4$–1	7	7	8
Large	$^3/_4$–1	12	12	10
Milling fixtures				
Small	5/8–$1^1/_2$	3	6	10
Medium	5/8–$1^1/_2$	6	9	12

2. Angular Base Plates

Dimension	Type		
	Base	Combination Hr	C-Angle
3 × 3 × 3	10	5	7
6 × 6 × 6	15	9	12

3. Two Base Plates for Box Jig

Dimension	Hours
$^3/_4$ – 1 × 3 × 3	11
$^3/_4$ – 1 × 7 × 7	14

4. Drill Plates and Support Blocks

Type	Dimension	Hours
Drill plate	$^3/_4$ × 3 × 3	3
Box jig	1 × 3 × 3	5
Support blocks	5/8 × $1^1/_2$ × 3	2
Collet, round	to 4 in. OD	7

5. Miscellaneous

	Hours
Hand knobs, each	1.5
Guide bushing, each	2.0
Hinge plate, each	12.0
Clamp with screw	7.5
Feet, each	2.5
Pin locator, each	1.8

tooling on a single run as opposed to amortizing tooling costs against probable future reruns.

But the best expedient is to add the amortized cost of the tooling to operation cost. The application of Equation (7.18) applies the amortized cost of the tooling to the product.

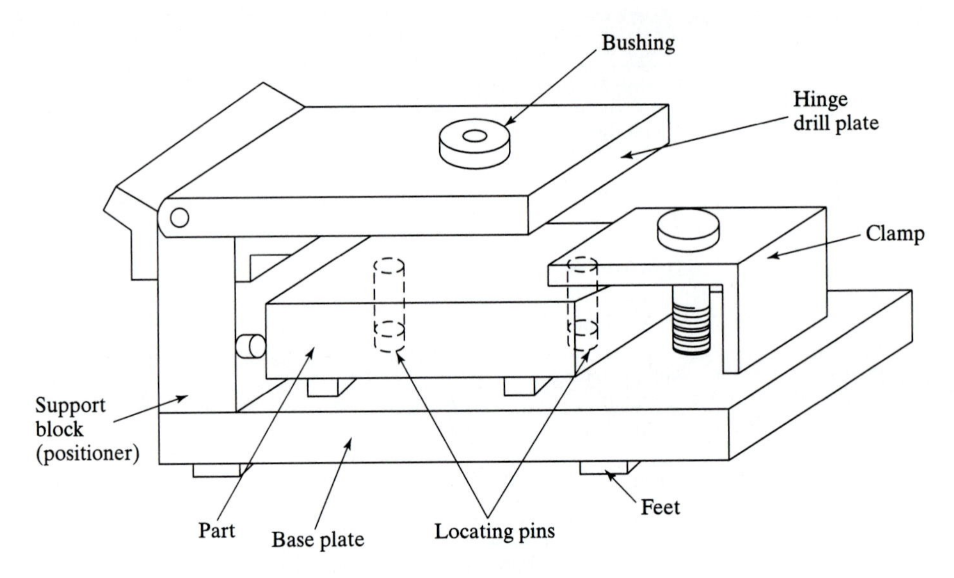

Figure 7.13 Basic components of a simple jig.

TABLE 7.12 Tool Estimate for Figure 7.13, Drill Jig

Tool type:	Drill fixture with hinge top, one bushing
Tool material:	Hardened high carbon steel alloy, 8 lb
Description of part:	Flat block with one accurately located hole
Description of Tool Elements	**Hours**
Angular base plate, C-angle	7.0
Guide bushing	2.0
Hinge plate	12.0
Clamp with screw	7.5
Feet, 3	7.5
Pin locators, 5	8.8
Total	44.8
Productive hour cost for tool making	$75
Cost of tool, 44.8 × 75	$3360
Material cost, $25/lb	200
Total cost of tool	$3560

In some business situations the manufacturer supplier may sell the tools directly to the original equipment manufacture (OEM) customer. Suppliers will design and build the tool and separately quote the cost of the tooling independent of the unit price to a customer. A separate tooling charge sidesteps the trade-off problem involved with amortization quantity, because the customer receives the tooling on delivery of the product. In other situations the engineer will add tooling costs to overhead, which spreads the cost to all products. This is not a recommended practice.

A part can be manufactured with or without tooling. This is another analysis that becomes necessary because of the high cost of tooling. For instance, tooling will allow less direct labor time in the operation but may increase initial costs. Break-even costs between the adoption of tooling versus increased direct labor can be evaluated by using

$$C_{nif} = \frac{Na(1 + p) - SU}{I + T + D + M} \tag{7.19}$$

where N = units manufactured per year, number

a = savings in labor cost compared to another operation, dollars per unit

p = percentage of overhead applied to labor saved, decimal

SU = yearly cost of setup, dollars

I = annual allowance for interest on investment, decimal

T = annual allowance for property taxes, decimal

D = annual allowance for depreciation, decimal

M = annual allowance for maintenance, decimal

This formula is used in interesting ways. A design can be manufactured by a number of methods. Is it cheaper to use a tool or employ direct labor that performs the same operation without a tool? This question, one that deals with comparative economics between competing methods, has many variations.

Example: Finding Tool Cost

Consider the tool that was previously estimated for the block with a single hole. There are 50,000 units to be produced, and a previous nontooled method indicates that $0.15 per unit will be saved if a tool were available that guides the drill. Savings in direct labor translate to 20% comparable savings in overhead. Setup of the tooled method is expected to cost $215. This company requires an annual interest of 12.5%, property taxes are 9%, depreciation is 25%, and annual maintenance of 18% for tooling is anticipated. The first cost of a tool that is based on these facts is found as

$$C_{nif} = \frac{50{,}000(0.15)(1 + 0.20) - 215}{0.125 + 0.09 + 0.25 + 0.18} = \$13{,}620$$

But the estimated cost of the tool as given by Table 7.12 is $3560, so the tool is paid back after 13,100(= 3560/13,620 × 50,000) parts are run. Tool cost is estimated as $3560 using Table 7.12 and is less than the tool cost necessary for the break even. The new tool recommended for the operation will reduce the cost of producing the part.

Once the separate cost components of direct material, direct labor, and tooling are known, they are combined in the operation cost.

7.6 OPERATION COST

Operation cost is found during the preparation of the operations sheet. Once the manufacturing operations are identified, along with the tooling, the calculation of cost starts with the computer entry of the setup and cycle times that are expressed in hours and hours per 100 units. (Obviously, other dimensions can be used, such as hours per 1000 units or cycle minutes.) Each operation has its own unique setup and cycle time.

Batch manufacturing will allow queues and storage between operations, and the layout of the factory is more suited to job shop style. Backtracking or spaghetti movement of material between operations is often pictured with job or lot production. The terms "batch" and "lot" are used interchangeably here.

In *flow-line production*, conveyors are usually the equipment used between stations. Serial movement connects adjacent operations and there may be virtually no storage between operations, though this is not a mandatory requirement for flow line production.

The calculation of cost for batch and flow-line estimating is different, as we now see.

7.6.1 Batch Manufacturing

For batch production, lot hours are found for each operation by using the relationship

$$\text{lot hours} = SU_b + N \times H_b \tag{7.20}$$

where lot hours = total time for each batch operation, hours

SU_b = batch operation setup, hours

N = lot quantity, number

H_b = cycle estimate for batch operation, hours per unit

For example, operation 10 in Table 7.6 shows 3.2 setup hours, 10.600 hr/100 units, and $N = 200$ units. Thus lot hours $= 24.4(= 3.2 + 2 \times 10.600)$.

The next step is to enter the productive hour cost rate (PHC) on the operations sheet, which is sum of the machine hour rate and the gross hourly cost rate for labor to operate the equipment. The PHC is different for each machine and work station. Using methods studied in Chapter 4 (Table 4.13), note that a PHC value of \$39.16 would be found for operation 10 in Table 7.6. The cost of a batch operation is found as follows:

$$C_{bo} = \text{lot hours} \times \text{PHC}_i \tag{7.21}$$

where C_{bo} = batch operation labor, equipment, and overhead cost, dollars

PHC_i = productive hour cost rate for labor, perishable tooling, equipment and other overhead for operation i, dollars per hour

$$C_{tbo} = \sum_{i}^{n} \text{lot hours}_i \times \text{PHC}_i$$

where C_{tbo} = total lot cost for all operations, dollars

$i = 1, 2, \cdots, n$ operations

In operation 10 of Table 7.6 the C_{bo} calculation is \$955(= 24.4 × 39.16), the batch cost for the operation. Also C_{tbo} is \$3262 for the four operations.

There are other ways to consider the PHC, and Equation (7.21) can be modified. In some cases an average value for the plant, department, or other logical grouping can be used, or PHC_{ave}.

Unit cost can be determined where labor, equipment, material, and tooling costs are summed for all operations and given by the following:

$$C_{bu} = \sum C_{bo}/N + C_{dm} + \sum C_{ot}/N \tag{7.22}$$

where C_{bu} = cost of batch manufacturing, dollars per unit

C_{dm} = cost of direct materials, dollars per unit

C_{ot} = operation cost for permanent tooling, dollars per unit

For example, in Table 7.6 the unit cost of pinion part number 4943806 is \$37.66 (= 3262/200 + 21.34 + 0).

It would seem that most of manufacturing cost estimating is concerned with batch production, as the emphasis has been on this role so far. We do a disservice to the student with that assumption. Indeed, most production is probably flow line if one considers commodity manufactured products, which will be consumer goods.

7.6.2 Flow-Line Manufacturing (Optional)

There are differences between flow-line and batch production. Flow-line production is an old method of continuous assembly that Henry Ford created in the early 1900s. In electronic fabrication, for instance, the flow line can have operations such as screen printing, automatic placement of components on a printed circuit board, manual placement of components, wave soldering, and testing, to name a few. The operations are sequential between adjoining production stations.

A flow line is serially arranged with automatic equipment, conveyors, benches, etc., where the machines are operator assisted or not. Bench work is dominated by operator controlled work, or manual, and there may be small assisting tools, such as nut drivers and pliers. Automatic equipment is commonplace in flow-line production.

A setup for electronic fabrication may include loading the component placement machine with the appropriate software program, stocking the materials to assemble to the board, and arranging part feeders, for example. If assembly is manual stuffing of components onto the board, the setup elements would include location of bins of components, obtaining the instructions, and arranging the assembly station. Conveyors of all kinds connect these operations.

In a flow-line production system, machines and benches are set up simultaneously and production is not initiated until every machine and bench are ready. In a flow-line system of production, unless the setup of each station is identical, there will be possible idle and lost time, unless other duties are found for the direct labor and equipment having the shorter setup.

The total time that the factory flow cell is working on setup due to changeover is the setup time of the station that takes the longest to prepare, or

$$SU_f = \max\{SU_i\}$$

where SU_f = maximum setup of flow-line stations, hours

$$SU_i = \text{setup of flow-line stations}, i = 1, 2, \ldots, n \text{ stations, hours}$$

$$\text{cost of setup for flow line} = \sum_i^n PHC_i \times SU_f, i = 1, 2, \cdots, n \text{ stations, dollars} \quad (7.23)$$

Production starts immediately after the flow-line setup is concluded and ends when the desired quantity is complete. With intermixed automatic equipment and manual-controlled bench work, and despite attempts of line equalizing of the separate operations into equal time length, the cycle time of each station is gated or bottlenecked by the station having the longest cycle time. The line is unable to produce product faster than the slowest operation. Unless the operations are perfectly balanced, idle time is likely. The "drop-off rate" for flow-line production is fixed by the longest cycle time, or

$$H_f = \max\{H_i\} \quad (7.24)$$

where H_f = maximum cycle time of flow-line stations, hours per unit

H_i = time estimate for flow line operations, $i = 1, 2, \cdots, n$ stations, hours per unit

$$\text{cost of cycle time for flow line} = N \sum_i^n PHC_i \times H_f, i = 1, 2, \cdots, n \text{ stations, dollars}$$

N = number of product units

The number of stations or operations n in the flow line directly influences the time each product unit is traveling. If the longest cycle time sets the gated time for the flow line, the total cost of producing N units is given by considering all operations and stations. The setup and cycle time for all units is composed as follows:

$$\text{lot hours} = nSU_f + nNH_f \quad (7.25)$$

where lot hours = total time for flow-line production for stations and quantity, hours

$$C_f = \sum_i^n PHC_i \times SU_f + N \sum_i^n PHC_i \times H_f \quad (7.26)$$

where C_f = cost of flow-line labor, equipment, and overhead, dollars

It is possible that the PHC can differ between setup and cycle work for any operation, though that nuance is not shown by Equation (7.26). Typically, it is more appropriate to find the unit cost once we have total cost using

$$C_{fu} = C_f/N + C_{dm} + \sum C_{ot}/N \quad (7.27)$$

where C_{fu} = cost of flow line manufacturing, dollars per unit

If N is very large and the significance of the setup cost is trivial, then the majority of the flow-line cost is composed of cycle time effects. In this case, the cost of setup is figured into the overhead. The adding of setup cost effects to overhead is not uncommon, though the practice is not as accurate.

The direct material C_{dm} for Equation (7.27) is added to the product cost as a lump sum independent of any operation. This practice varies widely. In electronic fabrication where many components are inserted into the printed circuit board, various operations may have direct materials added as the board progresses through the stages of flow-line production. Some flow-line cost models add the direct materials via the operation and then sum over all operations. This model may be more accurate, especially if there is a "yield" reduction of material and labor as the product progresses along the flow line.

In flow-line production the development of the operations sheet differs from batch planning. Symbols are added to the operations plan, as shown in Table 7.13.

Example: Flow-Line Cost Estimating of Computer Hard-Drive Casting Base

An aluminum casting for a computer disk drive is manufactured in Malaysia. The outside dimensions of the component are $3.000 \pm 0.008 \times 5.770 \pm 0.010 \times 1.128 \pm 0.004$ in. Worldwide demand for the base casting for the approaching quarter is 6 million units.

The critical manufacturing operation is the die casting of the base. The operation employs die casting equipment rated at 750 tons. The principle of a separate cold chamber is used.

The factory utilizes two shifts per day and just-in-time delivery of 40-lb aluminum ingots for the melting of the base. Enough ingot is stored and smelted for casting for the two 8-hour shifts. The setup for the morning shift will require melting of the aluminum in the pot and slow heating of the 8-cavity die, and at the conclusion of the second shift the caster and die is cooled. The time required for the daily setup, which includes both the setting up and the tearing down, is 2.7 hr. Material cost per unit is found as \$0.79, which includes waste due to the stack emissions and other melting losses.

Other operations follow the die casting operation in a flow line. The information is given in Table 7.13. In addition to the operations of casting, trimming, and machining, there are unadvoidable delays for storage, and direct labor for inspection. Setup and cycle times and productive hour costs are provided for all direct labor consuming operations. Tooling costs—estimated for the die, trim die, milling fixture, inspection tools and go/no-go gauges—are shown for the operations.

Operations 3, 4, 5, 7, and 8 are direct labor. Notice that inspection and test are considered direct labor, which is somewhat unlike batch operations where they are frequently classified as indirect labor and are a part of overhead computation for the productive hour cost rate.

Die casting is the bottleneck operation, both for the setup and cycle. Line equalization is such that all following operations have successively faster cycle times, this being done to avoid excessive interoperation storage. This illustrates a "pull" system. Each day 13.3 hr $(= 16 - 2.7)$ is available for component production. With a production rate of 0.00417 hr per unit for die casting, the line drop-off rate is 3189 units per day $(= 13.3/0.00417)$. The quarterly production per flow line is 287,000 units. For the quarterly production requirement of 6 million units over 90 days, 21 flow lines are necessary $(= 6{,}000{,}000/(90 \times 3{,}189)$.

Using Equation (7.23), the cost of setup $= 2.7(83.16 + 19.24 + 24.15 + 79.14 + 29.63) =$ \$635.36 per day. Using Equation (7.26), the cost of flow-line labor, equipment, and overhead $= 3{,}189 \times 0.00417\,(83.16 + 19.24 + 24.15 + 79.14 + 29.63) =$ \$3129 per day

TABLE 7.13 Flow-Line Operations Sheet for Computer Hard-Drive Casting Base

Workstation	Operation	Description of Operation	Setup, hr	Cycle, hr/unit	PHC, $/hr	Tooling, $
Jib crane	▽ 1	Material input to caster				
Jumbo pallets	▱ 2	Storage next to die caster				
750 ton die casting machine	▭ 3	Cast platter of 8 units	2.7	0.00417	83.16	27,500
200 ton trim press	▭ 4	Separate base from runners and gates	1.2	0.00385	19.24	8,500
Inspection bench	○ 5	Examine for blow holes, visual and ultrasound	0.5	0.00379	24.15	7,500
Robotic pallet with arm	▱ 6	Store next to milling machine				
CNC milling machine	▭ 7	Mill and drill to dimensions	0.8	0.00341	79.14	4,500
Inspection bench	○ 8	Plug gage holes, go/no-go for milled dimension	0.6	0.00323	29.63	4,300
	◇ 9	End process				

▽: Material input, ▱: Storage, ▭: Operation, ○ : Test or inspection, ◇: Termination

for 3189 daily units. Setup and cycle costs are $1.18 per unit. Tooling cost is amortized over the product demand of 287,000 units for the flow line, or $0.19 per unit (= 53,200/287,000). The total unit cost of the computer component is $2.16 (= 1.18 + 0.79 + 0.19).

The die casting of the computer base example focused on flow line production that was straight-line, or serial, from start to end. Other flow line types are possible and are shown in Figure 7.14. Figure 7.14 (a) is the serial flow model again, and this simple case has definite starting and ending workstations. The rectangles indicate the direct cost operations. Although not shown in the figure, additional symbols could be added for material input, storage, test or inspection.

The converging model is found frequently in industry, where two or more chains precede the serial-flow final assembly of a product. Look at Figure 7.14 (b), for example. The number of workstation links in the chains need not be identical. An example would be subassemblies that are joined at various points and are finally produced as an integral product.

Flow line production can be diverging, as described in Figure 7.14 (c). In case (c) imagine final products that have variation, and those different products are produced by one of the diverging chains after the common design is manufactured.

More complicated production systems can be configured, where flow combinations of serial, converging, and diverging can be joined in various ways. The production of the popular 12 oz. beverage can is an example. An early operation of blanking and cupping has a production rate of 3800 cans per minute. Later operations of necking or forming the top of the can are 615 cans per minute; so seven diverging systems produce 4305 cans per minute, which prevents a storage queue. (The student may want to work Challenge Problem 7.28 to understand this important "pull" production system.)

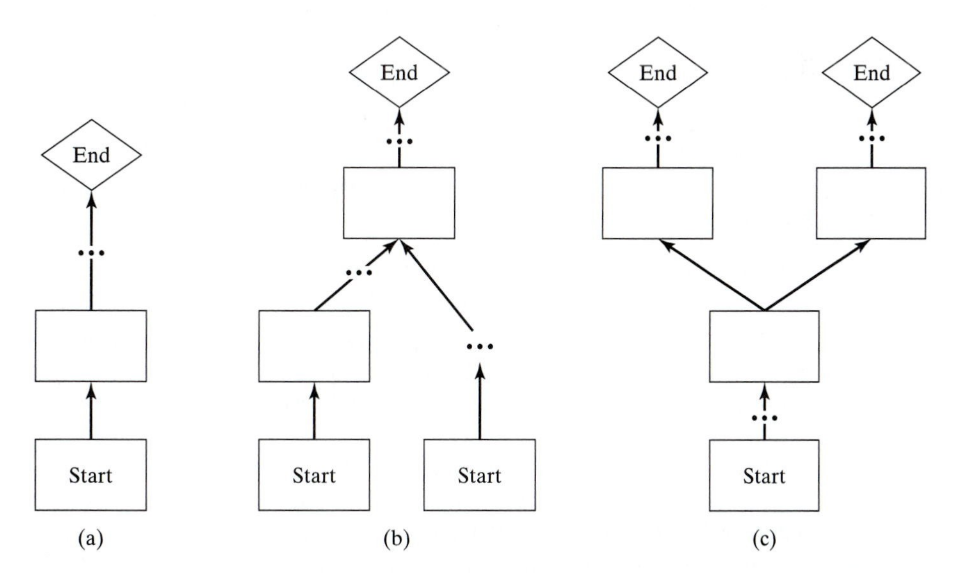

Figure 7.14 Illustrative production flow models: (a) Serial, (b) converging, and (c) diverging.

Cost calculations for the diverging or converging chains are similar to the basic serial flow model, as setup and cycle time of each link are found, and the bottleneck of each chain is determined. Although not given here, the cost equations need to reckon each workstation, and subsume that cost into the total cost. In the next chapter, discussion of the bill of material will elaborate on the necessary steps for cost calculation. The bill of material is useful for finding the cost of various flow systems.

SUMMARY

Cost is the major objective for operation estimating, and material, labor, and tooling are the principal components. If the labor is designated and engineering performance data are available, then time estimates are selected for work elements. The total time is subsequently multiplied by the wage or the productive hour cost. Tools can be estimated and appraised for their economic justification.

In the next chapter we consider product estimating. Manufacturing operation estimates are a prerequisite for product estimating and provide information to cost and price products.

QUESTIONS FOR DISCUSSION

1. Define the following terms:

Amortization
Approach
Cycle
Flow line manufacturing
Gating operation
Length of cut
Machining cost
Manufacturing
Microanalysis
Moderate production
Operation
Operation sheet
Permanent tooling
Process
Rough pass
Setup
Standard man hour
Tool changing
Tool life

2. What role does the RFQ have in the management of cost estimating?

3. Separate the types of production according to volume. How else can you classify production?

4. Why is setup—a fixed direct-labor cost—divided by the quantity of the lot?

5. Name examples of nonrecurring initial fixed costs. How are those costs handled? Does the act of separating a tooling charge from the unit cost of the product for a buyer eliminate the amortization problem?

6. Discuss the differences between batch and flow-line cost analysis.

7. List the pros and cons of trade-off policies dealing with the selection of quantity in the amortization of nonrecurring initial fixed costs.

8. Describe the limitations of the Taylor tool life model.

9. List the important steps in contemporary operations analysis for the finding of cost.

10. Check out various Websites for viewpoints of manufacturing.

PROBLEMS

7.1 An operation has a setup of 1.5 hr and a cycle time of 0.50 min. If the lot quantity is 75, find the lot hours and unit time expressed in hours.

7.2 An operation is estimated to require a setup of 10 hr and a cycle time of 1 min. If the productive hour cost is $75, and the run quantity is 210 units, find the operation cost.

7.3 A collet-type turret lathe will rough turn, finish turn, drill and tap using an end turret, and cutoff with the cross turret. The running of the 7-in. long stock includes the following metal-cutting times:

Estimate setup, pieces per hour, hours per 100 units, and lot hours for a run of 710 units. (*Hint*: There are five speed changes for machining RPMs; use Tables 7.2, 7.4, and 7.5.)

Element	Minutes
Rough turn	0.89
Finish turn	1.26
Drill	0.73
Tap	0.47
Cut off of stock	0.83

7.4 A slot milling operation is performed on a 13-lb gray-iron casting. The setup includes an angle plate with fixture mounting and a tolerance on the dimension being machined as ± 0.002 in. Machining time is 2.24 min. The shop advises that air blowing and brushing chips will be necessary. Estimate setup, pieces per hour, hours per 100 units, and lot hours for a run of 1783 units.

7.5 An operation is estimated for a Burgmaster NC turret drill press. Ten aluminum flat parts are stacked-drilled in 37 locations. Each part weighs 2.7 lb and is 1/8 in. thick and is loaded against a rail clamp-type fixture. Four turret stations are used. Drilling time through the stack is 0.19 min per hole. Estimate setup, unit time, pieces per hour, hours per 100 units, and lot hours for a run of 983 units. (*Hint*: Hole sizes and locations are not pertinent to the solution.)

7.6 A press operation for an automobile body fender production line has a setup of 30 hr for the form die, which costs $72,500. The operation is estimated to require 2.173 min for loading, drawing to size and trim, and unloading to the conveyor, which completes the cycle. The run for this model year is 247,500 units. Material cost for the corrosion-resistant steel sheet-metal blank including all losses included is $7.12. The productive hour cost for the press is $192.50.

Find the operation lot hours. Each press is expected to operate 4000 annual hours. How many presses are necessary? Find the unit cost then find the proportion of productive hour cost, material cost, and tool cost. (*Hint*: It is necessary to find the number of machines and die sets that are required for the production quantity.)

7.7 Permanent tooling is to be evaluated for an operation and product. Let the estimated unit savings in direct labor be $0.30 for a part if tooling is used, burden on labor saved = 40%, cost of each setup = $100, interest rate = 20%, allowance for taxes and insurance = 10%, allowance for depreciation and obsolescence = 50% and allowance for maintenance = 25%.

(a) The cost of a fixture is $4000. With one run per year, how many pieces must be made per year to have the fixture pay for itself?

(b) Let depreciation be 100%, because the fixture must pay for itself within a year. How large must that run be?

(c) By using the initial data, how much money can we afford for a fixture for a single run of 15,000 units at an estimated savings of $0.30 per piece?

(d) How many years for payback will a $4000 fixture require for an annual quantity of 20,000 units?

7.8 (a) An operator earns $15 per hour. Handling and other constant time elements total 1.35 min. What is the cost for the element?

(b) The machining length is 20 in. and the part diameter is 4 in. Standard velocity and feed for this material is 275 fpm and 0.020 ipr. What is the time for turning?

(c) The Taylor tool life equation is $VT^{0.1} = 372$. Find the average tool life for $V = 275$ fpm.

(d) Tool changing time is 4 min and other conditions are $VT^{0.1} = 372$, $V = 275$ fpm, $C_0 = \$0.25$/min, $L = 20$ in., $D = 4$ in., and $f = 0.020$ ipr. Determine the tool-changing cost.

(e) By using information in earlier parts of this problem and $C_t = \$5$, find the tool cost per operation.

7.9 (a) Operator wage and variable machine expenses are $60 per hour and handling is 1.65 min. Find the handling cost for this element.

(b) The length and diameter of a gray iron casting are 8.5 by 8.6 in. Surface rotational velocity of the casting is 300 fpm and feed is 0.020 ipr. Find the turning time.

(c) The Taylor tool life equation is $VT^{0.15} = 500$. Find the average tool life for 300 fpm.

(d) The time to remove a square insert and index it to another new corner is 2 min, tool life equation is $VT^{0.15} = 500$, $V = 300$ fpm, $C_o = \$1$/min, $L = 8.5$ in., $D = 8.6$ in., and $f = 0.020$ ipr. What is the cost to change tools?

(e) By using the information in part (d), and that an 8-corner insert for rough turning of cast iron is $24, find the tool cost per operation.

7.10 (a) Stainless-steel material is to be rough and finish turned. Diameter and length of the bar stock are 4 × 30 in. Recommended rough and finish cutting velocity and feed for tungsten carbide tool material are (V = fpm, f = ipr) or (350, 0.015) and (350, 0.007). Determine rough and finish cutting time.

(b) Medium carbon steel is to be rough and finish turned by using high-speed steel-tool material. The part diameter and cutting length are 4 in. and 20 in. Determine the total time to machine by using Table 7.2.

(c) Gray cast iron is to be rough and finish turned with tungsten carbide tooling. Part diameter and cutting length are 8.6 in. and 8.5 in. What is the part RPM for the rough and finish? Find the total turning time. (*Hint*: Use Table 7.2 for the turning velocity and feed. To find the time, use the relationship $t_m = L/fN$.)

7.11 (a) Find the time to rough mill a stainless-steel work material having a total-machining 4-in. flat surface length of 35 in. The plain milling cutter is high-speed steel material; it is 4-in. OD and 6-in. wide and has 16 teeth. (*Hint*: Assume only one pass; refer to Table 7.2 for information.)

(b) Find the time to finish mill a medium-carbon steel work material having a total-machining slot length of 11.5 in. by 1-in. wide. The end milling cutter is high-speed steel material; it is 1-in. OD and has 4 teeth.

7.12 (a) A gray cast-iron block has one hole that is drilled and tapped according to the drawing callout of 5/16-16 for a length of 1.5 in. Standard time data retrieval for setup are

2 hr and all handling requires 1.65 min. There are 93 units on the operation order. What is the cycle time? Find the lot hours for this operation. (*Hint*: The final tap diameter is 5/16 in. and has 16 threads per in. The drill diameter is smaller than 5/16 in. and thus $^1/_4$ in. is chosen, it being the only stated value in Table 7.2. The drilling length is typically longer than the tapped length by one thread length or so.)

(b) A medium carbon steel part has one hole that is drilled and tapped according to the standard drawing callout of 3/8-10 for a length of 0.75 in. Standard time data retrieval for setup is 1.22 hr and all handling requires 0.65 min. There are 273 units on the operation order. Find the lot hours for this operation.

7.13 (a) A 10-in. long block is to have a 1-in. wide slot machined on it. The HSS end mill is 1 in. OD and only one finish pass is machined over the slot. Calculate the approximate length of cut for the tool if the safety stock is 1/4 in. Repeat if the width of the slot is 2 in.

(b) A slab milling cutter is 6 in. in diameter. The design length for metal removal is 15 in., and one rough machining passes of 3/8 in. depth of stock removal is necessary. Safety stock is 1/4 in. Determine the approximate length of cut for roughing.

(c) A slab milling cutter is 6 in. in diameter. The design length for metal removal is 15 in., and two rough machining passes of 3/8 in. depth of stock removal and one finish 0.015 in. depth of cut are necessary. Safety stock is 1/4 in. Determine the length of cut for roughing and finishing.

(d) Find the length of cutting for a 1 in. in diameter drill in soft steel where the design length is 2 in.

7.14 (a) A stainless steel surface 1×10 in. is to be end milled with a 1-in. high-speed steel four-tooth end mill for a depth of 0.015 in. The end mill will pass entirely over the material and have a safety stock of 1/8 in. Velocity is 95 fpm, and chip load per tooth is 0.0015. What is the length of cut? How much time is necessary for machining? What is the cutter RPM and feed rate in inches per minute?

(b) A 1-in. slot is to be end milled in gray cast iron for a depth of 1/4 in. for a design length of 20 in. A rough and finish pass is required for the HSS four-flute cutter. A safety stock of 1/8 in. is necessary. What is the machining length of cut for the rough or finish machining pass? Determine rough and finish machining time. Find cutter RPM and cutting rate in inches per minute. If the nonmachining time is 1.73 min/unit and the setup is 3.24 hr, find the lot hours for 217 units for the operation. (*Hint*: Use Table 7.2 for machining data, where data are cutting velocity and tooth load. The cutter must pass entirely over the surface for the finish pass.)

7.15 (a) A gray cast-iron surface 6 in. wide by 30 in. long is rough milled to a depth of 1/4 in. The diameter of a 16-tooth cemented-carbide face mill is 6 in. Estimate the cutting time. Find the cutter RPM. (*Hint*: The computer controlled milling machine is performing the operation as if the cutter is an end mill in a vertical position. There is 1/2 in. for safety space. Use Table 7.2 for machining data and apply Equation (7.14). In exiting the surface, the cutter does not pass entirely over the surface for a rough machining pass.)

(b) A high-speed steel vertical milling cutter having 12 teeth is 4 in. in diameter and is used to mill a soft-steel surface 3 in. wide by 9 in. long with a depth of cut of 3/4 in. A cutting speed of 170 fpm and a feed of 0.008 in. per tooth-rev. are selected. Find the cutting time and cutter RPM. What is the operation lot hours if the setup is 3.25 hr and other cycle time is 14.17 min for 27 units?

7.16 (a) A stainless-steel part is drilled 5/16 in. which is followed by a 3/8-16 N.C. tap for a depth of 7/8 in. Find the drilling length and the total of drilling and tapping time. (*Hint*: Use Table 7.2.)

(b) Medium carbon steel is tap-drilled 1/4 in., which is followed with a tapping element of 3/8-24 N.F. for 1.3 in. There are 26 holes. Find the drilling length and the total of drilling and tapping time.

There are 172 units in a lot and the setup is 0.7 hr and the cycle time exclusive of the drilling and tapping of holes is 7.13 min. What are the pieces per hour. Find the lot hours. (*Hint*: Pieces per hour is exclusive of setup time.)

7.17 (a) Find the cutting time for a hard copper shaft 2 in. OD × 20 in. long. A surface velocity of 250 fpm is suggested with a feed of 0.009 in. per revolution. (*Hint*: This is a lathe turning operation.)

(b) An end facing cut is required of a 10-in. diameter work piece. The revolutions per minute of the lathe are controlled to maintain 400 surface feet per minute from the center out to the surface. Feed is 0.009. Find the time for the machining cut. (*Hint*: A facing cut is along the end of the bar stock, and the work piece is held by a 3-jaw chuck allowing the tool to move free of interference. As the tool post moves from the outside to the inside, the work piece RPMs will increase to maintain a constant 400 fpm. The length of cut for a lathe facing operation is $\frac{1}{2}$ × diameter.)

7.18 If the tool material grade is Kennametal 3H carbide, work material is AISI 4140 steel, depth of cut is 0.150 in., and the feed is 0.010 ipr, find the surface feet per minute for a 4-in. OD bar and a 6-min life if the tool life equation $VT^{0.3723} = 1022$. Find the RPM as well.

7.19 Find the ratio of tool life using tungsten carbide to high-speed steel tool materials for a rough-turning operation of cast iron. (*Hint*: Use Tables 7.2 and 7.3.) What is the average tool life increase using tungsten carbide tool material?

7.20 Using the classical Taylor tool life model, and for the work and tool materials, find the cutting velocity for a tool life of 20 min for each of these combinations..

Work Material	Tool Material
Stainless steel	Tungsten carbide
Medium carbon steel	Tungsten carbide
Medium carbon steel	High-speed steel
Gray cast iron	Tungsten carbide

7.21 A company is planning to "turn" thousands of bars for a long production run. As one step in the planning, it conducts a Taylor test and following a field study of the AISI 3140 work material, lathe, and a newly developed tool material, the tool-life curve in Figure P7.21 is determined. The conditions of the test are a feed of 0.013 ipr with a depth of cut of 0.50 in. Find the tool life for 100 fpm. What are the parameters for $VT^n = K$? (*Hint*: Use the curve to determine n and K.) How much time is necessary to machine the new-product design AISI 3140 bar stock if the diameter and length are 2 in. OD and 27.125 in. for a life of 60 min if the feed is 0.0125 ipr? What are the revolutions per minute?

7.22 A Taylor tool-life performance test is conducted for a rough-turning operation of newly developed carbon steel. The tool material is a recently developed grade of high-speed steel. The testing determines $VT^{0.1} = 372$. An operation requires planning for optimum

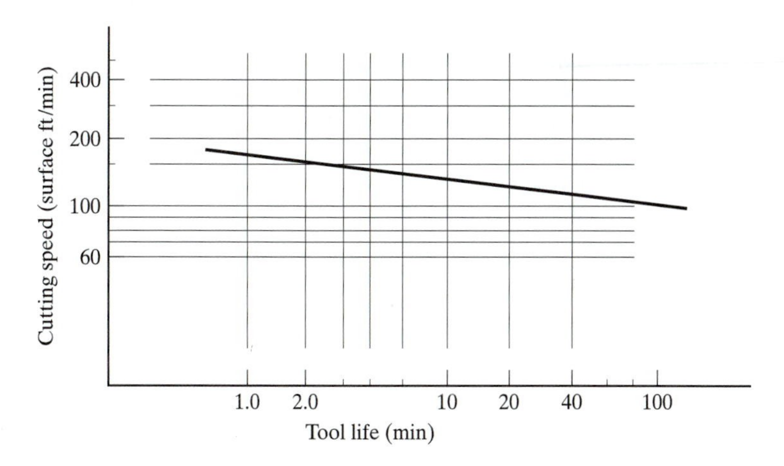

Figure P7.21

performance. The work material diameter is 4 in. OD and the total cutting length is 20 in. The tool point costs \$5 and time to change the tool after it wears out is 4 min. Part handling is 2 min and the operator wage is \$20 per hour. The feed of the turning operation is 0.020 ipr for a depth of cut = 0.5 in.

(a) Find the optimum cutting velocity analytically and calculate the RPMs.
(b) Plot the handling, machining, tool-changing, and tool costs versus feet per minute to find the total cost curve. Find the optimum cutting velocity graphically. What is the cost? (*Hint*: Transform the four equations (7.1, 7.3, 7.7, and 7.8) to have the cost expressed in units of V, and then substitute values of V from 50 to 350 in steps of 50 units. For example, handling cost = \$0.333/min × 2 min = \$0.667, which is a constant. Again, machining cost = $349/V$, and so on. You may want to use a spreadsheet.)
(c) Let the cost of $C_t = 0$ and compute V_{max}, and T_{min}.
(d) Find the cycle time, units per hour, and hr/100 units.

CHALLENGE PROBLEMS

7.23 Find the material cost and setup and cycle time for the fabrication of a pinion, part no. 4943806, similar in all respects to Figure 7.7 except for the following changes required by (a)–(e). (*Hint*: Work each subproblem separately; complete the operations process sheet and find the unit cost as demonstrated by Table 7.6. Many work elements are similar to existing worksheets.)

(a) Let the 18.750-in. dimension be 8.750 in., the 1.100-in. dimension be 1.40 in.; assume that there are no holes. (*Hint*: The 1.40 in. OD dimension is reduced to 1.75 in. length. Check it out!)
(b) Let the 4.75-in. flat dimension be 8.00 in. long, and the raw material be 2 in. OD instead of 1 3/4 in. OD.
(c) Let the 6-in. wide milling cutter be 3 in. in width for operation 30, and operation 10 will use high-speed tool material instead of tungsten carbide.
(d) Let the material be medium carbon steel instead of stainless steel. Medium carbon steel costs \$1.00/lb.
(e) Let the lot quantity be 1000 instead of 200.

7.24 Sometimes it is less costly to design and construct one multipurpose tool that will serve two or more similar designs rather than an individual tool that will satisfy only one design. There are, however, complications that can thwart this practice, such as simultaneous scheduling of parts or the need to use the designs that are so radically different that multipurpose tools are more expensive than using one tool for each design. Though multipurpose tooling may at first appear to offer cost benefits, the operation cost for different parts may cancel out many of those gains.

Consider the following situation: Two product designs are called Left Hand and Right Hand and are sufficiently similar to allow multipurpose tools. Some details are shown below.

	Left Hand	Right Hand	Both Hands, Multipurpose
Cost of tool	\$47,500	\$32,500	\$62,000
Productive hour cost per unit	\$1.92	\$2.17	\$2.03 (LH) \$2.65(RH)
Life time units	83,000	27,500	

Should two tools or one multipurpose tool be designed and constructed? (*Hint*: Productive hour cost per unit includes direct labor and machine cost. Ignore material cost in the analysis.)

7.25 A malleable-iron casting having an as-cast nominal diameter of 9 1/4 in. is rough turned to 8.600 ± 0.005 in. for a length of 8.500 ± 0.010 in. The feed for this turning operation is 0.020 ipr.

A renewable square carbide insert is used. The tungsten carbide insert has eight usable corners suitable for turning work and costs \$24. The time for the operator to remove the insert and install another new corner and qualify the tool ready to cut is 2 min. Operator and variable production costs are \$60 per hour. Taylor's tool life equation for work and tool material is $VT^{0.15} = 500$. Robot handling time is 1.65 min for a casting mounted in a fixture.

(a) Determine optimum cutting velocity analytically. Find the RPM.
(b) Construct individual cost and total cost curves similar to Figure 7.6. If the y axis is unit cost and x axis is fpm, then plot the curves to locate the minimum cost. What is the range of cost if shop performance can be expected to vary ±10% from the optimum velocity. (*Hint*: Transform the four equations (7.1, 7.3, 7.7, and 7.8) to have the cost expressed in dimension of V, and then substitute values of V from 50 to 350 in jumps of 50 fpm units. A spreadsheet analysis will be helpful. The diameter is the starting rough diameter of the casting.)
(c) Find the unit time T_u, pieces per hour, and hr/100 units.
(d) Assume that the cost of the tooling $C_{pt} = 0$ and find V_{max} and T_{min}.
(e) Find the time-versus-velocity curves if $C_{pt} = 0$.

7.26 Estimate the total lot and unit cost for the design given by Figure P7.26 for material, labor, overhead, and tooling. The quantity is 7500 units, which is produced in one lot. Follow the steps shown below.

(a) How many bars are necessary? Find the material cost for the lot order and unit. Raw material bar stock is 3 in. OD, 8 ft long, and is \$1.75/lb. Density is 0.285 lb/in.^{3} The cutoff and facing length is 1/8 in. and 0.015 in.

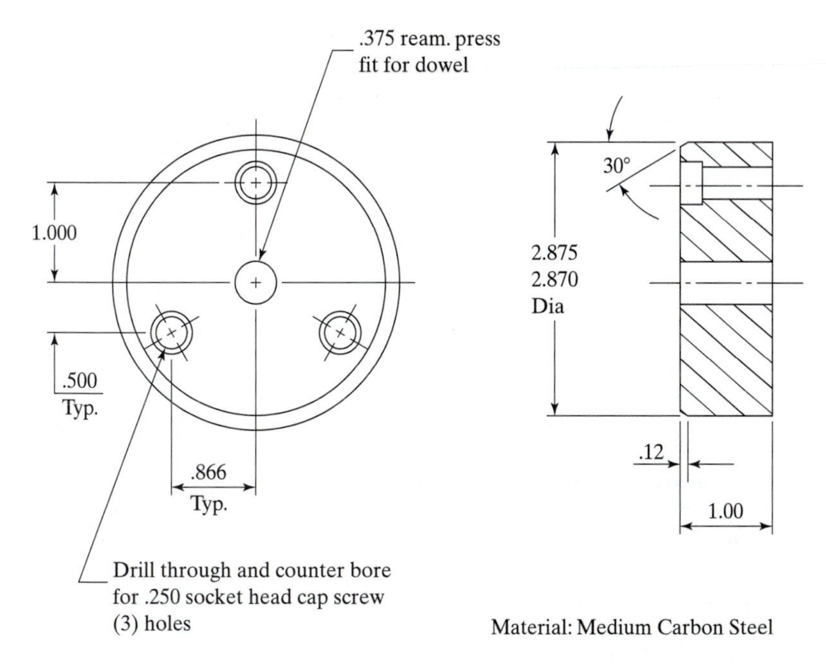

Figure P7.26

(b) Find the setup, cycle minutes, hours per 100 units for the two operations for entry to a operations process sheet. (*Hint*: Use a method similar to Table 7.6.) The operations area is as follows:

Operation 10: Face, turn, chamfer, drill, and ream the 0.375 in. hole and cutoff with a turret lathe

Operation 20: Drill and counter bore three holes with a drill press.

(c) Prepare an operations sheet and find the operation cost for the unit and lot. Productive hour costs for turning and drilling are $75 per hour. Use the tabular columns shown below.

Workstation	Op. No.	Description	Setup	Hr/100	Lot Hr	PHC	Lot Cost	Unit Cost

(d) Find the total and unit tool cost. A small drill jig with pneumatic clamping is necessary for the 0.25 in. drill and counterboring operation. Tool design is expected to cost $8000, and the PHC for tool building is $100 per hour. (*Hint*: The tool cost is estimated using the part print, rather than the tool design, which means that you should visualize the elements of a tool design as suggested by Table 7.11. There are three removable slip bushings that come out for the counterboring. Expect about 4 lb for the jig. Make any material adjustments for missing data based on local experience.)

(e) Using an operations sheet approach, estimate the total lot and unit cost of material, operations, and tooling.

7.27 Find the unit cost for flow line production, which consists of three stations for an electronic product assembly. There are 1750 printed circuit boards produced. The flow line starts with a predrilled printed circuit board obtained from a supplier costing $0.93 each. Designed tooling are holding fixtures, special grippers and a sled for wave soldering. Information is estimated as shown below.

Station	Machine Component Insertion	Manual Stuffing	Wave Solder
Setup time, hr	1.6	0.5	1.2
Cycle time, min/unit	3.75	5.13	0.65
PHCR, $/hr	$78.25	$27.25	$110.73
Direct materials, $/unit-operation	$17.62	$1.29	$0.85
Designed tooling, $/operation	$283	$18.46	$73.09

7.28 Can production of the ubiquitous 12 oz. beverage container is very large volume, and a plant will produce billions of cans per year. We are interested in finding the cost for material, labor, tooling, and equipment per can and for 1 million units.

A brief description of the 10 operations for the body of a three-piece aluminum can (e.g., body, lid, and pull tab) is shown in the table below.

Operation Description	Material Cost per Can for Operation	Yield of Operation, Decimal	Minute per Can	Productive Cost per Minute
1. Blanking and cupping	$0.0350	0.9800	0.00026	$5.42
2. Draw and iron body		0.9750	0.00026	4.58
3. Trim top of can to even height		0.9500	0.00500	2.33
4. Wash and dry	0.0001	0.9990	0.00020	3.00
5. Decorate and overcoat	0.0003	0.9988	0.00049	3.17
6. Oven cure for external coating		0.9988	0.00020	2.08
7. Internal spray and bake	0.0002	0.9975	0.00167	3.00
8. Neck and flange		0.9400	0.00163	4.67
9. Inspect		0.9990	0.00045	1.25
10. Palletize and storage		0.9925	0.00020	8.33

Initially, the can is blanked and cupped from coil stock in operation 1, and material is added in operations 4, 5, and 7.

Yield effects are given for each operation, and each operation reduces the quantity throughput because of waste losses due to processing. For example, a desired output of 1 million cans after operation 10 is 1,007,557 units (= 1,000,000/0.9925), which are required as input. Because of yield losses for operation 9, input is 1,008,000 units (= 1,007,557/0.9990). Each operation has yield losses, and the algorithm is to work upwards to operation 1 to find the incoming number of units to give a final output of 1,000,000 units. The yield effect hurts the system as the waste means that earlier operations lose not only the material but also the productive cost to produce the unit.

The production is a "pull" system where queues are avoided, and successive operations such as intermediate diverging and converging chains have higher throughput to prevent storage or additional equipment to keep the cans moving. The production rate for each operation is given by minutes per can. Setup is not considered for high volume. Conveyors, equipment, tooling and space cost are summarized by the cost per minute for each operation.

Find the cost of material and the cost of each operation, then find the total cost for one can and for 1 million cans.

PRACTICAL APPLICATION

There are many opportunities for a practical experience in operations estimating. Your instructor may suggest several. One possibility is to visit a contract production shop and interview the engineering manager responsible for cost analysis of the quotation procedure. Prepare a list of questions in anticipation of the visit. Be professional in the arrangements of the interview, and afterward write a report detailing the lessons that you have learned. Send them a courtesy letter expressing your thanks after the interview.

CASE STUDY: ESTIMATING A STAINLESS-STEEL PART

Estimate the unit cost of the design given by Figure C7.1 for material, tooling and labor and overhead. The annual quantity is 1500 units, and material is purchased and produced in three equal-sized lots.

Bar stock, stainless steel 430 Ferritic material, is 6-in. OD, 6-ft long, and is \$2.25/lb. Density is 0.275 lb/in.^{3} Two operations are necessary:

Operation 10: Face, turn the 5.88 in. dia, drill, ream and countersink the 0.375-in. hole, turn the 4.5000/4.995-in. dia, and cutoff and chamfer in a lathe. (*Hint*: Cutoff and chamfer are one tool.)

Operation 20: Drill, tap, and counterbore six holes. A drill jig is necessary for the drill, counterbore, drill-tap, and tap operation. The following hole schedule is given for operation 20:

Specification	Tool	Number of Holes
5/16 in. × 1 in. long	5/16 in. drill for cap screw	3
1/4 in. socket head cap screw × 0.25 in. long	3/8 in. flat bottom counter-bore for socket head	3
	3/16 in. drill full-length hole	3
	20 thread/in. tap	3

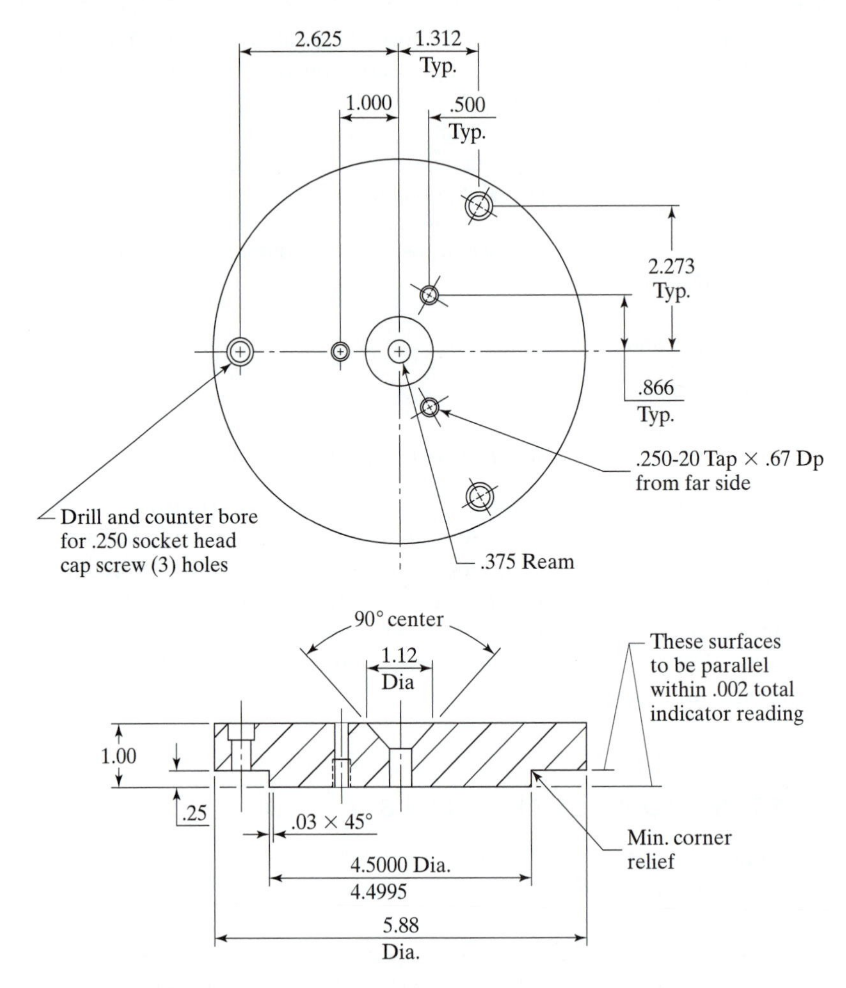

Figure C7.1

Productive hour cost rates for turning and drilling are \$75 and \$40 per hour. Tool design is expected to cost \$5000, and the PHC rate for tool building is \$100 per hour.

After the analysis and cost estimate are complete, discuss the following: What are the higher cost elements of the work, and where can cost reduction be suggested to the designer? What can be done to reduce the cost of the design? For this advice, what is the dollar and percentage reduction?

Chapter 8

Product Estimating

Product estimating begins with a request for proposal (RFP) or request for quote (RFQ) that contains the engineering requirements. The customer's package may be general or it may give specifications and CAD files or drawings. In another case, someone within the enterprise may need a cost analysis. The request may be from marketing for a new product idea, from engineering for a product improvement, or from management or manufacturing. Such concurrent engineering tasks require timely and accurate estimates.

To permit the preparation of a product estimate, sales, marketing, and operation estimates must be concluded or undergoing work—for the product estimate assists the pricing decision and may determine cash flow, rate of return, and profit and loss statements. This appraisal is made on new or continuing products.

En route to price setting various procedures may work on the task so as not to overlook the objectives of consumers and stockholders as well as management and engineering. Engineering will obtain the necessary data, calculate component and assembly labor and material costs, and compile the results into a total product cost. Outside sources such as supplier and commodity prices may be an important piece of the information. At the end of this process, a price for the product is established, which must be harmonized with the realities of marketplace.

Product estimating matters. These are important principles for every student to understand.

8.1 BACKGROUND

Development of a product pricing strategy calls for a wide assortment of decisions. This is a large undertaking for any firm. The risk can be high because, in general, only a few out of several hundred fresh ideas will congeal into successful new products. In addition to the cost of the product, opportunities exist concerning cash flow, rate of return, and meeting obligations to customers, employees, and investors. All aspects of the product cost need to be determined and then compared with the market conditions for similar products—both within the company and from competitors. A profit is then determined along with the price. Remember that price is equal to the cost plus profit, and is market driven as well.

What the buyer is willing to pay is influenced by the lowest competitive price and is determined at the point of sale, not in the factory. There is a school of thought asserting that "price is not related to cost." For example, this argument begins by assuming that houses A and B are identical as to neighborhood, appearance, and the like. Price A is $200,000 and price B is $180,000. Owner A argues that the cost for house A is $200,000. The buyer, however, considers this as irrelevant and so would choose house B. But what about owner B? If the cost is under $180,000, then owner B has earned a profit, but if the owner's cost is $180,000 or more, there is no profit.

If the expected attainable price is less than the product cost, then the company must take steps to reduce cost or abandon the product. Of course, cost is not the only factor that sets a price, but it is a vital one. For long-term survival it is imperative to recover the cost for the full consumption of all resources used to produce the product. A price strategy must accommodate this policy.

A cost estimate for a product undergoing redesign requires a different treatment than a new product. The components of the redesigned product are compared to the old design and classified as changed, added, or identical parts. The costs of the unchanged parts are found from records and may or may not be altered to reflect future conditions. Operation estimates are prepared for the altered and new parts. Methods of product costing and pricing vary in each of those several product conditions.

Figure 8.1 describes the elements in a product cost estimate with price as the ultimate objective. The bottom layer, operation costs, consisting of direct material, direct labor, and nonrecurring initial fixed cost, provides the foundation for an operation estimate (see Chapter 7). Addition of the upper blocks is achieved by the several methods described in this chapter.

Those costs classified as general, engineering, and selling are handled by overhead practices (see Chapter 4). There are many refinements that can be undertaken—most of them usually worth the trouble because the costs can be more accurately distributed to a product for eventual recovery.

Contingencies form an interesting category in Figure 8.1. Uncertain costs may be estimated here. Although this is not a desirable category, there are circumstances where it becomes necessary. For instance, government-imposed product disposal rules may be required for a product's hazardous material recovery after end-of-life cycle.[1] The firm, uncertain of this future behavior and unprepared to introduce this feature until the government has provided the specifications, uses this category for a visible cost. The contingency identification has the advantage of providing special provisions for legitimate future costs. The use of contingency to cover careless estimating practices is not encouraged. Products requiring extraordinary research and engineering may be candidates for adding contingency.

Conventionally, pricing is figured "from the bottom up." Determining the price using estimates and applying factors and then adding up the blocks, as shown in Figure 8.1, is an ordinary practice. But under competitive conditions, the price may be

[1]For example, the European Economic Community requires prepayment to a fund by the firm for an extensive list of hazardous material that are included in the product.

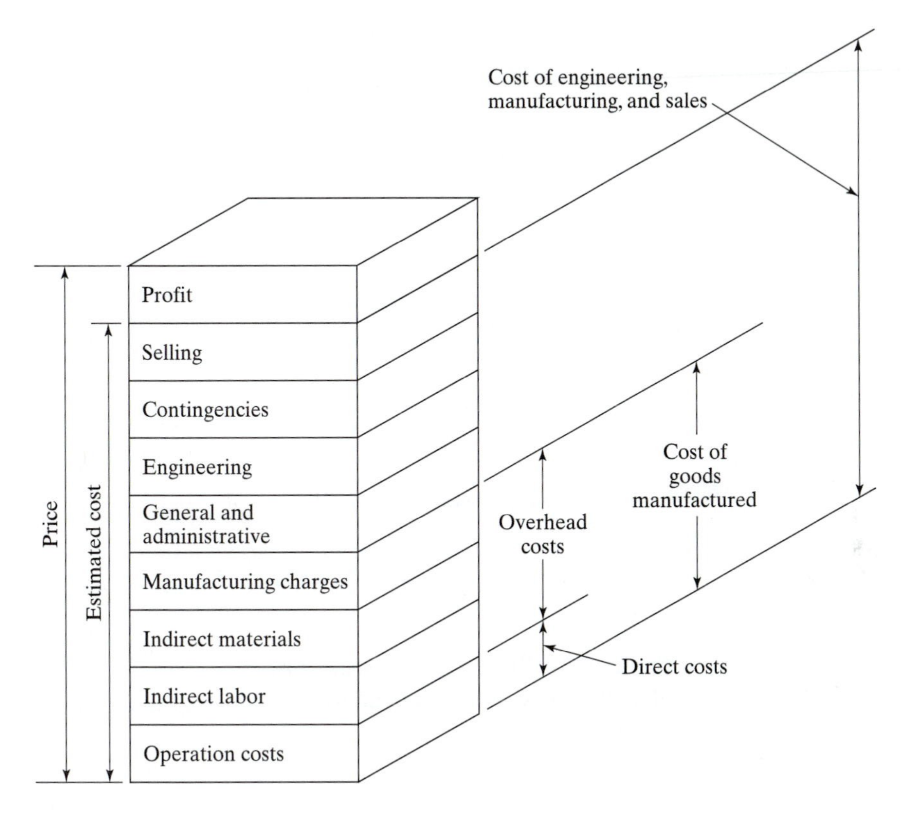

Figure 8.1 Elements in a product cost estimate.

limited by the market. In this situation the price and costs are set "from the top down." Design and manufacture to profit and cost principles are a popular engineering practice, and that is discussed in Chapter 11.

The elements of Figure 8.1 vary in importance depending on the business sector. Some industries may be labor intensive, and direct labor may be most important. For other products, the material, such as standard purchased parts, raw materials, commodities, and direct utilities, may dominate. For capital-intensive industries, recovery of capital money is vitally important.

The relative importance of the major elements is illustrated for a real product in Figure 8.2. The axes are quantity and percentage product cost per unit, Figure 8.2 is a layer chart and is read by finding differences between the chart lines to show the percentage of cost attached to each category. For instance, at any quantity the percentage cost of contribution is found by subtracting the lower y value from the higher y value. For example, at a quantity of 100,000, material contributes 30%, labor contributes $40\%(= 70\% - 30\%)$, capital accounts for $15\%(= 85\% - 70\%)$ and tooling contributes $15\%(= 100\% - 85\%)$. Labor is the most significant contributor to the product cost at this quantity.

It is possible to optimize the choice of the production system with product estimating methods. A series of envelope curves can be determined for a product that will indicate the best system and relate that to the regions of quantity. In Figure 8.3 the axes

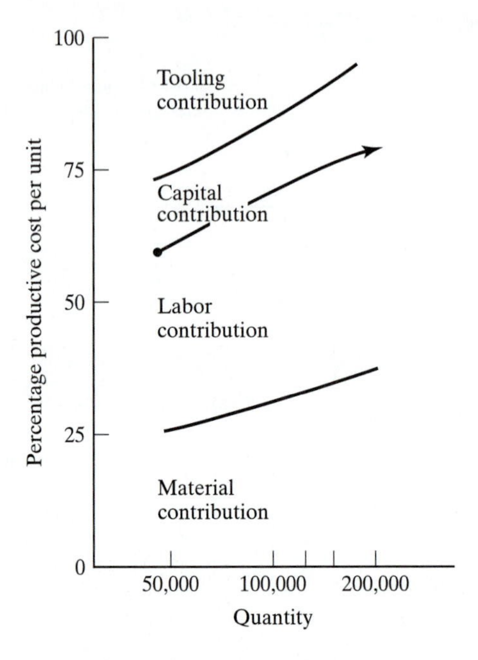

Figure 8.2 Illustrative product cost elements percentage for quantity range.

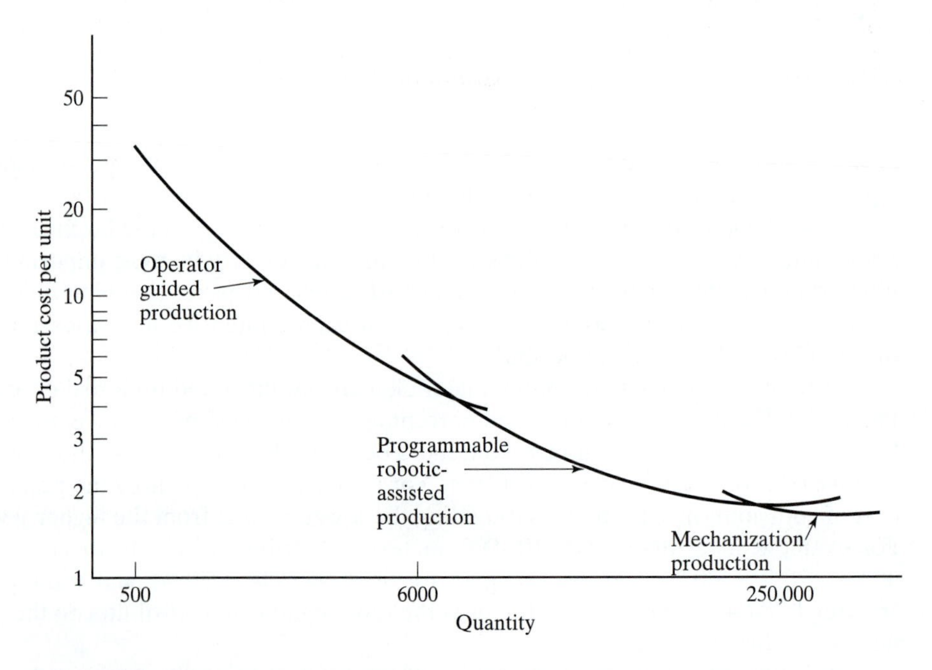

Figure 8.3 Optimization of production methods based on quantity using product estimates.

PICTURE LESSON Hoover Dam, an American Civil Engineering Icon

Bureau of Reclamation.

The Hoover Dam—the great pyramid of the American West—was the supreme engineering feat of its day. Although new generations of bigger and more sophisticated dams have risen, Hoover Dam remains the benchmark and inspires awe. It remains the highest concrete dam in the Western Hemisphere. The Boulder Canyon Project would be called the most ambitious government sponsored civil engineering task ever undertaken at the time.

With tentative plans and cost estimates, which were developed by the U.S. Reclamation Service, debate started in Congress and finally concluded in the signing of an act in December 1928 by President Calvin Coolidge. With the six-state ratification of the Colorado River Compact, the mining engineer President Herbert Hoover authorized $165 million to build the dam and the All-American Canal.

In January 1931, the Bureau of Reclamation released the plans and specifications to interested bidders at five dollars a copy. Each bid was to be accompanied by a $2 million bid bond and the winner was required to post a $5 million performance bond, enormous amounts in light of the depressed U.S. economy and banking system.

Scores of firms were interested, but their enthusiasm cooled as they realized magnitude of the job. Even the most aggressive of contractors were brought up short by the cost of the bid and performance bonds. It was simply too much money for any one company to handle alone.

Some companies banded together—Utah, Morrison-Knudsen, J.F. Shea, Pacific Bridge, MacDonald & Kahn, and Bechtel-Kaiser-Warren Brothers—and incorporated under the name Six Companies. W.A. Bechtel was named vice president. In short order, the key men met with the Bechtel superintendent of construction, engineer Frank Crowe, in the smoke-filled hospital room of the company's dying president William Wattis. They combed over the details of the bid items and, accepting the engineer's cost estimate, the executives added 25% for profit for this extremely risky venture.

The bid opening at the Bureau of Reclamation's office in Denver was tense, and movement in the crowded room was difficult. The first bid read "$80,000 less than the lowest bid you get" and was disqualified. A second bid was disqualified because it was not accompanied by the required $2 million bid bond. The last bid to be opened, at $48,890,955, with the required bid bond, was declared acceptable. It was only $24,000 more than the cost calculated by the Bureau of Reclamation engineers. Pandemonium broke out, camera flashes popped, and the crowd cheered. Six Companies had won the right to build the dam.

The view of Hoover Dam is just below Nevada (left) and Arizona (right) powerhouses. Water is released from Lake Mead in a regulated flow to farms, homes, and factories downstream. The water passing through Hoover's turbines generates hydroelectric energy for markets in Nevada, Arizona, and California.

are quantity and product hour cost per unit, an important metric. Note in this figure that programmable robotic-assisted production is preferred for the region of 6000 to 250,000 annual units. For low quantity, it is optimal to use labor intensive methods or operator-guided production.

8.2 ESTIMATING ENGINEERING COSTS

Engineering provides design of the product, specifications, bills of material, test plans, and perhaps prototypes. In addition, research and development for future product improvements and preliminary engineering are performed for jobs that may not lead to orders. Engineering liaison may extend to production, field service, and maintenance engineering during the life of the product. Sometimes engineering functions, in addition to design, are included in engineering costs. These functions include manufacturing, industrial, test, and quality engineering. Time or cost estimates for these functions are determined by similar methods to design engineering estimates.

Where engineering costs for a product are not covered by overhead, they are estimated separately. These engineering and companion estimates include research and development engineering costs that are added with the cost estimates for manufacturing. High-technology firms dealing with large systems calculate engineering, development, and design as a separate line-item cost. When products are similar to existing products, ordinary techniques useful in judging the costs of engineering are adopted. For those conventional situations, engineering does self-estimating on the basis of similar and historical records.

For those specific functions, the hours and a cost factor for hourly rates for wages and salaries of the group, charges for supervision, facilities, light, and heat are estimated. Engineering costs to develop proposals that do not lead to new orders are additionally collected into the conglomerate hourly rate and used as the multiplier for the estimated number of hours. The total of this cost is amortized to the product by

$$C_e = \frac{\text{total engineering expenses}}{\text{product quantity}} \tag{8.1}$$

where C_e = engineering cost per unit

Engineering costs can be large and if the product quantity is small, as in batch operations, can be a significant portion of the product cost. In mass production, however, with the quantity large individual part portion due to engineering is small.

Equation (8.1) is for exceptional charging not included in overhead practices. This model can be developed as a spreadsheet such as the one shown in Table 8.1. The engineering estimate is categorized by type of labor and rate per hour and is extended to total engineering labor.

Engineering effort may be a significant proportion of a contract. In special cases a firm will be hired for the engineering and, say, manage a large technically sophisticated job. Those costs may become a competitive bid among engineering firms. In addition, engineering costs may be negotiated on a lump sum package where the cost of engineering is included in the product cost for a production run. Sometimes engineering contracts are negotiated on some cost plus basis. Contractual variations are discussed in Section 8.7. No matter what the contract type, engineering contains these

TABLE 8.1 Worksheet for Estimating Engineering Costs

Design Engineering Estimate for X-152 Product

Description: Product Design Activities
Customer: Marketing
Inquiry or Quote No.: Date:
Based on build quantity of: 87 units During period:

Work Category	Hours	Rate per Hour	Extended Labor ($)
Scientist, research			
Engineer, senior design	40	$33.00	$1,320
Engineer, design	800	30.00	24,000
Technician, electrical			
Designer/engineering aide	160	18.75	3,000
Writer, technical			
Illustrator			
CAD operator	80	17.00	1,360
Procurement specialist			
Technician, model shop			
Total engineering labor	1080		$29,680

elements of total cost:

$$C_e = \Sigma S + \Sigma E + \Sigma OH + \Sigma F \tag{8.2}$$

where C_e = sum of engineering costs, dollars
S = salaries, dollars
E = variable expenses such as travel, living away from home, communication, dollars
OH = overhead, rent, depreciation, heat, light, clerical, supplies, and benefits, dollars
F = fees paid to other consultants and engineers, dollars

Salaries may be found on the basis of time estimates for the various job elements, or they may be calculated as a percentage of some other factor, such as the labor to support manufacturing, and they could be a percentage of the total direct manufacturing labor.

Factors will multiply expected salaries to arrive at the estimated total cost. Those factors vary from 1.0 to 3.0, depending on complexity, novelty, or secrecy of the work. In chemical and architectural work, the ratio of design drafting is on the order of two or three times other types of engineering. In electronics, this is reversed. Standards for engineering work have related size of drawing and design productivity. In design of tooling and dies, for instance, the size of the drawing can be associated to so many hours of design time.

Other similar rules of thumb have been established. In any one industry there seems to be a tendency toward standardization in CAD drawing size as well as the kind and quantity of information recorded on the print. For example, at one company their guidelines indicate that a CAD "D-size" file will require about 32 hours of CAD drafting time. A supervisor may be asked to determine the engineering cost by estimating the number of drawings that the product will require and thus giving the total time.

The cost of engineering is prorated to product charges like any other cost. Its impact may amount to little, or it can be a major factor, particularly when a product, new or revised, calls for a new process or plant, or the product is significant.

8.3 INFORMATION REQUIRED FOR PRODUCT ESTIMATING

Before the estimate is started, engineering needs answers to two major questions: (1) What does the product look like? and (2) How many will be made? The answer to the first question includes the shape, dimensions, tolerances, materials, and so on. The quantity helps determine the manufacturing processes and tooling, which is necessary to find the costs. In addition to determining processes, the quantities determine the production capacity and labor requirements, which in turn may lead to capital expenditures for increasing or even decreasing facilities and equipment.

For internal cost analysis, a request for estimate (RFE) will include much of the following, though all of this information may be unavailable for preliminary estimates.

- Engineering bill of materials, specifications, and CAD files
- Due date for completion of estimate
- Quantity, rate of production, and schedule for production

- Special test, inspection, and quality control requirements
- Packaging and shipping instructions
- Marketing information

The estimating of labor, material cost, engineering cost, tooling, and other important capital costs and extension by overhead calculations gives the important quantity *full cost*. This is then increased for profit to get price. This is shown symbolically by Figure 8.1.

8.3.1 Bill of Material (BOM)

The bill of material can serve as the structure on which to build a cost estimate. The costs of individual components are made and then added together as you progress up the bills of material to the top assembly—product level. A costed bill-of-material tree is shown in Figure 8.4.

The bill of material is an important document because other business functions use it as a central source of information. Engineering is responsible for its accuracy, completeness, and timeliness. The bill is sometimes entered as a table on a corner of the assembly drawing, but this gambit is simplistic because of data processing and its widespread use in the company. It is most common today to issue a separate document. Functionally, within the enterprise this responsibility is sometimes termed *configuration engineering*, as important products, say an airplane, require knowledge of the design and all parts and assemblies for a particular numbered airplane unit. Even individual parts may be coupled to a separate BOM.

With a complete BOM, engineering is assured that all materials, both purchased and fabricated, will not be inadvertently overlooked. Even the most unimportant of

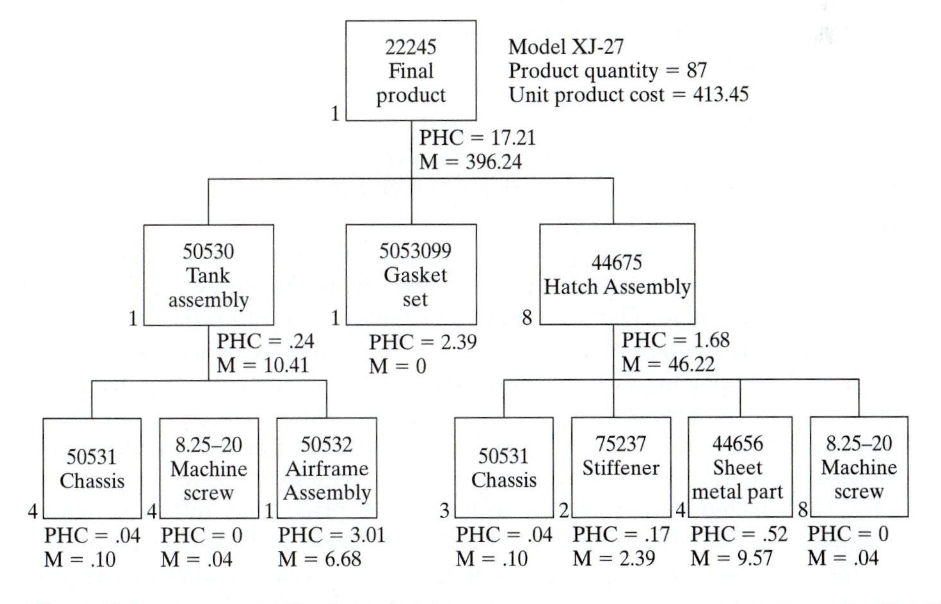

Figure 8.4 A costed bill-of-material tree.

materials or items must be attached to the BOM. It is used to identify purchase orders if the item is purchased or to start production.

A product tree can be constructed from the bills of material and is described by levels, three in the case of Figure 8.4. The top level is the final product and shows the sum of the costs of the components and assemblies below this top level. Each box of the tree contains a part number and part name. The box gives the number of units required for each part that then goes into the next higher assembly. In Figure 8.4 the number of units for the next higher assembly is shown at the left of each box. For example, one unit of final product (model XJ-27) requires one unit of the tank assembly, one gasket set, and eight units of a hatch assembly. Each hatch assembly contains eight machine screws, and so eight hatch assemblies require 64 machine screws. Assembly of the second level boxes sums to the level 1 box or the final product. Note that the gasket set does not have any lower level components.

Labor cost and material cost are indicated below each box. In some cases material cost (M) or productive hour cost (PHC) equals 0. A zero value for material could indicate consignment material, meaning that the material was supplied by the customer. A zero cost for labor indicates that the part is a purchased item. The number on the left side of the box is the quantity required, ultimately for one final unit.

Costs roll up from the bottom. For instance, in the tank assembly 50530, the estimated material cost of \$10.41 is the sum of the labor and material costs adjusted for quantity of the lower levels. A labor cost of \$0.24 is estimated to assemble the lower levels, which becomes the 50530 assembly. For example, material cost is $10.41[= 4(0.04 + 0.10) + 4(0 + 0.04) + 1(3.01 + 6.68)]$. Note that lower-level productive hour cost (composed of gross hourly labor cost and machine hour cost) becomes the value added to the material to transform it to higher-level material. The term $4(0.04 + 0.10)$ is the quantity 4 required for the next higher level, and \$0.04 and \$0.10 are productive hour costs and material costs. Those values are below the 50531 chassis box in Figure 8.4. Level 2 labor and material estimated costs roll up to level 1 material cost after adjustment for quantity.

The BOM tree may also show common materials. A purchased machine screw is used in two different locations in the third level. Each unit of final product uses four screws through the tank assembly and 64 screws as a result of the hatch assembly. A total of 68 screws become necessary for one unit of final product, and 5916 screws are required for 87 units of final product.

Though the tree shown here is a graphical illustration, more often such BOMs become too big and are awkward in practice. Then the BOM is provided as a table. The corresponding printout for Figure 8.4 is given in Table 8.2. Level 1 is the final product; level 2 indicates the next lower subassembly, and level 3 is the raw material or purchased parts for level 2. The headings for the costed bill of material indicate their purpose. The UM provides the unit of measure for the bill of material that can have a variety of dimensions. Next, assembly quantity and lot quantity are important calculations. Order policy (OP) indicates either make (M) or buy (B), but other situations are possible.

The machine screw that appears in two places has different next-assembly requirements, but for estimating and ordering purposes the quantity will be summed for the lot. Total lot hours and unit material cost are values that estimates provide and are input information to the costed bill of material.

TABLE 8.2 A Costed Bill of Material

Level	Part Number	Description	Unit of Measure (UM)	Quantity Next Assembly	Order Policy (OP)	Unit Material Cost	Unit Labor Cost	Total Unit Cost	Material Cost, Next Assembly	Lot Quantity
1	22245	Top Assembly XJ-27	Ea	N/A	M	396.24	17.21	413.45	—	87
2	505309	Gasket set	Ea	1	M	0	2.39	2.39	2.39	87
2	50530	Tank assembly	Ea	1	M	10.41	0.24	10.65	10.65	87
3	50531	Chassis	Ea	4	M	0.10	0.04	0.14	0.56	348
3	8.25-20	Machine screw	Ea	4	B	0.04	0	.04	0.16	348
3	50532	Assembly	Ea	1	M	6.68	3.01	9.69	9.69	87
2	44675	Hatch assembly	Ea	8	M	46.22	1.68	47.90	383.20	696
3	44656	Sheet metal part	Ea	4	M	9.57	.52	10.09	40.36	2784
3	75237	Stiffener	Ea	2	M	2.39	.17	2.56	5.12	1392
3	50531	Chassis	Ea	3	M	0.10	0.04	0.14	0.42	2088
3	8.25-20	Machine screw	Ea	8	B	0.04	0	0.04	0.32	5568

The control and assignment of numbers for various engineering data such as drawings, bill of material lists, and engineering instructions are a function of design engineering. The design numbering system is comprehensive enough to serve engineering, manufacturing, and sales and yet unique enough to indicate readily specific conditions. Careful control of this numbering system is required for *configuration management*, a critical task in controlling the production of products. Configuration management affects the interchangeability of parts, product quality and safety issues, and is a topic too big to fully discuss in this text.

BOMs provide an abundance of information leading to other existing documents, products, and cost estimates. New product designs incorporate some parts, subassemblies, or major subassembly units of other manufactured products, and engineering uses the bill to find previously prepared estimates. Those older estimates are updated by using indexes or new productive hour cost rates and the like. It may not be necessary to estimate the elements of a product from scratch, because using data from similar parts can reduce the estimating requirements.

8.4 THE PRODUCT ESTIMATE

Costs in terms of dollars or other currency are determined for all materials, parts, subassemblies, commodities, labor and overhead for all of the operations that are required to manufacture a product. These costs may be estimated by individuals and functions apart from engineering. To find the total product cost and determine the price, all cost estimates must be collected, verified, adjusted to eliminate any overlap or to uncover oversight, and hourly estimates must be converted to dollars, and finally the collected facts are summarized into the estimate. All costs are collected, and we term this compiled number of the cost as *full cost*—nothing is unintentionally overlooked, nor are costs carelessly added. A preliminary estimate may have preceded the detailed product estimate. The preliminary estimate, though not complete, can provide useful organizational information for the final product estimate.

The product estimate is a formal document. The format for the final estimate is generally a standardized company document and its preparation is reviewed by key people. Usually approved by high-level enterprise executives, it is the underpinning upon which the decision to proceed or not to proceed with engineering and production. It is the single most important document of the product-estimating effort.

Once the sublevel and other contributing estimates are collected, there are many different methods in compiling the product estimate. These maneuvers are not rocket science, and each company will have their special approach. Millions of worldwide firms exist that manufacture products. There are over 250,000 firms in the United States. Variations in product-estimating methods can be expected. These distinctions are summarized into the three popular methods that we now study:

- Productive hour cost
- Activity-based costing
- Learning

8.4.1 Productive Hour Cost Model

Most of the labor estimates produced for determining product cost are first developed in hours. This activity occurs for both in-house or outsourced materials and products. At some point the cost rate for the labor hours must be determined. As we have seen, the cost rate is more complicated than the wage rates of the workers. Overhead costs need to be distributed across the productive machinery and operations in some logical manner. The concept of *productive hour cost*, which was discussed in Section 4.9.5, is one way to handle a large portion of the overhead costs. The productive hour costs vary by machine, operation, department, or some other subset of the company. The productive hour cost rate includes the basic wage of any direct labor operator(s) involved that can be directly tied to the operations—including fringe benefits plus the overhead to cover the indirect labor (quality inspectors, production control, supervision, etc.) equipment utilities, maintenance, and consumables (indirect materials). Once the productive hour cost is calculated for each operation unit, then determining the overall product estimate is straightforward.

Manufacturing operation sheets are prepared for each part, subassembly, and assembly. The operation sheets provide the manufacturing procedure and give the operation setup and cycle times, lot hours, productive hour cost rates, and material costs. The total unit cost for a part (and hence a part number) is given as

$$C_u = \sum_{i}^{n} PHC_i\left(\frac{SU_b}{N} + H_b\right)_i + C_{dm} + C_t \tag{8.3}$$

where C_u = unit cost of batch manufacturing operations for one part or assembly on BOM, dollars per unit

PHC_i = productive hour cost rate for labor, perishable tooling, equipment, and other overhead for operation i, dollars per hour

SU_b = batch operation setup, hours

N = lot quantity, number
H_b = cycle estimate for operation, hours per unit
i = 1,2, $\cdots$, n operations for one part or assembly on BOM
C_{dm} = cost of direct materials, including scrap and waste, dollars per unit
C_t = prorated cost of permanent tooling for one part or assembly on BOM, dollars per unit

Better practice dictates that setup and cycle standards for operations be determined from any of the methods of time measurement discussed in Chapter 2. Those data are improved if the information were screened by the methods of engineering performance analysis given in Section 6.3.2. Each operation is estimated for setup hours and cycle minutes and lot hours are found by using

$$H_l = SU + (N \times H_c/60) \qquad (8.4)$$

where H_l = lot hours
H_c = cycle minutes per unit

The engineering-performance-data hours per unit (H_s) are found by

$$H_s = H_l/N \qquad (8.5)$$

If the lot quantity N is large, H_l and thus H_s will not include setup time considerations, because SU/N becomes negligible on a per unit basis. Under those circumstances, cycle time H_c is the only time considered. Setup costs for large volume products are included in the overhead distribution in the productive hour cost rate.

It is customary for the cycle H_s to include personal, fatigue, and delay (PF&D) allowances. Composition of this standard is discussed in Chapter 2. On the other hand, performance against those standards invariably requires adjustment based on actual time. The adjustment is either up or down and divides the operational cost by efficiency.

Total product cost is found by

$$C_p = \Sigma C_u + \Sigma C_e + C_c + C_s \qquad (8.6)$$

where C_p = total product cost of all BOM items, dollars
C_e = engineering cost, dollars
C_c = contingency cost, usually zero, dollars
C_s = sales and marketing costs, dollars

Example: Estimating Product Cost Using Productive Hour Cost Model

Table 8.3 illustrates the operation process sheet for the manufacture of the final assembly of model XJ-27, the product given by the graphical bill of material in Figure 8.4. Except for purchased parts, each item on the bill of materials will have a manufacturing process sheet. The operation process sheet in Table 8.3 covers the assembly of the tank assembly, gasket set, and the hatch assembly. The incoming material has a total cost of \$396.24, which is the value-added costs of direct labor, machine, and assembly costs and purchased material costs for all levels below it. This is shown in the graphical bill of material (Figure 8.4) and the operations process sheet.

The final operation—to bench assemble the tank assembly, gasket set, and the hatch assembly is estimated to require a setup of 0.75 hour and cycle minutes of 31.561. Those estimates were found from engineering performance data, the most accurate way to estimate.

TABLE 8.3 Worksheet for productive hour method of product estimating

Part no. 22245	Part name XJ-27	Subassembly Part no.	Material cost
Quantity. 87	General notes	50530	$10.65
Estimator TM	Three Subassemblies	505309	2.39
Date 3/12		44675	383.20
Estimate expires on 6/9		Total →	$396.24

Work Station	Operation no.	Description of operation (list tools and gauges)	Setup hours	Cycle minutes	Lot hours	PHCR	Cost of lot
Bench assembly 10	Assemble		0.75	31.561	46.51	32.19	1497.27

1. Total productive hour cost (for top assembly)	1497.27
2. Material cost (Fig. 8.4 top assy, M = 396.24 × 87)	35,970.15
3. Engineering cost (Table 8.1)	29,680.00
4. Contingencies	____
5. Selling costs	____
6. Total lot cost of manufacturing, engineering, and sales	$65,650.15
7. Unit cost	$754.60

Using Equation (8.4) for 87 units, the lot hours are $H_l = 0.75 + 87(31.561/60) = 46.51$ hours. The lot hours are then multiplied by the productive hour cost rate for bench assembly, \$32.19. This rate is composed of the gross hourly direct labor cost (see Chapter 2) and the machine hour cost (see Chapter 4). The productive hour cost rate will vary, and more expensive manufacturing cells can be expected to have rates that reflect the cost of efficient operation of these assets. Multiplying the lot hours and the productive hour cost rate gives the cost for the operation or lot hours multiplied by PHC = 46.51 × 32.19 = \$1497.27.

The unit productive hour cost for this operation is 1497.21/87 = \$17.21. The total unit manufacturing cost of the model XJ-27 product is the sum of the labor and material or, 17.21 + 396.24 = \$413.45.

The product cost estimate is completed by using Equations (8.3) and (8.6) and the data in Figure 8.5. The engineering design cost is found from a companion estimate, Table 8.1, and is itemized in Figure 8.5. Both a total cost for 87 units and 1 unit are given.

$$C_u = 1497.27 + (396.24 \times 87) = \$35{,}970.15$$
$$C_p = 35{,}970.15 + 29{,}680.00 = \$65{,}650.15$$

The per unit cost is simply 65,650.15/87 = \$754.60. Note that there is no tooling, contingency, or selling costs used in this example.

There are many advantages for the productive hour method, which is sometimes called the *engineering method.* The method has good accuracy because it combines the machine hour rate with gross hourly rate. Some machine rates can be several times more costly than others—for example, heavy machining as compared to simple bench assembly. The productive hour cost method gives the ability to consider many systems of production, such as cell, programmable automation, flexible machining centers, and flow line. Finally, with this method it is possible to consider operations without direct labor, the so-called "automatic machines."

This discussion developed the productive hour cost model with batch manufacturing. But there is another similar approach that uses flow-line manufacturing, a concept discussed in Section 7.6.2. Flow-line manufacturing is probably more common, when one considers all manufacturing, such as large volume and commodities. The student is given the opportunity to consider what the differences might be.

8.4.2 Activity-Based Costing Model (Optional)

The productive hour cost method allocates indirect and overhead costs to each hour of manufacturing processing. In that method overhead costs may be assigned to products and parts of processes that do not benefit from all of the overhead activity. Activity-based costing (ABC) is an accounting method that attempts to allocate indirect and overhead costs to units of product based on what is actually used. It sets up predetermined rates at the beginning of an accounting period and applies those rates to all products made during the period based on the units of activities used. The estimated or budgeted rates are used for the period regardless of the actual overhead and indirect costs experienced.

In Chapter 4, overhead and indirect costs of an activity are forecast along with an expected usage for the activity during the period. The expected costs are then divided over the units of activity to obtain a per unit cost. When the activities are established, the units are also determined. The units can be based on the number of hours the activity is used or the on some other division, such as the number of purchase orders required for the job, number of setups, or number of times per product. When a product engages in an activity, the volume of the activity is multiplied by the rate to determine the applicable overhead. Remember that most products do not use every one of the activities in the facility.

Many costs in a production system are driven not by the volume of product produced but by the number of times an activity is performed. For instance, labor-hours is a volume-based cost driver. The more labor hours in a product, the more units are generally produced. But it may not be labor hours that cause overhead levels to vary. It may be an activity like stocking part numbers. Two products could have identical labor hours but one uses 32 unique parts whereas the other one uses 8. The product that uses 32 unique parts will have more indirect costs related to picking raw material inventory and moving it around the factory. If labor hours were used as the cost driver, then the two products would have the same overhead cost assigned. However, using activity-based costing, the first product will have four times the overhead cost assigned from the activity that accumulates raw material overhead costs than the product that only uses eight unique parts. The key with activity-based costing is that the cost driver (the denominator) is related to an activity, not to the volume of production.

If the company implements ABC, then the costs of all of the activities to make a product can be accumulated into a product cost estimate. Engineering will determine which activities are required to produce the product and how much (how many units, such as hours or number of occurrences) of each activity are used. Accounting furnishes the costs of the units for each activity. In manufacturing, the activities are based on the number of hours per individual part or product, number of hours per batch (setup time) or the number of times an activity is used per batch (purchase orders, scheduling actions, design changes).

Products for each batch can be costed in the ABC method by

$$C_{uabc} = \Sigma(H \times WR)_i + \Sigma(SU \times WR)_i/N_d + \Sigma(H_d \times WR_w)_j/N_d + \Sigma(H_s \times WR_s)_k + \Sigma C_e/N_a + C_{dm} \quad (8.7)$$

where C_{uabc} = unit cost of product using ABC method, dollars per unit
H = cycle estimate for operation activity, hours per unit
WR = wage rate for operation activity, dollars per hour
SU = setup for daily batch for operation activity, hours per day
N_d = daily batch number
H_d = warehouse activity, hours per day
WR_w = wage rate for warehouse activity, dollars per hour
H_s = inspection time activity, hours per sample unit
WR_s = wage rate for inspection activity, dollars per hour
C_e = engineering, tooling, management, dollars
N_a = annual units
i = 1,2, . . . , operation activity, number
j = 1,2, . . . , warehouse activity, number
k = 1,2, . . . , inspection activity, number

These formulations are unique to each product situation, and Equation (8.7) is not intended to be general.

In order to determine the complete product cost including the indirect labor and overhead, according to the ABC system, all of the activities used must be accounted for. Some overhead costs occur once for each unit. In this case a unit- or volume-based driver like labor hours is needed. This would be the case for equipment costs. The utilities, consumables, and maintenance depend on the hours of operation. However, many overhead costs occur once for each batch (for example, inspection). Then the cost driver is the number of inspections and the cost will be applied to the batch of units that is inspected and then divided over the number of units in the batch. For instance, if you have two batches of units that require inspection and you have developed a predetermined rate of $100 of inspection overhead per batch, then the units would receive inspection cost as follows:

Batch 1 (50 units in the batch): $100 inspection cost assigned to the batch or $2 per unit.

Batch 2 (25 units in the batch): $100 inspection cost assigned to the batch or $4 per unit.

ABC is designed to tell managers where they could improve costs. An example of product level costs is engineering change orders, and an example of facility level overhead cost is depreciation. No matter which way the activities are divided, the units of measure must be held consistent and all activities used by the product must be considered.

Example: Activity-Based Costing of Die-Cast Aluminum Part for Off-Shore Plant

A small die-cast aluminum casting for a computer disk drive is manufactured in Mexico in a single-component dedicated plant. Worldwide demand for the casting for the approaching year is 1,148,000 units. The plant produces only this product, and the enterprise is consigned to engineering, warehousing, logistics, and management.

Material cost per unit is found as $0.32, which includes waste due to the stack emissions and other melting losses. Other operations follow the die-casting operation in a flow line. In addition to the operations of casting, trimming, and machining, there are delays for storage, and direct labor for inspection. Setup and cycle times are necessary for all direct-labor consuming operations. Tooling costs are estimated for the die, trim die, milling fixture, inspection tools, and go/no go gauges, and are shown for the operations. Inspection is done on a sampling basis. Management and depreciation are additional annual charges. A tabular summary follows below.

Item	Activity	Activity Rate	Measure
Material input	Warehouse	Daily	3189 units per day
Store material on flow line	Warehouse	Daily	3189 units per day
Die cast	Direct labor	Operation, each	0.00417 hr per unit
Separate parts	Direct labor	Operation, each	0.00385 hr per unit
Inspect	Inspect, indirect labor	Every 100 units	15 min per sample unit inspected
Store parts	Warehouse	Daily	3189 units per day
Mill	Direct labor	Operation, each	0.00341 hr per unit
Inspect	Inspect, indirect labor	Every 10 units	20 min per sample unit inspected
Tooling design, construction	Engineering	Total units	$75,000 for design and construction
Engineering, management		Total units	$225,000 for engineering and management
Space, utilities	Plant	Total units	$60,000 for space and utilities
Annual depreciation	Equipment	Total units	$300,000 annual depreciation for equipment
Direct material cost		Unit	$0.32 for good material, waste and scrap

A worksheet analysis of the single-product plant can be summarized as follows:

Activity				
Direct Labor Operations	Hours/unit	Dollars/hr	Hours/unit × $/hr	
1. Cast platter	0.00417	20	0.083	
2. Separate part from trim	0.00385	22	0.085	
3. Machine	0.00379	25	0.095	
Total for direct labor			0.263	
Setup Cost for Labor	Hours/day	Dollars/hr	Hours/day × $/hr	Hours/day × $/hr/ units/day
1. Setup	2.7 hr	20	54.00	0.017
2. Setup	1.2 hr	22	26.40	0.008
3. Setup	0.8 hr	25	20.00	0.006
Total setup cost for labor				0.031
Warehouse per day	Hours/day	Dollars/hr	Hours/day × $/hr	Hours/day × $/hr/ units/day
JIT receipt of ingots	1.7	19	32.30	0.010
Storage next to flow line	1.3	19	24.70	0.008
Total warehouse cost				0.018
Inspection	Minutes/unit	Hours/sample unit	Dollars/hr	Dollars/hr × hr/ sample unit
Inspect per 100 sample parts inspected	15 min/unit	0.0025	17	0.043

Inspection	Minutes/unit	Hours/sample unit	Dollars/hr	Dollars/hr × hr/ sample unit
Inspect per 10 sample parts inspected	20 min/unit	0.0333	17	0.566
Total inspection cost				0.609
Leadership and Engineering	Product Total	Total Units	Product Total/ Total Units	
Engineering, manage, tools	360,000	1,148,000	0.314	
Annual depreciation			Total depreciation/ total units	
Equipment, plant annual depreciation	300,000	1,148,000	0.261	
Direct material cost			0.320	
ABC total unit cost			$1.816	

The ABC total unit cost = $1.816 for the annual volume of over 1 million units in a plant engineered and dedicated to this one product.

When ABC is implemented in a system with appropriately divided and defined activities, and that has costs associated with the activities, there is opportunity for a good estimating and cost control tool. Objectively, the best purpose for the tool is cost control. Moreover, its cost-estimating success for complex engineering systems has not proven superior to the productive hour cost method.

8.4.3 Learning Model (Optional)

Costs often decrease during the product production life cycle due to improvements in engineering design and manufacturing and the skills of the operators. We lump these improvements into a concept called *learning*. In Section 6.4.1 we saw how to quantify these improvements and determine the costs of any number of units of production given certain estimates or actual values.

The learning function is defined if the number of direct labor hours or cost required to complete a particular unit of the original production is known and if the rate of improvement is estimated. Alternatively, the learning curve is defined if direct labor hours for a downstream unit, such as the first unit of the follow-on order and the learning curve rate are estimated. Other possibilities for defining the learning function can be selected. The number of direct labor hours required to complete the first production unit depends on the following circumstances:

1. Experience of the company with the product: If the company has little or no experience, then the first unit time is greater than for a product having greater company experience.
2. Amount of engineering, training, and general preparations that the organization expends in preparation for the product: In some cases the first several units are custom-made and tooling is not designed and constructed until more sales can be ensured. This "hard way" production inflates first unit time.
3. Characteristics of the first unit: Large complex products are expected to consume more direct cost resources than something less complex.

To compile a complete and comprehensive product cost estimate, it is often necessary to include learning in the calculations. One approach is given as

$$\text{cost element} = C_f + C_v \times T_c \tag{8.8}$$

where cost element = individual calculation subject to learning, dollars
C_f = fixed cost element, dollars
C_v = variable cost element, dollars per unit
T_c = cumulative learning factor, number

Equation (8.8) deals with the separate elements of the product. It is not considered as accurate to use only one learning rate Φ for all the components of the estimate. This nuance improves the accuracy of the model, so it is supposed. For instance, CNC machining is generally thought to have 90 to 98% learning because of the maturity of the operation, but labor intensive assembly may have a range of 60 to 75% because of its potential for improvement. Furthermore, if a cost element is truly fixed, then the learning concept would not ideally apply, and thus the separation of fixed from variable learning costs is made.

The operation estimates consist of direct labor, direct material, and nonrecurring initial fixed costs. The estimate is made for the first unit, or some other specified unit, which subdivides the cost into fixed and variable portions. In this model direct labor and direct material are considered variable only. In Chapter 7 nonrecurring initial costs—such as tooling—were considered fixed. Now we acknowledge that tooling and test equipment may have a variable component, such as perishable tooling that depends on the production quantity. In the example, though, its proportion to the nonrecurring initial fixed cost sum is small for Table 8.4.

Learning slopes for this method are usually estimated from historical evidence. These learning rates or Φ should be carefully determined. Principally, the reduction is due to direct labor learning and engineering design as well as process improvements and cost reduction programs. The direct labor learning process assumes that as a worker continues to produce, it is natural that he or she requires less time per unit with increasing production.

The engineering programs improve production, encourage quality, reduce design complexity, create technology progress, and foster product improvement. Those programs inspire time and cost reduction beyond what direct labor could do by themselves.

What is this improvement? Pundits suggest that the operator is responsible for approximately 15% of the total learning reduction, with engineering and their programs contributing the remaining 85%. In the manufacturing industries, engineering's 85% is broken down into 50% due to engineering design endeavors, and 35% as a result of manufacturing and industrial engineering activities. Furthermore, it is necessary to have an active engineering improvement program that is sustaining during the production cycle, because the learning effect does not naturally happen.

Large products such as ships, aircraft, machine tools, and specialized capital equipment have in common high cost, low volume, and discrete item production and are candidates for treatment by learning. Although the same principle applies to consumer goods mass produced in huge quantities, the learning effects per unit are negligible and may take years to uncover because of the large production volume.[2] The learning curve is usually not applied to high volume or low cost consumer products. Commodity and single-item products would not use the learning model.

The learning method is often required for major contractors selling to agencies of the U.S. government and is also used for expensive low-quantity industrial products such as turbines, boilers, and airplanes.

Example: Using the Learning Model to Estimate a Generator for a Wind Farm

The Chilean government engages an engineering construction firm as its prime contractor in the erection and manufacture of a wind power project in the Andes. The RFQ for the prime contractor requires 18–1200 kW separate wind generator units in the wind farm. The three-blade assembly will rotate at subsonic speed on the propeller tips. The major components of the project are (1) erection of the 90-m high tubular tower and three 30-m blades and gearbox generator, (2) design and manufacture of the nacelle, gearbox, and generator, (3) construction of the electrical grid, and (4) manufacture of the blades.

The prime contractor approaches a custom design and build firm specializing in gearbox generators. The request for quotation is for a new design, and while it is roughly similar to other designs, an estimate is necessary to respond to the RFQ.

The prime contractor defines the performance requirements. The propeller assembly is to rotate at 30.25 RPM, meaning that the blade hub of the propellers is attached to the slow shaft end of the gear box. The cast-steel gear box increases the rotor to 1500 RPM in a field of heavy copper conductors of the generator. The gearbox and generator are housed in a formed steel nacelle. The gearbox houses a direct gear drive to the induction generator. The unit includes an electronic controller that monitors yaw and any generator overheating, which then locks the wind blades preventing damage to the unit.

The custom job shop recognizing that it is one of three bidders for the 18 generators, develops estimates on the basis of experience for the first unit and does the estimating for direct labor for the process categories of machining, casting of steel shroud and gear reduction box, electrical winding of the field and rotor, and assembly. The learning model is used. Direct material estimates are made for raw, standard, and subcontract materials on the basis of history for similar generators.

Various slopes are estimated for the cost elements. Reliable values from similar experiences are used. A spreadsheet shows the estimate. The cumulative learning factors are found from tables similar to those shown in the appendix.

A plant overhead rate of 75% is applied to the total operation estimates. The sum of direct costs and plant overhead gives the cost of goods manufactured. General and administrative and selling overheads of 20% are applied on the basis of the cost of goods manufactured. Engineering costs require a separate estimate. The total of the five major costs leads to the full cost, or the cost of engineering, manufacturing, and sales.

The total cost of $4,405,557 proceeds to the pricing step, as the response to the prime contractor. That development is covered later.

---

[2]In an interesting contrast to the accepted theory that learning does not happen for large quantities, a research study was made of the production of Ford cars. The study demonstrated that learning consistently happened for over 75 years and over millions of units. Indeed, while there were upward spikes in the time per car for assembly, which were caused by strikes, the depression of the 1930s, and the massive shift to change the car marketing appeal in the late thirties, the time per car has declined, and is still declining today. This decline is due to engineering efforts. Arguably, it can be asserted that while car assembly declined, did not off-assembly time increase to make up the difference? The research did not answer that issue.

TABLE 8.4 Learning Method of Product Estimating for 18 Wind-Powered High-Energy Generators in the Andes

Cost Element	First Unit Cost Estimate: Fixed	First Unit Cost Estimate: Variable	Slope, Φ	Cumulative Learning Factor	Cumulative Cost at Unit 18	Total Element Cost
Direct labor						
Machine shop		$8,500	90	11.4055	$96,949	$96,949
Foundry		7,000	75	8.2453	57,717	57,717
Electrical winding		6,500	80	9.7162	63,155	63,155
Assembly		4,500	85	11.4055	51,325	51,325
Direct material						
Raw		6,800	90	13.3344	90,674	90,674
Standard		15,000	95	15.5249	232,874	232,874
Subcontract		24,000	95	15.5249	372,598	372,598
Nonrecurring						
Tooling	$17,500	500	95	15.5249	7,762	25,262
Test	20,500	100	90	13.3344	1,333	21,833
Subtotal						$1,012,385
1. Plant overhead at 75% of subtotal of total element cost						759,289
2. Total cost of goods manufactured, calculated as sum of total element cost and plant overhead						1,771,673
3. General and administrative cost, calculated at 20% of line 2						354,335
4. Engineering cost						375,000
5. Selling cost, calculated at 7.5% of line 2						132,876
Total cost of engineering, manufacturing, and sales for 18 generator units						$4,405,557

8.5 EXTENSIONS FOR LEARNING MODEL (OPTIONAL)

Applications using the learning model are found in procurement, production, and the financial aspects of an enterprise. In purchasing, a learning function helps to negotiate purchase price or may be used for make-versus-buy decisions. In estimating, decisions related to cost are based in part on the concept that the production cost will decrease with increasing units produced. Contract negotiation is sometimes reopened after a first satisfactory model is produced. The contract for later units is based on learning reductions with time or cost known from the prototype unit.

We next consider the following topics:

- Follow-on procurement
- Engineering change order
- Break even

8.5.1 Follow-On Production

In many cases, products are not produced on a continuing basis, for an indefinite period. Production runs are made to produce a specified quantity and are then discontinued. All of the initial fixed costs such as engineering and tooling may be amortized

PICTURE LESSON The First Computer

Used by permission of the University of Pennsylvania School of Engineering and Applied Science.

The ENIAC (Electronic Numerical Integrator and Computer) was unveiled to the public in February 1946 at the Moore School of Electrical Engineering at the University of Pennsylvania. The primary objective of the project, which was supported by U.S. Army, was to build a machine that would speed up calculations for ballistic ordnance. However, the inventors wanted to make the ENIAC as flexible as possible, able to solve not only numerical integration but a wide range of problems involving various numerical operations, and store and retrieve intermediate results. The design performed these operations consecutively or concurrently, with automatic transfer of data from one step to the next.

Inventors J. Presper Eckert and John W. Mauchly were familiar with electric/mechanical calculators, punched card and tape machines, and the differential analyzer. The differential analyzer was particularly well suited to solving ballistic equations, but it was not general enough to handle a greater variety of numerical calculations. It is more likely that ENIAC was conceived in the tradition of the mechanical adding, multiplying, and dividing machines of that era. Manufacturers of that time were working on electronic ring counters and scalers for experimental physics, and those developments were known to the engineers.

The ENIAC contained 17,468 vacuum tubes, 70,000 resistors, 10,000 capacitors, 7200 crystal diodes, and 6000 switches. It had a footprint of about 33 $\times$ 1 m, occupied a room of

170 m^2, and weighed 30 tons. The basic operation of addition/subtraction (or transmit/receive) required 200 μs. Programming the ENIAC was very different from what we consider programming on a present-day stored-programmed computer. The data-flow architecture of the ENIAC required setting switches and making wire connections between units. The replacement of the vacuum tube by the transistor and the integrated circuit was one of the first reasons for decrease in computer size, weight, and power.

over the first production run. In addition, much of the decrease in production time due to learning has been accomplished. If sometime after the end of the first production run, another order for the product comes in, the cost for this *follow-on production* needs to be re-examined. Costs for follow-on production are noticeably lower than are original costs.

Example: Follow-On Estimating with Learning

Figure 8.5(a) describes follow-on estimating for procurement, a problem that can be stated as follows: Given that historical values that are available for some specific unit number and the associated direct labor hours, find the value for the cumulative quantity of a follow-on estimate. This example is shown in Figure 8.6(b) for a 10-kW product. A follow-on estimate is necessary at unit 18 for a two-year bid. The follow-on bid quantity will be for 482 units (= 500 − 18). Because the estimate is for future conditions, a history of the product's performance up through unit 18 is only one consideration.

If at unit 18 the hours are 73.5 and learning is 95%, then the first step finds the first unit cost (K) by using Equation (6.6), $K = 73.5/18^{-0.074} = 91.0$

We next want to find the average cost per unit for the follow-on units (19–500). The lot time for the follow-on shipment is found by determining the cumulative costs at units

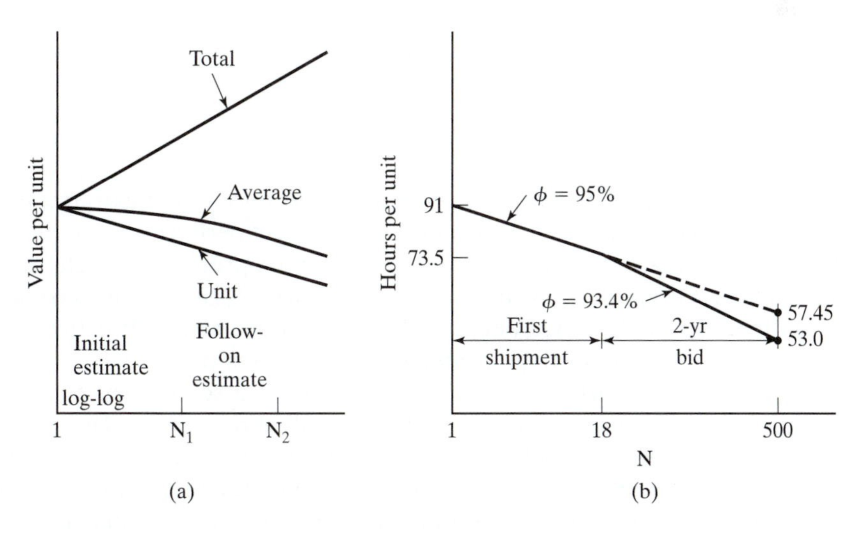

Figure 8.5 (a) Follow-on procurement with learning effects, and (b) follow-on production with a "challenge" to improve the learning.

18 and 500 and then subtracting to determine the portion of the total cumulative cost at unit 500 that applies to the follow-on 482 units. In order to find the cumulative total, the average cost is found from Equations (6.11) and (6.12) and then multiplied by the number of units to get the cumulative totals.

The cumulative total for the first 18 units is

$$T_a = \left(\frac{1}{1 - 0.074}\right)(91.0)(18^{-0.074}) = 79.349 \text{ hr per unit}$$

$$T_c = 18 \times 79.349 = 1428 \text{ hr}$$

The cumulative total for the whole lot of 500 units is

$$T_a = \left(\frac{1}{1 - 0.074}\right)(91.0)(500^{-0.074}) = 62.045 \text{ hr per unit}$$

$$T_c = 500 \times 62.045 = 31{,}022 \text{ hr}$$

Time for the follow-on 482 units is $31{,}022 - 1428 = 29{,}594$ hr and the average time is $29{,}594.21 \div 482 = 61.4$ hr per unit.

If the follow-on estimate indicates a value too high for competitive reasons, management may want to reduce the cost even more. For example, using the information above the time for the 500th unit can be projected as

$$T_{500} = 91 \times 500^{-0.074} = 57.45 \text{ hr}$$

Management may choose to "challenge" engineering and manufacturing to reduce the time for the 500th unit to 53.0 hr. Using the known time at the 18th unit, management's desired time for the 500th unit and Equation (6.13) yields a slope for the learning curve of

$$s = \frac{\log 73.5 - \log 53}{\log 18 - \log 500} = -0.0984$$

Applying Equation (6.9), $\varphi = 2^{-0.0984} = 93.4\%$. We have changed the learning slope from 95% to 93.4%. Engineering will assess if the new slope is doable and realistic.

8.5.2 Engineering Change Order

During the course of production, it may be necessary to incorporate a design change. Design changes are made to improve product performance or manufacturability or in some cases to correct a safety or reliability problem. Designs are changed using an instruction called an *engineering change order (ECO)*, which consists of the new design, revised drawings, and the authorizing documents. Direction is provided about the incorporation of the changes. In most cases the change applies only to those units produced after the release of the ECO. In some cases, previously produced items require modification in a process called *retrofitting*. The time required to produce the new design will probably be different from the previous product, and there will be time required to perform the retrofit on the previously built items. It is straightforward to determine the time for single units, but when learning is involved, determining the cost of a production run, including any retrofit, entails interesting calculations.

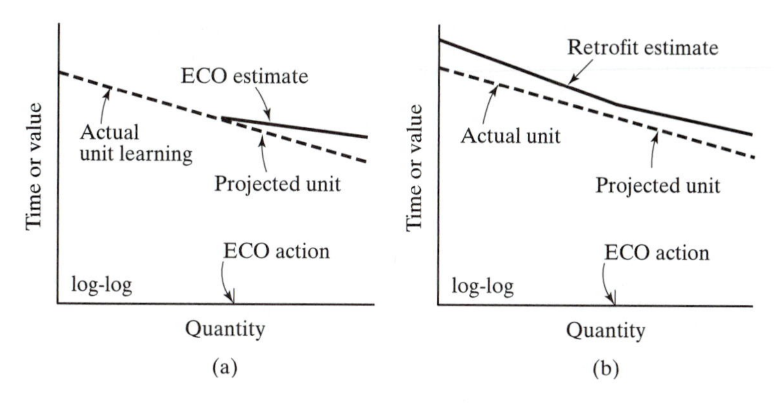

Figure 8.6 Engineering change order learning curves: (a) nonretrofit, and (b) retrofit.

The effects of ECOs can be evaluated by learning theory. ECOs may occur after the product is estimated, priced, and perhaps after a contract exists between a buyer and seller. Production and delivery may even have started.

It is necessary to estimate an equitable adjustment to product cost and price, if the contract terms allow. Now, consider Figure 8.6(a) and (b), which show curves for the nonretrofit and retrofit cases. From the point of incorporation of an ECO, the learning line could be above or below the continuation of the line for the product without the change. The cost of retrofitting will always be above (added to) the curve for previously produced items.

In the nonretrofit case, shown in Figure 8.6(a), no adjustment is required for products supplied prior to ECO action. In Figure 8.6(b), retrofit work is required for units before the ECO point. Additional time or value is also required after the ECO action, and an ECO learning line with different slope is shown. At the unit where the ECO is issued, the estimated ECO and actual learning lines do not intersect. The difference between the two lines at the ECO point is the unit time or cost to affect the ECO on the unit at the point of issue.

The solid lines represent average and total learning before and subsequent to ECO action. Note the change in slope for the retrofit and the ECO-configured product. At the ECO point there is an incremental cost for the change.

Example: ECO Cost with Retrofit

A product has been in production when an ECO is issued at the 100th unit. The 90% learning curve for the original design is $T_u = 1066\,N^{-0.1520}$ hr.

The ECO will require an additional 75 hours for the 100th unit and includes a retrofit requirement. The learning rate for the retrofit is estimated to be 95% and a new learning rate of 93% will be applied to subsequent units. This situation is depicted in Figure 8.6 (b). The steps include calculating the cost of producing units 101–500 with the ECO, including the original cost and retrofit for the first 100 units, then determining the cost of the ECO by comparing the cost of the 500 units produced with the ECO and what the cost would have been to produce all 500 units without the ECO.

The cost to build the initial 100 units is figured by determining the average time using Equation (6.12) and then multiplying by the build quantity:

$$T_{a=100} = \frac{1066}{(1 - 0.152)} 100^{-0.152} = 624.25 \text{ hr}$$

$$T_{c=100} = 624.25 \times 100 = 62{,}425 \text{ hr}$$

The ECO time for the retrofit is computed knowing the time at the 100th unit and the learning curve, $\Phi = 95\%$, or $T_u = K(100^{-0.074})$ and $K = 105.45$. Now the average and cumulative times for the retrofit can be determined:

$$T_{a=100} = \frac{105.45(100^{-0.074})}{1 - 0.074} = 80.99 \text{ hr}$$

$$T_c = 80.99 \times 100 = 8099 \text{ hr}$$

The time for producing units 101–500 with the ECO incorporated is calculated as follows:

(a) Determine the time for unit 100. Use the original design and learning curve $T_{100} = 1066\,(100^{-0.152}) = 529$ hr. With the additional 75 hours for the ECO $T_{100} = 529 + 75 = 604$ hr.

(b) Determine K for the new $\Phi = 93\%$ learning curve for the subsequent units $s = \log 0.93/\log 2 = -0.1047$, and $T_{100} = 604 = K(100^{-0.1047})$ and $K =$ 978.2 hr.

(c) Find the cumulative time for units 1–500 and subtract the time of units 1–100 (previously handled above). For units 1–500:

$$T_{a=500} = \frac{978.2(500^{-0.1047})}{1 - 0.1047} = 570 \text{ hr}$$

$$T_c = 570 \times 500 = 285{,}000 \text{ hr}$$

For units 1–100:

$$T_{a=100} = \frac{978.2(100^{-0.1047})}{1 - 0.1047} = 674.62 \text{ hr}$$

$$T_c = 674.62 \times 100 = 67{,}462 \text{ hr}$$

Time of units 101–500:

$$285{,}000 - 67{,}462 = 217{,}538 \text{ hr}$$

Sum the times with the ECO:

Total time of 500 units including the ECO	
Original build units 1–100	62,425
Retrofit first 100 units	8,099
Build units 101–500	217,538
	288,062 hr

Compare this to the time of 500 units built to the original design:

$$T_{a=500} = \frac{1066(500^{-0.152})}{1 - 0.152} = 488.785 \text{ hr}$$

$$T_c = 488.785 \times 500 = 244{,}392 \text{ hr}$$

ECO time is: $288{,}062 - 244{,}392 = 43{,}670$ hr

8.5.3 Break-Even Analysis

In a conventional break-even analysis the production cost for each part, C_v, (sometimes called the variable cost) and the initial fixed costs, C_f, are compared with the selling price P to find the quantity of parts, n_{be}, that must be sold to recover the costs and "break even." The conventional break-even analysis can be studied in Section 9.3. Now the conventional mode simply finds the quantity of product where the total cost equals the net sales revenue and is shown in Figure 8.7(a) and described by

$$Pn_{be} = n_{be}C_v + C_f \tag{8.9}$$

This model assumes that the costs C_v for the product remain constant throughout the production run. In learning cases the production costs decrease at a rapid rate early in production. We can take the learning model and use it in place of a constant C_v to get a new break-even model, as shown in Figure 8.7(b).

Generally, the learning model allows the break-even quantity to be attained sooner, depending on how C_v is set. If we ignore the fixed costs for simplification, the break-even point can be determined by using a variation of the learning equation, using the cost and price in units of money.

$$P = KN_{be}^s \tag{8.10}$$

K is the cost of the first unit and s is the slope of the learning curve.

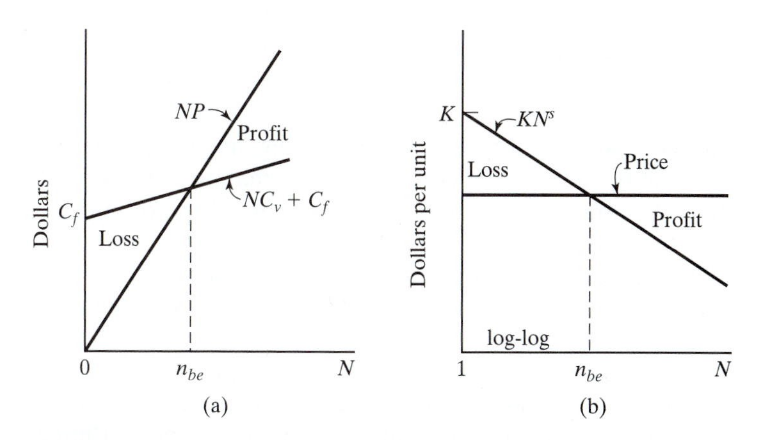

Figure 8.7 Break-even situations (a) conventional with a constant variable cost, and (b) learning curve for variable cost, with fixed cost not shown for simplicity.

Example: Break-Even for Learning Theory

Let the price of a product be $4820 with a learning equation of 12,500 $N^{-0.2138}$. Setting price equal to the learning equation $4820 = 12{,}500\ N^{-0.2138}$ yields $n_{be} = 86$. Profit develops if $N > 86$.

8.6 PRICING PRINCIPLES

Price matters. In Chapter 1 it was pointed out that the driving force behind desire of products was "an economic want," or what the consumer is willing to pay. Two factors dominate the economic want: (1) the degree that the customer wants or needs the product and (2) how many products and at what prices are available to satisfy the want (in other words, what does the market look like?).

But as with many other concepts, there is no single price. For example, if a customer purchases directly from the manufacturer raw materials or component parts, then the price that the customer pays is the same as the price the manufacturer receives for the product. As the distribution channel expands, the price paid by the customer includes shipping costs, mark-ups (costs and profits) for distributors or wholesalers, and the retailer's costs and profits. From the customers' viewpoint the only price that matters is the one that they pay.

So far there has been no discussion of a formal definition of *profit*, and its commonly understood meaning has sufficed. For the purpose of this discussion, profit is defined for engineering estimating as the expected excess of the planned price over the estimated cost. These figures are as put forward by the product's manufacturer. The actual or historical profit eventually realized is of more importance to accounting than engineering. Estimated and actual profit may not be equal.

In the real estate business, the agent who is asked by home sellers how much they can ask for their house will tell them that they can ask whatever amount they want. But realistically (based on the market) the agent will tell them that they can only get so much. It is the same for any company's product. The price can be set for whatever the company wants, but there is no guarantee that customers will pay that much. It is the task of the company's marketing department to determine the price that the company can get for a product.

Other company groups, such as engineering, manufacturing, and finance, determine the cost of the new product through the estimating process. An estimate provides information for comparison to marketing's estimated price or in the absence of a forecasted price to determine a price.

In small companies an engineer may be charged with price setting. Engineering tends to produce elegant (thus costly) designs. Manufacturing tends to overestimate, or "pad" costs to cover estimating errors or other production inefficiencies. Finance and management tend to inflate overhead and profit requirements. And finally, marketing tends to lower price estimates in order to ensure a large market and unit sales volume in the face of the competition. The difference between the (un)intentionally inflated price set within the company is contrasted with the price determined realistically attainable by marketing. The reconciling of the two will improve both and finally determine if the product is financially feasible.

The attainable price does not remain constant over the life of the product. In a neutral inflation-deflation milieu, and as the product progresses through the stages of its life cycle, price usually decreases. In addition, a number of practices exist and are closely related to price setting and add to the complexities. For instance, promotional pricing, premiums, coupons, trade-ins, extras, fire-sale gimmicks, volume discounts, repeat discounts, geographic price differentials, lease-buy arrangements, and reciprocal agreements are deals that squarely affect the price choice.

The pricing situations presented to the price setter are diverse and may range from a price for a one-of-a-kind product to one where there are identical units offered to many buyers. The firm may have only a single product or it may have a multiple-product line. The product may be brand new to the firm or on rare occasions may be a product of research or invention. The product may have been manufactured for decades in the same form, or minor modifications may be introduced every so often. Commodity products are another circumstance, as these common products are market driven. Engineered products for a undefined market present different opportunities.

Market distinctions exist. There is the open market in which products are sold to an unknown buyer. Competition among products is based on price and nonprice factors. Then there is the bid or order market in which manufacturers produce an item for a specific buyer. The type of contract, specification, and engineering design, in addition to price, are factors. In some cases, price is the single means of competition. In others, price may be relatively unimportant. Different pricing methods may be better for these different situations. It is important to choose the best method for the conditions. There are four pricing concepts to consider:

1. *Prices proportional to cost:* If the concept uses the same percentage profit for all elements of cost, then it is full cost plus a markup. Another variation would have different markup percentages on the several cost elements.
2. *Prices can be established proportional to conversion cost:* This concept ignores the effects of the several kinds of material cost in its calculation. A conversion cost concept emphasizes value added by manufacturing, or adds a percentage onto direct labor plus overhead.
3. *Prices can be proportional to variable cost—i.e., resulting in the same percentage contribution over variable costs:* This concept can lead to different proportions for different designs; that is, fixed costs and profit. Direct labor and direct material (and variable overhead in some situations) are used as the base, and marginal cost of producing additional units is emphasized.
4. *Prices can be systematically related to the stage of market and competitive development of the product:* Price and cost-estimating relationships are necessary features of the concept. Price is largely determined by marketing. Manufacturing is challenged to reduce costs to accommodate the price reductions.

Four methods of product pricing are studied:

- Opinion, conference, and comparison
- Markup on cost

- Contribution
- Price-estimating relationships

8.6.1 Opinion, Conference, and Comparison

Opinion, conference, and comparison are nonanalytic methods that involve experienced people who know the product's market and its price and understand the technical factors. These people meet to determine the price for a new product driven by real or anticipated demand, or for existing products, as a consequence of competitors' actions, or someone may notice that the price of an item may be out of line. Perhaps costs have gone up or down and adjustments are felt necessary. Discussion regarding the volume effect that each of the alternative prices would have and the product's profit as a result of alternative prices is undertaken. Data on competitors' prices are usually available as are comparisons of strengths and shortcomings of competing products. Cost estimates, past sales figures, and the history of price changes are discussed.

To assess the number of units that might be sold and their price, we try to imagine the customer response to each alternative. Furthermore, the responses of distributors and sales staff, changes in the sales of other products, and probable response of competitors, and the effect on the share of the market are evaluated. The predictions will be shaky, and we must recall their shortcomings. Though we forecast the future actions of our customers and competitors, we can be certain that reactions in the future will differ from those in the past. Analytic methods to determine a price are overlooked. Despite the absence of analytic methods, opinion, conference, and comparison remain essential to price setting.

8.6.2 Markup on Cost

The markup on cost method, variously called "cost plus" or "markup," is probably the most popular method of price setting because it identifies the types of costs and adds an additional percentage of those costs as a markup.

Simple as this appears, there are variations. For instance, the total cost can be used or the cost can be broken down into material, labor, overhead, and engineering, with each item having its own markup. This second method is found in government contracting where limitations on the markup of the cost components are sometimes required. Some companies use the same add-on percentage year after year. Others use markups reflecting the preceding year's sales results. For the most part, percentages vary with business conditions. This capability, and an ease of understanding, are desirable features. When companies use these procedures, it is usually as a starting place for a price decision. In some companies the sales managers in the territories eventually decide what price they will actually quote, and the markup percentage serves as background.

If the total cost of manufacturing, development, and selling is used, then it is a full-cost base. The formula for a full-cost markup is

$$P = C_t + R_m(C_t) \tag{8.11}$$

where P = unit price, dollars

C_t = total cost of manufacturing, development, and sales, dollars

R_m = markup rate on cost, decimal

Look again at the "total unit cost" at the top assembly for the X-125 product shown in Table 8.2, where $C_t = \$413.35$. If we want a mark-up $R_m = 13\%$, then, $P = \$467.09$.

A second method, conversion cost pricing, emphasizes value added or direct labor plus overhead as the base for a markup calculation. In some cases, material will be provided as a direct pass through to the customer, because it may not be competitive to add markup on materials. We can define a price by using the following formula:

$$P = \Sigma C_{dl}(1 + R_{oh})(1 + R_m) + C_{dm} \tag{8.12}$$

where C_{dl} = direct-labor cost for each operation, dollars
R_{oh} = general overhead rate, decimal overhead dollars to direct labor dollars
R_m = markup rate, decimal
C_{dm} = direct material cost, dollars

If the cost base is on direct labor and direct material only and the markup is not made on overhead, then the price objective is the incremental cost of additional units, or

$$P = (\Sigma C_{dl} + \Sigma C_{dm})(1 + R_m) + C_{oh} \tag{8.13}$$

where C_{oh} = costs of overhead, dollars

To get the same profit, Equation (8.13) requires a larger markup on a smaller base than is the case of the full-cost base (Equation 8.11).

It is fundamental to good business that if a new venture or a major improvement is to be undertaken, then the expected net return from it must exceed the cost of capital required. This philosophy provides a pricing method that depends on invested cost and requires that the cost of operation and the consumption of fixed assets and an acceptable rate of return be estimated. The acceptable rate of return on investment varies with numerous economic factors, but overall cumulative values have emerged and range from a low of 10% to over 50%. A simplified markup pricing model based on investment cost is

$$P = \frac{\left(\dfrac{iI}{N_y} + C_f + C_v N\right)}{N} \tag{8.14}$$

where i = desired return on investment, decimal
I = capital investment, dollars
N_Y = number of years for payback of investment
C_f = product fixed costs, dollars
C_v = product variable cost unit, dollars per unit
N = number of units sold

In this model the return on investment substitutes for the markup rate. To completely cover all costs, the overhead must be included as part of C_v.

Another cost-based approach to pricing includes methods to achieve a mark-up on sales value. It may be formalized as

$$P = \frac{C_t}{1 - R_s} \tag{8.15}$$

where C_t = total cost of engineering, manufacturing, and sales, dollars
R_s = markup on sale values, decimal

Refer to Figure 8.4 again. $C_t = \$413.45$, and if a mark-up of $R_s = 13\%$, is desired then $P = \$475.23$. This compares with \$467.09 for a markup of 13% of total cost. Terms found in pricing have discount synonymous with margin, list with selling price, and net with cost.

8.6.3 Contribution

Contribution is the amount left over from revenue after paying the variable costs and is used first to pay the fixed expenses. Any overage is a contribution to profit. As we have said, some costs vary with changes in production quantity, and others do not. Pricing can be related to the variable costs as a percentage of the full variable cost. In the contribution pricing method we find the principal items of variable cost, including labor, material, marketing, and administrative costs. This C_v value is raised by an amount of "contribution" to obtain a sales price, or

$$P = \frac{C_v}{1 - R_c} \tag{8.16}$$

where P = list price for manufacturer, dollars
C_v = full variable cost, dollars
R_c = contribution rate, decimal

For example if the variable cost of a product is \$125 per unit and a contribution percentage of 35% is assumed, then $P = \$192.31 (= 125/0.65)$.

Obviously, the percentages for the several kinds of pricing rates (i.e., markup, contribution, etc.) are not comparable. The percentages have meaning only in relationship to a pricing method. The method used is normally established by company management, as is the amount of markup, contribution, etc.

8.6.4 Price-Estimating Relationships

Price-estimating relationships (PERs) are mathematical models or graphs that estimate price. In these models, price is plotted with respect to some quantity of interest. Time is usually chosen, although production quantity or other variables are possible. A variety of functional forms are available; see Chapter 6 for details.

Consider PERs in the application of a price ceiling and a price floor. The marketplace determines the price at which products will sell and sets a price ceiling. The cost, profit, and price determined by the best firm establishes a price floor. In this context we define an *opportunity margin* offered—the difference between the ceiling and the floor—by the market. If the policy is to accept the marketplace as determining the price at which products will sell, then when the two meet the firm may choose to stop manufacturing the product or to reduce profit or cost. In economics, the price for a product will remain static in constant dollars if production (total from all suppliers) and demand remain in equilibrium. Excess demand tends to increase prices, and excess supply suppresses prices. This situation is demonstrated by the cyclic changes in the semiconductor industry. If profit and opportunity margin are high, more competitors are apt to enter the market.

These market conditions can be modeled for the opportunity margin as shown:

$$M = P_o e^{-k_m t} - P_f e^{-k_f t} \tag{8.17}$$

where M = opportunity margin, dollars
P_o = price ceiling of the initial unit at the beginning of production, dollars
k_m = decay experience for the product, decimal
t = time, typically years
P_f = price floor initially, dollars
k_f = decay experience of the price floor, decimal

A typical graph for a product-selling price is as shown in Figure 8.8(a). The floor price is the point at which the most efficient producer will make a reasonable profit. With a fairly stable floor price, the margin over this floor price decays rapidly as more competitors attempt to capture a share of the market.

Example: Opportunity Margin

A new product has been developed. Novelty will sustain early sales but long-range estimates have concluded that competing designs with a constant retail price of \$9 will provide a floor. The initial price of the item is \$21, and a decay of 0.25 is expected on the basis of experience. The number of years for the intersection is

$$9 = 21e^{-0.25t}$$
$$t = 3.4 \text{ years}$$

The opportunity margin can be used to set the product price and to estimate the life cycle under constant conditions and market expectations. In the example above, the life cycle is only a little over 3 years (a long time in the computer or the new toy

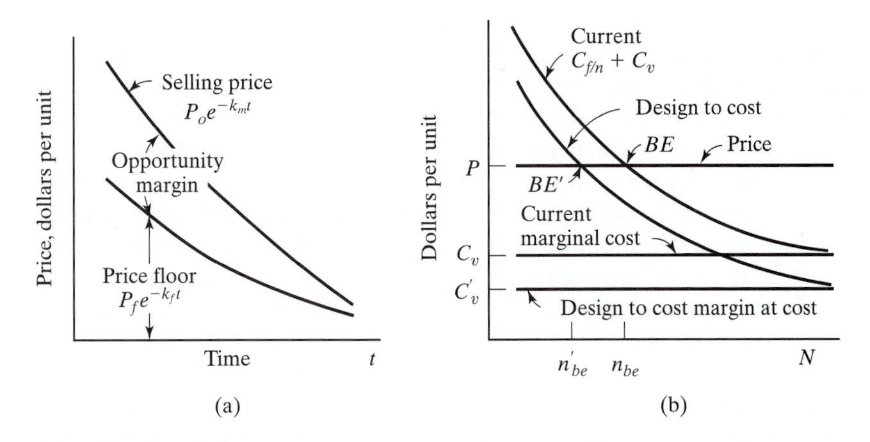

Figure 8.8 Application of price-estimating models: (a) Opportunity margin for the model $P_o e^{-k_m t} - P_f e^{-k_f t}$ and (b) fixed price with traditional and design-to-cost average costs.

industry). If a longer life cycle is desired, additional effort will be required to reduce the production costs in order to reduce the price floor and thus continue the opportunity margin for a longer period.

If the opportunity cost model exposes the product to a time decay, traditional models show the break-even under $C_f/n + C_v$ where engineering cost reduction programs drive the product cost to lower break-even locations through programs known as "design to cost," a topic covered shortly. Figure 8.8(b) demonstrates that average cost asymptotically approaches variable cost (also known as a *constant marginal cost*) when appropriate programs are encouraged.

8.7 BASIC CONTRACT TYPES (OPTIONAL)

An engineering or product sale agreement between a buyer and a seller are many times governed by legal documents generically called *contracts*. Placing and acceptance of a purchase order based on a published or quoted price by a supplier constitutes a contract with the customer.

Contracts are important to the engineer. Though the engineer may not be lawyer, he or she needs a fundamental understanding of the contract practice because the estimate is sometimes construed to have legal importance. At the very least, the results of the estimate may be part of the contract. This brief discussion is meant to familiarize engineering students with aspects of procurement contracts.

There are two basic types of contracts used in the sale of products: (1) firm fixed-price and (2) cost reimbursable. Firm fixed-price arrangements have in common that one party, a supplier, is to deliver a product, project, or system, or perform an operation in accordance with the terms and conditions of the contract, and an agreement by a buyer, firm, agency, or government to pay a price equal to that specified by the contract. In contrast, cost-reimbursable contracts pay the supplier for costs incurred plus an agreed-upon profit, subject to restrictions and special negotiated understandings. Firm fixed-price (FFP) contracts are used for low technology, well-developed production of low-cost consumer products, large volume and/or short time period contracts. Products requiring highly complex development or new technologies may use a cost-reimbursable type of contract. Cost-reimbursable contracts transfer the economic risk to the customer, and fixed-price contracts place the economic risk on the supplier.

To illustrate these concepts, suppose the product is a computer hard disk drive. This is a mature technology product with a very high production. The market determines a maximum (ceiling) price and the supplier has a solid knowledge of the cost to manufacture the drives. A firm fixed-price contract (quote and purchase order) would be executed between the hard drive manufacturer and the computer maker. On the other hand, many times the government (think about NASA or the Department of Defense) will desire a very large, high technology product or system that will require development of complex new designs and technologies. This would impose risk to the supplier with all of the unknowns and the difficulties of determining the cost accurately. In this case the customer and supplier would develop a risk-sharing, cost-reimbursable contract. Both types of contracts may depend on competitive bidding, but the bid and award process is more complicated in a cost-reimbursable contract.

8.7.1 Fixed-Price Arrangements

The fixed-price contract requires that the design (operation, product, etc.) be delivered as described on a predetermined schedule. The price is fixed for the life of the contract precluding changes allowed by the contract. Though the fixed-price contract provides the greatest risk to the seller, it also offers incentive and opportunity to realize the greatest profit. The vendor or contractor fully recovers savings due to any cost reductions. Thus, if the actual costs are less than estimated costs, then greater profit materializes. The fixed-price or lump-sum contract is popular from the customer's viewpoint because the total project cost is known in advance.

The parties agree to the price before a firm fixed-price contract is awarded. An example for a $1 million contract under varying consequences is given as

Contract price	$1,000,000	$1,000,000	$1,000,000
Actual cost	900,000	1,000,000	1,100,000
Realized profit (loss)	$100,000	$0	($100,000)

The supplier either gains or loses based on performance.

Some vendors bid for work on the basis of a productive hour cost rate. For example, the manufacturing company specifies the following schedule:

Heavy machining: $182.84/hr

Light machining: $63.89/hr

The time it then takes to complete the work is multiplied by the appropriate productive hour costs, which include profit. Note that the values for these PHCs from Table 4.13 were increased by 25% using the markup method. Obviously, the buyer needs to be satisfied that a fair and reasonable time is charged against the contract.

A *time and material contract* is used between a customer and supplier for work at a fixed and specified rate (hourly, daily) that includes direct labor, indirect costs, profit, and materials. This is often the form used by home repair contractors, or small manufacturers working for a larger product producer. The materials may be at cost or cost plus profit. This contract is suitable for operations where the amount or duration of work is unpredictable or insignificant.

There are other variations on the firm fixed-price contracts. One type is the quote or price-in-effect (QPE) contract. These are also known as escalation contracts (introduced in Section 3.5). Long-term uncertainties that result from inflation or deflation effects are the reason for their use. The supplier estimates costs and then establishes a mutually agreed upon benchmark or index. Adjustment is mostly upward, but downward adjustment is possible. The price adjustment clause must identify a base or benchmark period and one or more indexes to measure changes in price level in relation to the reference cost and reference period. If the contract provides for economic price adjustment, then the contingency element of the product cost estimate is ignored if it deals with inflation effects. Remember that contingency may be for considerations other than inflation. The contract may also allow the supplier to pass purchased material cost increases through to the customer.

8.7.2 Cost-Reimbursable Arrangements

If the project cannot be accurately estimated, then the fixed-price contract may not be suitable. The cost-reimbursable contract places the risk on the customer and is used whenever research, development, design, or urgency is necessary. The customer has to be sufficiently knowledgeable in the product and manufacturing and be allowed (by the contract) to sufficiently monitor the supplier performance and records to ensure that the costs charged are fair and accurate. It is as necessary to estimate accurately for cost-reimbursable contracts as it is for fixed-price contracts.

Cost-reimbursable contracts have provisions for payment of allowable, allocable, and reasonable costs incurred in the performance of the contract. Certain formulas, to be described shortly, permit adjustments for fees, incentives, and penalties.

In one type of a cost reimbursable contract, a negotiated pricing formula can be agreed to that motivates and rewards the supplier on the basis of performance. In these fixed-price incentive (FPI) contracts, the process involves an estimated cost, target profit, target price, ceiling price, and profit (loss) sharing formula for costs incurred above or below the estimated cost. The example above with the $1 million contract price uses a 0/100 sharing formula (customer/supplier). This means that the customer does not share in improved or poor performance and the contractor accepts 100% of the difference between price and cost.

The FPI contract represents joint responsibility for the ultimate cost. In this arrangement the supplier and customer share in any dollar difference between the estimate and final cost. In an 80/20 example, the supplier is responsible for 20% of the difference either as an addition to or a deduction from target profit. Though shares always total 100%, the proportions can vary because of uncertainty, amount of target profit, and the spread between estimated cost and ceiling price. Expressions of the customer/supplier shares are 60/40, 75/25, 50/50 and so on.

Example: Fixed-Price Incentive Contract

Assume a 70/30 sharing proposition for the following cost facts:

Price ceiling	$11,500,000	
Target price	10,850,000	
Estimated cost	10,000,000	
Target profit (8.5%)	850,000	
Final cost	9,600,000	
Difference	400,000	Under run

The supplier receives $120,000(= 0.30 × 400,000) as an increase in profit. This is added to a target profit.

Target profit	$850,000
Contractor's share	120,000
	970,000

The customer receives 70%, or $280,000 (= 0.70 × 400,000) difference, as a reduction in price.

Final cost	$9,600,000
Final profit	970,000
Final price	10,570,000
Target price	10,850,000
Price reduction	280,000

Now, assume that the performance was not satisfactory and instead of the cost reduction, a cost over run is incurred so that there is a final cost of $10,500,000.

Estimated cost	$10,000,000	
Final cost	10,500,000	
Difference	500,000	Over run

The supplier share of the loss of 30%, or $150,000 (= 0.30 × 500,000) is a decrease in profit.

Target profit	$850,000
Less 30% of over run	150,000
Final profit	700,000

The customer accepts 70% of the over run, or $350,000 (= 0.70 × 500,000) as an increase in price.

Final cost	$10,500,000
Final profit	700,000
Final Price	11,200,000
Target price	10,850,000
Price increase	350,000

If the performance was even worse and the final cost is $12,000,000, or $500,000 in excess of the ceiling, then the ceiling of $11,500,000 is the final price and the supplier incurs a $500,000 loss.

In extreme cases of uncertainty, a full "cost plus" contract can be negotiated. These are used only when there is very little experience with a large new product and/or new technology—for example, a major new system for the space station. In these cases the customer (many times the government) is usually significantly larger than the supplier and better able to shoulder the unknown costs. In a total cost plus contract the customer assumes 100% of the risk for a 100/0 sharing. The supplier will provide their best estimate given the limited amount of information at the start, and specify a fixed amount or a percentage of the actual cost that will represent the profit (cost plus profit).

The customer must be sure there are provisions for monitoring and controlling the suppliers' costs (so that there is responsible management of costs) and some incentive may be included to encourage the supplier to pursue cost savings. This starts to approach the fixed-price incentive type of contract except that the customer will share some of the savings, but will assume all of the cost increase. Cost plus contracts are not as common as they once were, somewhat due to abuses by suppliers. Customers are very cautious about entering into cost plus contracts and the negotiations and subsequent monitoring are usually intense.

8.8 BENCHMARKING AND COMPETITIVE ANALYSIS

In this chapter we have studied the essential procedures for product cost estimating. There were two objectives—finding full cost and price. Of course, the accuracy, reliability, and timeliness of the analysis and estimates need to be professional and conducted with modern tools. With this as background, what else can be done? The answer is a lot. There are many opportunities for product estimate extensions, such as the following:

- Make versus buy
- Break-even
- Value analysis engineering
- Design for assembly and manufacturing
- Design to cost or design for profit[3]
- Target costing
- Concurrent engineering

There are more applications for product estimating, but space limitations precludes their discussion.

In make versus buy analysis the engineer determines if it is cheaper to self-manufacture ("make") the product or to purchase it from an external source ("buy"). The engineer compares the price as promised by the vendor's quotation to a calculated make value. If a purchased part is analyzed as cheaper, then the firm will buy the design from a supplier. If the calculated make cost is less, then the company will manufacture the product. The objective of the analysis is selection of the cheapest source. The analysis is complicated by the effect on the overhead structure of the company if the make scenario significantly changes the business. In addition, the suppliers', qualification, management, and sustaining costs must be factored into the buy scenario. These principles are discussed in Chapter 11.

Break-even analysis is one of the significant tools used in engineering cost inquiry. The models balance the negative and positive effects of money transactions and the break-even chart dramatizes the relationships that exist between costs, price, volume, and profits. (Discussion is deferred to Chapter 9.)

Value analysis/engineering (VA/E) involves application of tools that identify the functions of the object, finds a monetary or nonmonetary value for the function, and then recommends reengineering of the functions for the lower cost. Though simple to define, the thrust of the work—cost reduction—is the general aim. Defining "functions" is not a trivial pursuit. In VA/E the analysis breaks down functions in simple noun-verb combinations, and a tree decomposes them from the highest level to individual components. Common tools include quality function deployment (QFD), which is a hierarchical decomposition of requirements to features and functions. Moreover, the object may be a product, service, system, or operation. More often, the application deals with existing objects—those that are already made and in existence. Many of these goals and practices are discussed in this book. There is a major difference though. Engineering cost analysis and estimating deal in the future. Generally, the product in

[3]The name "design to cost" was introduced in the early 1970s. Another choice was "design to production," but it was considered as too restrictive. Shortly after the name christening, the U.S. Department of Defense added the practice to its purchasing requirements, and the technique expanded quickly. Later the name "target costing" was introduced, because "design to cost" was pictured (incorrectly) as too engineering. Target costing practice has flourished in military contracting and with some commercial products.

its final form is unavailable. The student is referred to the references for additional information on VA/E.

Design for assembly and manufacturing uses computer tools to analyze the potential design for ease of assembly and manufacture. Software for this work is listed in the references. Principles for these activities are discussed throughout Chapters 7 through 11.

A primary step in design to cost or target costing is to have a cost before the product is designed. For the most part engineers "build up" an estimate starting with operation estimates and concluding with product cost and price. This activity begins with minimal information if preliminary estimates are made. If bills of materials and designs are complete, then detail estimates are prepared. Most detail estimating is "bottoms up."

Occasionally a reverse direction ("top down") is preferred. This practice is called *designing for profit*, or *design to cost*. The procedure begins with a market price, which is used as the target price. The engineer calculates downward from a market price to find the cost for design assemblies and components. Much of the these costs are rough-order-of-magnitude numbers. Those targets become goals for engineering, sales, procurement, and production. If the cost goals are realized, then a profit is ensured because the procedure began at the top, at a price that ensures market success, so it is reasoned. Design to cost is discussed in Chapter 11.

Concurrent engineering is an integrated approach to design of products, and product manufacture. This attitude encourages the enterprise, from the outset, to involve all business elements in the product life cycle from concept through disposal, including quality control, cost, scheduling, and customer requirements. Also called integrated product development, it promotes teaming of cross functional disciplines to satisfy the customers' needs, and these activities begin with the product's first kickoff meeting.

Pundits have stated that the benefits of concurrent engineering include 30% less development time, 65% fewer engineering changes, 20% less time to market, higher quality, and improved white-collar productivity. This comparison is to straight-line and traditional engineering, where design initiated the business sequence, and other functions follow in their order. For instance, cost work does not begin until drawings are completed. But concurrent engineering philosophy changes this approach, and cost analysis and estimating are now involved in the design process from the beginning. Business experience has confirmed this paradigm, and these principles have been a part of engineering education ever since the 1970s.

The design-first-and-cost-estimate-second premise is reversed in concurrent engineering. Design does not drive cost—instead, cost drives design. While design works toward product performance as its primary goal, and it is the imperative one, concurrent engineering presents the notion that cost and its surrogate profit are equally important as performance.

There are many nuances to concurrent engineering, and the student will want to review the references for additional reading. Principles, jargon, and management style have made this field exciting. That said, some principles are illustrated with the following example.

Example: Concurrent Engineering of Lawn-Tractor Wheel Rim

A firm designs, manufactures, and markets lawn tractors and mowers. It has noticed that competition is increasing both domestically and internationally, eroding income in this

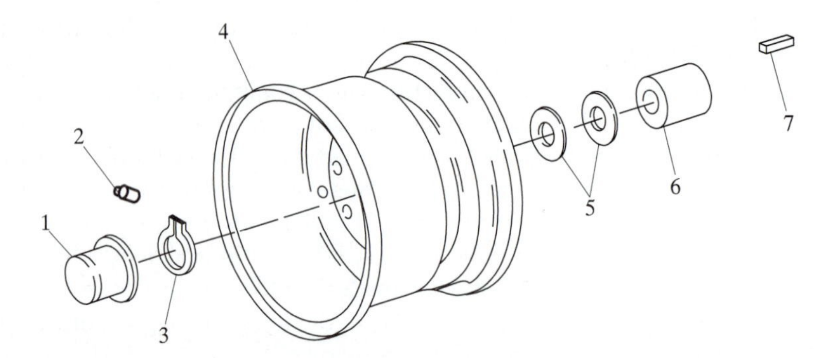

Figure 8.9 Wheel rim assembly of lawn tractor.

product. Sales of lower-quality lawn tractors have migrated to super grocery outlets, discount stores, and Wal-Mart. At one time, lawn tractor products were sold only through specialty stores. As a consequence, the enterprise is planning to add new models for exclusive sales to large estates and affluent home owners by upscale hardware stores and boutique tractor store outlets. It thinks that a superior quality machine will command a price and margin advantage. This quality reengineering should have a 10-year head start over other competitors who may wish to follow, so it is reasoned.

Upscale engineering will concentrate on the molded exterior, leather seat, engine, wheel rims, and a much tighter turning radius—mostly a fashion redesign. The frame, tires, and ignition system and power train will be repeated from existing models.

Two sizes, the 13.5 and 16 hp models, are selected. There are no patent restrictions for this effort. The wheel rim, along with the body shell and seat, are designed with improvement in appearance as an objective.

The wheel rim is selected for concurrent engineering. The wheel rim is a two-piece hydroformed steel metal arc-welded component. A view of the wheel assembly is shown in Figure 8.9. The tires size specifications are 18×8.50-8 and 13×6.50-8 in. with maximum inflation of 10 psi., which will remain as before.

Marketing and engineering collect the information given in Table 8.5. Although the Internet is useful for searches of this kind, such information is actually inadequate because not all engineering details can be found or communicated to suppliers. Capital equipment, manufacturing process representations, CAD integrations, process times, labor rates, and

TABLE 8.5 Marketing and Engineering Factoids for Wheel Rim

Item	Factoid
Global units of metal wheel rims produced	More than 400 million annually
Number of global suppliers of metal rims for lawn tractor	More than 3000
Number of basic manufacturing schemes	More than 30, excluding minor variations
OEMs for lawn tractors globally	More than 370
U.S. marketing outlets	1 per 14,500 population
Forecast tractor units sales, U.S. domestic market	20,000 first year; 300,000 for 10 years
Number of rims for 13.5 and 16 hp tractor sizes, 3 sizes	80,000 first year
Number of supplier-engineering firms providing a dual design/make role for wheel rims	25

TABLE 8.6 Bill of Material for Lawn-Tractor Front-Wheel Rim

Description	Unit of Material	Quantity Next Assembly	Order Policy	Unit Material Cost, $	Unit Labor Cost, $	Material Cost, Next Assembly, $
1. Cap	Ea	2	M	$0.03	$0.01	$0.08
2. Tire valve stem	Ea	2	B	0.09	0	0.18
3. Snap ring	Ea	2	B	0.07	0	0.14
4. Rim, left hand and right hand	Ea	2	M	1.93	5.89	15.64
5. Washer	Ea	4	B	0.04	0	0.08
6. Spacer	Ea	2	M	0.08	0.11	0.38
7. Shaft key	Ea	2	B	0.11	0	0.22
Total				$2.35	$6.01	$16.72

geographic information for outsource analysis are unavailable or unreliable, though it is initially helpful. Direct contact is necessary.

Table 8.5 indicates the global production of metal wheel rims for the total market—for cars, trucks, ATVs, motorcycles (but not toys or lightly-loaded recreational bicycles). Furthermore, the search turns up the number of suppliers able to collaborate on design as well as manufacturing. Engineering design may be unnecessary for the new design of the wheel rim, since there are many suppliers who would cooperate with design and manufacturing. Wheel rims are a commodity product, and competition is intense.

Existing information is summarized in Table 8.6, where BOM information is entered. Each tractor has two wheel rim sizes, and for the 18×8.50-8 in. size, the unit cost is $8.36 and for the tractor, $16.72.

Our firm buys competitor tractor units, as listed in Table 8.7, on the open market. Intelligence is uncovered from the teardown of a competitor's units, doing reverse engineering and estimating.[4] The results of this practice is an unordered ranking of competitors, count of parts in assembly, and cost. Additional evaluations can be made, for example, quality and appearance rating, though not shown here.

This benchmarking is of the cost and quality of the competitors' products. When benchmarking is done, engineering studies the "best of the class" and determines several kinds of metrics about their competitors. This benchmarking can be about the product, which is the type described here, or it may deal with the process.

An exclusion chart—a scattergram of prices, costs, indexes, or other measured variables—can be drawn using this information. In Figure 8.10(a) note that there is an area on the chart that is free of data points. This area is a rectangular or convex polygon after lines are roughly drawn to circumscribe the data. The abscissa is designated with a definable measure, rim diameter, while the ordinate is cost. The exclusion chart suggests that the cost of a future wheel rim should avoid that zone.

[4]The Ford Motor Company conducts teardowns of competitors' products. As an example, in a full-scale mock-up shown to the authors, Ford disassembled Ford Ranger, Mazda, and Dodge truck of the same size. The parts, subassemblies, engine, frame and so on were unmanufactured as much as possible. The unit body was stripped down and sections were cut into the roof to disclose the internal construction. In another example, the dash assemblies of the three models were disassembled and mounted to a vertical 4×12 ft plywood panel. Wheels were attached to the bottom of the panel allowing easy rolling and storage, access, and viewing. Engineering was able to examine the parts, minor assemblies, etc. They also made experienced cost estimates, by determining the processes and conducting reverse engineering, allowing product improvement. Through such processes, Ford is able to estimate the cost that competitors pay for these products.

TABLE 8.7 Cost Benchmarking of Competitors' Wheel Rim Assembly

	Teardown	
Competitor	Count of Parts	Unit Cost, $
Sears	10	8.32
Murray	11	5.29
MTD	5	7.21
Scotts	7	6.27
John Deere	7	7.36
Simplicity	5	4.29
Snapper	8	8.31
Yard Man	12	5.21
Kubota	14	11.69
Ariens	8	9.27
White	9	10.32

Engineering needs to appreciate the perception by the customer of the product. If *features* are the physical or aesthetic attributes of the product wanted by the customer, then *function* is the physical action associated with a component or subassembly of a product with a possibility for engineering design. The feature may have an indirect connection to function. A customer desire may be appearance, but that is difficult to map to function. If it is possible to uniquely map to a function, then a cost estimate can be coupled to the function for design engineering. Obviously, a cost estimate is not plausible for a customer's desired wants. Difficulties arise. For example, if the features map to more than one function, separable and mutually exclusive cost estimates may be questionable. Note Figure 8.10(b) where customer features are arrowed to engineering functions.

The customer appearance feature is mapped uniquely to the wheel assembly, then manufacturing sequences are compiled. For this case, seven manufacturing plans for the wheel assembly are listed with painted, polished, machined, and chromed surfaces as appearance choices. Furthermore, an escutcheon or acorn lug nuts are possible. Continuing, preliminary designs are estimated, and the results shown in Table 8.8.

TABLE 8.8 Estimated Cost of Metal Rims and Appearance Ranking

Manufacturing Process of Wheel Rim with Final Appearance Feature	Raw Material	Estimated Cost, $/unit	Ranked Appearance
Two piece, blanked, stamped, formed, spot welded, painted	Cold rolled steel, AISI 1025	$5.14	7
Two piece, cold forged, welded, painted	Cold rolled steel, AISI 1025	7.22	6
Cast one piece, machined	Aluminum, alloy 201 T4	9.94	5
Cast one piece, polished	Aluminum, alloy 201 T4	9.63	4
Cast one piece, chrome	Steel, AISI 1151 for rim	11.27	3
Cast one piece, painted with chrome insertable escutcheon of company logo on plastic end cap	Steel, AISI 1151 for rim	10.23	2
Cast one piece, painted, 3 chrome acorn lug nuts to axle plate	Steel, AISI 1151 for rim	12.92	1

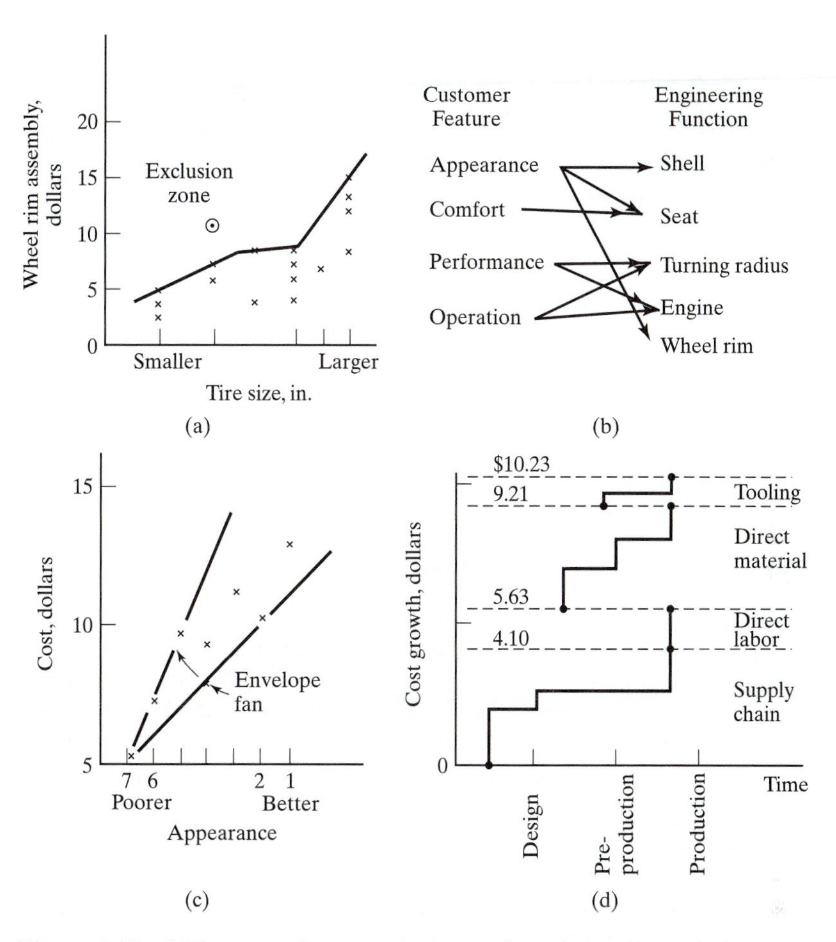

Figure 8.10 Milestones for concurrent engineering cost analysis: (a) Exclusion chart of the manufactured cost of competitors' wheel rims; (b) customer feature to engineering design mapping; (c) envelope fan of choices of cost and appearance rank; and (d) monitoring of design-to-cost targets.

A jury is convened to rank the alternatives for appearance. Rules given to the jury, such as not disclosing the costs of the alternative products, can be wrongly persuasive. Wheel rim products are openly available, and samples from other applications are gathered in a design studio and comparison is easily achieved. Final ranking is shown in Table 8.8. The envelope fan of choices is seen in Figure 8.10(c).

The jury chooses the process of steel casting a one-piece rim with the cap end-piece (item 1 in Figure 8.9) that is refined with a chrome company logo (which is thermoplastically injection-molded as an insert in the cap, a common procedure). The selection has a ranked appearance of 2 in Table 8.8. Estimated cost is $10.23, an increase of $1.87 per unit, and an additional $150,000 annually. The data point appears as a circle in the restricted area of the exclusion zone, Figure 8.10(a), a situation that engineering and marketing will need to resolve.

At this early point, engineering continues until the design alternative 2 is finalized. Throughout these phases, cost analysis and estimating continues, but the focus turns to production launch tools that monitor and update the cost. Activities include supply chain engineering, process planning for direct labor and material, and design and construction of the tooling. Figure 8.10(d) shows the step-cost growth, where the major cost contributors are shown as dashed horizontal lines. While the cost target is $10.23 per unit, the supply chain contributes $4.10 to product cost, for example. The analysis continues until the objective of the cost target of $10.23 or less is verified. Greater cost growth is not a good option, and efforts are made to prevent that from happening.

Cost estimating and analysis continues on with post-launch tools after production begins. Cost-reduction efforts, learning-curve philosophy, waste analysis, quality function deployment, and supplier assistance are a few of the many possibilities, and are collectively called *value analysis tools*, which are employed after production begins.

8.9 TRENDS IN PRODUCT COST ESTIMATING (OPTIONAL)

Computing for cost work began in the 1960s, though it was mostly data processing, and little if any comparative cost analysis and estimating was found until later. The handy data books with engineering performance information for manufacturing, as discussed in Section 6.3.2, did not appear until the early 1980s,[5] though in construction they existed since the 1920s. Specialized DOS software programs that estimated the lowest-level BOM or component cost materialized,[6] but were limited to estimating one-piece parts, and simplified assemblies. BOM software extensions were a proprietary feature of some companies, and even then the software only rolled up the costs, as shown by Figure 8.4.

The common tool used in business and for engineering cost work is the computer spreadsheet. Computing tools are more advanced in other areas of engineering. Certainly the spreadsheet approach is helpful, but it is limited for future applications. A strategy of engineering management is to have the accuracy, reliability, and timeliness of the analysis and estimates be professional and conducted with modern tools. So, what does the future hold for product estimating?

8.9.1 Designing, Estimating, and Producing in a Global Economy

The theme for future engineering and management strategy, so it seems to us, is to prepare for designing and cost estimating in a global and domestic economy without borders. Trade treaties indicate that seamless movements of product across international borders will expand. Today, virtually all manufacturing can be contractual, with suppliers available throughout most of the world. This is also true for design services.

[5]Phillip Ostwald, *American Machinist Manufacturing Cost Estimating Guide* (New York: McGraw-Hill, 1982), 382 pp.
[6]Phillip Ostwald, *AM Cost Estimator Software*, (New York: McGraw-Hill Publications, 1985).

TABLE 8.9 Future Directions for Product Cost Analysis and Estimating

Future Directions	Future Directions
Web browser based, collaborative system providing real time access to cost data across the enterprise with application service provider	Part estimating along with a manufacturing process tied to the BOM and specification
Supplier-design collaboration	People coalitions across the enterprise have role-specific access to real time data
Commodity team collaboration	Forecasting of labor rates, material costs domestically, internationally
Contract manufacturing assessment	Configurable user access
Integration with data from enterprise information systems such as resource planning, manufacturing execution systems, supply chain, and product data management	Product cost visibility and forecasting from initial concept through end of life, life cycle costing
Make versus buy analysis	BOM cost roll-up and cost tracking
Cost saving/increase tracking	Product portfolio optimization by standardization on components and sub-assemblies
Blended cost from multiple suppliers	Process times, labor rates under preliminary or detail levels
Supplier access for quoting	Overhead extensions for productive hour cost or activity based costing principles
CAD integrations from RFP to final design release	Specification management
Service costs of warranty, repairs	Capital-equipment total cost of ownership

If small companies, who serve in a supplier role, choose to concentrate only on the domestic market, then they will discover that buyers hunt for good service, on-time delivery, and Internet site convenience in addition to these engineered products. Certainly, for the company who chooses to focus on the domestic market, there is international competition. In contrast, large companies, which stress international markets for their goods, lose the advantage of having customers receive and hold the physical product and gaining feedback from a local experience.

Consolidation of information from international and domestic sources into an application service provider (ASP) and accessing the application through a Web browser is a likely future trend. Information of many kinds will be provided (see Table 8.9).

For these and many other reasons, trends suggest that a full-bodied menu of computing services is the future—where designing, estimating, and manufacturing will be global.

SUMMARY

In this chapter we constructed the estimating system for a complete product. The product is estimated with various techniques that were introduced in earlier chapters. Direct costs are assembled around a product structure and indirect and overhead costs are added. The impact of learning theory was applied to the situations of engineering change orders, follow-on procurement, and break-even analysis. Finally, methods for determining the price and establishing contracts between the customer and supplier were addressed. Concurrent engineering procedures work on the product to drive the cost to lower levels and improve quality. The material covered in this and preceding chapters give the methods for obtaining an estimate for total product cost, large or small, high or low quantity, planned or in production.

QUESTIONS FOR DISCUSSION

1. Define the following terms:

Activity-based costing	FPI
BOM	Full cost
Break even	Make and design anytime and anywhere
Concurrent engineering	Markup on cost
Contingencies	Price
Costed bill of material tree	Product cost
ECO	Productive hour costing
Engineering cost	Retro-fit
Envelope curve	RFE
Fixed price arrangements	RFQ

2. What are the ways a product estimate may be used? List others not included in the chapter.
3. What kinds of information are required for a product estimate? Which are internally determined by the design engineer?
4. In the sharing arrangement for FPI, what are the pros and cons of a sharing plan such as 90/10 versus 50/50 to the contractor and owner for cost reductions below estimated cost?
5. Describe the purpose of learning for a cost estimate. In addition to production quantity, what other things show learning correlation? Where in the cost-estimating process is the learning theory applied?
6. How may the parameters of the learning theory be defined? What factors contribute to this definition?
7. Segregate the methods of product costing. Outline those differences.
8. Indicate the complexities of pricing. Rank several objectives of a pricing policy.
9. When is the full-cost method of pricing appropriate? What are its failings? Devise a new model that uses the full-cost and investment method of pricing.
10. Contrast fixed price versus cost plus types of contracts for (a) a 10 MW gas turbine generator (very low volume, semistandard product) and (b) prototype manufacture of an ultra high vacuum chamber environmental test unit.
11. When would an owner prefer an FPI contract? Why would a supplier desire an FPI arrangement?
12. Each sketch of Figure Q12 is defined by a horizontal axis of time and a vertical axis of dollars per unit. Market and engineering studies have calculated a price and a cost estimating relationship. Indicate the type of mathematical functions and their parameters that give these curves. Discuss the engineering and business implications of the product estimate that leads to these curves.

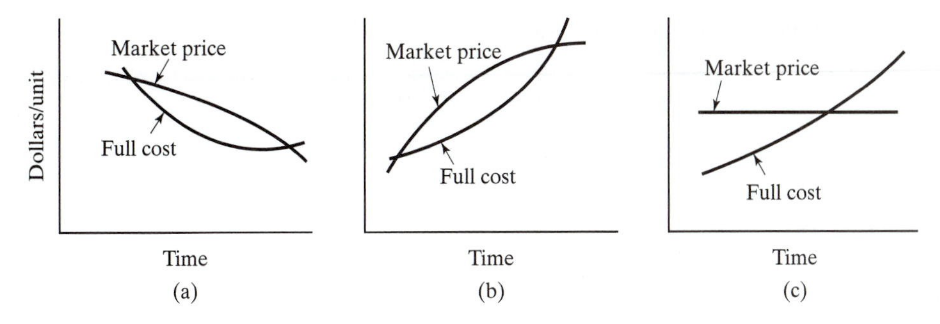

Figure Q12

PROBLEMS

8.1 Find the price for the product if markup is 50%. Use Figure P8.1.

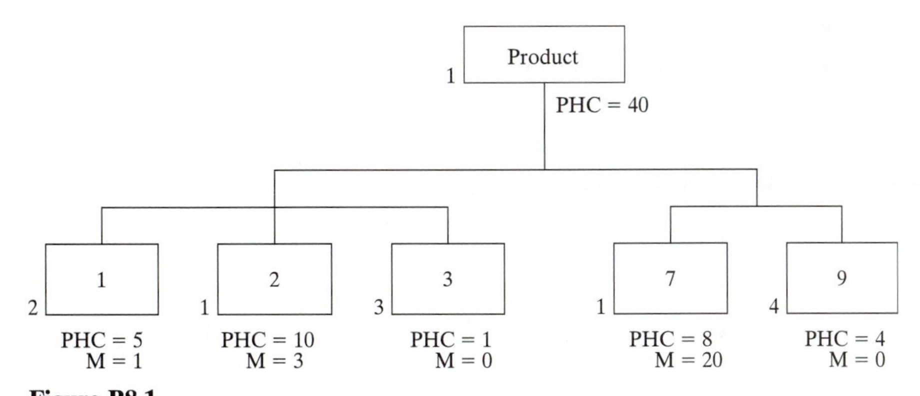

Figure P8.1

8.2 (a) If the markup on sales is 20% and cost is \$160, find the selling price.
(b) If the margin on sales is 30%, find the markup percentage.

8.3 Engineering has found the following costs per unit:

	Cost
Direct labor	\$0.10
Direct material	0.20
Overhead	0.06

(a) Find the price if the markup of full cost is 50%.
(b) If the markup of direct labor and material is 100%, then determine the price for an objective of incremental cost of additional units.
(c) Disregard the cost of materials for a markup of 200% on conversion cost. Find the price.
(d) Find the price if the margin of sales is 50%.

8.4 An estimate is as follows: direct labor, \$4.50 per unit; direct material, \$8.00 per unit; and overhead, \$12.50 per unit. Find the price for a margin on sales as 25%.

8.5 A product is priced at three levels:

Unit Price	Projected Units
$4	40,000
5	30,000
6	15,000

The full variable cost is $3. For each unit price find the following:

(a) Contribution and contribution percentage
(b) Total revenue and contribution
(c) Which price level is preferred? (*Hint*: Find the maximum contribution.)

8.6 A revolutionary product has been developed. Novelty will sustain early sales, but long-range estimates have concluded that competing designs with a constant retail price of $30 will provide a floor. An initial price of $70 with a decay of 0.25 is expected. How many years will elapse until the new product intersects the competing price of $30 per unit?

8.7 A price strategy has the following information: initial price = $10, decay = 0.25, floor price = $6, and decay = 0.15.

(a) By using those estimates, at what year will there be an intersection of the selling price with competitive floor price?
(b) At a floor price of $3.50, what is the opportunity margin? At what year will this level be reached?
(c) Discuss what your pricing policy should be after the intersection of prices.

8.8 Product PN 8871 has the following operational times:

Operation	Setup	Hours/100 Units	PHC
Shear	0.1	0.0048	$36.50
Punch press	0.4	0.150	49.00
Punch press	0.4	0.150	49.00
Tumbler		0.010	26.05
Degreaser	0.1	0.100	29.45

Material unit cost = $0.095. The 2500 quantity is to be shipped in one order. (*Hint*: Assume no learning for this product.) Find the full cost for manufacturing, development, and sales by using the productive hour cost (PHC) method. Find the price if the markup is 20%.

8.9 A single-component product is estimated for cost and price for a lot of 175 units. The operation estimates for this product are as follows:

Operation	Setup (hr)	Cycle (hr/unit)
Shear	0.1	0.001
Pierce	0.5	0.038
Countersink	0.3	0.043
Brake	0.6	0.010
Weld	0.2	0.035
Grind	0.1	0.020
Deburr	0.4	0.006

Average shop labor is \$17.25 per hour. Factory overhead is 75% of direct labor, material cost and overhead per unit is \$0.075, and G&A is 25% of the cost of goods manufactured. There is no selling or contingency cost. The markup on full cost is 20%.

(a) Develop a system of cost estimating, and find the unit and lot cost and price.
(b) Which operation is most costly? If a 50% reduction can be achieved in this operation, what addition to profit would be expected at the same selling price? (*Hint*: Assume that profit is increased only by direct-labor cost reduction.)

8.10 A universal projection screen mount, Figure P8.10, is to be estimated. This product suspends the screen from the ceiling and eliminates the tripod support. The device permits rotation of

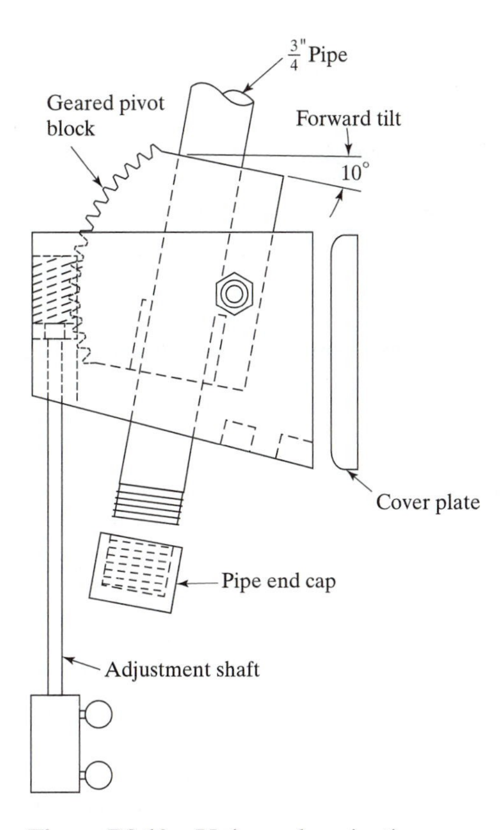

Figure P8.10 Universal projection screen mount.

360° and tilting of a maximum of 30°. Vendors supply parts to the company. The bill of material below summarizes cost facts. Provide solutions using a spreadsheet program.

(a) Find the direct costs for labor and material.
(b) Define the cost of manufacturing as labor and material. Let the overhead costs for the plant be at 200% of the cost of manufacturing. Find the full cost.
(c) If markup is 25% of full cost, find the profit and price.
(d) Repeat part (c) if the margin is 25% of the sales price.

Number Required				Part Name	Material Cost ($/unit)	Labor Estimate (hr/unit)	Gross Hourly Cost ($/hr)
1				Complete assembly		0.01	20.05
	1			Gear pivot assembly		0.03	20.05
		1		Pivot, geared	2.25	0.005	18.75
		1		Bolt	0.25	0.004	16.00
		2		Nut	0.10	0.002	15.00
	1			Adjustment shaft	0.75	0.01	20.05
		1		Worm	1.15	0.01	18.75
			1	Shaft	0.60	0.01	16.00
		1		Connector	0.50	0.02	17.75
			2	Thumbscrews	0.20	0.01	15.00
			2	Rod	0.40	0.03	16.00
	1			Housing		0.02	20.05
		1		Cover plate	0.60	0.005	16.00
		2		Side plate	0.30	0.015	16.00
		1		End plate	1.40	0.010	16.00
		1		Top plate	1.75	0.010	16.00
		1		Bottom plate	1.65	0.020	16.00

8.11 A vendor is considering a fixed-price contract where the productive hour cost rates will be quoted. The vendor uses a markup of sales value method of pricing. Markup rates depend on the machine center.

Machine Center	PHC ($)	Markup (%)	Estimate (hr)
Light machining	39.16	10	80
Heave machining	90.98	20	40
Assembly	32.19	15	20
Finishing	42.33	5	15

(a) What are the fixed-price quotations for each center?
(b) Given the customer's estimate for the job times, what is the quote for the job?

8.12 For the engineering change proposal A, the cost at unit 14 is $8000 and the learning rate is 80%. For EC proposal B, the unit 14 cost is $6000 and the learning rate is 90%. Which proposal is best at unit 15 or 30?

8.13 If the two-year bid in Figure 8.5(b) is estimated to decline to 52 hr, what is the slope? What is the reduction in total hours for the two-year bid as compared to 53 hr?

8.14 The actual and follow-on learning equation is $T_u = 1000\ N^{-0.322}$ An ECO is planned for $N = 4$, and the slope parameter $\Phi = 75\%$ is estimated. Find the total hour effect for units 4 and 5. Retrofit is not required. Repeat for $\Phi = 85\%$. (*Hint*: Units 1, 2, and 3 are unchanged.)

8.15 A company that produces engine-generator sets is planning follow-on bidding with the learning model. At the 980th unit, a 5-kW product required 68 hr of direct labor with a slope of 95% learning.

(a) Follow-on will involve 2460 more units. Extend the constructed line to include this second shipment. What is the value of unit 3440?

(b) Engineering now believes that the 5-kW product must reduce to 56.3 hr at the 3440th unit. Find the new slope and the lot time for the second shipment. What is the average time per unit for the follow-on lot?

8.16 The actual and follow-on learning equation is $T_u = 1000\ N^{-0.322}$. At $N = 4$, an engineering change is planned with a slope of 85%. At $N = 4$, an additional 50 hr will be necessary. The slope for the retrofit portion is 90%. Find the total time for the ECO for units 1–5.

8.17 The actual and projected hours for a design are $T_u = 35{,}000\ N^{-0.322}$. An ECO is planned at the 50th unit. Retrofit is scheduled for units 1–50 and units 50–100 supplied according to ECO configuration. From a similar experience, 90% and 85% are estimated as learning before and subsequent to the ECO action, and 150 hr are estimated at the ECO point. Find the number of hours resulting from the ECO for the ECO and projected units 50 to 100.

8.18 A company uses the learning approach to estimate its product.

	First Unit Cost Estimate		
Cost Item	Fixed	Variable	Learning (%)
Direct labor	—	450 hr	90
Direct material	—	$2,000	95
Manufacturing support	$2,000	$50	95

Labor costs $23 per hr, based on direct labor and manufacturing support, fixed and miscellaneous overhead is 100%, distribution and administrative overhead is 20%, and selling costs are at 2%. The company uses the full-cost method with profit at 10%. Construct a method of estimating. Find the average and total price for 100 units.

8.19 (a) What is the supplier's profit for a 75/25 sharing arrangement if there is a $135,000 over run where estimated cost is $10 million? Use the information in the example in Section 8.7.2.

(b) What is the supplier's profit if there is a $150,000 under run?

8.20 (a) Find the cost plus profit for a cost reduction of $80,000 if the sharing arrangement is 80/20. Consider the side of the supplier. Use the information in the example in Section 8.7.2.

(b) What is the cost plus profit if there is a cost overrun of $70,000 and the sharing arrangement is 75/25?

8.21 A supplier for a major system estimates and negotiates a fixed price incentive contract:

Estimated cost	$10,000,000
Target profit	850,000
Target price	10,850,000
Price ceiling	11,500,000

There is a 70/30 sharing arrangement below the ceiling and a 0/100 above the ceiling.

(a) Find the profit for a $250,000 cost reduction and for a $250,000 overrun.
(b) Find the profit for a $400,000 cost reduction and for a $400,000 overrun.
(c) What is the profit/loss if the supplier experiences a 10% cost overrun?

8.22 A print shop implements an ABC system. Printing activity costs $0.023 per page for ink and maintenance of equipment (paper is a direct material cost and is estimated separately.) Bindery costs are $1.42 per "book" for bindings, glue, etc. (covers are a direct cost.) (*Hint*: Labor for printing and binding are a direct hourly cost and are estimated separately.) Finally, the delivery cost is $20 per order.

The Woodland Trading Company orders 400 copies of a 50-page catalog. On the same day, Kelso Bridge Works orders 50 copies of a 400-page manual.

Activity	Cost Driver	Cost per Unit of Activity ($)
Woodland Trading Company		
Printing	No. of pages	0.023
Binding	No. of copies	1.42
Delivery	No. of orders	20.00
Kelso Bridge Works		
Printing	No. of pages	0.023
Binding	No. of copies	1.42
Delivery	No. of orders	20.00

Compare the indirect and overhead costs for the two orders. (*Hint*: To complete the estimate for these jobs, the direct labor and material costs would be included, but that is not required.)

CHALLENGE PROBLEMS

8.23 Electronic components must be protected during manufacture. For example, spurious electric charges can destroy microelectronic chips. To avoid those problems, a tote-box liner is designed that uses a static-free plastic material. An assembly sketch is given in Figure P8.23. A supplier has provided a quotation for the per unit costs of items 3 and 4 with quantity discounts:

Quantity	Item 3	Item 4
500	$4.10	$5.22
1000	3.51	4.47
2500	3.28	4.17
Material	0.28	0.33

Items 1, 2, and 5, assembly, and packaging are estimated by the company. Additional cost factors are as follows:

Item	Unit Material Cost	Tooling Cost	Direct Setup (hr)	Labor (hr/unit)	Gross Hourly Cost
1	$1.22	$3500	6.5	0.025	$19
2	0.92	1750	4.0	0.013	19
5	0.67	1200	3.0	0.012	19
Assembly			0.5	0.005	19
Packaging	0.2		0.5	0.005	19

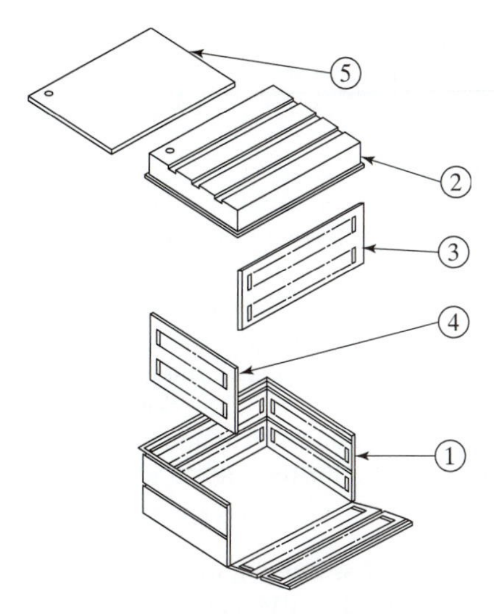

Figure P8.23 Tote-box liner for electronic components.

General administrative and sales costs are at 100% of the company's direct labor costs. Determine the total and unit costs for quantities of 500, 1000, and 2500.

8.24 A product consists of three manufactured parts with one operation each. Material for each part needs to be ordered, received, and inspected. The parts are assembled into the final product, inspected, and packed. Order processing, scheduling, and accounting are required for each batch. R&D and engineering costs are apportioned to each unit produced. An order for 250 of the product is received. A summary of the ABC cost estimate for the batch is shown.

Activity	Cost Basis	Hourly Rate	Cost/ Activity
Log order	1/batch		$250.00
Schedule	1/batch		176.00
Order material	3/batch		125.38
Receive material	3/batch		121.41
Inspect material	3/batch		287.43
Setup Part A	4 hr/batch	$18.65	
Setup Part B	7.35 hr/batch	23.88	
Setup Part C	8.50 hr/batch	25.43	
Fab Part A	0.35 hr/part	27.94	
Fab Part B	0.86 hr/part	29.11	
Fab Part C	0.51 hr/part	31.86	
Assemble	0.14 hr/part	30.12	
Inspect Assembly	0.08 hr/part	30.46	
Packing	0.3 hr/part	26.34	
Accounting	1/batch		151.78
R & D	1/part		1.75
Engineering	1/part		4.14
Material Cost	1/batch		1578.00

(a) Find the total ABC cost per unit.
(b) What is the total cost of the order?
(c) Which item(s) should be considered for cost reduction studies based on the cost per unit?

8.25 (a) Engineering Services has information on the average time for the second and the 35th units as 125 and 90 hr. What is the average time for the 300th unit? Find the cumulative time.
(b) A product is estimated with a learning equation as $T_u = 1000 N^{-0.322}$. After the initial design, an engineering change is required that will be effective for the fourth and any follow-on units, with an estimated learning slope of 85%. It is estimated that 50 hr will be necessary for retrofit at $N = 3$. The slope for the retrofit line will be 90% for units 1, 2, and 3. Unit 4 will require 47.9 hr additional to the original product-learning equation. Find the total time for the retrofit and new production owing to the ECO for units 1–5.

8.26 An estimate finds a cost of $5 million, with a profit set at $600,000 and a maximum and minimum fee of $800,000 and $450,000. A sharing arrangement of 80/20 is negotiated. This arrangement is known as a cost plus incentive fee (CPIF).

(a) Find the supplier's profit and the buyer's cost for a final cost of 15% above the estimate.
(b) Explain what happens if the final cost is 10% below the estimate.

8.27 A company's experience in manufacturing custom gear transmissions indicates that the first unit will consume 100 hr of fabrication and assembly time. A learning rate of 75% is anticipated. A contract is being estimated in which 40 units will be supplied. Labor cost of $24/hr covers labor time, and indirect manufacturing expense and material expenses should be $25/unit. (*Hint*: Material and indirect labor are not subject to learning.)

If a 20% add-on for profit is historically applied, what should the bid price be? How much must the time estimates be off to consume the profit?

8.28 A prototype unit is constructed with 450 hr of work recorded on job tickets. The average direct labor hourly rate is $24, and the overhead rate on the basis of labor is 100%. Raw material costs charged to the job are $2000 for the first unit. After a period for design changes, the company will build units 2–1000 for sale. They anticipate a learning rate of 90% for direct labor and 95% for material. What are the total estimated costs for this product, average, and the 1000th-unit cost?

8.29 A multinational company is considering a "twin-plant" project where two plants are adjacent in the sense of low-cost transportation but are separated by a national border, and the material is nonbulky and transportation costs both ways are negligible on a per unit basis. On the U.S. side, labor rates and productivity conform to typical values, but across the border labor rates vary from one-tenth to three-quarters as much.

The U.S. plant will initially process the material, transport the semifinished material to the twin plant, where the foreign plant adds labor value to the product and returns the completed product to the U.S. plant. The U.S. manufacturing unit cost for the semiprocessed material is $1.

At the border, custom fees amount to 5% and return custom fees are 5%. In the foreign plant the labor work value is $0.25 for an Equivalent $1.25 per unit of U.S. work. An ad valorem of 20% of value added is assessed against nondomestic labor.[7]

[7]An ad valorem is a phrase applied to duties levied on goods as a percentage of their value.

Once back in the United States, a 35% contribution is added to give a final price. Find the product price for a twin-plant and a single-plant operation. (*Hint*: Set up a side-by-side comparison of a single U.S. plant versus foreign twin plant. Use manufacturing costs and apply the rates. Start with the cost for the semiprocessed material of $1. Use Equation (8.16) to find price with the contribution method.)

Discuss the implications of an international trade policy that sends goods for intermediate processing to other countries. What must the labor cost of the non-U.S. labor be to make the decision indifferent between the twin plants?

8.30 The company has produced 50 units so far, and the learning rate is calculated as $\Phi = 90\%$, where the first unit time is 750 hr. Thus $T_u = 750\,N^{-0.152}$. An ECO requires retrofitting of the first 50 units. From experience the learning rate for the retrofit is $\Phi = 95\%$, and the retrofit requires 25 hr at $N = 50$. In addition, the ECO made manufacturing improvements that will save 10 hr at $N = 51$. For the remaining 200 units (units 51 through 250), $\Phi = 85\%$. This information is shown graphically by Figure P8.30.

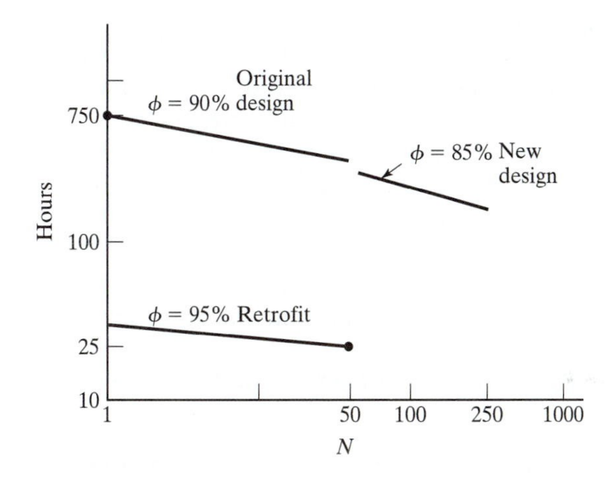

Figure P8.30

Find the time of the production run with the ECO (which includes the original production time of the first 50 units), the retrofit to those units, and the time of the remaining 200 units.

PRACTICAL APPLICATION

Understanding product estimating is important, and many opportunities exist for a practical application using the principles of this chapter. Consider these choices.

A. Business between a buyer and seller, who are involved with engineering products, is a formal arrangement, which is concluded with a contract. But before this point is reached, there is the opening procedure, i.e., a RFQ from a customer to a supplier. The RFQ process is detailed and oftentimes complicated.

Between those two epochs, RFQ and the contract, are a number of steps, perhaps only a few for the smaller jobs, but many for the more complex engineering work. Your assignment is to understand this process and to form a team to handle this project.

First, find a cooperative company willing to provide these details. When your inquiry is made to the company, reassure them that you are interested in the process, and that you do not wish to know any confidential information. A supplier and a buyer are secretive about cost numbers and proprietary arrangements.

To better understand their practices, prepare your questions before the interviews. For example, does the RFQ package include drawings, BOM, and other specifications? Is product scheduling all at once or on a just-in-time, and how is delivery accomplished? What about first article inspection? Will the RFQ be delivered to more than one supplier? Is there open or closed competition? In your judgment, are there any hidden agendas or prior agreements or past understandings that anneal the arrangements?

The report should include flowchart, obligations and legal promises, and statements regarding penalties or bonuses for performance, shipping requirements, quality inspection criteria, and special engineering information. Your instructor will provide additional suggestions.

B. The opportunity for a concurrent engineering application can follow interesting paths. Frequently, students are involved with a senior design project in another class. The principles of concurrent engineering can be a co-course project. Another possibility is to study some simple commercial product. A choice could be "bird feeders," and product information could be found in catalogs, magazines, Internet, and bird stores is available. Bird feeders, for instance, are not well engineered. Your instructor will give additional instructions.

When the project report is concluded, remember to send a thank-you letter to your helping company. They will appreciate seeing a copy of the report.

CASE STUDY: UNIJUNCTION ELECTRONIC METRONOME

Ray Enterprises is releasing a new design for an electronic metronome and plans to market it through a well-established chain of catalog houses. President Ray says that the key to the quality of the product lies in its new electronic circuitry, shown in Figure C8.1. The bill of material is given by the parts list, which appears on the circuit diagram.

It is assumed that increased sales will lower the potential market price. Marketing and sales provide the following price-volume information:

Annual Volume	Potential Catalog Price
100,000	$100.00
200,000	91.00

Standards for production are determined from operation estimates as follows:

	Hours/100 units	Dollars/hr
Machine shop	7.00	25.00
Assembly	6.25	15.00
Inspection	2.80	16.20

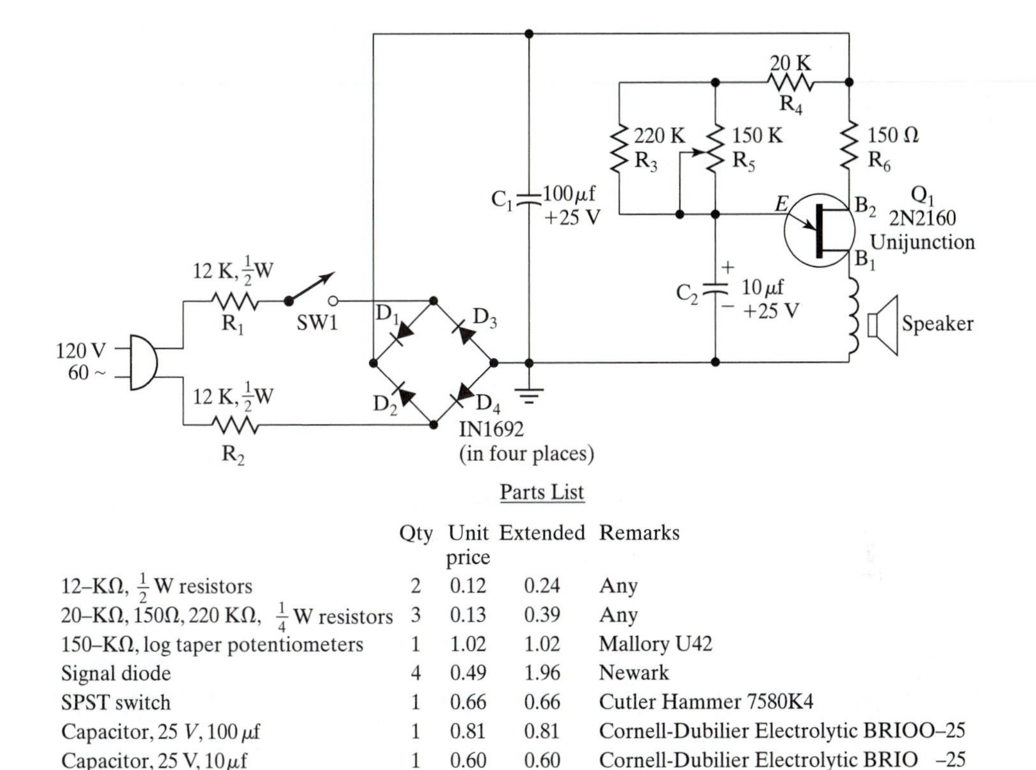

Parts List

	Qty	Unit price	Extended	Remarks
12–KΩ, $\frac{1}{2}$ W resistors	2	0.12	0.24	Any
20–KΩ, 150Ω, 220 KΩ, $\frac{1}{4}$ W resistors	3	0.13	0.39	Any
150–KΩ, log taper potentiometers	1	1.02	1.02	Mallory U42
Signal diode	4	0.49	1.96	Newark
SPST switch	1	0.66	0.66	Cutler Hammer 7580K4
Capacitor, 25 *V*, 100 μf	1	0.81	0.81	Cornell-Dubilier Electrolytic BRIOO–25
Capacitor, 25 V, 10μf	1	0.60	0.60	Cornell-Dubilier Electrolytic BRIO –25
Unijunction transistor, 2N2160	1	1.49	1.49	Allied
Speaker, 3.2Ω, 2 W	1	1.85	1.85	Quam 30A05

Figure C8.1 Unijunction electronic metronome.

The manufacturing burden is 100% of direct labor cost. New investment requires $20,000 in tooling and the amortization policy for tooling is two years. Engineering and marketing burden costs amount to 50% of direct labor costs.

A practice of adding 25% for profit to all costs has been followed. The distributor's charge for a product of this sort is usually 40% less than list price.

Construct a cost estimate and determine a price, comparing it to the potential catalog price. In view of the preliminary marketing data on price and volume and estimating data on cost, what recommendations on price and cost can you make to President Ray? Should this product be made and sold?

Chapter 9

Cost Analysis

This chapter presents methods for applying cost analysis to engineering designs. Unfortunately, strict procedures cannot be set out beforehand in some recipe fashion that will standardize the cost analysis approach. If that were possible, then by learning those procedures and adhering to them, we would be confident of the result. Although analysis is special for each design, there are, however, principles that are helpful, and the student will find them worth learning.

Cost analysis is performed during the engineering and manufacturing "tradeoff" period. These events happen early in important engineering products—perhaps 60 to 80% of the consequential cost/tradeoffs are made before and during design. Even older products can be reengineered and tradeoffs found. Figure 9.1 illustrates the major events that influence cost and profit.

9.1 FIRST PRINCIPLES FOR TRADEOFF STUDIES

An important engineering task is the cost analysis for "tradeoffs." A typical design tradeoff is "increase the diameter of the gas engine cylinder for improved performance," as against another alternative of "a hotter spark with the existing diameter and engine." What is the cost of either alternative? How do the technology differences influence the performance of the engine? The cost analysis that answers these questions is a large undertaking. It may be a team effort, which depends on the scope of the product definition.

Engineering determines the *comparative economy* of the alternative choices and makes the decisions in the tradeoff. As an aside, we employ the term "economy" in its original sense, meaning *thrift*. Software, company reference books, standards and codes, design, and engineering skills are the underpinnings for these comparisons.

This tradeoff activity affects the profitability of the product. The following are typical steps for tradeoff studies:

- Search for design and manufacturing alternatives.
- Determine the cost estimates, profit and loss, and cash flow statements.

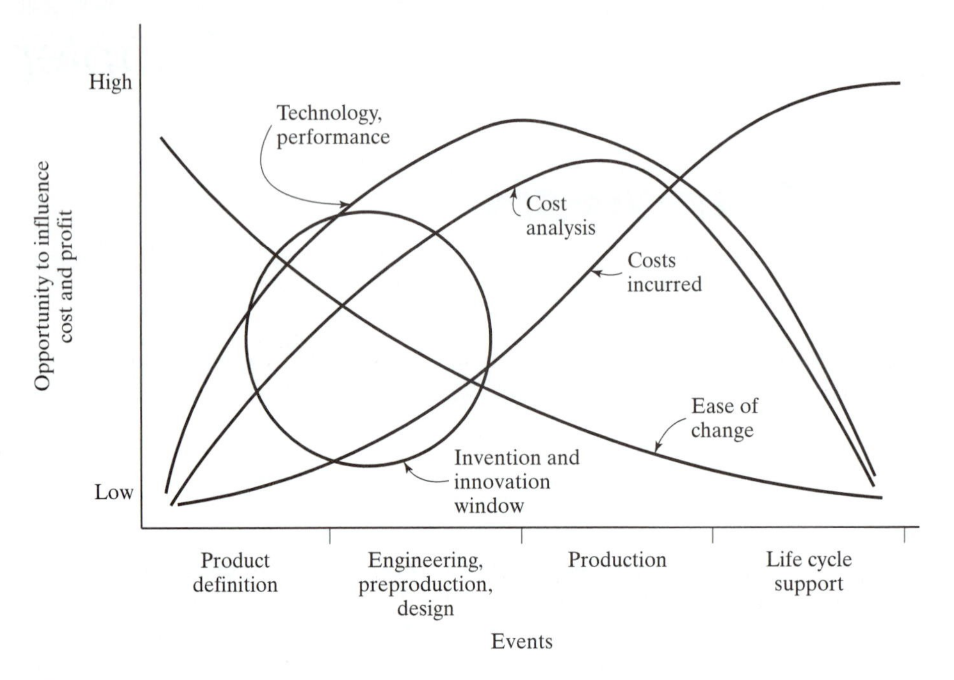

Figure 9.1 Influence of product life-cycle events.

- Apply the analytical methods in comparative economy.
- Consider the constraints on the cost analysis.
- Understand the irreducible factors.
- Select the design.

The beginning of cost analysis is the *search for design and manufacturing alternatives*.[1] The need for imaginamachina in the forming of alternatives cannot be overstated. No matter how good an analysis and selection among two or more alternatives may be, if another and unidentified alternative is superior, then the selection is suboptimal.

An engineer is expected to consider a broad range of alternatives while remaining within the realm of practicality. After a reasonably exhaustive list of alternatives is formed, there is a pruning of those that are less feasible within the constraints of time, effort, and money. Preliminary estimating methods may abridge the list to a few that are more promising. The converse is also true: as the analysis continues, opportunities may present themselves that were not known at the start.

This search for alternatives goes on during the estimating phases. In large-scale projects, for instance, such as a major change to a gas combustion engine, there are many estimates that follow the technical events, as illustrated in Figure 9.1. Should an insertable ceramic liner be used for the enlarged engine cylinder, or can the existing

[1]A wide variety of resources are available on the Internet. A useful one for patent searches involving invention and design in engineering and manufacturing is www.USPTO.gov.

gray cast-iron engine block be bored to the newer diameter, 135.21/135.19 mm? There are thousands of questions—literally—that might be asked.

To choose between two cylinder material alternatives for an improved engine by listing the things we value—for example, satisfaction, dependability, and lack of pollution—is not very helpful. Meaningful cost analysis cannot be conducted on such lofty levels. More specific measures for selection between the alternatives are desirable, such as cost, heat-loss cost through the wall annulus, waste-heat recovery cost, and equal-marginal fuel cost. There is a greater chance for successful analysis using narrow measures of *design to cost.*

The word *cost* is meaningless when used alone. An adjective is needed in order to clarify the comparative economy of the design-to-cost alternatives. There are many modifiers that can be used, such as absolute, annual, current, direct labor, discounted, equipment, escalated, first, full, marginal, material, overhead, plant, project, relative, salvage, short or long term—and the list continues. Refer to the index of this book for pages where these and other modifiers are found. Using precise language for the word *cost* aids the understanding of the constraints and conditions on the cost estimate.

These precise modifiers are indicated in the preparation of estimates. Once the cost facts (identified as facts because this information is the best that we have at this time) are available, various analysis methods are used for the cost tradeoffs. Cost-estimating encyclopedias, as described in Chapter 7, are applied for the simpler cases of manufacturing-work tradeoff. For more complex levels of the product, Chapter 8 gives the organization for the collection of the product data. These chapters describe the methods to find the estimates. Estimates are made for the alternative choices.

The principles for doing cost trades do not depend on the magnitude of the dollars. Whether the comparative economics are for several thousand or millions of dollars, the analytical principles are similar. Several analytical methods are presented in this chapter.

An engineer may be unable to convert some aspects of a design into ordinary monetary units. Function, beauty, safety, legal exposure, and quality of life are difficult to evaluate and are called *intangible* or *irreducible* factors. Although nonfundamental units for a scale of measurement or ranking might be forced on the intangibles, the engineer usually believes that it is not worth the effort and that the intangibles are truly inscrutable in terms of ordinary units. In that event, then there are other ways to consider them. If the stakes are not high, then it may be convenient to ignore them. Also, considering that "intangibles" advantageous to some may be disadvantageous to others, a simple listing of the intangibles, with descriptions both pro and con, may suffice. In addition, management may be the court, judge, and jury on the merits of those intangible factors. Perhaps, the prominence of the intangible factors may be so significant that they overshadow the economic comparison.

The step of *selection between the alternatives* is the last one in tradeoff analysis. For example, the recommendation is to use a centrifugal cast thin-wall cylinder liner, and this selection becomes a design feature on the engine working drawings, or a listing on the tabulated material specifications, and the inclusion of other engineers that implement these design-to-cost decisions expands in many ways. Eventually, all of the many selections are confirmed in this *mega* cost analysis, and they become an inseparable part of the business of engineering and manufacturing.

PICTURE LESSON Aerial Locomotion: The First Soar

Library of Congress.

It was in 1901 that Samuel Langley built a gasoline-powered version of his tandem-winged model, the first to propel an aircraft, and launched large unmanned steam-powered models on many successful flights. But in 1903 it was the Wright brothers' *Flyer* that made the first successful manned and powered flight.

The first powered, sustained, and human-piloted flight carried Wilbur Wright 120 feet in 12 seconds. Later the same day, Orville Wright flew 852 feet for 59 seconds. The Wright brothers' success was built on the pioneering work of German engineer Otto Lilienthal with gliders, and Octave Chanute, an American who worked on multiplane gliders. The race to be the first to fly successfully was a worldwide competition with numerous prizes, and pioneers like Langley and Gustave Whitehead were serious contenders. Though the Wright brothers' first historic flight in 1903 is recognized as the true beginning of aviation, the feat was largely ignored at the time. Despite pictures of the Wright's first flight, the British were skeptical and wondered if the claims could be substantiated with a public exhibition of formal testing, to confirm this "aerial locomotion." The photo shows Orville Wright on a cold December day lifting off the earth and for 12 seconds flying less than the length of an airliner.

The engineering and science of aviation soon exploded, and the grim needs of World War I encouraged rapid development in aviation. In 1927 Charles Lindbergh became the first person to make a solo nonstop crossing of the Atlantic and As the aviation industry began to mature, planes such as the B-17 Flying Fortress and DC-3 appeared on the scene during the 1930s. In 1936 the German Focke-Achgelis Fa-61 rotary-winged aircraft was demonstrated. Sir Frank Whittle in England and Hans von Ohain in Germany constructed the first turbojet propulsion engines. The Heinkel He-178 experimental aircraft, powered by Hans von Ohain's centrifugal-flow HeS-3b engine made the world's first turbo-jet powered flight in 1939, and that same year Igor Sikorsky invented the first helicopter.

Perhaps the most astonishing achievements in aviation were at the beginning of the twentieth century, when little was known about flight. This was a time of intuitive creation and experimentation—with wing size and shape, with materials, and with power systems.

Early flight was a risky and daring undertaking. Fliers were called "aeronauts," a title that reflected the dangers they faced. Planes were noisy, low-flying craft made of cloth or wood. The cockpit was completely open, leaving the pilot unprotected from weather and other hazards. Fuel was still unrefined and therefore not always reliable, and crash landings were not uncommon. There were no guidance systems or satellites to warn of storms or other weather hazards.

Perhaps the greatest outcome of this technology has been the expansion of personal horizons. In contemplating what this century of flight will bring, Orville Wright's early prediction seems appropriate. He said, "I cannot answer except to assure you it will be spectacular."

The remainder of this chapter continues the development for important cost analysis procedures. Earlier, techniques were given for doing the cost-estimating procedures. In Chapter 8, the purposes for the cost estimate were for product estimating and the pricing step. But this chapter gives another purpose: doing tradeoffs for alternatives engineering designs.

9.2 CASH FLOW

Cash flow analysis may be compared to a reservoir receiving a stream of water. At certain times more water is received than at other times, and occasionally there may be greater outflow than inflow. In a company money behaves like this metaphor and thus is frequently called a *cash stream*.

The meaningful considerations in cash flow analysis are as follows:

- Income and expenses
- Depreciation
- Taxation
- Inflation and deflation

Even if the cost of the business venture opportunity is small in comparison to the inflow or accumulated cash surplus, a cash flow document is necessary. A small business may require operating capital before manufacturing revenue is received. It would find a cash flow document necessary. Nevertheless, if the manufacturing venture is a big one, the company evaluates its cash position to meet its obligations. It makes "cents" to determine the cash flow analysis. Bankers, venture capitalists, suppliers, owners, and others with a financial stake in the business want the reassurance that a cash flow analysis gives.

In Chapter 4, we considered the profit and loss statement, and the student may want to scan that material again. As one step in tradeoff studies, the finding of estimates and profit and loss statements for product or process alternatives is necessary for competent cash flow analysis.

Profit is the goal, but for a comprehensive analysis the necessities of taxation are frightful—frightful because of the details that a thorough examination requires. The

IRS Code is large, and specialists are necessary,[2] but the engineering student needs to have an awareness of how cost analysis is influenced for the resolution of engineering product and project studies. The next section gives an overview of how engineering cost analysis is influenced.

9.2.1 Taxation Effects on Cash Flow

We study three important types of taxes:

- Income taxes
- Property taxes
- Sales taxes

Income taxes are taxes on pretax income of the firm in the course of ordinary operations. Income taxes are also levied on gains on the disposal of capital property. Income taxes are usually the most significant type of tax to consider in design-to-cost analysis.

Tax codes are complex, and their minutiae are beyond the needs of this book. But one needs to recognize that income taxes paid are just another type of expense, while income taxes saved (through business deductions, expense, or direct tax credit, and so forth) are similar to other kinds of reduced expenses, such as savings.

Property taxes are based on the valuation of property owned, such as land, equipment, buildings, inventory, and so forth, and the established tax rates. They do not vary with income, and are usually much lower in amount. Still, they can be significant in terms of money or impact.

Sales taxes are taxes imposed on product or material sales, usually at the retail level. They are relevant in cost analysis only to the extent that they add to the cost of the purchased direct material or equipment.

At the end of each tax period (usually a year, but more frequent periods may be necessary) a business must calculate its taxable income before finding the profit or loss on ordinary business operations. An approximate word relationship showing this maneuver is given as

$$\begin{aligned}\text{Taxable income} = {} & \text{gross income} \\ & - \text{all expenses (except capital expenditures)} \\ & - \text{depreciation deductions}\end{aligned} \tag{9.1}$$

Taxable income is often referred to as gross income *before* taxes, and after income taxes are deducted, the remainder is called net income *after* taxes. There are two types of income taxes for computation purposes: ordinary income (and losses) and capital gains (and losses). We do not discuss capital gains or losses in this book.

Ordinary income is the net income before taxes that results from routine manufacturing business. For federal income taxes purposes, all ordinary income adds to taxable income and is subjected to a graduated scale, which has higher rates for higher taxable income. Table 9.1 presents an example of a corporate tax table.

[2]As a small glimpse of the magnitude of understanding required for tax work, the student may want to visit www.IRS.Gov.

TABLE 9.1 Illustrative Corporate Federal Income Tax Schedule

If Taxable Income is		The tax is	of the Amount Over
Over	But Not Over		
$0	$50,000	15%	0
50,000	75,000	$7,500 + 25%	50,000
75,000	100,000	13,750 + 34%	75,000
100,000	335,000	22,500 + 39%	100,000
335,000	10,000,000	113,900 + 34%	335,000
10,000,000	15,000,000	3,400,000 + 35%	10,000,000
15,000,000	18,333,333	5,150,000 + 38%	15,000,000
18,333,333	—	6,416,667 + 35%	0

Example: Finding the Income Taxes for a Corporation

Suppose that in the last fiscal year, a business had a gross income of $5,527,000, expenses (excluding capital expenses) of $3,290,000, and a depreciation of $1,650,000. Taxable income = 5,527,000 − 3,290,000 − 1,650,000 = $587,000. The federal income taxes on this amount are demonstrated in Table 9.2.

TABLE 9.2 Calculation of Federal Taxes on Income of $587,000

	Taxable Income	Income Taxes
Income taxes = 15% of first	$50,000	$7,500
+25% of next	25,000	6,250
+34% of next	25,000	8,500
+39% of next	235,000	91,650
+34% of remaining	252,000	85,680
Total	$587,000	$199,580

Income taxes are collected by the federal and state governments. While there is similarity of the state practices to the federal government, there is variation in the tax rates. State tax rates are much lower than federal rates but remain significant. The net effect of taxes is to make the profitability and attractiveness of manufacturing more risky.[3] The corporate tax rates for the states typically range from $2\frac{1}{2}\%$ to 10%, and 5% is a rough average. The state and federal income tax rates can be consolidated with the following approximation:

$$t \cong \text{federal rate} + \text{state rate} - (\text{federal rate})(\text{state rate}) \quad (9.2)$$

where $t \cong$ effective corporate income tax rate on increments of taxable income of importance to engineering economy studies, decimal

[3]Taxes are important for a democratic society and necessary for its continuance to provide a broad range of needs and to protect its citizens. The tax rate is a lively subject for debate, but it is not discussed in this book.

$$\text{federal rate} \cong \text{approximation of corporate tax rate, decimal}$$
$$\text{state rate} \cong \text{approximation of corporate tax rate, decimal}$$

For instance, if a corporation has a federal tax rate of 34% and a state rate of 6%, then the consolidated rate $t \cong 38\%$. Federal taxes allow a deduction for payment of state taxes from gross income, though the reverse is not true.

9.2.2 Inflation or Deflation Effects on Cash Flow

Price *inflation* or *deflation* is a general increase or decrease in the prices paid for materials, labor, subcontracts, and other goods and services, and it affects the purchasing power of the monetary unit, the dollar. When inflation occurs, the purchasing power of the dollar decreases, and in the case of deflation it increases.

The frequency of inflation is much more common. The last general decrease in the cost of labor occurred in 1931 in the United States. In contrast, deflation is seen with imported materials and with improved products. Inflation and deflation are charted with indexes (see Section 5.4). The student will want to review that material.

There are three basic factors of inflation-deflation that influence the cash flow model:

- Real dollars
- Actual dollars
- Inflation or deflation rate

Real dollars are expressed in terms of the same purchasing power relative to a particular benchmark time, or base year. For example, if the future material, labor, and subcontract prices are changing rapidly, they are estimated in real dollars relative to some base year to provide a consistent floor for comparison. Sometimes real dollars are called *constant dollars*. This maneuver eliminates inflation effects by converting all cash flows to money units that have constant purchasing power. This approach is more suitable for before-tax analysis and when all cash flow components inflate at a uniform rate.

Actual dollars are the actual amounts received or spent and not adjusted for inflationary factors. For instance, engineers anticipate their salaries two years hence in terms of actual dollars. The purchasing power of an actual dollar, at the time the cash flow occurs, includes the effect of price inflation or deflation. Sometimes actual dollars are referred to as *then-current dollars* or *escalated* or *future dollars*. These money units are actually exchanged at the time of each transaction. The actual-dollar approach is generally easier to understand and apply, and it is more versatile than the real-dollar method.

Naturally, there are issues with either method. The account depreciation is a non-cash expense that depends on the initial amount of the capital asset and the method of depreciation. The capital asset money that is spent is at a certain point in time, and to inflate the depreciation value for future increases is somewhat artificial. Furthermore, it is probably incorrect to assume that all business money transactions will inflate at the same rate, if the money is declared as real.

The inflation or deflation rate is a metric of the change in the purchasing power of a dollar during a specified period of time. This metric is known as an *index*, and it is a ratio of a market basket of goods and services over extended periods of time relative to a benchmark period. Indexes such as the Consumer Price Index (CPI) and the Producers Price Index (PPI) were discussed in Section 5.4. These inflation or deflation rates are developed by the Bureau of Labor Statistics.[4] The rates are dynamic and are influenced by many factors.

The relationship between actual dollars and real dollars is given as

$$D_r = D_a\left(\frac{1}{1+f}\right)^{n-k} \tag{9.3}$$

where D_r = real dollars at a point in time, dollars

D_a = actual dollars as of the time it occurs, dollars

f = inflation or deflation rate, decimal

n = point in time, period, and usually years

k = base time period used to define the purchasing power of the real dollar, usually 0.

This relationship may be used to convert actual dollars into real dollars, or vice-versa. Further, it assumes a constant f, which is not usually true, even over a few periods. While k is a general index of time and can be any period, for tradeoff studies between competing designs, it is usually zero.

Example: Converting Real Dollars into Actual Dollars

Consider the construction, manufacturing, and operation of a plant to cast thin-wall cylinder-liner products in the diameter range of 35 to 210 mm. The estimated values are considered "real." Inflation is expected to be 5.3% per year for this period. Actual values are calculated in Table 9.3.

TABLE 9.3 Conversion of Real Dollars into Actual Dollars for Construction and Operation of Ceramic Cylinder-liner Product Plant

End of Year, n	Real Dollars	$(1.053)^n$	Actual Dollars
0	−\$775,000	1.0	−\$775,000
1	60,000	1.0530	63,180
2	125,000	1.1088	138,601
3	185,000	1.1676	216,000
4	250,000	1.2295	397,364

[4]The BLS home page can be found at www.BLS.gov. Earnings, worker safety, productivity, unemployment rates, PPI, CPI, etc., and other indexes can be found there.

For cash flow analysis we define

$$F_c = (G - D_c - C)(1 - t) + D_c \quad \text{or equivalently,} \tag{9.4}$$

$$F_c = (G - C)(1 - t) + tD_c$$

where F_c = total cash flow of money, dollars per year

G = gross income, dollars per year

D_c = depreciation charge, dollars per year

C = costs not estimated elsewhere, dollars per year

t = effective income tax rate, decimal

This basic relationship shows the major factors for cash flow. A tax reporting period will follow the firm's fiscal year reporting period, which may not correspond to the calendar year (January 1–December 31).

Depreciation is introduced in Chapter 4 and those calculations to find D_c apply here. Depreciation is defined as a *noncash tax expense*. Depreciation is the recovery amount for the money that originally purchased the capital expenditure, which has been already spent. As a noncash tax expense, it acts to reduce the taxable income, but it is added back to the cash flow. Note this manipulation in Equation (9.4).

Example: Cash Flow Statement

A ceramic products plant will produce industrial products for automotive, aerospace and aviation, gas turbine and reciprocal engines, and many items that are subjected to high temperatures, gas flow, abrasion, and severe shock and thermal loading. Products are also planned for power generating equipment components and for chemical, mineral, and metal processing. These ceramic-matrix small-grain but high density products have in common raw materials such as alumina, zirconium oxide, aluminum oxide and silicon oxide. Patented processes increase the concentration of intergrain contact, thus influencing electrical and super conductor performance, for instance. The vision for the plant is the consolidation of the mineral processing of the ores and intermediate supplies to the final shaping of the products—net shape with little, if any, final grinding and machining to precision dimensions and complex geometry.

Now consider an example of the cash flow for a ceramic products plant. An owner[5] will analyze this cash flow carefully as the financial performance for the construction and operation is essential to the decision of whether to proceed with the design and build the plant. Assume that the ceramic products plant designs and estimates, which are expressed in real dollars and are indexed to the year 0, are summarized by the cash flow analysis, as given by Table 9.4. This table consolidates capital expenditures, revenue, expenses, depreciation, and allows the net cash flow to accumulate over several years. This analysis indicates when the revenues will exceed the capital and operating expenses, an important fact.

Cost estimates for construction bid and manufacturing, schedule, and marketing and sales rates, and capitalization costs for new equipment and plant enlargement are necessary to construct a cash flow statement. Construction is shown as a lump sum amount at point-in-time 0. The analysis concludes with year 6.

The ceramic products, while differing in design features and materials, can be conveniently modeled by the metric "ton." The ton measure is the best activity variable for plant capacity. Thus, growth expressed in terms of 10,000 tons per year reaches 100% in year 4, but design improvements will allow learning and greater capacity for years 5 and 6.

[5]The term "owner" is representative of proprietor, shareholder, or owner's representative, and implies one who is responsible for "due diligence," a scrutiny of the economic facts that protects the owner's interests.

TABLE 9.4 Consolidated Cash Flow Statement for Construction and Operation of Ceramic Products Plant with Real Cash Values Expressed as $\$10^3$

Item	0	Year 1	Year 2	Year 3	Year 4	Year 5	Year 6
Capacity, percent		25	50	75	100	110	116
Production, 10,000 ton/yr		12	25	37	50	55	58
Construction cost of plant, $	−700,000						
Startup expenses, $	−75,000						
Gross revenue, $		60,000	125,000	185,000	250,000	269,500	282,500
Less expenses, $		15,000	31,250	46,250	62,500	68,750	72,500
Less depreciation, $		70,000	93,750	71,250	70,000	70,000	70,000
Gross income before taxes, $		−25,000	0	67,500	117,500	130,750	140,000
Less income taxes @ 40%		0	0	27,000	47,000	52,300	56,000
Net income after taxes (1–40%)		NOL*	0	40,500	70,500	78,450	84,000
Add back depreciation, $		45,000	93,750	71,250	70,000	70,000	70,000
Net cash flow, annual, $	−775,000	45,000	93,750	111,750	140,500	148,450	154,000
Cumulative cash flow, $	−775,000	−730,000	−636,250	−524,500	−384,000	−235,550	−81,550

*Net operating loss.

Project estimates provide the $700 million investment cost for plant construction. Preoperating expenses and investment costs are first-year cash out, lumped at time zero for simplicity. Production quantity and price are estimated with information from the marketing components of the business.

Expenses for the operation for the production quantity for the design flow sheet of the ceramic products plant are entered. These expenses involve working capital, which is made up of accounts receivable, raw material inventory, work in process, and finished goods inventory. Direct labor and overhead are included. Increases in inventory require immediate cash outlays that delay cash flow from generating sales revenue. Ordinarily, a higher requirement for cash-on-hand occurs during periods when operations are increasing. Determination of what constitutes working capital is usually meant to be incremental capital (i. e., differential capital between present and prior year.)

Depreciation is a noncash tax expense, and a tax credit (a product of the depreciation times the firm's tax rate) helps to provide the inflow of funds. For this example, depreciation is straight line over 10 years or $70,000 (= 700,000/10). Typically, straight-line depreciation is not used, as the modified accelerated cost recovery system (MCARS) is the more common method. But notice in Table 9.4 that there is inadequate revenue to allow even the full depreciation until year 3, and this is a *carry forward* process until sufficient revenue exists for allowable and full depreciation. For instance, in year 1, gross income of $\$60{,}000 - 15{,}000 < 70{,}000$, leaving only $45,000 of depreciation that is allowable for that year. In year 2, the $25,000 shortage of depreciation in the previous year can only be allowed to the amount of no taxable income, or $93,750. The carry forward provision of the tax code affects future tax computation.

Consider the application of Equation (9.4) for year 4 where we want to find the yearly cash flow.

Item	Year 4
Gross income, G	\$250,000
Working capital, C	−62,500
Depreciation, D_c	−70,000
Subtotal of gross income before taxes	\$117,500
Taxes at 40% of gross income	−47,000
Net income after taxes	\$70,500
Add back of depreciation, D_c	+70,000
Cash flow of money for year 4	\$140,500

The cash flow for year 4, or \$140.5 million, becomes the yearly amount that is consecutively added to the previous years and the running total of a negative \$384 million to that point. The ceramic products plant at the end of year 4 has not returned sufficient cash flow to pay off the capital investment.

Once expenses and depreciation are deducted from sales revenue, we have the gross income before taxes. The effective tax rate of $t = 40\%$ is applied to this quantity. Net income percentage after taxes is $(1 - t) = 60\%$.

The arithmetic for net and cumulative cash flow is evident in Table 9.4. The payback time calculates the number of years required to regenerate, by means of profits, depreciation, and tax credits, the total investment of the fixed assets, and the preoperating expenses that are required to launch the product. It is seen that there is insufficient cash flow to recoup the capital investment and pretax operation expenses within 6 years. The cash flow looks at two factors simultaneously: (1) product profitability and (2) ensuring that there is sufficient cash to see it through its initial design and construction and money for the ramp-up production. It must pass both tests.

The results of this cash flow analysis may give a negative decision to build the plant and produce the product. This may not be a good investment. This example illustrates an economic tradeoff study for engineering design. Other designs will be examined to see if a more profitable ceramic liner plant can be modeled that is satisfactory to the owner.

In our example, we consider the ceramic products plant as the only investment of the business. Otherwise the losses of this operation are distributed to other parts of the business, and the total corporate income is taxed at another rate. Not knowing income from the other parts of the business, we assume that this plant is the only operating entity for this corporation. This is a good practice, because it requires the new opportunity to be a freestanding enterprise, thus showing its own warts or beauty marks.

The cash flow of Table 9.4 is based on a *nondiscounted* dollar (i.e., the face value of the real dollar cash flows for each year). This cash flow did not deal with the concepts known as *time value of money*. Methods discussed in Chapter 10 show how these very important principles are applied to cash flow analysis.

Cash flow logic is important to much cost analysis conducted in engineering. Its understanding allows us to consider break-even models, prolific methods for economic testing of design opportunities.

9.3 BREAK-EVEN MODELS

Engineering provides estimates about designs to make future decisions. There is the premise that the action recommended by the estimate will add to the profit of the enterprise to make it worth the trouble. Break-even analysis is one of the significant tools used in engineering cost inquiry. The essence of the model is more than 80 years old, and though there have been improvements since those pioneering times,[6] the early models remain incisive in balancing the negative and positive effects of money transactions. The development of the break-even chart dramatizes the importance of knowing the relationships that exist between costs, price, volume, and profits. Its very simplicity is the reason for its widespread use. Additional titles for break-even models, illustrated Figure 9.2, are profit graphs or cost-volume-profit charts. Students need to understand these principles.

In engineering planning, many important questions arise: How much do I have to reduce centrifugal-cast ceramic liner cost to reach the profit goal? How will a change in fixed costs affect net income? How much does direct labor costs need to decrease to cover a planned increase in equipment costs? These are typical of the many questions that can be answered by using break-even study. Versatility of the model allows solution to a wide range of problems.

Break-even analysis applies simple mathematical formulas to determine the sales or production level at which the business neither incurs a loss nor makes a profit. The break-even point, in language terms, is the juncture where

$$\text{Total costs} = \text{net sales revenue} \tag{9.5}$$

Equation (9.5) is accomplished by three basic break-even models:

- Linear cost
- Semifixed cost
- Nonlinear cost

These cases are illustrated in Figure 9.2.

In the cases shown in Figure 9.2 it is assumed that total costs and total revenue depend on the level of activity units, n. In our discussion, we are thinking of units produced and sold, but other descriptions for activity units are valid. Surely, every

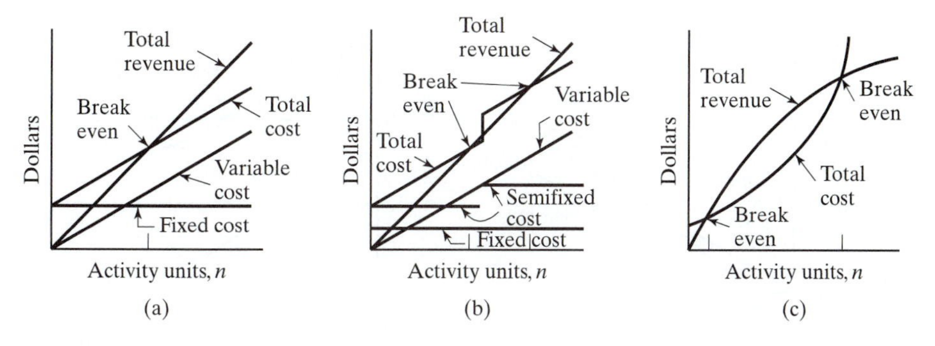

Figure 9.2 Break-even models: (a) linear cost, (b) semifixed cost, and (c) nonlinear cost.

[6]See Walter Rautenstrauch, *The Successful Control of Profits* (New York: B.C. Forbes, 1930). This is a landmark book by an early engineering pioneer in break-even chart analysis.

company wants to do better than break-even. Thus the estimating of the information to conduct the comparison has the objective of increasing profit.

9.3.1 Linear Cost Case

Linearity assumes variable cost per unit, C_v, remains the same over an intended range of the activity variable. Furthermore, fixed costs are constant throughout the same relevant range, and the plant has capacity to produce the units without adding additional plant or equipment. Costs, which change with volume, are usually called direct costs and those which do not are referred to as fixed costs. There is a distinction, however, as how these costs vary in either total costs or unit cost. Direct costs are shown to vary in total with the number of units made and sold. Thus the basic diagram of Figure 9.2(a) can be expanded into two graphs, Figure 9.3(a) and (b).

Fixed costs, C_f, are those expense items that generally do not change in the short run, regardless of how much is produced or sold. These costs do not go up or down with production or sales level. Examples of fixed costs include general office expenses, rent and space, equipment and plant depreciation, utilities, telephone, property tax, and the like. Obviously, all expenses vary over the long run. Rent and property taxes may increase every so often. But in break-even analysis, calculations are based on short-run information in order to reveal the current profit structure of the business.

Chapter 4 focused on the composition of the machine hour and activity-based costing rates—information that after some reconstruction will give the fixed costs of equipment and plants. The period of the fixed costs is tied to a time constraint, and it corresponds to the period for the production of the activity range of n.

Variable costs are those expenses that change with the level of production or sales. In retail firms, variable costs include the cost of goods, sales commissions, billing costs, etc. In manufacturing, variable costs are oftentimes restricted to direct labor and direct material. The principles studied in Chapters 2 and 3 are required

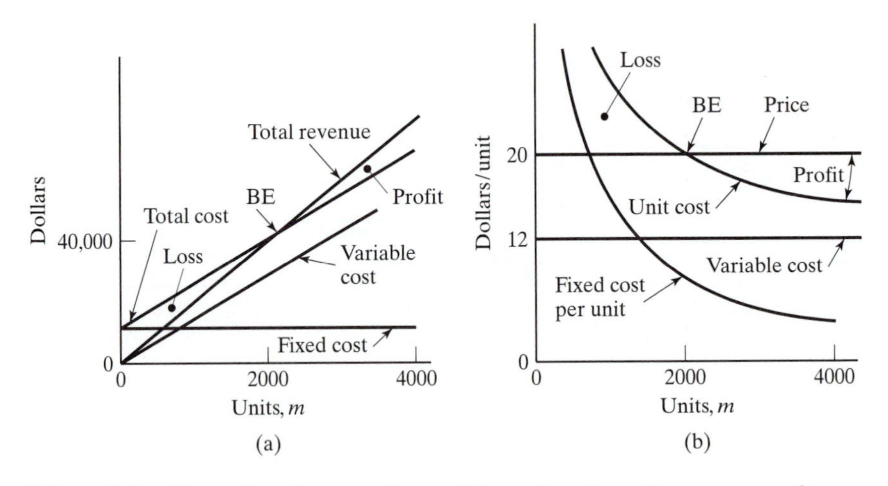

Figure 9.3 Linear break-even charts: (a) total costs and revenue, and (b) unit costs and price.

now; the student will want to review those chapters. The labor costs start with productivity, say hours per unit, and when multiplied by the gross hourly labor cost, the identified term direct labor cost results, and that is what is required here. These costs increase with increased production or sales because they are directly involved in either making the product or making the sale. The unit direct cost is constant regardless of the number of units sold.

Material costs are the shape requirements from the engineering drawing, which allow for waste and scrap. When multiplied by the raw material unit cost, the resulting material cost per unit is found. Those two items are direct cost, and vary linearly with the activity of production. As variable costs, they do not exist unless there is production.

Total costs, C_T, may appear on the income statement or be given by the estimate worksheet. With this description we can define

$$C_T = C_f + nC_v \tag{9.6}$$

where C_T = total cost at n units or activity, dollars per period

C_f = fixed cost, dollars per period

n = number of units or activity

C_v = variable cost, dollars per unit

$$\text{and } C_u = \frac{C_T}{n} \tag{9.7}$$

where C_u = unit cost or average cost, dollars per unit

The total cost C_T is easily found as the sum of the fixed and variable costs. A period of time is necessary—for example, weekly or monthly.

The next step is to match revenue against the total costs, as shown in Figure 9.3. The versatility of the method allows a number of approaches. The amount of sales revenue could be available from the income statement as net sales or from the product estimate. Net sales revenue is all sales revenue, often called *gross revenue* less any returns and allowances. Even tax effects will allow net revenue to account for an after-tax revenue case. For new products, an estimated sales number is given, (see Chapter 8). Revenue is found by

$$R_T = nR_v \tag{9.8}$$

where R_T = total revenue at n units or activity, dollars per period

R_v = revenue or price, dollars per unit

The constant value R_v can be found for a new product from the estimate worksheet, marketing information, or even from competitor prices. Theoretically, after fixed costs are covered, each dollar of sales will have to cover only variable costs. The break-even point which makes no profit or sustains no loss can be calculated by

$$n_{BE} = \frac{C_f}{R_v - C_v} \tag{9.9}$$

where n_{BE} = break-even point, number

$R_v - C_v$ = contribution, dollars

In Figures 9.2 and 9.3, the break-even points are shown. For the linear cases, profit and loss are wedges that have their apex at the break-even point. A break-even point will have a corresponding dollar value, as graphically found, or n_{BE} may be back-substituted to find the dollars. As the number n increases above the break-even point, the profit wedge increases. Similarly, the loss zone is given by a number less than n_{BE}.

Example: Linear Break-Even Charts

A product is estimated to have a fixed process cost of $16,000 for a year. The variable cost is $12 per unit, and the expected production and sold units will range from 0 to 4000 units. The estimating worksheet discloses a price of $20 per unit. The chart is shown as Figure 9.3(a). The total revenue line, $20n$, and the total cost line, $16,000 + 12n$, intersect at the break-even point of 2000 units and $40,000. Profit and loss wedges are shown.

The break-even chart can be shown differently, as now given by Figure 9.3(b). The emphasis is on unit cost and unit price. When depicted graphically, fixed cost per unit is a rectangular hyperbola, or C_f/n. To obtain the full cost per unit, unit cost is added, and becomes $C_f/n + C_f$. At break-even point, unit sales revenue is equal to unit full cost. A new term is coined as *contribution* and is graphically shown as the vertical distance between unit revenue and the unit variable cost functions.

With break-even charts management is naturally interested in increasing profit. From an engineering standpoint, that may mean lowering the break-even point by reducing direct material or direct labor or fixed costs. Lowering direct costs will then raise the gross margin. In addition, there may be a necessity to raise prices. These break-even charts can be studied to see these effects. Figure 9.4 illustrates four situations.

Though the narrow meaning of break-even analysis is that revenue equals costs, another application is applied to the comparison of two or more different processes to find the most economical process. The point at which the costs of two processes are equal is called the *isocost point*, or equal cost point. No revenue is considered in this case because the goal is to select the lowest-cost process rather than maximizing profit. In Figure 9.4(a) FC_A, or fixed cost for alternative A, and VC_A, or variable cost for process A are compared to process B. The intersection of TC_A and TC_B gives the isocost point. In the graphical illustration, process A is preferred to the right of the isocost point.

The make versus buy situation—a special instance of isocost analysis—can be modeled in the following way. From Figure 9.4(b) we see that for the buy alternative there is no fixed cost, that being subsumed in the price that the supplier is asking. The make variable-and-fixed costs are set equal to the sellers variable cost of the buy alternative. Depending on the actual number of units, the decision is made to either buy or make. The decision is to buy if the quantity is to the right of the isocost point. Make or buy considerations are amplified in Chapter 11.

A savings in direct material cost is a immediate contribution to the profit. These savings may be achieved by reducing waste or improving the reject experiences by better product design, material substitution, less expensive sources of supply, processing methods, and so on. A dramatic area for savings, which improves every facet of the company's profit, is the reduction of purchase cost of materials. Two material cost situations are described in Figure 9.4(c). A lowering of the variable cost (VC)

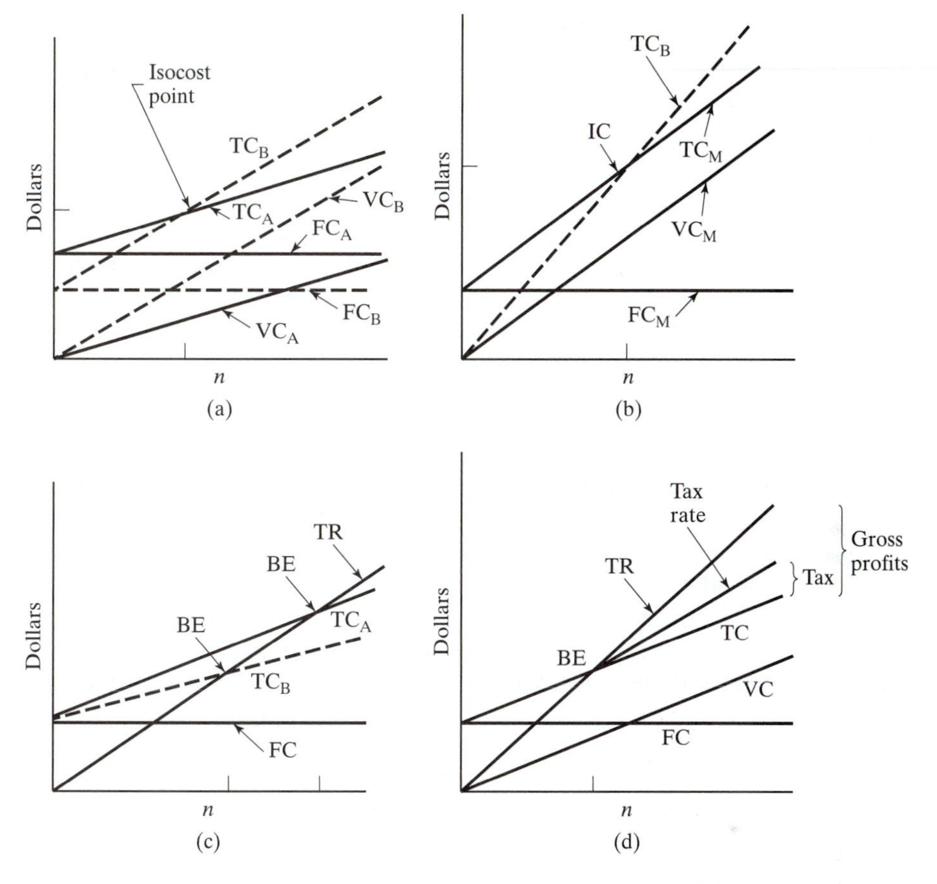

Figure 9.4 Examples of linear break-even chart versatility: (a) process selection, (b) make versus buy, (c) reduction of material costs, and (d) tax effects on profits.

with a constant total revenue (TR) line results in a break-even that is less than the previous situation.

Another viewpoint of break-even charts is the imposition of taxes on the product revenue. In Figure 9.4(d), the typical break-even chart is shown. A tax line begins at break-even, as taxes are not paid on losses. For this example, we are assuming a one-product business, or one where the chart is isolated for a single product experience. The gross profit is reduced for taxes, leaving net taxes as a smaller wedge.

Some costs are *periodic* (time based and fixed) while other costs are *object* based, such as direct materials or direct labor, and there are some costs that are *joint*. Some joint costs seem to have a periodic fixed cost basis, and have a portion that can be further separated logically. This additional subdivision is called semifixed costs, and we now look at that case. Costs, in an estimating sense, are often identified as to behavior, either as a function of time or as a function of the object of the manufacturing or service performed.

9.3.2 Semifixed Cost Case (Optional)

We have defined costs in a number of ways. In another interpretation, costs are fixed, semifixed, and variable. *Fixed costs* are incurred regardless of the differences in levels of manufacturing activity. Generally, they do not change in the period as a result of more or less manufacturing activity. The decision to build a plant in a city under a tax system makes the costs of carrying the depreciation and taxes on the plant uncontrollable and fixed. The governmental agency controls the taxes. Only the closing down or removal of the plant can affect these continuing charges against the property. Assuming the ongoing business convention, other fixed costs are salaries of the president, engineers and management, etc., who are employed a full period, say a month or for the duration of the specific job being studied.

Sometimes fixed costs step up or down as operations vary beyond a certain level. These are called *semifixed costs* or sometimes *programmed* fixed costs, because they are forecast ahead according to a level of business activity. Once the decision is made to commit to a level of manufacturing activity—say make 15 assemblies instead of 10 during a period—these costs usually remain fixed unless significant revisions are made in schedules during the period covered by plan. The modifier "semi" clarifies their difference from fixed costs.

Workers' compensation, social security taxes, unemployment taxes, energy rates, major building or process alterations, and employee training are in the noncontrollable portion of semifixed costs. Energy rates are "blocked" with declining costs as consumption increases above the several steps. Income tax rates are set by law and are blocked. Notice the corporation tax rates given earlier. Within the semifixed category of costs, there are a number of controllable items such as maintenance, overtime pay, and supervision. It is possible that these costs will vary with levels of manufacturing activity but not with the same sensitivity as found with "pure" variable costs. Blocking or stepping is a characteristic of semifixed costs.

Semifixed costs may result from the purchase of a machine tool during a period, and the graphical result is seen as a spike or step. This spike is a discontinuity, and analysis needs to handle it with an appreciation. Note the sawtooth full-cost line in Figure 9.5.

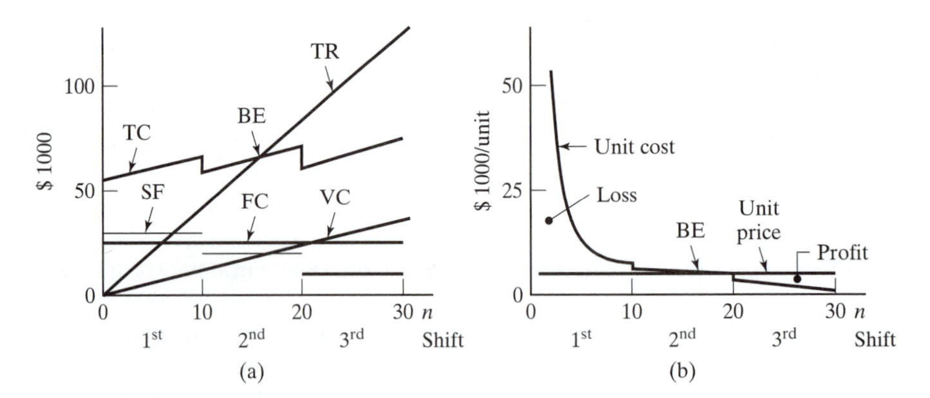

Figure 9.5 Break-even locations for aircraft subassembly manufacture: (a) fixed, semifixed, variable costs, total revenue and total cost, and (b) unit cost and price.

Theoretically, variable costs in the very short term can be classified as fixed, because of the inability to terminate the variable costs instantaneously, and fixed costs in the long term can be viewed as variable, because of the ability to liquidate those charges. So the length of the period has an influence on the nature of costs. The behavior of fixed, semifixed and variable costs is described in Figure 9.5.

These costs can be modeled in the following way:

$$C_T = C_f + C_{sf} + nC_v \tag{9.10}$$

where C_T = total cost at n units or activity, dollars per period

C_f = fixed cost, dollars for period

C_{sf} = semifixed cost for level of n or activity, dollars per period

n = number of units or level of activity

C_v = variable cost, dollars per unit

and $C_u = \left(\frac{C_T}{n}\right)$

where C_u = unit cost or average cost, dollars per unit

The cost per unit, C_u, can be thought of as *average cost*, since it considers fixed, semifixed, and variable costs per unit. Furthermore, if the range of the activity variable n is reasonably stable and known, the semifixed costs can be combined with the fixed costs. For this relationship, n is used, indicating that quantity is implied.

Example: Semifixed Cost Case

The manufacture of airframe subassemblies for the aileron of a business jet has the possibility of one-, two-, or three-shift operation. There are 83 separate components that are assembled into an aileron unit of product.

The programmed costs depend on the number of units that are sold and the planning of production. Monthly scheduling of production can be accommodated by one, two, or three shifts. A single 8-hour shift will have a larger plant area and more equipment and workers than a two- or three-shift operation to produce the same number of product. In this shift-cost analysis, the mobilization of supervision, inspectors, and other indirect shift employment is guided by the practical considerations that the second and third shifts do not require proportionate indirect hires as the first shift, which is the indirect cost that is studied. Production output is scheduled for identical structural frames for this example, as those considerations are limited by assembly fixtures and tooling. The planning that follows estimates the cost for three-shift operation for one month of operation. Other analysis would be necessary for a one- or two-shift operation.

Type of Airframe Cost	Cost
Variable	
Direct labor	\$250 each
Direct material	\$1,000 each
Fixed costs	\$25,000 per month
Semifixed costs	
First shift	\$30,000 for units $0 \leq n \leq 10$ per month
Second shift	\$20,000 for units $10 < n \leq 20$ per month
Third shift	\$10,000 for units $20 < n \leq 30$ per month

The price is expected to be $4200 per unit. Sales and production are unlikely to be above 30 units for the period of 1 month. Three shifts are necessary for the 30 production units.

Fixed and semifixed costs are necessary before any airframe assembly can begin, and thus are charged at $n = 0$. Semifixed costs are estimated according to the level of tools, assembly fixtures, portable cranes, and equipment to move materials, indirect inspection labor, etc. Figure 9.5(a) and (b) provide graphical break-even results. The total and unit cost curves are erratic because of the way in which the semifixed stair-step costs are defined. While the fixed costs are increasing at the stairs, unit costs are declining.

Break-even is graphically found at 16 units, but remember that the analysis period is a month, and when 16 units are produced during the period, and after that point, the remainder of the month is operating at a profit. Until the costs that underlie these estimates are changed, each period will have similar results for the order of airplane assemblies. Once the next month starts, there is a loss until 16 units are reached.

Contrariwise, the development for the analysis could be resolved by using the entire sales order, meaning that once the break-even location is reached, the profit wedge begins. The selection of the period and the number of units is important to the interpretation of the break-even. The student may want to ponder this dilemma.

This analysis is compared against one- and two-shift production, and whichever shift pattern that has the more favorable break-even and total profit is a better choice. If planned production is greater or less than 30 units per month, a different schedule of equipment and direct labor manning is required, and fixed and semifixed costs will change and the break-even analysis is recalculated. Break-even analysis is valid only through a relevant range of normal activity for a predetermined period as shown by this set of estimates.

Cost and revenue functions do not always follow linear stereotypes. More often than not, realistic break-even analysis requires a nonlinear interpretation.

9.3.3 Nonlinear Cost Case

Not all future decisions ensure that the firm will be better off when a cost reduction opportunity presents itself. Consider our ceramic product again. The firm produces an identical automotive ceramic valve and guide assembly in two separated production lines, called A and B. The manager of valve and guide production has permission to add another operator to either line A or B. Which one shall she choose? In line A the unit average cost of the valve and guide assembly is $0.82 at $n = 20{,}000$, while in line B the average cost is $0.88 at $n = 15{,}000$ units. Our manager chooses to assign the operator to line A, "because it is already lower in average cost." But it is possible that the difference in the costs occurred because the crew A was more efficient. If so, the new associate may add little to cost reduction, and in B the associate may lead to a greater decrease in cost. Reasoning based on average cost decisions does not give the entire story. Concepts other than average unit cost are important to break-even analysis. The nonlinear case allows an expanded approach.

Example: Nonlinear Cost Case Break Even

An illustration of the nonlinear case starts with the estimate of the total cost for product units, shown by Table 9.5. For the identified period, say a day, for units of output of the

TABLE 9.5 Estimated Cost Schedule for Product

Units, n	Fixed Cost, C_f	Total Variable Cost, nC_v	Total Cost, C_T	Average Cost, C_T/n
0	$4,000		$4,000	
1	4,000	$800	4,800	$4,800
2	4,000	1,120	5,120	2,560
3	4,000	1,340	5,340	1,780
4	4,000	1,600	5,600	1,400
5	4,000	1,900	5,900	1,180
6	4,000	2,520	6,520	1,087
7	4,000	3,520	7,520	1,074
8	4,000	4,800	8,800	1,100

TABLE 9.6 Cost, Revenue, and Profit Analysis for Product

Units, n	Average Fixed Cost C_f/n	Average Variable Cost, C_v	Marginal Cost, $\Delta C_T/\Delta n$	Total Revenue, R_T	Average Revenue, R_T/n	Total Profit, $R_T - C_T$	Marginal Revenue, $\Delta R_T/\Delta n$	Marginal Profit $\Delta R_T/\Delta n - \Delta C_T/\Delta n$
0								
1	$4,000	$800	$800	$2,000	$2,000	$(2,800)	$2,000	$1,200
2	2,000	560	320	4,000	2,000	(1,120)	2,000	1,680
3	1,333	447	220	6,000	2,000	660	2,000	1,780
4	1,000	400	260	7,800	1,950	2,200	1,800	1,540
5	800	380	300	9,300	1,860	3,400	1,500	1,200
6	667	420	620	10,500	1,750	3,980	1,200	580
7	571	503	1,000	11,600	1,657	4,080	1,100	100
8	500	600	1,280	12,700	1,588	3,900	1,100	(180)

product, estimated fixed cost, C_f, and total variable costs, nC_v, are added to give total cost, C_T. Average total cost is the final extension. Average cost (arithmetic mean) is determined, which usually concludes the calculation.

Straightforward extensions can be made on Table 9.5. For instance, fixed cost can be divided by the quantity n. This shows a declining reduction on the fixed costs. Since fixed costs are independent of output, their per-unit amount reduces as output increases. This widespread recognition is confirmed when the sales staff speaks of "higher sales spreading the overhead." These calculations are shown in Table 9.6.

In addition, total variable costs can be divided by the quantity n. Notice that the unit variable cost, C_v, is no longer constant as assumed by previous linear and semifixed break-even discussion. Notice that the average variable cost declines to $n = 5$, and then increases. A number of logical explanations can be given for this nonlinear response.

The variable cost is zero before any activity units. Marginal cost, by convention, is also zero, and average cost is undefined at this point. At two units, marginal cost is the amount added by total cost from one to two units, or $4800 - 4000 = \$800$, or $\Delta C_T/\Delta n$. *Marginal cost* is the added cost to make another unit, or it is the extra increment of cost to produce another increment of output. If the last increment of cost is smaller than the average of all previous costs, it pulls the average down. Thus average total cost declines until it just equals marginal cost. Equivalently, the rising marginal-cost curve also pierces the average variable cost curve at its minimum point (Figure 9.6) Minimum point on the average-cost curve occurs where it crosses the marginal-cost curve.

The term marginal cost is interchangeable with differential cost or incremental cost. More broadly, marginal cost and marginal revenue results from generating an engineering change order for a product. In a "sweeping" statement all estimates are marginal cost/marginal revenue estimates, because they report change from a current course of action.

The cost side of the analysis is only half of the job to study the effect of break-even on profits. Next, marketing and engineering determine the total revenue, which is joined to Table 9.6. Average revenue is found by dividing total revenue by the unit number and is thus seen to decline. This occurrence can happen for a variety of real reasons. The student may want to think of examples.

For the second unit, total revenue minus total cost gives a loss of \$1120. After that total profit is positive. Marginal revenue is the additional money received from selling one more unit at a level of output. For a linear revenue function, the marginal revenue is a constant price. Thus for each additional unit sold, the total revenue is increased by one unit of price. When this assumption is replaced by the more realistic nonlinear assumption, the situation is not as obvious.

The hypothetical variable revenue is zero before any production units. Marginal revenue, by convention, is also zero, and average revenue is undefined at this point. At two units, marginal revenue is the amount added by the activity from one to two units,

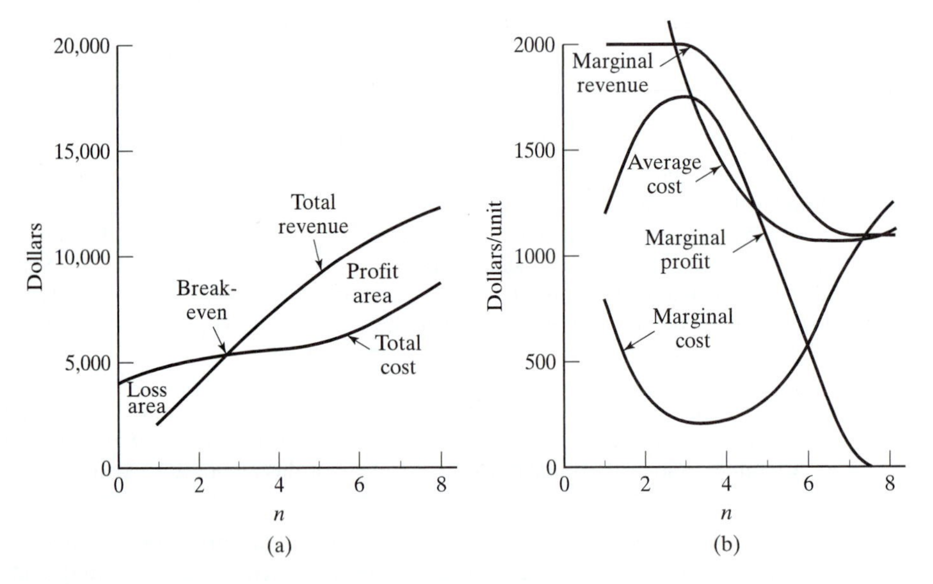

Figure 9.6 Cost and profit analysis curves for product: (a) total revenue and total cost, and (b) marginal revenue, marginal cost, average cost, and marginal profit.

or 4000 − 2000 = \$2000, or $\Delta R_T/\Delta n$. Marginal revenue is the added revenue to sell another unit. It is the extra increment of revenue to sell another increment of output. The tabulation shows a decelerating revenue rate, and thus price is reducing with increasing quantity.

In Figure 9.6(a) break-even is shown for three units for the period. Loss occurs before this point. In Figure 9.6(b) dollars per unit are shown for the marginal analysis of costs and revenue.

The intersection of marginal cost and marginal revenue gives the maximum profit point. As marginal cost increases above this point, each additional unit costs more to make than the sale will provide, and thus total profit is declining. The table shows that unit profit at $n = 8$ is negative, though total profit remains positive.

Consider another nonlinear break-even example of the valve assembly, which is illustrated in Figure 9.7; (a) illustrates the cold formed stages of the valve stem, which starts as round bar stock, and (b) illustrates ceramic assembly replacement components. All components are ceramic—i.e., zirconia and alumina materials (with the exception of the steel spring). The retaining clip housing is not shown.

The enumeration approach of Tables 9.5 and 9.6 is a brute force way to analyze revenue and costs. An exhaustive directory of the possible estimates is required before the arithmetic could begin. A preferred approach to enumeration is to have a scattering of cost-estimated experiences and then fit linear and nonlinear regression lines. This is shown in Table 9.7, where total annual cost data are estimated for the future period. Each activity value of n is separately estimated for total annual cost, C_T. Thus there are six estimates of cost for Table 9.7. Similarly revenues are shown by Table 9.8.

In a similar way, revenues are estimated for the product and are scattered for a variety of the activity variable n. For our purpose here, a total revenue function is estimated with knowledge about input price and demand and that it increases or

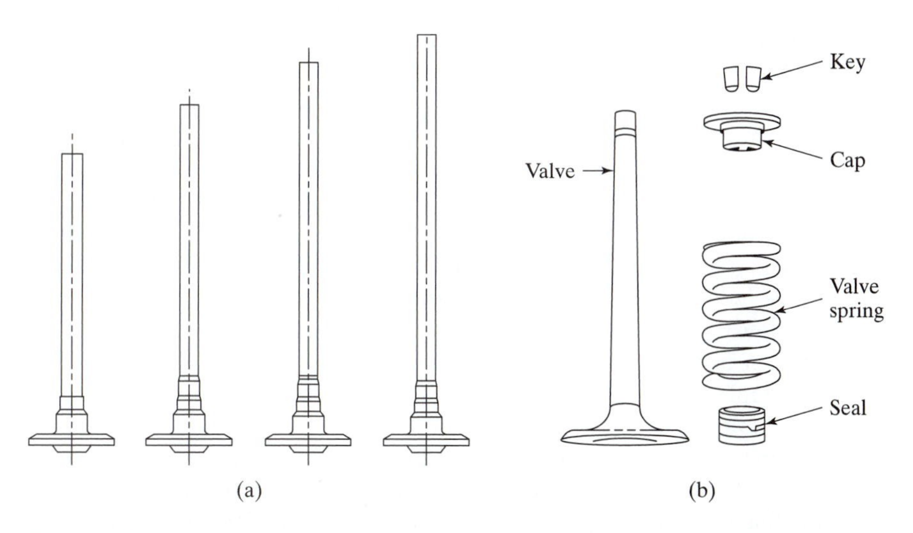

Figure 9.7 (a) Progress of the cold forming of steel valve stem, and (b) ceramic valve and guide assembly.

TABLE 9.7 Estimated Total Cost For Ceramic Valve and Guide Product for Future Year

n	Total Estimated Cost, C_T
5,000	$6,875
10,000	10,460
15,000	13,260
20,000	16,400
25,000	21,125
30,000	29,250

TABLE 9.8 Estimated Total Revenue for Ceramic Valve and Guide Product for Future Year

n	Total Estimated Revenue, R_T
0	0
5,000	$6,000
10,000	11,125
15,000	15,200
20,000	19,300
25,000	23,500
30,000	27,000

decreases with increasing output according to whether the demand is elastic or inelastic. Elastic demand implies a demand that increases in greater proportion than the corresponding decrease in price, and vice versa.

Costs are estimated as lump sums, as we assert that it is too complex to decouple fixed costs from variable costs and list those quantities separately and add them together, as done in Table 9.5. This development of total cost and total revenue estimates may arise from complicated multiproduct issues, or from a company's commingling of the information. Thus the action to lump the information as inseparable is a major consequence, and it leaves no other method than regression analysis to continue the analysis.

It is useful to remind ourselves of the assumptions for regression or least-squares. A cause-and-effect relationship is necessary and there are no extraneous variables that make the regression of little value. The deviations of the dependent y values of "effect" are mutually independent, as the forcing variable x_j "causes" changes. The student will want to review the listed assumptions for regression given in Section 5.2.5.

We then use regression models to analyze the break-even properties of those data. It is simple to fit a polynomial regression equation of low order and much software is available.

The first regression of Table 9.7 is linear and is given as $C_T = 0.84n + 1527$, where $C_v = 0.84$. The fixed cost C_f is $1527, which is the vertical-axis intercept for the plot of the linear model. If C_v is the same at various values for n, we have the linear cost model. If C_v varies, then we are concerned with nonlinear models. For the constant C_v, nC_v is a straight line increasing (or decreasing if C_v is negative) at a constant rate per unit, and C_v is the slope of the variable cost line. If the estimates are independently made for the activity variable, then preknowledge of the slope is unknown, as we assume for this example.

In Table 9.8 zero income for no activity n is a trivial addition, but it aids the development of the data by a least-squares fitting. The total revenue column of Table 9.8 contains no reference to the price for the sale of a valve and guide assembly unit, but price is found by dividing total revenue by the units of output. We require that units made and sold equals sales volume. In a similar fashion to the cost models above, we fit polynomial regression curves to those data and analyze those determined models. We modify the revenue model given by Equation (9.8) to the form

$$R_T = nR_v + R_f \tag{9.11}$$

where R_T = total revenue per period, dollars per period

R_v = variable revenue or price, dollars per unit

R_f = fixed revenue, dollars per period

$$\text{and } R_T = \begin{cases} 0 & \text{if } n < 0 \\ nR_v + R_f & \text{if } n \geq 0 \end{cases}$$

This is a linear or nonlinear model if R_v is a constant multiplier or a nonlinear term. If we consider the income arriving from product sold, then revenue is not obtained until $nR_v > 0$. For a no fixed income condition, $R_f = 0$. For a product business this is the sales price. For average revenue, or average price, the model becomes

$$R_a = \frac{nR_v + R_f}{n} = \frac{R_T}{n} \tag{9.12}$$

where R_a = average revenue or price, dollars per activity unit

First, the linear regression gives $R_T = 0.887n + 1284$. It might be assumed that at $n = 0$ there should be no revenue, which is frequently correct. However, a displaced origin is not uncommon with a least-squares approximation of data points. Additional regression methods, which are not given in this book, can force the linear equation to pass through the origin. In contrast, for some product and manufacturing enterprises, there is often an initial "earnest" payment given for product development by the buyer, and the R_f conveys that advanced payment.

The break-even for the general linear case of regressed linear equations where the fixed revenue R_f exists is given by

$$n_{BE} = \frac{C_f - R_f}{R_v - C_v} \tag{9.13}$$

where n_{BE} = "break even" number of activity units per period

When using linear models $C_T = 0.84n + 1527$ and $R_T = 0.887n + 1284$, the break-even point is

$$n_{BE} = \frac{1527 - 1284}{0.887 - 0.84} = 5170 \text{ units}$$

The linear lines for cost and revenue and the break-even point are given by Figure 9.8(a). There is a linear profit wedge beginning at 5170 units, which continues indefinitely upward to the right.

One approach to finding the break-even point is to plot the original data of Tables 9.7 and 9.8 and determine their intersection. A preferred method is to fit the data to a k-order polynomial, $y = a + b_1x + b_2x^2 + \cdots + b_kx^k$, and plot that equation because regression has the purpose of smoothing the errors or any quirks of the estimates at the various values of n. For the data of Tables 9.7 and 9.8 we have third-order polynomials $C_T = 840 + 1.527n - 7.418 \times 10^{-5}n^2 + 1.827 \times 10^{-9}n^3$ and $R_T = -48 + 1.23n - 1.64 \times 10^{-5}n^2 + 1.833 \times 10^{-10}n^3$.

Once the cost and revenue equations are set equal for break-even analysis, *two* relevant intersections are $n = 9{,}237$ and 28,015 units. This nonlinear situation gives a lower and an upper limit break-even intersection, unlike the linear case. Whereas the linear approximation suggests a continuing profit zone once the break-even is reached, the nonlinear approximation is a better evaluation (see Fig. 9.8 [b]).

The analysis continues similar to that shown for the tabulated example above, except that the regressed functions are susceptible to calculus treatment. Marginal cost and marginal revenue analysis depend on the calculus, Equations (9.14) and (9.15). The differentiated marginal models are for the most part derived functions from regressed estimated and engineering data. The functions C_T or R_T are assumed continuous (despite the fact that they are routinely determined from arbitrary and selected activity levels) and differentiable. A good deal of information is uncovered from these basic models. The simplifying assumptions are necessary, but it is important to remember cases where the assumptions may not hold.

By similar reasoning, if the output n is increased by an amount Δn from an established level n, and if the matching increase in cost is ΔC_T, then the increase in cost per unit increase in output is $\Delta C_T/\Delta n$. Marginal cost is the limiting value of this ratio as n

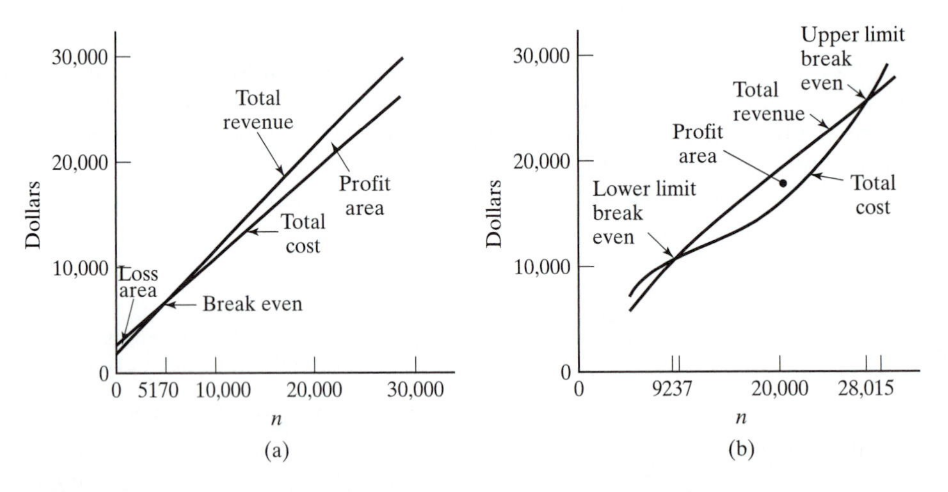

Figure 9.8 Break-even graphs of ceramic valve product: (a) Linear and (b) nonlinear.

gets smaller (i.e., marginal cost as the derivative of the total cost function). It measures the rate of increase of total cost and is an approximation of the cost of a small additional unit of output from the given level. Sometimes marginal cost is called the slope or tangent of the total cost curve at the point of interest. We define *marginal cost* as

$$C_m = \frac{dC_T}{dn} \tag{9.14}$$

where C_m = marginal cost, dollars per activity unit

$\frac{dC_T}{dn}$ = derivative of total cost function with respect to activity variable n

The marginal cost function becomes $C_m = 1.527 - 14.836 \times 10^{-5}n + 5.481 \times 10^{-9}n^2$ from the third-order polynomial regression and a marginal curve is plotted directly. The marginal cost curve is a parabola, as shown in Figure 9.9

Note that the marginal cost line intersects the lowest point of the average cost curve C_T/n. At this particular value of n, average cost is minimum and equal to marginal cost. This is shown in Figure 9.9.

The parallel to marginal cost is marginal revenue. Marginal revenue is defined as

$$R_m = \frac{dR_T}{dn} \tag{9.15}$$

where R_m = marginal revenue, dollars per activity unit

$\frac{dR_T}{dn}$ = derivative of total revenue function with respect to activity n

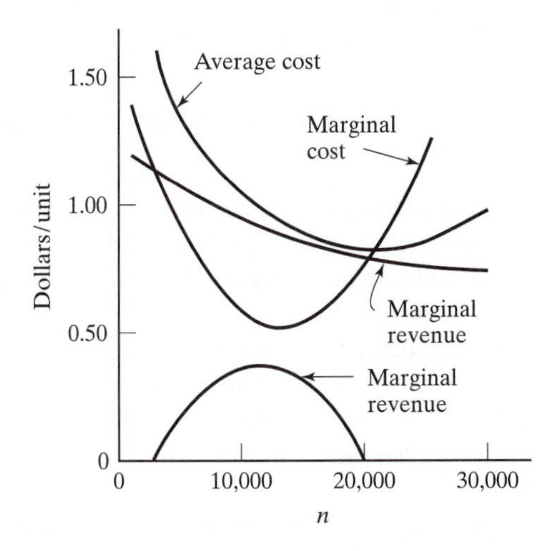

Figure 9.9 Average cost, marginal cost, marginal revenue and marginal profit for ceramic valve and guide product.

For a third-order polynomial fit, the marginal revenue function is $R_m = 1.23 - 3.28 \times 10^{-5}n + 5.499 \times 10^{-10}n^2$ from which a marginal return parabolic curve is plotted in Figure 9.9. If the total revenue curve is linear, then the marginal revenue is a constant and equals R_v. In the usual circumstance, the second-order marginal revenue model is considered more accurate than a first-order model.

Reduced revenue or unit price can occur because of a variety of economic happenings. In addition, increased unit costs beyond that which is normal are possible as activity increases indefinitely. Increasing unit costs of production as n increases can be found; for example, a process of manual insertion of components into printed circuit boards is uneconomical above a limited number of production units as the costs increase instead of declining.

C_v is an increasing function and R_v is a decreasing function. With this dual intersection of the revenue and cost lines, additional analysis of the meaning of optimum profit is called for. Thus the indefinite profit above the profit break-even point of linear models can be misleading.

With curves and mathematical models, such as those described for the example, it is possible to optimize sales revenue and production cost. The question is asked, "What shall we optimize?" Maximizing sales revenue may not guarantee maximum profit, nor does minimizing cost balance other factors for maximum profit. In some cases those actions may reduce profits. There may be more than one point of break-even operation as several points of profit-loss neutrality are possible.

The intersection of the slopes of the total cost and total revenue function is given as

$$\frac{dR_T}{dn} = \frac{dC_T}{dn} \tag{9.16}$$

The finding of this intersection is the point at which marginal profit equals zero, or

$$\frac{dZ}{dn} = \frac{dR_T}{dn} - \frac{dC_T}{dn} = 0 \tag{9.17}$$

where $\frac{dZ}{dn}$ = marginal profit, dollars per activity unit

If n increases beyond this point, then the cost of each unit exceeds the revenue from each unit and total profits begin to decline. Total profits do not become zero until the profit limit point, as they do at the upper break-even point in Figure 9.8(b), which is the beginning of a loss operation.

When the marginal profit is zero, we have the critical activity rate and maximum profit (about 12,200 units), which is the level spot on the marginal profit dome. This critical activity does not necessarily occur at the rate corresponding to the minimum average unit cost (about 21,000 units), nor does it necessarily occur at the point of maximum profit per unit.

A plot of $dR_T/dn - dC_T/dn$ is given in Figure 9.9 and is called marginal profit. The point at which $dR_T/dn = dC_T/dn$ is also zero for marginal profit or dZ/dn. The critical production rate may be determined directly from differentiation of the profit function Z. This critical activity rate is zero at n = 20,800 units, which is the point of maximum gross profit. But it is not the rate corresponding to minimum average unit

PICTURE LESSON First Television

RCA and NBC Announce

Television Broadcasting Begins on April 30th in the New York Area

...and RCA Victor Television

Receivers Are Ready Now!

Years of patient effort in RCA Laboratories—millions spent on research—now convert a fantastic dream into a splendid reality . . .

April 30th marks the birth of a new industry—television. On that day sight will join with sound to bring you a wealth of new experiences.

Television offers something everyone wants. If you live in the New York metropolitan area you can have it right now. No prediction can be made as to how soon television will be available nationally, but RCA is bending every effort to meet popular demand.

When television becomes a nationwide service it should provide new opportunities for workers. Think how recently radio was an experiment and a toy. Swiftly it became a great industry. Today, radio is a source of livelihood to thousands. RCA hopes to help in a similar growth of television in the future.

The development of television has required much research. To insure success RCA gathered in Camden, a distinguished group of scientists and engineers. A long step forward was their development of the Iconoscope, the "eye" of television, and the Kinescope, the "screen." These are the bases of RCA electronic television, and have been made available to the entire industry.

Television also had to be proved in the field. RCA has spent more than two million dollars in practical field tests of television in New York. RCA and its various subsidiary companies have been, and are, engaged in every phase of television research, engineering, manufacturing, installation, broadcasting and service. This experience is unmatched anywhere. Now the great day has arrived. A new era begins. Through RCA Victor Television Receivers you can take part in one of the greatest adventures in all scientific history. It is an adventure you will never forget.

The development of television is one more example of the ceaseless research of RCA and its various subsidiary companies. By always looking ahead, RCA seeks not only to improve the general services of radio, but to produce equipment of highest standards at moderate prices for home, industry and laboratory. That's why, in radio and television . . . it's RCA All the Way.

FACTS YOU'LL WANT TO KNOW ABOUT TELEVISION

Indications point to the operation, in the near future, of three stations in the New York area; also one at Schenectady, N. Y., and one at Los Angeles, California . . . At the average electric rate it will cost about one cent an hour to operate a television receiver. Sizes of pictures are shown in set descriptions on these pages. Beginning about April 30, 1939, NBC will provide two one-hour programs per week, plus special event pick-ups of sports, visiting celebrities, etc.

Commercial television began in New York City on April 30, 1939. The photograph shows an inaugural advertisement announcing the date of a brand new industry. "On that day sight will bring you a wealth of new experiences," the advertisement proclaimed. "A great day has arrived. A new era begins," read the joint communication by RCA Victor and NBC.

The Radio Corporation of America (RCA) was the incubator of research, engineering, manufacturing, installation, and service. At first, three television models were available. A small table "vision" model connected to a separate radio for sound. The set featured a screen size of 3 3/8 by 4 3/8 inches. A floor model was highlighted by a 7 3/8 by 9 inch white screen—the Kinescope—allowing indirect viewing via a mirror. These two models were dependent on vacuum tube technology, five television "channels," and numerous scientific breakthroughs.

There were two regular one-hour programs per week. The announcement pledged that the cost of operation would be one cent per hour to operate.

cost or the maximum profit per unit. The maximum profit per unit may be estimated from the level dome of the marginal profit curve, or approximately 12,200 units. By using Figure 9.9, minimum marginal cost is estimated as 14,500 units, and maximum marginal revenue is at the minimum sales projection of 5000 units.

Our earlier discussion about the addition of an associate to the ceramic valve assembly line, either A or B, was based on the average cost per unit as the decision-making factor. Decisions based on an average cost metric can be misleading. Look at Figure 9.9 again. The additional resource of an extra worker was assigned to line A because the cost per unit in line A at $n = 20{,}000$ units was cheaper than line B at $n = 15{,}000$ units. The nonlinear marginal cost and marginal profit, as it turns out, is better with line B than with A. The results of marginal cost (A = \$0.75 vs. B = \$0.54 per unit) and marginal profit (A = \$0.04 and B = \$0.33 per unit) underscores the usefulness of the methods of nonlinear break-even as a way to understand choices between two opportunities. Production line B is preferred for the additional worker resource because of the marginal cost and profit benefits.

Although several points of operation can be obtained from minimizing total or unit cost, maximizing total or unit revenue, or maximizing gross or unit profit, we should not conclude that all is lost if those exact objectives are never precisely realized. There is a redeeming feature if the actual and the ideal do not coincide. Fortunately, most optimums are relatively flat and not V-shaped, as would be seen with graphical plots. For those flat domes, operation on either side of the optimum may be insensitive and not influence cost or profit very much. Naturally, a converse example can demonstrate a sharp singularity as optimum. For this case, operation at this point is important.

Several comments about linear break-even analysis are worth noting. Break-even analysis is considered short term. To establish arbitrary rules about what is short or long term is conjectural unless specific cases are known. Short term for one activity may be only weeks, while another firm may view short term as an extended period of months and even years. In the next chapter, these concepts are extended with what is called engineering economy analysis where the important matters of discounting and time value of money concepts are introduced.

9.4 LIFE CYCLE COST (OPTIONAL)

Life cycle cost (LCC) analysis is a process of estimating all of the costs likely to be incurred over the economic life of a product, equipment, or a facility and expressing those costs in terms of an equivalent single value that allows comparisons among alternative designs, operating strategies, and external conditions that may influence costs. Examples are numerous, such as alternative designs of equipment, say one that is more reliable and expensive as compared to a cheaper model; or with a building having increased expenditures for materials that require less energy, which may be economical in the long run as compared to other plausible alternatives. Pay now or pay later is what LCC is about.

Intuitively, individuals have used LCC principles for economic evaluation of cars when they concern themselves not only with initial cost (sticker price), but with operating and maintenance expenses (gas mileage, worn parts, insurance, and license) and residual value (resale price). Broadly, life cycle cost is the summation of all estimated cash flows from concept, design, manufacturing and or construction, operation, maintenance, and disposal of the equipment or facility at the end of its useful life. LCC attempts to estimate all relevant costs, both present and future, in the decision-making process for the selection among various choices.

Operating cash flows can easily be greater than the original R&D or the investment. It has been shown that for military hardware systems approximately two-thirds of life-cycle costs are unalterably fixed during the design phase.[7] Some states have legislation requiring LCC in the planning, design, and construction of state buildings.

LCC gives an important role to the influence of total costs rather than initial costs of ownership. The selection is then focused on the total cost of ownership rather than the first cost of a building or equipment, for example. LCC does not consider revenue and profit as was discussed with break-even analysis. LCC is without revenues from the equipment choices, although such additions are possible for the general problem.

What is presented in this chapter is an approach to understanding the ideas behind LCC. As discussed throughout this chapter, the time horizon plays an important part. The practice of short-term versus long-term conditions are meaningful here too, and they are easily accommodated.

There are no universally accepted procedures that are standardized to do LCC analysis. In its stead, we give an overview of the several types of cost estimates that are required for LCC.

- Design, development, and engineering
- Initial capital investment and financing
- Operation, maintenance, and functional use
- Replacement
- Alteration, refurbishing, and improvement
- Salvage and retirement

Figure 9.10 illustrates engineering, project investment, operation, and maintenance costs as separate cash flows. In LCC analysis, the estimates are scheduled as

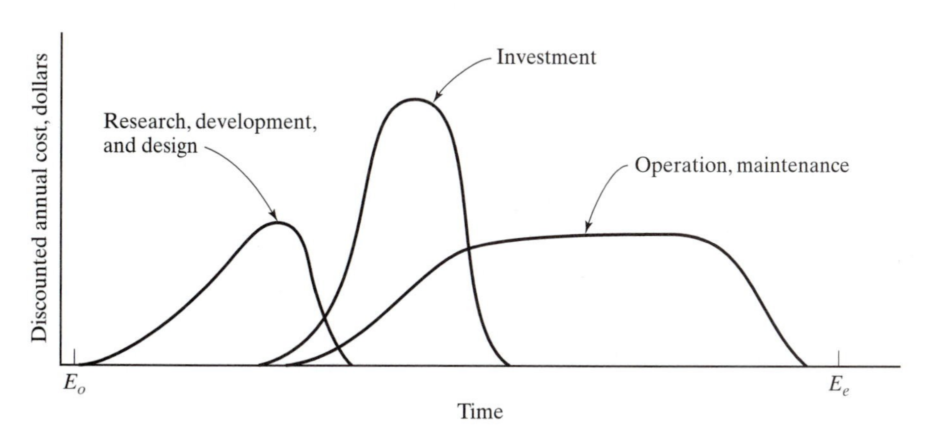

Figure 9.10 Cost-time phasing for a product life cycle.

[7]The U.S. Department of Defense is the pioneer in the popularization of the terminology of "life cycle costs," though it was engineering practice that had understood the concepts for over a century.

period cash flows, starting at the moment of preparation of the estimate, E_o to E_e, end of costing and life cycle.

The length of the life cycle is sensitive to wear-out, casualty or destruction, obsolescence, economic, or technology factors. When obsolescence is a factor, qualified opinion is required, because the life of the equipment may be suddenly terminated by a change in company policy, buying habits, government legislation, competitive pressures, or new designs.

When confronted with notification of obsolescence of electronic components within an airplane product, the avionics supplier is faced with a series of alternatives. The basic decision is whether to make a last-time buy of the components necessary to last through the life cycle of the airplane, or to initiate redesign of a replaceable unit, or printed circuit assembly in order to design the obsolete component out of the product. In complex avionics, especially those involved in flight-critical systems, redesign to give product emulation of the obsolete component, qualification, testing, software, and certification is very expensive. In this illustration we see some of the complexities of components/products with a separate life cycle that are embedded in another system.

There are two major concerns in forecasting the life of an asset: annual deterioration and obsolescence. Where physical life or annual deterioration establishes the life, statistical data from past records become the basis for future prediction. If life is regarded as economic, then statistical methods find the probability that the economic life of a proposal will terminate during each year of its service life. In either of those two cases, deterioration or obsolescence, the life of a particular proposal terminates because it is worn out or because the equipment or service is no longer profitable.

In the first instance, the physical condition has deteriorated and does not produce the desired quality, or the cost of maintenance exceeds the cost of replacement. In the absence of statistical data, reliance is placed on the opinion of engineers, operators, and the manufacturers producing the equipment and the constructors. In the case of obsolescence, competition introduces substitute products, processes, or equipment with better prices, qualities, or services.

This raises the point: "What is the life cycle?" For long-haul commercial truck tires, for instance, a life cycle may be 30,000 miles, while in public works 30 to 100 years can be used for buildings, bridges, etc. In a weapons system the horizon may terminate within a period of months to 20 years. In commercial enterprises a system may logically extend from a fad period to an enduring life of 20 to 40 years. All kinds of priorities—political and social—affect the length of time.

Salvage value is another bewildering prediction. An experienced appraiser may give opinions of future land and building values. If the life is not expected to be great, then the engineer is in a position to trust this source of information. For the longer-lived equipment, information may be unavailable. Despite the uncertainty of distant predictions, errors in evaluating salvage value are not normally serious. Error in the present or the near future is given more study because the effects are greater.

Note that in Figure 9.10 the cost of design precedes the investment of the project, which precedes the operating and maintenance costs. It is not necessary that the curves be symmetrically shaped. Also discrete spikes of lump-sum cash flows are possible—for example, salvage value. Costs conclude at the end of the life cycle.

In preparation for the LCC analysis, the following procedures and equations are required:

- Operating profile
- Maintenance schedule
- Repair data

The operating profile has a repetition time and contains all the operating and nonoperating modes of the equipment or building. It is sometimes possible to have operating profiles internal to other operating profiles. For tradeoff studies, candidates are evaluated with the same profile. The profile says when or in what way the equipment is operating.

Two parameters are mean time between failures (MTBF) and mean time to repair (MTTR). Time between overhaul, power consumption rate, and preventive maintenance routines, such as cycle and the preventive maintenance rates, are required information.

Labor for preventive and corrective maintenance is calculated by using

$$PM \text{ actions} = \frac{SOH}{PM \text{ cycle time}} \tag{9.18}$$

PM = preventive maintenance cycle time, units consistent with SOH

SOH = scheduled operating time, usually hours

$$C_{cm} = \frac{SOH}{MTBF}(MTTR)C_m \tag{9.19}$$

where C_{cm} = cost for corrective maintenance, dollars

$MTBF$ = mean time between failures, usually hours

$MTTR$ = mean time to repair, usually hours

C_m = cost of maintenance labor, usually dollars per hour

Maintenance costs need to be framed into annual costs, if the dimensions deal with hours or maintenance cycles. Typically, most life cycle costs are dimensioned into annual payments.

Example: Life Cycle Cost Analysis for Concrete-Pumping Equipment

Consider concrete-pumping truck equipment, which pipes semiliquid concrete into wall and floor forms at the construction site, a popular method in recent years. An alternate method can be hand hauled or motorized buggies, for example, which is labor intensive. The portable concrete-pumping equipment has a maximum capacity of 195-yd^3/hr and is purchased for $1,200,000 delivered cost. It has a service life of five years before the first overhaul, which restores the equipment nearly to its initial performance, and this major maintenance cost is $125,000. The restored equipment will have a life of an additional four years, and it is retired at the end of this period. The refurbished equipment is eventually sold for 15% of the initial price.

Cost analysis is not rocket science. Variability and difference in approach are normal. But it is important to write down the assumptions that influence the collection and

TABLE 9.9 LCC Information for 195-yd^3/hr Capacity Concrete-Pumping Truck Equipment

Information Element	Amount
Equipment purchase cost	$1,200,000
Service life before first overhaul	5 yr
Overhaul cost	$125,000
Remaining life after overhaul	4 yr
Concrete travel and pumping operating hours, annual	2,600 hr
Standby hours, annual	3,000 hr
Salvage value, percentage of purchase cost	15%
Periodic and unscheduled maintenance cost, percentage of first cost, annual	2.5%
Spare parts and tire replacement, annual	$10,000
Gross hourly labor costs (available during all standby hours)	$21/hr
Diesel fuel, oil, and truck operating costs, annual for 3000 hours	$45,000
Cleanup material (sacks of Portland cement, etc.), annual	$7,800
Minimum attractive rate of return, i %	20%

analysis of the data. For example, we use the start-of-year convention for expenses of labor, equipment operating costs, and investment costs as if they were prepaid. This means that the counting of the costs starts with zero time. The costs are considered discrete, inflation effects are ignored, and values are *constant dollar*. These and other facts are collected and are shown as Table 9.9.

Table 9.10 shows the LCC analysis of the equipment estimates. An interest rate of 20% discounts all future cash flows.

The results of using the discounted annual and cumulative cost values from Table 9.10 are shown in Figure 9.11. Unlike Figure 9.9, which shows individual element graphs, the LCC curves in Figure 9.11 are consolidated into annual and cumulative values. Two pumping capacities are compared, and it is seen that the smaller capacity, 175 yd^3/hr, while it has a smaller first cost, is less attractive from a life-cycle-cost examination. When considering the cost analysis of Table 9.10, and comparing that to a table for the 175 yd^3/hr capacity truck, the larger capacity equipment is better for the cumulative and annual cost models. Only a LCC can comprehend these intricacies.

As with other predictive methods, LCC analysis requires assumptions about economic life and the discount rate. Long time horizon LCC uses *discounting*. This discounting allows a systematic method of comparing streams of costs and revenues that have differing moneys and times of payment. Generally, the discounting or present worth model is used. If engineering desires an annual rate of profit as a 5% return, then a dollar received 1 year from now is really worth $(1 + 0.05)^{-1}$ dollars, and in n years $(1 + 0.05)^{-n}$ dollars right now. Or if you presently had $(1 + 0.05)^{-n}$ dollars and lent it at 5% interest compounded, you would have back 1 dollar n years from today.

A longer time horizon will lower the annual savings required to justify an increased initial capital investment. Consequently, the shorter the life of equipment or a building, the less worthwhile it is to invest in actions to reduce future operating or maintenance costs. Higher discount rates require larger annual savings to justify an initial capital investment. Consequently, the higher the discount rate, the less worthwhile

TABLE 9.10 LCC Analysis for Concrete-Pumping Truck with 195-yd^3/hr Capacity

Cost Element and Year	0	1	2	3	4	5	6	7	8	9
Equipment	$1,200,000									
Labor	63,000	$63,000	$63,000	$63,000	$63,000	$63,000	$63,000	$63,000	$63,000	
Equipment operation	45,000	45,000	45,000	45,000	45,000	45,000	45,000	45,000	45,000	
Maintenance	30,000	30,000	30,000	30,000	30,000	30,000	30,000	30,000	30,000	
Spare parts, tire replacement		10,000	10,000	10,000	10,000	10,000	10,000	10,000	10,000	
Cleanup material	7,800	7,800	7,800	7,800	7,800	7,800	7,800	7,800	7,800	
Overhaul						125,000				
Income from salvage sale										(180,000)
Total	1,345,800	155,800	155,800	155,800	155,800	280,800	155,800	155,800	155,800	(180,000)
$(P/F\ 20\%, n)$	1	0.833	0.695	0.579	0.482	0.402	0.335	0.280	0.233	0.194
Discounted annual value	$1,345,800	$129,781	$108,281	$90,2081	$75,096	$112,882	$52,193	$43,624	$36,301	$ (34,920)
Discounted cumulative value	$1,345,800	$1,475,581	$1,583,862	$1,674,071	$1,749,166	$1,862,048	$1,914,241	$1,957,865	$1,994,166	$1,959,246

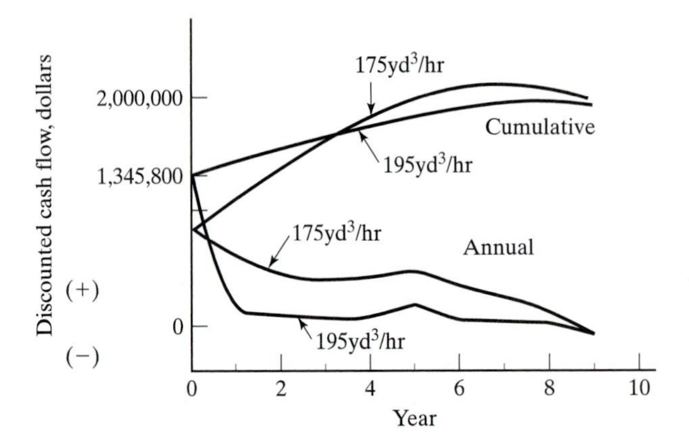

Figure 9.11 Life cycle cost comparison of two rated capacities for concrete-pumping equipment.

it is to invest in initial capital costs in order to reduce future operating and maintenance costs. As the system life cycle becomes longer, the actual length becomes less crucial, because the discounting factor drops rapidly if either i or n or both increase.

These thoughts are amplified in the next chapter. For the present, we understand that discounting will reduce future cash flows by an interest rate, which we call the minimum attractive rate of return.

In LCC, the notion is to experiment with several design configurations under various scenarios, which are exercised for future events in identical ways. Then the LCC analysis allows for visibility of "tall poles," a jargon implying significant cost elements. In LCC it may not be necessary to find *total* or *absolute* or *full* cost, as *relative* costs may separate the better alternatives from those of dubious merit.

Much software is available to conduct LCC comparisons. There is software for machine tools, equipment, concrete pipe, bridges, highways, building designs (with allowances for geographical location, materials of construction, maintenance requirements, and so forth), and other applications. The software handles estimates, inflation effects, taxes, and various discount rates as well as many of the engineering design features.

SUMMARY

This chapter presented several fundamental techniques for cost analysis. Those methods are applied to the important "tradeoff," where competing candidates in design and manufacturing are compared side by side under similar rules of calculation and engagement. The eventual objective of this work is selection of the optimal design.

As each design and manufacturing project is unique, so must the technique be adapted to the circumstances. Estimates are prepared that give a plethora of costs for those alternatives. There are intangible factors that must be logically considered and evaluated, for they too are significant.

In this chapter the time horizon for these tradeoffs is considered as short term, but long-term consequences are important, a topic to be studied in the next chapter.

Chapters 9 and 10 can thus be considered complementary, in the sense they are both necessary for effective cost analysis.

QUESTIONS FOR DISCUSSION

1. Define the following terms:

Average revenue	Minimum average cost
Break-even point	MTBF
Carry forward depreciation	Non cash expense
Cash stream	Operating profile
Escalated dollars	Real dollars
Isocost point	Taxable income
Life cycle cost	Tradeoffs
Marginal cost	Upper break-even point

2. How does a scattering of estimated points aid break-even analysis?
3. Discuss the distinction between break even and isocost points.
4. List several objectives for break-even analysis.
5. Develop the rationale for the events of tradeoff study of engineering design and manufacturing. Where in the strategy of product development is the greatest opportunity for profit?
6. Which is more important for cost analysis—real dollars or actual dollars?
7. Why is a scattering of cost estimating experiences necessary for fitting cost equations to the data?
8. List several cost break-even opportunities in engineering, and define the "activity variable" for its analysis.
9. Discuss how marginal-revenue and marginal-cost principles parallel one another.
10. Several points of optimum location are possible with break-even analysis. Describe the nature of these points.
11. Give the types of situations that lend themselves to life cycle costing.

PROBLEMS

9.1 An investor is evaluating an inventor's design, and cost and revenues are estimated for each year. Calculate the net cash flow table for this investment opportunity. Find net cash flow and cumulative cash flow. (*Hint*: Disregard taxes.)

Year	Cost	Revenue
0	$50,000	0
1	5,000	$2,500
2	5,000	10,000
3	2,500	25,000
4	0	25,000
5	0	20,000

9.2 A business earns $1,050,000 of gross income and incurs operating expenses of $825,000. There are interest payments on borrowed capital for equipment, which amount to $85,000. Depreciation for the year is $115,000. Find the taxable income for this firm. If interest had been $120,000, what is the taxable income? Discuss the consequences of operating losses.

9.3 A corporation has a gross income for the year of $3,125,000, expenses of $2,100,000 which excludes capital expenditures, and depreciation of $510,000. Find the taxable income and federal income taxes.

9.4 A company has a gross income of $5,250,000, expenses of $2,900,000 (excluding capital expenditures), and a depreciation of $1,870,000. Find the taxable income and federal income taxes. Determine the average tax rate. Discuss the tax rate schedule. State the marginal and the maximum tax rate.

9.5 Suppose that the federal tax rate is 34% and the state tax rate is 9% on ordinary business operations. Find the effective income tax rate.

9.6 A manufacturer has the following yearly facts for its business operations: sales revenue, $200,000; depreciation, $40,000; and other expenses, $130,000. The income tax rate for this firm is 20%. Find the gross income before taxes, net income after taxes, and net cash flow.

9.7 An owner forecasts the following cash flow for a commercial development:

Year	Cost	Revenue	Amount for Depreciation
0	$60,000,000	0	0
1	5,000,000	$20,000,000	$9,000,000
2	5,000,000	20,000,000	9,000,000
3	3,000,000	15,000,000	9,000,000
4	4,000,000	15,000,000	9,000,000
5	5,000,000	10,000,000	9,000,000

The first year cost includes $45,000,000 as fixed assets and $15,000,000 as initial operating expenses. If the company has an effective tax rate of 40%, determine the annual net cash flow. Find the period when positive cumulative cash flow starts. (*Hint*: Assume that this investment is the only asset for the company.)

9.8 The estimated after-tax cash flow in actual dollars for a design alternative is given:

End of Year, n	Cash Flow
0	−$72,500
1	−20,500
2	75,500
3	82,000
4	110,000

If the inflation rate is expected to average 5.2% per year during this period, what is the real dollar equivalent of these actual amounts? (*Hint*: Use a convention called *end of year*, which says that the dollar amounts are seen as lump sums that occur precisely at this point in time. The base or reference time period is year 0.)

9.9 Three years ago a firm purchased computer servers for $25,000 each. Nowadays, a similar server, but with improved processing speeds and storage features, costs $22,000. Ignoring the technical improvements, what is the rate of deflation for the computers? During the same period the Producers Price Index increased from 123 to 131. Find the differential rate of inflation for the computer. (*Hint*: Use the PPI in finding the differential rate.)

9.10 A quotation is dimensioned as a *unit-price contract*. The buyer requires a quotation expressed in current dollars, but allows *escalation* of the unit prices based on evidence of inflation as demonstrated by a reliable index.

(*Hint*: A unit price contract means that the number of units that are manufactured are not a quoted part of the offer as the number can vary, and a verified count is determined after all product is delivered.)

End of Year, n	Unit Price Estimate
0	$538
1	538
2	538
3	538
4	538

The Producer Price Index is expected to average 4.7% per year for the next several years. Find the actual dollars per product year for the unit-price contract.

9.11 A supplier has estimated that the direct material and labor costs are $10 per unit. For this product, fixed cost is $30,000 per year, and the selling price is $16 per unit. Find the break-even. What is the profit if 10,000 units per year are made and sold? If 4000 units are sold per year?

9.12 An injection molding company produces toys, and its practices are similar to a commodity business, as production quantity is driven suddenly by changes in demand. This firm isolates the production equipment for one toy product, referring to this separation as a "cell." The fixed costs for this cell are $9 million, and the direct labor and material are $12 per unit. These costs are for a production of one million units per year for a two-shift per day basis. Find the costs of production and the current unit cost.

The company is able to sell the toys at $32 per unit. Find the total revenue and total profit from the two-shift operation.

Business is very good, and the company is considering the addition of a third shift for the cell. If the fixed costs and variable costs remain the same, what is the new unit cost if the third shift adds 350,000 units per year to the output? What is the total revenue and profit if the selling price of the incremental output is the same at $32 per unit?

Realistically, fixed costs increase due to increased management costs for the third shift, and there are other additions too. Now assume that the fixed costs increase to $10 million. Further, with the increased volume, price is expected to lower to $29 per unit while variable costs remain the same. For this increased business opportunity, find the total revenue, total cost, and total profit.

Plot the total revenue, total cost, variable cost, and fixed cost against quantity per year for the realistic planning model. From the graph, what is the approximate break-even point?

9.13 Printed circuit board light-directed stuffing machines aid the operator in the insertion of electronic components into a printed circuit board. These semiautomatic machines are often compared to situations that are more labor intensive. In the semiautomatic case, fixed costs are higher, but direct labor costs are lower when compared to bench stuffing.

For the selection of the most economical process, revenues and operation quality are identical and are ignored. Direct material costs are equal between the two operations.

For a printed board product, the labor cost is $1 per unit and the fixed cost is $2500 per year for the semiautomatic machine. When manually stuffing the board, the labor cost is $1.50 per unit and the fixed costs are $1800 per year. Find the quantity where the two processes are equal in cost, or referred to as the *isocost* point. Plot the comparison.

9.14 Manufacture of an electronic product requires $10,900 for annual fixed cost for its processes. The direct materials, which are mostly purchased, are $5 per unit, and labor costs for assembly are $5.50 per product unit. The unit can be purchased from the Far East, and on-shelf cost for inventory stocking to the United States is $14.50 per unit. The first-year quantity is 5000 units. Should this company make or buy the product?

(*Hint*: When purchasing a product, the price per unit is considered a variable cost, and the buyer is unaware of any fixed costs that the supplier encounters. Nor is the buyer aware of the profit markup, so the analysis compares the buyers unit price against the make cost.)

Consider the two situations identical in quality and specification. What is your recommendation as to whether the unit should be manufactured in the United States, or purchased?

9.15 Product planning for a new design has developed three methods of manufacture. A batch method can be devised to make the product to meet design specifications and delivery requirements using existing equipment and available floor space. A second alternative is to use automated equipment, and a small inside factory will satisfy requirements. The third is to outsource the product from suppliers; bids are received, and the lowest quotation is known. Information is given as shown below.

Technology/Policy	Fixed Cost	Variable Cost/Unit
Batch factory layout and make	$80,000	$17.00
Flow line layout and make	180,000	9.80
Purchase from supplier		29.00

Plots can be used to determine the isocost points for the three technologies. Give the range of product quantity and the policy for obtaining the product. (*Hint*: Revenues are not necessary to find the optimum ranges for selection of the policy of satisfy the requirements. A purchase decision will not involve a fixed cost, that being a part of the price given by the supplier in this situation.)

9.16 An existing process is at a maximum rate because of limitations due to production. Potential process improvements would add robotic handling, increasing the output from 8000 to 9000 annual units. It is expected that variable costs would be unchanged, though there are additional fixed costs for the process improvements. Marketing believes that the price under the new quantity plan would be discounted from the current price because of greater product availability. The estimates are summarized below:

	Fixed Cost	Variable Cost/Unit	Price/Unit	Sales, *n*/year
Existing process	$25,000	$13.80	$21.50	8,000
Incremental improvement	10,350	13.80	20.00	9,000

Does it make sense to improve the process for the additional quantity that will be available for sales?

9.17 There are two methods to treat fouled processing water, A and B. Information is as follows:

Method	A	B
Labor, $/year	140,000	190,000
Depreciation, $/year	100,000	$56,000
Chemical and treating materials, $/m^3	0.15	0.10

Find the annual amount of water needed to make the two methods equal. (*Hint*: Evaluate this problem as short term.)

9.18 An automotive supplier has net sales of $600,000 annually. The annual fixed costs are $350,000 and the direct costs are 31.5% of the net sales dollars for this level of business. Find the gross profits. What is the break-even point in terms of the sales dollars? Find the annual sales to have a gross profit of $80,000.

9.19 Work is estimated to have a fixed cost of $25, variable cost of $1/unit, and a semifixed cost as shown below:

Number of Units	Semifixed Cost
1–10	$5
11–20	10
21–30	15

Find the total cost for 15 units and for 30 units. What is the average cost for these quantities?

9.20 A cost schedule of electrical rates is given as

Energy	Cost
First 50 kWh or less/month	$25.00
Next 50 kWh/month	$0.15/kWh
Next 100 kWh/month	$0.09/kWh
Next 800 kWh/month	$0.06/kWh
Over 1000 kWh/month	$0.25/kWh

What is the cost for 500 kWh? For 1250 kWh? Find the average cost per kWh for these power consumptions. Discuss how "block" rates relate to fixed, semifixed, and variable cost concepts. Because this is a power company committed only to rural and farm service, how is it discouraging new business to commercial users?

9.21 A contractor specializes in concrete airport/road and commercial parking surface paving. The engineer determines monthly fixed and variable costs as $10,000 and $2.50/ft^2, respectively. Semifixed costs are estimated as follows:

Construction Rate, 1000 ft^2	Monthly Semifixed Costs
0–4,999	$40,000
5,000–8,999	60,000
9,000–15,999	140,000
16,000–20,000	200,000

Determine two plots, total cost versus construction rate, and cost per ft^2 versus units of area. If the construction engineer anticipates that the concrete paving rate will be in the range of 12,000–14,000/1000 ft^2 for the forthcoming summer months, what simplifications can you make to the cost analysis?

9.22 An opportunity sale of an additional 200 units is received above an existing production activity. The buyer is prepared to pay $35,000 for 200 units. Cost for the current 500 units is $95,000. Total cost for producing the 700 units is $127,700. What is the total marginal cost? Find the unit marginal cost? What is the profit or loss? Find the minimum opportunity sale price to break-even?

9.23 A sphere has the following weight and cost information, (lb x, cost y): (2, 2), (3, 3), (4, 5), and (5, 8). Sketch the cost curve, average cost curve, and marginal cost curve. Do those data comply with the *economy of scale* law?

9.24 The total cost for manufacturing a product varies according to the number of units and is given as follows:

Product Units	Total Cost	Product Units	Total Cost
0	$ 0	5	$ 19,200
1	10,000	6	20,500
2	15,000	7	23,000
3	17,500	8	28,000
4	18,700	9	38,000

Sketch the total cost curve, average cost curve, and marginal cost curve. (*Hint*: Use one graph for the three curves, but the graph will have two different vertical axes.)

9.25 A price for a product is given as $P = \$21{,}000n^{-1/2}$/unit. On the cost side, variable cost per unit is $1000 per unit and fixed costs are $100,000 per period. Find the marginal revenue function. Determine the break-even quantities. Find the point at which profit is a maximum indicating the amount and quantity.

9.26 A marginal revenue function = $500 + 0.10n$; marginal cost = $1000; point of maximum profit = 5000 units and $5000; and fixed costs = $10,000. Find the demand function for price, quantity per period, and total cost function for the same period.

9.27 The firm's cost profile is given as follows:

Product, n	Total Fixed Cost, C_f	Total Variable Cost, nC_v
0	$3000	$0
1	3000	700
2	3000	1300
3	3000	1800
4	3000	2400
5	3000	3100
6	3000	3900
7	3000	4900
8	3000	6200
9	3000	7800

At what product n are minimum marginal cost and minimum average total cost found? Plot the average fixed cost, average variable cost, marginal cost, and average total cost. Where does the marginal cost curve pierce the average variable cost curve?

9.28 Consolidated revenue and cost functions are found by regression methods and are given as $R_T = 100n - 0.001n^2$ and $C_T = 0.005n^2 + 4n + 200{,}000$ where n is the number of units made and sold. The selling price of finished units varies according to $P = (100 - 0.001n)$ dollars per unit. Fixed costs are \$200,000 per month. The variable cost is found to be $(0.005n + 4)$ dollars per unit. The plant is efficient at about 12,000 units per month.

Find the quantity for the greatest profit, the least average unit cost and the break-even points. Show the curves for average unit cost, marginal cost, marginal revenue, and marginal profit. (*Hint*: A spreadsheet will be helpful for the plots.)

9.29 An original equipment manufacturer announces that a popular type of equipment is up for bid and an LCC approach is mandatory. The firm publicizes that an LCC model to select the winning bid is based on the following formula:

$$\text{Cost per unit} = \frac{\text{unit price} + \text{unit qualifying and stocking cost}}{\text{bid MTBF}}, \text{ dollars per unit}$$

The equipment qualifying and stocking cost of \$11,000 per unit is the total cost for storage, transportation from central inventory, mobilization at the factory floor, setup and final disposal. Each bidder supplies the following information:

Bidder	Unit Price	Bid MTBF
1	\$100,000	800
2	125,000	615
3	117,500	917

Which bidder wins the contract? Discuss the merits of this sort of analysis. Propose another scheme to evaluate manufacturing equipment using LCC principles.

9.30 The U.S. Army Corps of Engineers issues purchase instructions informing potential suppliers that roadway graders will be evaluated using the following life cycle cost model:

$$\begin{aligned} \text{LCC} &= \text{unit operating cost, dollars} \\ &= (\text{unit price} + \text{logistic cost}) + \text{service life} \end{aligned}$$

The instructions require that the selected equipment supplier demonstrates service life in a post-award reliability acceptance test. If the reliability test does not meet the level guaranteed by the contract, then a penalty function deducts from the unit price as

$$\begin{aligned} \text{penalty cost} &= (1 - \text{test value MTBF/quoted MTBF}) \times (\text{unit price} + \text{logistic cost}) \\ \text{where } MTBF &= \text{mean time between failures, service life, hours} \\ \text{logistic cost} &= \text{value supplied by U.S. Army Corps of Engineers, dollars} \end{aligned}$$

The following bids are received:

Supplier	Unit Price	MTBF, Hr
A	\$30,000	1000
B	35,000	1200
C	60,000	1635

The buyer determines that the logistic cost is $25,000 and supplies this information to qualified suppliers. Find the winning supplier for the road equipment. Now suppose that the supplier fails to meet the quoted MTBF by 10%. What is the penalty cost and the final price for the roadway grader?

9.31 A national transport truck firm has 17,135 53-ft trailer and tractor rigs. These 16-wheel rigs will wear through a number of tires annually. The company establishes a life cycle cost model to evaluate tires used for its trucks. The model is given as

$$\text{LCC} = (\text{number of tires})(\text{unit price} + \text{shipping cost} + \text{maintenance cost}), \text{dollars}$$

Three tire manufacturers are invited to bid. Each supplier provides a sample for simulated working tests to determine the number of miles per tire. Then the truck firm determines the number of tires for each company's quote based on a tire mile index, which is given by number of required miles divided by best performance. Explosive loss of tires is not a consideration here. The number of annual miles for a tractor-trailer is 150,000. Shipping costs are evaluated from the supplier's factory to a central inventory. Maintenance cost to change a tire is $75. Make an evaluation to determine the LCC price and recommend the source of purchase.

Supplier	Tested Miles/Tire	Shipping Cost/Tire	Bid Price/Tire
A	31,000	$40	$180
B	30,500	80	170
C	29,500	30	160

9.32 A manufacturer announces that electronic fabrication equipment is evaluated using the following LCC model:

$$C_M = \left(\frac{SOH}{MTBF}\right)(MTTR)C_m + \text{Unit Price, dollars/unit}$$

where SOH = scheduled operating hours

$MTBF$ = mean time between failures, hours

$MTTR$ = mean time to repair, hours

C_m = cost of maintenance labor, dollars/hour

The following bids are received. Which supplier do you choose?

Supplier	Unit Price	SOH	MTBF	MTTR	C_m($/hr)
A	$15,000	50,000	10,000	150	$75
B	28,000	45,000	15,000	140	75
C	39,000	40,000	20,000	130	75

CHALLENGE PROBLEMS

9.33 A product investment cost is estimated as $100,000. Future inflation for the next five years is expected to be 6% per year. The company has a real interest rate of 4%. Find the equivalent worth of the product investment cost five years from this base point for the following conditions:

$$i_r = \text{real interest rate} = 4\%$$

$$f = \text{inflation rate} = 6\%$$

(*Hint*: This problem indicates two rates, interest and inflation, and asks for a future equivalent value adjusting for inflation and interest rate. The real and the inflation rate have a joint compounding effect.)

$$i_c = (1 + i_r)(1 + f) - 1$$

where i_c = combined interest rate, percent/period

and $F = P(1 + i_c)^n$

where F = future amount, dollars

P = present amount, dollars

9.34 There was a time from the early 1900s to the mid-1920s when horse wagons and internal-combustion engine trucks competed for economical haulage of basic materials.[8] Consider the following choices:

Comparison Item	Single Horse Wagon	700-lb Capacity Light Delivery Truck
Initial cost	\$430	\$910
Estimated life, year	10	2.5
Total operation cost/working day (300 days/year)	\$3.40	\$7.10
Total cost/idle day	\$1.74	\$1.21
Average mileage/day	20	60

Construct a break-even analysis and make suggestions for operation during this period. (*Hint*: Assume that the load carrying capacity are identical between the competing methods (which incidentally was nearly true), except that distance does vary.)

9.35 A product plant is being considered for a new booming market. Total fixed cost for design, land, and plant construction is $\$10^5$. The variable cost of producing product is a function $C(n) = 100n - 3 \times 10^{-3} n^2 + 10^{-7} n^3$ where n = number of product produced.

(a) Determine the total cost function *C(T)*. (*Hint*: It includes the cost of producing the product and fixed cost.)
(b) Find the marginal cost function C_m for producing the product.
(c) Find the number of products for minimum marginal cost. What is the marginal cost at this point? What is the average cost per unit at this level of production?
(d) Total revenue is given by $S(n) = 250n$. Find the marginal price. Determine the functions for total profit, and marginal profit.
(e) Find the point to maximize profit.
(f) Graph these cost equations. Your instructor will give additional instructions.

[8]Incidentally, fresh milk was delivered to homes until the late 1930s with horse wagons. The reason that horse wagons were chosen over internal-combustion trucks was that during the milkman's walk to the house with the milk, the horse would knowingly move to the next location and wait for the milkman there. The horse would also respond to distant voice commands. The truck was no cost match to the intelligence of the horse for this application.

9.36 Engineering studies of three plant sites, located at mile markers 103, 111, and 123, along a to-be constructed highway show a daily demand of 20,000, 7,000, and 6,000 yd^3 of concrete in each location for 6 days per week. Two plans, A and B, meet this demand.

In plan A, one central and large plant having a daily capacity of 40,000 yd^3, is located equidistant between the locations and produces concrete with a fixed cost of $20,000 per day, and a variable cost of $1.50 per yd^3.

Alternatively, plan B has portable and distributed operations in each location with capacity of 24,000, 8,000, and 8,000 yd^3. Fixed costs are $15,000, $7,000, and $7,000 per day. Because of lower transportation costs, variable costs per yd^3 are $1.00, $1.10, and $1.20 per yd^3.

(a) Which alternative, A or B, is more desirable based on demand requirements? (*Hint*: Consider this tradeoff study as "short term.")
(b) If sales were to increase to production capacity, does the best choice change?
(c) What variable cost per cubic yard may plan B increase to for an isocost equal to A at demand capacity? At full capacity?

PRACTICAL APPLICATION

There are many opportunities for a practical experience in cost analysis for design and engineering. Your instructor may suggest several. This is a possibility.

Contact and visit a design firm and inquire about cost analysis that they do. Prepare a list of questions in anticipation of the visit. What popular methods do they use? Be professional in your arrangements, and afterward write a report detailing the lessons that you have learned. Send them a courtesy letter expressing your thanks after the visit.

CASE STUDY: OPTIMAL INJECTION-MOLDING TOOL COST

The Don Boyle Company makes zinc die castings. It has one 300-ton machine with a trim press located at the end of the quench conveyor. The machine is a hot chamber type and is capable of running automatically.

Andy James, Inc., manufactures leather goods. The purchasing agent for Andersons, Inc., calls Andy James (and several other manufacturers) for a quote of a 3-in-wide belt. The belt is to be made with an adjustable buckle. The James engineers design a zinc die-cast chrome-plated buckle, and call Don Boyle (and several other die casters) to quote on the buckle. Andersons has estimated a volume of 75,000 belts annually for the next three years, when the item will be phased out.

Don Boyle must quote Andy James a unit price and a tooling price. He has a 300-ton machine capable of casting up to 16 buckles at one shot. His trim press is also capable of handling a shot this size. How does he quote this part to stay competitive? A 16-cavity die can minimize the unit price, but the tooling cost could be enormous. A single cavity die maximizes the unit price but minimizes the tooling price. Which alternative should Don choose?

A single-cavity casting die, Don estimates, would cost about $4500. A trim die would cost an additional $900. Direct labor and overhead to cast one shot, no matter how many parts it has, is 6 cents. In other words, to cast the required 225,000 buckles in the next 3 years, the lowest cost for tooling is $5400. However, the unit cost (excluding tooling) is 6 cents per part, or $13,500 for the total of 225,000 units (no allowance is made for waste or scrap, as waste and scrap are returned to the pot and used again as good product.)

Now, the marginal costs and marginal revenues must be examined. Marginal revenue is defined as the increment of total revenue (plus or minus) that results when the number of cavities is increased by one unit (or fraction thereof). Marginal cost likewise is defined as the increment of tooling cost that results from a similar increase in cavities. (Total revenue does not refer to the amount of money to be paid by Andy to Don. That would be a pricing problem, whereas this discussion concerns the problem of an efficient design.) Total revenue here is the total dollars saved by using a two-cavity die compared with a one-cavity die, or three cavities against two, and so forth. This total revenue may be called unit savings.

To cast the buckle for Andersons, the minimum starting point is single-cavity tooling. Total revenue at this point is zero. Don has not yet experienced a saving as a result of his tooling choice. A two-cavity design, Don estimates, would cost an additional $600 for the casting die and $200 for the trimming die. The marginal cost is $800, and this does not include unit cost. Unit costs are considered only in respect to changes that create a savings or loss. A two-cavity die will produce two parts per shot, reducing the unit cost to 3 cents per part, or $6750 for the three-year requirement. This is a reduction from $13,500 to $6750, or a unit saving of $6750.

For each additional cavity added to the tooling Don estimates that the tool cost will increase by an additional $800. With 16 buckles per shot, the maximum capacity of the machine has been reached. Anything over 16 buckles per shot would require a larger die casting machine.

Where does Don maximize the profits from a tooling standpoint? How many cavities does this decision call for? What is the full cost for this decision? How does raw material affect your decision?

Hint: To answer these questions develop a spreadsheet with the following column heads:

Cavities per Die	Tooling Cost	Unit Cost	Total Revenue	Marginal Cost	Marginal Revenue	Marginal Profit

Finally, develop a plot of marginal profit, marginal revenue, marginal cost, and full cost against number of cavities per die.

Chapter 10

Engineering Economy

This chapter provides a useful discipline for making choices among engineering alternatives where the project is dominated by an investment. Attentiveness to these principles will contrast decisions that should not be made on unsupported guesstimates. The units for the decision-making evaluation are in money equivalences, but the added feature at this point is that some amounts of money are future revenues or costs. Their timing influences the selection of the alternative.

Knowledge of earlier chapters is crucial to employ the principles. Operation, product, and project cost estimates are necessary for the models. Accuracy and thoroughness in this phase are important to the success of the decisions that are recommended by these engineering economic methods. Additionally, accounting information is important.

The analysis typically starts with the physical environment where engineering proposals encourage the development, building, making, or producing materials into desired buildings, consumer products, highways, machine tools, cars, computers, electronic products, airplanes, equipment, plants—the list is endless. A systems view is an incisive way to appreciate the intricacies of these maneuvers.

Engineering economy is a course found in many colleges.[1] Numerous textbooks are available and some are listed in the References. This chapter is a condensation of the important principles. Future engineers and managers need to understand engineering economy.

10.1 IMPORTANCE

In the business world, the purpose of converting money into designs, buildings, plants, and equipment is to return an amount of money that exceeds the investment. This statement assumes that capital is productive and earns a profit for its owner. In

[1]The American Society of Engineering Education has a division, Engineering Economy devoted to the teaching and encouragement of these principles. Similarly, *The Engineering Economist* is a journal that provides the latest information on research activities.

PICTURE LESSON Alexander Graham Bell and the Early Telephone

Library of Congress.

During the middle of the nineteenth century, news-bearing messengers traveled by train, horse, stagecoach, or foot. Even the telegraph, which could transmit information long distances "instantly," relied on people to decode and hand-deliver the message. An "urgent" telegram could take five minutes or even weeks to arrive.

The true pioneer in the field of telecommunications was Alexander Graham Bell (1847–1922), an inventor who changed the character of sending messages. His lifetime interest in the education of the deaf led him to invent the microphone, and in 1876 his "electrical speech machine," which we now call a telephone. By 1878 Bell had set up the first telephone exchange in New Haven, Connecticut. Long distance connections were made between Boston and New York City by 1884. Bell is shown opening the long-distance line from New York to Chicago in 1892.

Long-distance technology had arrived in 1892 with the invention of a signal amplifier, but many innovations were necessary before telephoning would truly cover long distances. The triode vacuum tube, invented in 1906 by American Lee De Forest, further amplified electrical signals to make long distance feasible, and the use of the loading coil, which connected to the cable every mile or so, increased speaking ranges to approximately 1000 miles.

Plans to connect the east and west coasts of the United States began in 1914 and the job was completed a year later. It had taken 39 years to go from a single telephone in Boston to 11 million nationwide. Today, estimates put that number closer to 200 million, plus an additional 100 million cell phones.

efficiency terms, productive capital is related to a ratio of output to input. Unlike physical processes, the economic efficiency of capital, assuming long-term success and a capitalistic society, must exceed 100%. The productivity of capital arises from the fact that money purchases more efficient procedures for supplying services and making goods than people can employ themselves. Those services and products are offered to the public at attractive prices that return cost and profit to the investor.

In earlier chapters, cost and price are emphasized as the metric of importance. For purposes of capital investment, however, the owner of the capital employs a term called *return*.[2] Like cost and price, return is expressed in several ways, including total dollars, percent of sales, ratio of annual sales to investment, or return on investment. Return on investment is favored by engineers and management.

The following methods are used for calculation of return on investment:

- Average annual rate of return
- Payback period
- Engineering-economy rate of return

In this book we favor the compound-interest-based investment computation that considers the time value of money. However, as a background let us look at methods that are used because of their simplicity.

10.2 AVERAGE ANNUAL RATE-OF-RETURN METHODS

In some economy studies, return on investment is expressed as an annual percentage basis. The yearly profit divided by the total initial investment represents a fractional return or its related percent of return. This recognizes that a good investment not only pays for itself but also provides a satisfactory return on the funds committed by the firm. This is one of several variations:

$$\text{Return} = \frac{\text{earnings per year}}{\text{investment value}} \times 100, \% \tag{10.1}$$

Earnings are after tax, and deductions for depreciation usually represent some average future expectation.

Example: Return on Investment

The investment for new equipment is $175,000, salvage will provide $15,000, and an average earnings of $22,000 after taxes is expected. Then

$$\text{Return} = \frac{\$22{,}000}{\$160{,}000} \times 100 = 13.8\%$$

Now, assume that an investment opportunity of $25,000 has come to your attention. This is broken down to $20,000 for the investment and $5000 for initial working capital (cash, accounts payable, and so forth). Annual operating and other expenses are estimated at $10,000, and income is estimated as $15,000 per year. Return is then equal to $5000/25{,}000 \times 100 = 20\%$.

---

[2]The word "owner" can imply shareholder, proprietor, or titleholder, for example.

Another variation is expressed as

$$\text{Return} = \frac{\text{average earnings} - (\text{total investment} \div \text{economic life})}{\text{average investment}} \times 100, \% \quad (10.2)$$

The earnings in the formula are the average annual earnings after taxes, plus appropriate depreciation charges. The original investment is recovered over the economic life of the proposal by subtracting the factor of total investment divided by economic life from average earnings. This difference denotes the average annual economic profit on the investment.

The *average* investment is defined as the total investment times 0.5, acknowledging that the life of an investment for tax purposes and its true economic life are not the same.

The original investment is based on the normal physical life or as legally defined by the Internal Revenue Service code.[3] Average investment represents the profitable life of the investment, which is frequently a different period. If engineering desires, it may incorporate a risk element by further shortening economic life.

Example: Return with After-Tax Earnings and Depreciation

An average after-tax earning of $22,000 is expected from an investment of $175,000 having an economic life of 10 years. Straight-line depreciation is assumed for a period of 12 years, and salvage is $15,000.

$$\text{Return} = \frac{22{,}000 + (160{,}000/12) - (175{,}000/10)}{160{,}000 \times \frac{1}{2}} \times 100 = 22.3\%$$

These "return" methods are popular and are framed in a number of ways. But another well-liked method is "payback," which we study next.

10.3 PAYBACK-PERIOD METHOD

The payback-period method is easy to understand and is widely adopted. Essentially, the method determines how many years it takes to return the invested capital. The formula is normally given as

$$\text{Payback} = \frac{\text{net investment}}{\text{annual after tax earnings}}, \text{ years} \quad (10.3)$$

The payback method recognizes liquidity as the important feature for the economic worth of capital expenditures. Payback dumps proposals of doubtful validity from those that call for additional economic analysis. Obviously, payback signals the immediate cash-return aspect of the investment, which may be desirable for corporations where a high-profit investment opportunity and limited cash resources exist. In

[3]See www.IRS.gov for information. This Website is a megasite and it offers an opportunity to broaden your understanding of the IRS Code.

some situations, the payback metric is used for those investment situations where the risk does not warrant earnings beyond the payback period.

Example: Years Payback

The installed cost for new equipment is \$175,000, and old equipment will be sold for \$15,000. Better productivity of the new equipment will earn \$29,400 additionally. This assumes constancy of the earnings. For a composite 25% corporate tax rate, after-tax earnings amount to \$22,000 (= 29,400 × 0.75).

$$\text{Years payback} = \frac{175{,}000 - 15{,}000}{22{,}000} = 7.3$$

Thus the investment requires about seven years before the earnings have liquidated the first cost. Now, consider two investment opportunities:

Investment	Equipment A, \$60,000	Equipment B, \$60,000
Revenue		
Year 1	20,000	30,000
Year 2	20,000	30,000
Year 3	20,000	30,000
Year 4	20,000	
Year 5	20,000	
Total annual after-tax earnings	\$100,000	\$90,000
Payback period	3 years	2 years

In this case, equipment B is preferred over A because of the shorter payback period. Notice that payback for equipment A is 3 years (= 60,000/20,000). If sufficient resources to purchase investments were available, then a management fiat could approve for spending any investment where payback was under some arbitrary cutoff level, such as five years.

The average annual rate-of-return method is acknowledged to have faults. It assumes equal distribution of earnings throughout the economic life of the asset. Even if this assumption held, there is a significant difference between the value of the dollars earned in the first year and those earned in later years. The time value of money is ignored. A project yielding savings in early years of its life is more beneficial, because those funds become available for additional investment, or for alternative use. Those early funds are subject to less risk than savings projected many years ahead. Furthermore, differences in salvage values and their relation to the time element are overlooked. Nor is interest on borrowed money in any way reflected in the above equations.

The payback method suffers similarly. The life pattern of earnings is ignored in payback formulas. In the example, equipment B had a shorter payback period than equipment A, yet A will return \$10,000 more. New equipment may be profitable during the early part of the payback period. On the other hand, new equipment may be quite profitable in the future. Payback does not provide for a technique of ranking with other investment possibilities. Nor does it take into account depreciation or obsolescence or the earnings beyond the payback period. For example, the payback method does not recognize that one investment earning $10,000 the first year and $2000 the second year is more desirable than another that earns $6000 in each of the two years. The use for which payback is suited, and then only provisionally, is as a rough measure of evaluation.

The payback and rate of return are used provisionally, and it is necessary to morph these linear methods into advanced quadratic procedures. The engineering-economy method of determining return overcomes those shortcomings. It is applicable to every possible type of prospective investment and it yields answers that permit valid comparisons between competing projects.

10.4 TIME-VALUE-OF-MONEY METHODS

The time-value-of money method applies compound-interest formulas to the additional cash flow produced by the investment. This concept enables engineering to place a value on the money that becomes available for productive use in the future as well as for the money available today. Fundamentally, the time value of money begins with *simple interest*, or

$$I = Pni \tag{10.4}$$

where I = interest earned, dollars

P = principal sum, dollars

n = number of periods

i = interest rate, decimal

This formula can be restated as the amount including principal and simple interest that must eventually be repaid, or

$$F = P + I = P(1 + ni) \tag{10.5}$$

where F = principal and interest sum at some future period, dollars

Payment of simple interest is made at the end of each time period, or the sum total amount of money is paid after a given length of time. Under the latter condition there is no incentive to pay the interest until the end of the contract time.

If interest is paid at the end of each time unit, then the lender could use the money to earn additional profits. Compound interest considers this point and requires that interest be paid regularly at the end of each interest period. If the payment is not made, then the amount due is added to the principal, and interest is charged on this converted principal during the following time unit.

An initial loan of $10,000 at an annual interest rate of 5% would require payment of $500 as interest at the end of the first year. If this payment were deferred, then the interest for the second year would be ($10,000 + $500)(0.05) = $525, and the total *compound* amount due after two years would be $10,000 + $500 + $525 = $11,025.

When interest is permitted to compound, the interest earned during each interest period is permitted to accumulate with the principal sum at the beginning of the next

TABLE 10.1 Derivation of Basic Compound-Interest Formula

Year, n	Principal at Start of Period	Interest Earned During Period	Compound Amount F at End of Period
1	P	Pi	$P + Pi = P(1 + i)$
2	$P(1 + i)$	$P(1 + i)i$	$P(1 + i) + P(1 + i)i = P(1 + i)^2$
3	$P(1 + i)^2$	$P(1 + i)^2 i$	$P(1 + i)^2 + P(1 + i)^2 i = P(1 + i)^3$
—			
n	$P(1 + i)^{n-1}$	$P(1 + i)^{n-1} i$	$P(1 + i)^{n-1} + P(1 + i)^{n-1} i = P(1 + i)^n$

interest period. This compounding is shown in Table 10.1. The resulting factor, $(1 + i)^n$, is referred to as the single-payment compound-amount factor. Notice the arithmetic of the progression of the exponents.

Reconsider the quantity called F, as defined above, implying an interest and principal sum at a future period. Typically, the total amount of principal plus compound interest due after n periods is defined as follows:

$$F = P(1 + i)^n \qquad (10.6)$$

$$\text{future amount} = (\text{present amount})(1 + i)^n$$

where F = future amount, dollars

i = interest rate per interest period, decimal

n = number of compounding periods

The single-payment compound-amount factor is used to solve for a future sum of money F, interest rate i, number of interest periods n, or a present sum of money P when given the other quantities.

Engineering-economy methods are preferred because they depend on time-value-of-money principles. But do not conclude that all methods employing interest computations are useful for all occasions. Some have limited applicability.

When those methods are given correct information and properly understood, their answers are equally valid. Though the answers that arise from these methods are not identical, of course, they are *equivalent*, given the number of compounding periods and interest rate and amounts. Thus equivalency implies a different idea when compared to equality. We present four distinct and equivalent variations:

- Net present worth
- Net future worth
- Net equivalent annual worth
- Rate of return

Each of those methods measures a different factor of the investment. They give different evaluations. Nonetheless, they generally lead to the same recommendation for consistent decision making.

Each of the four methods is demonstrated with the same standard numbers. (Cents are dropped from calculations for ease of understanding.) The $1025 can represent the cost of manufacturing equipment, for example. Notice that in this simple problem the revenues vary each year.

Year	Cost	Revenue
0	$1025	$0
1	0	450
2	0	425
3	0	400

Simplifying procedures are adopted for these engineering-economy methods. While the investment and revenues may be actually distributed throughout a year, we assume that nonuniform or annual revenues are instantaneously received at the end of year. Investment is scheduled at time 0, or the origin of the project.

10.4.1 Net-Present-Worth Method

The *net-present-worth method* is also known as the *net-present-value* or *venture-worth method*. It compares the present worth of future revenue with initial capital investment, assuming a continuing stream of opportunities for investment at a preassigned interest rate. The procedure compares the magnitude of present worth of all revenues with the investment at the datum time 0. A decision about the investment is made based on the magnitude of this comparison.

We may define net present worth as the added funds that will be required at the start of a proposed project, invested at a preassigned interest rate, to produce receipts equal to, and at the same time as, the prospective investment. For a given interest rate of 10%, the net present worth of the previously given problem is computed by discounting all revenues to year 0 at this rate and subtracting the proposed investment.

Example: Net-Present-Worth Method

Period	n	$1/(1+i)^n$ Present-Worth Factor at 10%	Cash Flow	Amount
Year 1 to zero	1	0.9091	$450	$409
Year 2 to zero	2	0.8264	425	351
Year 3 to zero	3	0.7513	400	301
Total				$1061
Less proposed investment	0	1.0000	1025	1025
Net present worth				$36

The $36 is the amount that must be added to the $1025 to set up the amount that would have to be invested at 10% to achieve receipts equal to and at the same time as those predicted for the recommended investment, or

$$(\$1025 + \$36) \times 1.1 = \$1167$$
$$\text{less payment } \underline{450}$$
$$\$717$$
$$717 \times 1.1 = \$789$$
$$\text{less payment } \underline{425}$$
$$\$364$$
$$\$364 \times 1.1 = 400$$
$$\text{less payment } \underline{400}$$
$$0$$

10.4.2 Net-Future-Worth Method

Assets and revenues can be invested at the preassigned interest rate where there is a continuous exposure of investment according to opportunities given by $F = P(1 + i)^n$. A comparison of investment of the original sum plus reinvestment of revenues at the preassigned interest is made against the standard alternative of investing only the original asset value. The calculation results in the added amount obtained at the end of the project's economic life. A common comparison uses the same stipulated interest rate i.

Example: Net-Future-Worth Method

For 10%, the net future worth is computed by compounding future revenues to the terminal year, then subtracting from this the amount that would have resulted from the other alternative of investing the original asset at the same preassigned interest rate to the terminal year:

Period	n	$(1 + i)^n$ Compound Amount Factor at 10%	Cash Flow	Amount
Year 1 to 3	2	1.2100	$450	$545
Year 2 to 3	1	1.1000	425	468
Year 3	0	1.0000	400	400
				$1413
Less disbursements compounded to terminal year at 10%.				
Year 0 to 3	3	1.3310	1025	1364
Net future worth				$ 49

The calculations point out that if the project is funded and if the revenues materialize as estimated, then a surplus of $49 will be expected over the simple alternative of investing only the asset of $1025. The same period of time and equal interest rates are parts of this comparison.

10.4.3 Net-Equivalent-Annual-Worth Method

Management often wants a comparison of annual costs instead of, say, present worth of the costs. Here we refer to net costs—that is, the net difference between any cost and revenues or credits. This method considers a supply of opportunities for investment of both assets and receipts at the predetermined interest rate plus a supply of capital at the same interest rate.

Example: Net-Equivalent-Annual-Worth Method

The sample problem does not have uniform annual receipts, and the receipts first must be converted to total present worth and then to annual equivalents. The total present worth at time zero is $1061 (from Section 10.4.1). The annual equivalent is found by dividing by the sum of the present-worth factors, or

$$\text{Annual equivalent amount} = \frac{1061}{(0.9091 + 0.8264 + 0.7513)}$$
$$= \frac{1061}{2.4868} = \$427$$

The $427 amount is the equal annual equivalent of the irregular series of receipts, or $450, $425, and $400 which are spaced at 1, 2, and 3 years for an interest rate of 10%. Next we find the net annual equivalent worth, which now includes the investment amount of $1025 at time zero. The amount $1025 is converted to an annual equivalent by dividing by the sum of the present worths, or 2.4868.

$$\text{Net annual equivalent worth} = 427 - \frac{1025}{2.4868} = \$14$$

This $14 is the amount by which the anticipated revenues depart from the proposed investment, rated at 10% interest, and exceed the annual equivalent of the proposed investment:

		Amount
Anticipated equal annual receipts		$427
Less equal annual equivalent worth at 10%		14
Equal annual receipts to be generated by investing		$413
	1025 × 1.10 =	$1128
Less payment		413
		715
	715 × 1.10 =	$787
Less payment		413
		374
	374 × 1.10 =	$413
Less payment		413
		0

Starting with the investment that earns interest and subsequently subtracting payments leads to a balance of zero dollars at the end.

10.4.4 Rate-of-Return Method

The rate-of-return method calculates the rate of interest for discounted values of the net revenues from a project to have the present worth of the discounted values equal to the present value of the investment. The rate-of-return method thus solves for an interest rate to bring about this equality.

Other titles also exist, such as *ROI* (*return on investment*), *true rate of return, profitability index*, and *internal rate of return*. The adjective "true" distinguishes true rate of return from other less valid methods that have been labeled rate of return—for example, Equations (10.1) and (10.2).

For this method there is no assumption of an alternative investment and no predetermined interest rate as was required for the previous methods. We define this interest rate at which a sum of money, equal to that invested in the proposed project, would have to be invested in an annuity fund in order for that fund to be able to make payments equal to, and at the same time as, the receipts from the proposed investment.

Example: Rate-of-Return Method

The solution for the interest rate is by repeated trials or by graphical or linear interpolation. Typically, two interest rates bound the value of the investment. For the sample problem, the two trial values are 5% and 15%, and the calculation is given as shown below.

Year, n	Revenue	$\frac{1}{(1+i)^n}$ PW Factor at 5%	Discounted Amount
1	$450 ×	0.9524 =	$429
2	425 ×	0.9070 =	385
3	400 ×	0.8638 =	346
Total			$1160

Year, n	Revenue	$\frac{1}{(1+i)^n}$ PW Factor at 15%	Discounted Amount
1	$450 ×	0.8696 =	$391
2	425 ×	0.7561 =	321
3	400 ×	0.6575 =	263
Total			$975

The two trial values bound the initial asset value of $1025. Note that the higher value, 15%, discounts the annual revenues into smaller values.

Interest Rate, %	Present Worth, $
5	$1160
To be determined	1025
15	975

Figure 10.1 shows a linear line between the two points, and the value of $i = 12.3\%$ is found. A better approximation occurs when the neighboring interest boundaries are narrower than that shown, as the line is nonlinear.

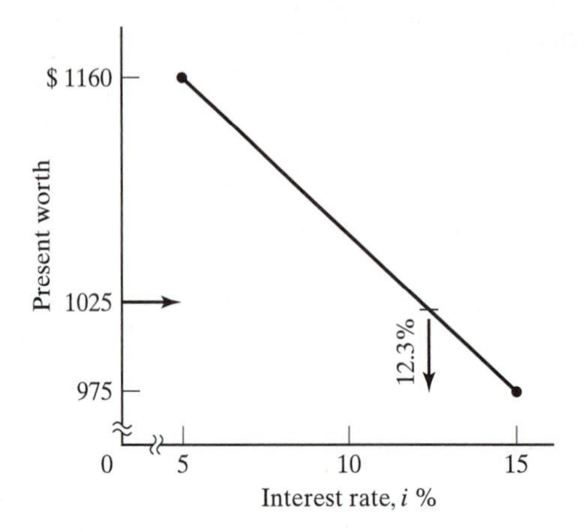

Figure 10.1 Graphical approximation for rate of return.

This rate of return is the interest rate at which the original sum of $1025 could be invested to provide returns equal to, and at the same time as, the three revenues of the prospective investment. The calculation below demonstrates this equivalence.

		Amount
	$1025 \times 1.123 =$	$1151
Less payment		450
		$700
	$700 \times 1.123 =$	$787
Less payment		425
		$362
	$362 \times 1.123 =$	$400
Less payment		400
		0

This calculation shows that the earning rate is true and is the actual return of the invested money.

This method has the important advantage of being directly comparable to the cost of capital, which is the lending rate charged by banks and investors for the project.

10.4.5 Comparison of Methods

The results of the four different methods are summarized below.

Method	Amount
Net present worth at 10%	$36
Net future worth at 10%	$49
Net annual equivalent worth at 10%	$14
Rate of return	12.3%

We can demonstrate that the first three methods give equivalent answers. For instance, the net present worth of \$36 can be compounded to the terminal year by using Equation (10.6), or

$$F = 36(1.10)^3 = \$49$$

The preassigned interest rate, 10%, yields the net future worth or \$49. The net annual equivalent worth of \$14 can be found by dividing the net-present-worth value of \$36 by a sum of present-worth values of 10%, or

$$\frac{36}{2.4867} = \$14$$

On the other hand, the rate of return cannot be calculated from the foregoing answers. It is only found directly from the data. Equivalency among the first three methods can be found for any arbitrary interest rate, and if those methods are equivalent to the rate of return, then they are equivalent at only one preassigned interest rate.

The differences in the first three methods are based on the choice of an interest rate. This arbitrary selection of an interest rate makes the three methods little more than decision tools for comparing projects. The rate of return is the only method that can provide a consistent measure of the extent of the economic productivity of prospective investments. The student needs to know that the four methods remain the most popular of all the engineering-economy methods, but no one single method or criterion of profitability analysis is preferred in all business situations.

10.4.6 Standard Approaches to Engineering-Economy Methods

The foregoing algebraic methods are instructive, but computer spreadsheets and software, calculators, equations, and factor tables are the techniques that are used for engineering-cost tradeoff appraisal. Table 10.2 summarizes the equations and key symbols for engineering-economy calculations.

The six relationships are divided into two groups: single- and equal-payment series. The notation consists of five symbols (P, F, A, i, and n), and their definitions are given. Cash flows are implicitly assumed to be at the start or the end of the year and are considered lump sum or discrete.

Functional notation is preferred over the writing of the equations. An example of functional notation is (F/P, $i\%$, n). Rather than write $F = P(1 + i)^n$, we write $F = P(P/F, i\%, n)$, and a table factor is substituted. The meaning of functional notation, for example, F/P, says that we wish to find F given P.

These symbols are the definitions for the formulas below.

where i = effective interest rate per interest period, decimal

n = number of interest periods, usually having units of years, but other units are allowed

P = present sum of money, dollars

F = future sum of money is an amount n periods from the present time that is equivalent to P at interest rate i, dollars

A = end of period cash receipt or disbursement in an equal payment series continuing for n periods, and sometimes called an "annuity," and the entire series is equivalent to P or F at interest rate i, dollars

TABLE 10.2 Summary of Periodic Compounding Formulas for Engineering-Economy Methods Using Functional Notation

Factor Name	Given	To Find	Functional Notation	Equation	Equation Number
Single Payment					
Compound-amount factor	P	F	$(F/P, i\%, n)$	$F = P(1 + i)^n$	(10.7)
Present-worth factor	F	P	$(P/F, i\%, n)$	$P = F(1 + i)^{-n}$	(10.8)
Uniform Payment Series					
Sinking-fund factor	F	A	$(A/F, i\%, n)$	$A = F\left[\frac{i}{(1 + i)^n - 1}\right]$	(10.9)
Capital-recovery factor	P	A	$(A/P, i\%, n)$	$A = P\left[\frac{i(1 + i)^n}{(1 + i)^n - 1}\right]$	(10.10)
Compound-amount factor	A	F	$(F/A, i\%, n)$	$F = A\left[\frac{(1 + i)^n - 1}{i}\right]$	(10.11)
Present-worth factor	A	P	$(P/A, i\%, n)$	$P = A\left[\frac{(1 + i)^n - 1}{i(1 + i)^n}\right]$	(10.12)

There are other engineering-economy equations, such as continuous interest or continuous flow of funds, but for the purpose of early tradeoff of engineering design and projects, our sense is that they are overkill when compared to the accuracy of preliminary estimates.

Now observe the abbreviated 5% sample factor table, Table 10.3. Clearly, computer spreadsheets, Internet browsers,[4] and calculators do similar things and are preferred for practice, but published tables are helpful for instructional purposes.

$(F/P, 5\%, 4)$ is the single payment, and given a value of P to find F and for $i = 5\%$ and $n = 4$ periods, examination of Table 10.3 gives a value of $(F/P, i\%, n) = 1.216$ as the factor. There are numerous tables, such as this one, but we only supply tables for 10% and 20%, which are given in the Appendix. Preprogrammed calculators or computers are used for other interest values.

The solution to engineering economy alternatives, which have a simplified presentation in this book, involves several assumptions:

- Discrete cash flows occur at year end
- Inflation effects are ignored, inasmuch as they affect the alternatives similarly (methods are provided in Section 9.2.2 that allow inflation analysis)
- Taxes are evaluated in special ways

[4]There are Internet sites that are available for solving the problems in this chapter. Entry of "engineering economic calculation" will provide several Websites from search engines.

TABLE 10.3 Factor Table for 5% Interest Rate for 1 to 5 Years

	Single Payment		Equal Payment Series			
	Compound-Amount Factor	Present-Worth Factor	Compound-Amount Factor	Sinking-Fund Factor	Present-Worth Factor	Capital-Recovery Factor
n	To find F Given P F/P i, n	To find P Given F P/F i, n	To find F Given A F/A i, n	To find A Given F A/F i, n	To find P Given A P/A i, n	To find A Given P A/P i, n
1	1.050	0.9524	1.000	1.0000	0.9524	1.0500
2	1.103	0.9070	2.050	0.4878	1.8594	0.5378
3	1.158	0.8638	3.153	0.3172	2.7233	0.3672
4	1.216	0.8227	4.310	0.2320	3.5460	0.2820
5	1.276	0.7835	5.526	0.1810	4.3295	0.2310

- Constant interest rate during the lifetime of the project
- Nonquantifiable effects are ignored until the analysis is ended
- Predetermined minimum acceptable rate of return (MARR)

A minimum attractive rate of return (MARR) or i is predetermined before project analysis, and it meets or exceeds the benchmark of economic performance for the business.

Those assumptions are employed because the information that is available in the early stages of finding the tradeoffs for a project are only dimly seen, and simple calculations expedite the decision making.

These discrete cash flows can be diagrammed where upward and downward arrows are given different algebraic signs, such as positive cash flow as a revenue can be specified as upward, while an investment, meaning "money out" would be negative or down (see Figure 10.2).

Now that formulas are given and the tables explained, we consider their applications. These practices are especially important for engineering management and personal economy. The student will want to understand that these applications are far more extensive than illustrated here.

10.5 ADVANCED APPLICATIONS (OPTIONAL)

For the most part, engineering economy applications deal with comparison of engineering alternatives where long time factors and large amounts of investment are significant. Furthermore, the replacement of equipment—a capital intensive operation involving a revenue or cost stream that would make the selection nonobvious—are candidates for these methods.

10.5.1 Contexts of "Interest" in Engineering Economy

The term "interest" is used in many engineering economic contexts. Here we restrict the meaning of the word *interest*, or *rate of return*, as the *cost of using capital*. The

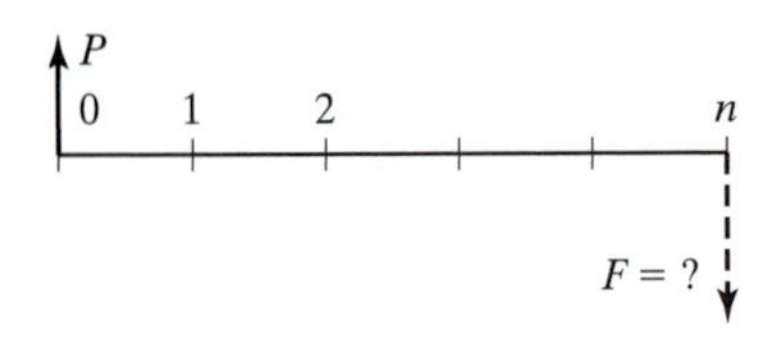

(a) Single Payment—Compound Amount Factor

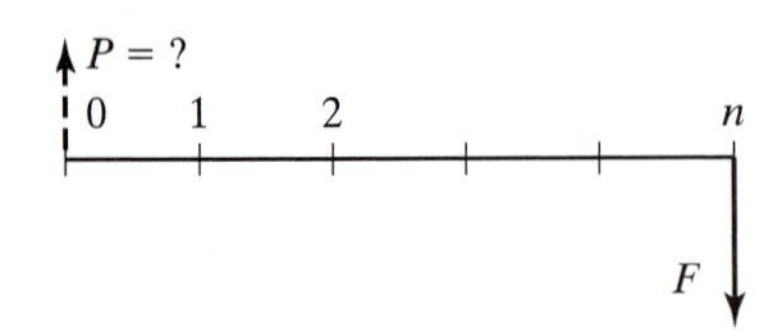

(b) Single Payment—Present Worth Factor

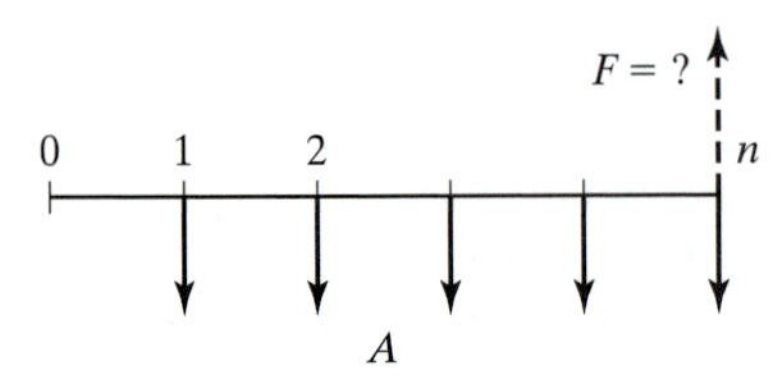

(c) Equal Payment Series—Compound Amount Factor

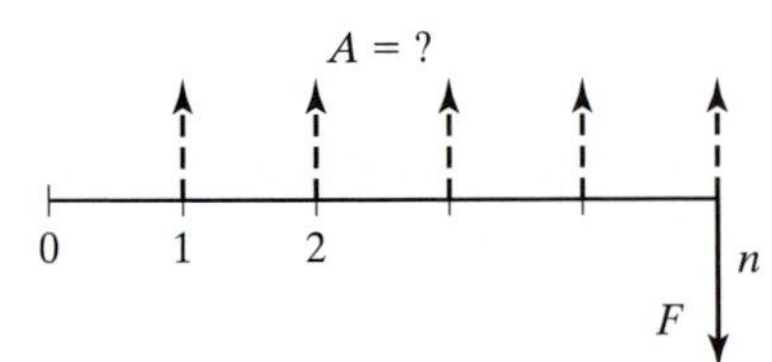

(d) Equal Payment Series—Sinking Fund Factor

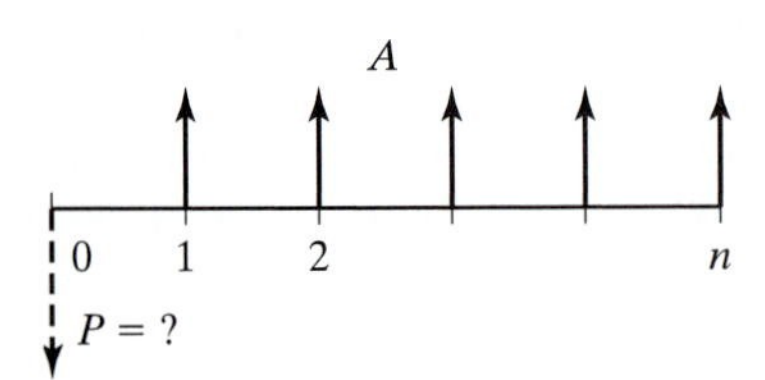

(e) Equal Payment Series—Present Worth Factor

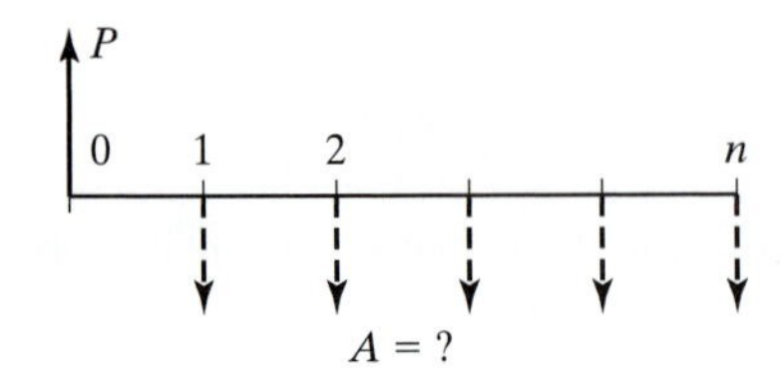

(f) Equal Payment Series—Capital Recovery Factor

Figure 10.2 Diagrams for simple compounding interest models.

term has a long history. In this brief discussion we now direct our attention to the following:

- Internal rate of return (IRR)
- Minimum attractive rate of return (MARR)
- External rate of return (ERR)

Whatever the terminology, it provides a percentage that shows the yield on different uses of capital. Interest rates are well understood throughout engineering and business, and there is general agreement on how the arithmetic is applied.

The IRR can be calculated for projects that involve only costs, as well as income-producing proposals. Consistent results are obtained between EAW, PW, and IRR comparisons.

The internal rate of return is sometimes known as *true rate of return* or *return on investment* (*ROI*). The term IRR is used here because it emphasizes that the rate is indeed *internal* and based solely on the project's cash flow, and that no external influences are used directly.

The net present worth of a cash flow is shown in Figure 10.3(a), for different interest rates. For our purposes here, consider the analysis to be before-tax cash flow. When the present worth is 0, we find the IRR. At any required interest rates less than the IRR, the present worth of the project is positive, and therefore acceptable. For a required interest rate greater than the IRR, the project shows a negative present worth and is rejected. It is pointed out that the IRR is earned only on the capital originally invested, and how the capital recovery payments are reinvested has no impact on the IRR.

In Figure 10.3(b) two intersections of the net present worth are found. In some cash flow patterns, there may be a sign reversal between positive and negative present worth cash flow, and two possible IRRs, such as "Low" or "High," are possible. Since a sign reversal is sometimes accompanied by dual rates of returns, a trial and error solution, or the programming code may need interpretation. An artifice is to imagine that the capital recovered is reinvested at a rate either near L or H where there is a choice to test the IRR for practicality. If the artificial and temporary interest rates reduce the

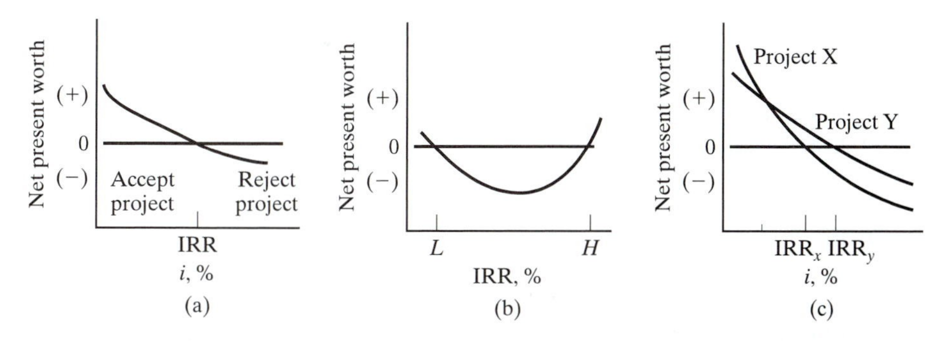

Figure 10.3 Behavior of present worth cash flows for interest rates: (a) zones showing acceptance or rejection of a project, (b) dual rates of return; and (c) switching and selection of projects with rates of return.

solution to either L or H, that selection is preferred. Furthermore, the value H may be so high in these dual-rate solutions that the higher rate is not realistic.

In Figure 10.3(c), consider the case where two independent projects have the same initial investment, but the subsequent cash flows differ. There may be an interest rate where the present worths of Projects X and Y are equal. To the left of the intersection, Project X is preferred, it having a greater present worth cash flow. To the right of the intersection, Project Y is preferred, but only up to the IRR at which the present worth becomes negative. Figure 10.3(c) shows that ranking of projects can switch depending on the interest rate i.

10.5.2 Minimum Attractive Rate of Return

For engineering purposes, an interest rate is sometimes called *minimum attractive rate of return*, (MARR)—the lowest level of interest that an alternative is still considered acceptable if the present worth of the cash flow is positive. MARR will vary between businesses and even within the same firm. It is generally agreed that it should be no lower than the cost of capital, and most likely considerably higher than the cost of capital. How much higher depends on the objectives of the organization. It can also vary depending on the risk, or the events of the economy that are current or projected for the future. As an arbitrary number, it intends to define the lower limit for the best possible use of money, which is a limited resource. For instance, the MARR may be lower for a cost reduction proposal than for a undefined new product development having less certainty.

The lower bound is the cost of capital. Even so, there is disagreement on what this is, as it could mean the costs of acquiring funds from various sources. A small company strapped for funds and without reserves or much profit will be required to ask angels or venture capitalists for money, who may justifiably insist on robbery returns for their risky investment. Or a larger company may be able to fund the cost reduction from the annual profit for the firm. These larger firms may be prone to use the idea that the money for the alternative can be used in an external development such as the purchase of another firm, and thus be put to a better application. The effect of setting a MARR is to ration capital. The purpose is to avoid unproductive investments in marginal activities, thus protecting the firm from squandering and allowing more opportune investments at a later time.

The *external rate of return* (*ERR*) is the investment rate outside of the firm, a parameter of less importance to engineering. It is the minimum attractive rate for reinvestments that are apart from internal project evaluation.

Engineering economy is used in many practical applications. We now consider the following situations:

- Comparison of alternatives
- Replacement
- Taxation

10.5.3 Comparison of Alternatives

Most engineering projects have several feasible alternatives, including the *do-nothing option*. When the selection of one of these alternatives excludes the adoption of another option, the alternatives are called *mutually exclusive*. These alternatives will have differing amounts of investment, revenues, and annual costs along with variation in technical performance. But the economic choice of the alternatives recognizes that the slate is of *feasible* candidates, and despite technical variation, the alternatives are inherently workable.

Because these alternatives give a range of economic consequences, an analysis is necessary to declare a winner. These alternatives will have differing amounts of investment capital, and this results in increased capacity, increased quality, increased revenues, decreased operating expenses, or increased life of the equipment.

The engineering-economy methods are able to make this comparison, and when correctly applied using the MARR as an economic decision criterion, these methods result in the suitable selection from a set of mutually feasible alternatives.

If two or more mutually exclusive alternatives are compared, and receipts or savings (cash inflow) as well as costs (cash outflow) are known, the feasible alternative that has the highest net equivalent worth should be selected as long as that equivalent worth is greater than or equal to zero. If only costs are known or are being considered when revenues for the alternatives are the same, that feasible alternative that has the least negative equivalent worth of costs should be selected.

The annual-cost method is frequently used because people are more familiar with the nature of annual cost than with the concepts of present worth or future worth or rate of return on the investment. Furthermore, the annual cost method is easier to explain. The annual-cost method is merely the cost pattern of each alternative converted into an equivalent equal series of annual costs at the minimum required rate of return i. The alternative with the lowest cost quantity is the preferred selection.

If the purpose of an engineering economy comparison is the pick of one alternative over the other, we are not necessarily interested in the total cost of the alternatives, but in the fact that one alternative has more or less cost than the other alternative. We are interested in the *relative* differences, not the absolute differences, or in the context of engineering economy that difference in annual, or difference in present worths or in the percentage earned on the difference. Those costs that do not contribute to the difference have no effect on the preference and may be omitted.

Equal-Life Alternatives The comparison of alternatives for mutually exclusive opportunities can be subdivided into either identical or nonidentical lives. When the mutually exclusive alternatives for a project have the same useful life, and that matches the analysis period, selection of the optimum alternative is straightforward.

Example: Annual Cost Comparison for Two Alternative Processes Having Equal Life

A wave soldering machine is necessary for the application of a layer of molten solder to a printed board surface. The solidification of the solder secures the inserted components into place. There are two candidates for the requirement: automatic and semiautomatic. The essential manufacturing difference is that in the semiautomatic equipment, the board is returned to the loading operator through a return carriage. In the fully automatic one, the circuit boards continue on their path of assembly without returning to front end of the equipment. The design, quantity, and quality of the printed circuit product and the amount of solder material are identical for either case. The minimum required rate of return is

10%, which is a before-tax requirement. The analysis needs to recommend which equipment is desirable and a comparative annual cost method is required for the decision. Estimates are given below:

	Automatic	Semiautomatic
Investment cost	$1,250,000	$800,000
Fuel, heating, and utility requirements, annual	570,000	480,000
Labor, annual	50,000	700,000
Insurance for first cost, annual	1%	1%
Supervision, annual	150,000	150,000
Floor space, annual	80,000	50,000
Maintenance, annual	140,000	90,000
Salvage	0	0
Life of equipment, years	5	5
(A/P, 10%,5) factor value	0.2638	0.2638

Notice that supervision is the same between the alternatives, and thus it could be ignored in the analysis. The automatic waver soldering machine has a correspondingly lower labor costs. Furthermore, the choices give the same quality of the loaded circuit board. See Figure 10.4 for the cash flow diagram.

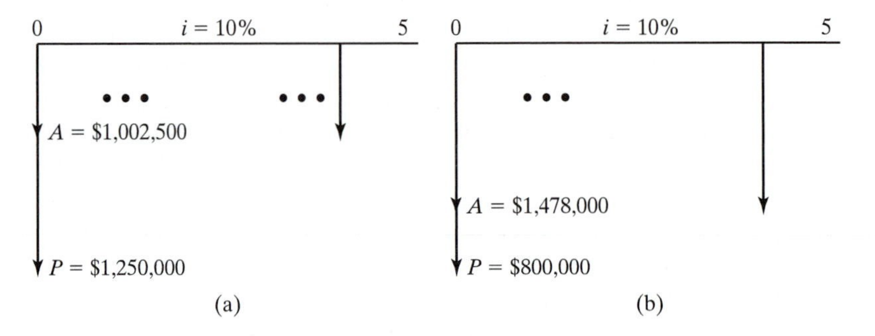

Figure 10.4 Cash flow diagram for wave soldering equipment for manufacturing operation. (a) automatic equipment, and (b) semiautomatic equipment.

The example proceeded from the estimates of all costs for the two alternatives to the cash flow diagram. The cash flow diagram summarizes the costs into a first cost or investment (P) and annual components (A). Because the comparison is on an annual cost

TABLE 10.4 Annual Cost Comparison of Equipment with Equal Life

	Automatic	Semiautomatic
Annual capital recovery cost, $P(A/P, 10\%,5)$	$329,750	$211,040
Annual cost	1,002,500	1,478,000
Comparative annual cost	$1,332,250	$1,689,040

basis, that alternative which gives the minimum quantity is selected. The automatic process is preferred. See Table 10.4.

A principle for mutually exclusive alternatives is that the comparison for replacement must be made over equal outputs. Each alternative must be producing equal quantities. Of course, the inputs will be different, because each project has a differing engineering efficiency. The input dollars must not reflect differences in output. Input dollars must purchase equal units of output. If comparisons are not made over equal outputs, one alternative will be charged with extra operating costs arising out of the extra productive capacity. At first glance, extra capacity beyond what is necessary is pointless, except that reserve is considered an important irreducible factor.

Now we consider the case where alternatives that have larger investments of capital than others may have higher annual revenues or lower costs when revenues are equal, or they may have different lives. Different useful life for the alternatives adds a special wrinkle to the analysis. We adopt procedures that allow the engineering economy procedures to be formed, placing the alternatives on a comparable basis. Two assumptions are used, repeatability and cotermination.

The repeatability assumption requires the analysis period be equal to a common multiple of the lives of the alternatives. For instance, if the two alternatives have lives of two and three years, then the common multiple is six years. Thus the two-year asset is replaced two more times. The three-year life asset is replaced once more with the initial selection. Furthermore, the circumstances that are estimated in the initial formulation occur in the repetitions of the replacements of the alternatives, meaning that the subsequent end-of-life of the alternate are replaced with the initial choice of the investment. Alternatives must be compared on the basis of equivalent outcomes.

Actual practice with this assumption suggests that it is effective so long as the multiple of lives is short. Longer multiple lives give the opportunity for a equipment derivative that is not similar to the replaceable model, and thus this practice is challenged in its reality.

Example: Unequal-Life Alternatives

Investments I1 and I2 are feasible designs for new production. Investment I2 has an initial cost of $320,000 and an expected salvage of $40,000 at the end of its 4-year economic life. Investment I1 costs $90,000 less initially and has an economic life of three years, and it has no salvage value and its annual operating costs exceed those of I2 by $25,000. The required rate of return is 10%. Which alternative is preferred when using the repeated project life method? The estimate of the investments is given by the following:

Investments	I1	I2
Initial cost	$230,000	$320,000
Annual operating cost	25,000	0
Salvage value	0	40,000
Economic life, year	3	4
Multiple of lives	12	12
Repeated investments	3	2
MARR before taxes	10%	10%

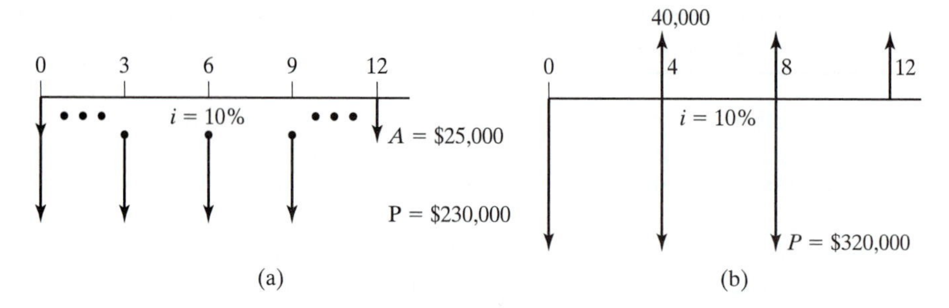

Figure 10.5 Cash flow diagram for equipment having unequal life using repeated life assumption: (a) three-year life asset, and (b) four-year life asset.

These estimates are summarized in Figure 10.5.

Equations to model the cash flow diagram of Figure 10.5 are given below.

$$\begin{aligned} PW(\text{I1}) &= -230{,}000 - 230{,}000(P/F\ 10\%, 3) - 230{,}000(P/F\ 10\%, 6) \\ &\quad - 230{,}000(P/F\ 10\%, 9) - 25{,}000(P/A\ 10\%, 12) \\ &= -230{,}000 - 230{,}000(0.7513) - 230{,}000(0.5645) \\ &\quad - 230{,}000(0.4241) - 25{,}000(6.8137) \\ &= -\$800{,}520 \\ PW(\text{I2}) &= -320{,}000 - 280{,}000(P/F\ 10\%, 4) \\ &\quad - 280{,}000(P/F\ 10\%, 8) + 40{,}000(P/F\ 10\%, 12) \\ &= -320{,}000 - 280{,}000(0.6830) - 280{,}000(0.4665) + 40{,}000(0.3186) \\ &= -\$629{,}116 \end{aligned}$$

The present-worth advantage of I2 over I1 for 12 years of investment is \$800,520 − \$629,116 = \$171,404. The repeatability method has limited use in engineering because actual situations may differ dramatically from the assumptions.

Reconsider the above example where the life of the assets is limited by operational requirements of the life of the product, which will conclude in two years. Now view the investments as restricted to a two-year study period. The coterminated assumption implies that the investment is disposed of, or retired to secondary service, at that time. If possible, estimates of the worth of the assets are helpful to the quality of the analysis. These salvage values could be large in view of the brevity of the service. If reliable estimates of the salvage cannot be found, minimum resale levels can be calculated to make the alternatives equivalent. Then only a judgment of the value of the salvage need be made to assess whether the market resale value will be above or below the minimum level.[5]

For instance, assuming salvage = 0 for alternatives I1 and I2 after two years of service,

$$\begin{aligned} PW(\text{I1}) &= -230{,}000 - 25{,}000(P/A\ 10\%,2) \\ &= -230{,}000 - 25{,}000(1.7355) = -\$273{,}388 \\ PW(\text{I2}) &= -\$320{,}000 \end{aligned}$$

[5]Information about new and used equipment, such as machine tools, on-location auctions, and wanted equipment, is available at www.MachineTools.com. Prices and history on age and performance specification can be found to help the salvage value judgment, making the "aspiration level" model businesslike.

Investment 1 has a lower worth of costs for the two-year service period and is preferred at this point. On the other hand, the salvage value for I2 that would make PW(I1) = PW(I2) is

$$273{,}388 = 320{,}000 - F_s(P/F, 10\%, 2)$$

and the salvage value is

$$F_s = (320{,}000 - 273{,}388)/(P/F\ 10\%, 2) = 46{,}612/0.8265 = \$56{,}400.$$

Investment 2 is preferred to investment 1 at the end of year 2 if the resale value of investment 1 is more than $56,400 greater than the resale value of I1 at the same time. Calculating this aspiration level avoids making an estimate of the salvage values. Only a judgment is necessary about whether F_s will exceed a certain amount, the aspiration level of $56,400 in this case.

10.5.4 Replacement

Principal reasons for replacement include wear-out, obsolescence, reduced performance, new requirements, costs, light-bulb failure, or cost-effective leasing alternatives.[6] Surveys suggest that approximately 70% of production equipment is greater than five years old. Physical impairment may lead to a decline in the worth of the service, increased operating costs, or increased maintenance cost. Even with numerical control over 50 years old, and mechanization nearing a century of age, automatic production is about 20% while manual control is 40% of the production processes. Thus replacement analysis is popular.

Practitioners view replacement analysis as an economic contest between a defender and challenger. The *defender* is the older asset that is being considered, while the *challenger* is the proposed asset. In replacement analysis there is already equipment that is delivering the service or product. So there is a difference in replacement analysis when compared to the study of mutually exclusive alternatives, which was discussed earlier.

Because the economic facts of the defender and the challenger are dissimilar, special attention is given to their appraisal. For instance, the magnitude and duration of the cash flows of the defender and challenger are quite different. Historical costs are available for the defender. Estimates are required for the challenger. New assets have high capital cost and low operating costs. The reverse is usually true for the defender. In addition, the remaining life of the defender is shorter and can be estimated with greater certainty. The defender's usable life may be different than its remaining depreciation or economic life. Eventually, the costs of keeping and maintaining the defender next year will be more than for this year. When this happens, the end of the economic life may be approaching, and the equipment is a candidate for replacement.

Viewpoints on equipment life include physical, accounting, useful, product, or economic factors. But we view equipment life as the time remaining until the defender's value or cost is greater than the challenger's value. *Equipment life*, then, is defined

[6]The Internet has Websites for engineering-economy calculators that answer questions such as Is it worthwhile to buy or lease or to repair or upgrade? Is it cost effective to purchase the next best asset?

as the period of time that will elapse before the equipment—either present or proposed—is displaced from the intended service by more economic equipment. Restated differently, it is the period of time that the equipment will continue to have the lower annual cost compared to any contender for the service.

There are opinions about how to consider the defender's historical cost facts. How does the fact that equipment, which was purchased for $50,000 three years ago, affect the analysis? When the equipment was purchased, the accounting system recorded that the cash account, an asset, decreased by $50,000 and the equipment account, also an asset, increased by $50,000 to balance the accounting equation. (Review Chapter 4 on this point). This entry transferred dollars from one asset account to another. That entry occurred three years ago. The accounting system recognizes the decline in the asset's value, and applies depreciation expense to the asset. *Depreciation* reduces the value of the equipment as shown on the balance sheet, and simultaneously increases costs as depreciation expense on the income statement. The net effect is the reduction of the asset over time and increased costs on the income statement, which in turn reduces profits and may reduce income taxes for the period. The *book value* of assets is the value of the asset as determined by depreciation entries that reduce the initial value of the assets as originally purchased. The "book" is the accounting record. Book value is not generally used as a measure of the assets value for replacement analysis. Finally, *sunk costs* are money spent in the past that has no influence on the current analysis. Theoretically, at least, this sunk cost cannot be recovered by current or future transactions. Current and future cost values are preferred.

Market value for the defender can be determined by an expert's appraisal, or by offering the asset for sale through a competitive system. Salvage value or future market value is the preferred choice for the defender's final value. If there are tear-down, removal, disposal or other related costs, it is probable that the salvage value could be negative. There are resellers or liquidators that remove the asset and offer it for sale. The reseller's offer can be used as the value of the salvage, for instance.

Trade-in value by a potential seller of the challenger may not be an accurate reflection of the defender's value. The trade-in value may be offered as an incentive, thus suggesting that the purchased price is exaggerated. Anyone who has dickered with used-car sales people can attest to this.

The *nonowner viewpoint* of the cost values of the defender is often used as an arbiter in valuation of the asset. The nonowner would argue that first cost, trade-in, and book value of the defender be neglected. These costs are ignored because they are not relevant to the nonowner's viewpoint.

Once the defender's and challenger's equivalent annual cost have been determined, a decision can be made on the replacement. The choice may be to keep the defender and defer the replacement. But the decision "not to replace" can be reversed at any time in the future.

Cash flows and the assumption of annual cash flows are at start of year, similar to that described earlier. Popular equations for expressing the capital recovery of an investment are given below:

$$EAC = (P - F_s)(A/P, i\%, n)F_s i \tag{10.13}$$

$$EAC = P(A/P, i\%, n) - F_s(A/F_s i\%, n)$$

where EAC = equivalent annual cost or worth, dollars
$(A/P, i\%, n)$ = capital recovery factor, number
$(A/F_s, i\%, n)$ = sinking fund factor, number
P = investment or first cost, dollars
F_s = salvage or resale value at end of equipment life, dollars

Consider an asset that costs \$60,000 and has a \$20,000 salvage value. The net amount to be recovered from the annuity payment $P - F_s$ = \$40,000. The residual of the purchase price is returned by the receipt of the salvage value, \$20,000. However, the owner of the asset is deprived of the \$20,000 during the life of the asset, so interest is owed on this amount because it represents unrecovered capital. The term $F_s i$ term accounts for this interest payment.

Example: Defender and Challenger with the Equivalent Annual Cost Method

An electronic component insertion machine was purchased three years ago. This machine is used for inserting microchips that have been packaged in a case size of 0.3 in. or 0.6 in. into printed circuit boards. The price then was \$400,000. It meets benchmark quality requirements even now, but an improved version is available for \$350,000. The challenger promises to reduce operating expenses and cut defect rates. Costs and salvage values for the two machines are as follows:

Year	Defender (D)		Challenger (C)	
	Operating Cost	Salvage Value	Operating Cost	Salvage Value
0		\$120,000		\$350,000
1	\$34,000	70,000	\$3000	310,000
2	39,000	40,000	10,000	270,000
3	46,000	25,000	12,000	240,000
4	56,000	10,000	15,000	210,000
5			20,000	170,000
6			31,000	120,000

Should a replacement be made if the required rate of return is 20% and the equipment will be needed for only four more years? Use the annual cost method.

The calculation determines the equivalent annual cost (EAC) when replacement is made immediately. This allows consideration of only the first four years of the challenger's life. The salvage value after four years is \$210,000. The defender's current salvage value is its market's worth now and the original purchase price of \$400,000 has no bearing on the replacement decision.

$$EAC(D) = [120{,}000 + 34{,}000(P/F\ 20\%, 1) + 39{,}000(P/F\ 20\%, 2) + 46{,}000(P/F\ 20\%, 3) + 56{,}000(P/F\ 20\%, 4) - 10{,}000(P/F\ 20\%, 4)](A/P\ 20\%, 4) = \$86{,}618$$

$$EAC(C) = (350{,}000 - 210{,}000)(A/P\ 20\%, 4) + 210{,}000(0.20) + [3000(P/F\ 20\%, 1) + 10{,}000(P/F\ 20\%, 2) + 12{,}000(P/F\ 20\%, 3) + 15{,}000(P/F\ 20\%, 4)](A/P\ 20\%, 4) = \$105{,}208$$

As the challenger has a higher equivalent annual cost than the defender, a replacement is not recommended at this time.

PICTURE LESSON Agricultural Mechanization

Wisconsin Historical Society.

At the beginning of the twentieth century it took a team of farmers weeks to plant and harvest one crop. Four farmers were required to feed 10 people. Today, machinery allows the Midwestern corn crop to be planted in 10 days and harvested in 20; one U.S. farmer can produce enough food to feed 97 Americans and 32 people in other countries.

Twentieth century engineering has made the difference. The tractor, reaper, combine, and hundreds of other machines gave farmers the mechanical advantage they had long needed to ease their burdens and make their lands profitable. Agricultural mechanization enormously increased farm efficiency and productivity.

Engineering began to affect the farmer late in the nineteenth century, with steam-powered tractors and various tools for drilling seed holes and planting. Still, most field work was done with hand tools like the spade, hoe, and scythe, or with oxen-, mule-, or horse-driven equipment. The photograph shows a team of 11 horses pulling a McCormick reaper in Eastern Washington in 1923. A farmer's day was labor-intensive, beginning well before sunrise and ending at sunset.

Mechanization did not advance rapidly until the twentieth century, with the advent of the internal combustion engine. As the chief power source for vehicles, it began replacing both horses and steam for planting, cultivating, and harvesting equipment. It made the evolution of the tractor possible, and led to sweeping changes in agriculture. The number of tractors increased dramatically. In 1907 there were some 600 tractors in use, and by mid-century the number had grown to almost 3.4 million.

Many inventions were created by the farmers themselves before commercial development. It would soon follow that inventions would be produced near their conceived origins. Interestingly, all 50 states today manufacture agricultural equipment.

Example: Defender and Challenger with Unequal Life Choices

The firm is considering the purchase of new equipment. The defender equipment cost \$135,000 four years ago. It is expected that six years of life are remaining and that removal costs will equal any salvage value at that time. Annual operating costs are \$5000, current book value is \$30,000, and market value is \$60,000.

Challenger equipment can be purchased for \$125,000 with an estimated life of eight years. Annual operating costs are \$3000, and salvage value is \$8000 in eight years. Other product-related costs, such as direct labor and direct material, are similar between the competitors. The MARR is 20% before taxes.

Year	Defender (D)		Challenger (C)	
	Operating Cost	Salvage Value	Operating Cost	Salvage Value
0		\$60,000		\$125,000
1	\$5,000	48,000	\$3,000	110,375
2	5,000	36,000	3,000	95,750
3	5,000	21,500	3,000	81,125
4	5,000	15,500	3,000	66,500
5	5,000	10,500	3,000	51,875
6	5,000	0	3,000	37,250
7			3,000	22,625
8			3,000	8,000

It is seen that the book value is disregarded as an entry in the following calculations. Book value is affected by original cost and cumulative depreciation, and is important to cost accounting but not to engineering-economy methods. Nor is the original price that is paid for the defender of use to the analysis.

$$EAC(D) = 60{,}000(A/P\ 20\%, 6) + 5000 - 0 = \$23{,}042$$
$$EAC(C) = 125{,}000(A/P\ 20\%, 8) + 3000 - 8{,}000(\text{A/F}\ 20\%, 8) = \$35{,}090$$

On the basis of the above calculations and the estimates of the operating costs and salvage value, the decision to keep the old equipment is the best solution, assuming that the existing equipment lasts six years.

10.5.5 Taxation Effects on Engineering Projects

Taxes are a major factor for any profit-seeking enterprise. Income, property, sales, and excise taxes affect net returns for both individuals and corporations. *Income taxes* are progressive—that is, they increase with greater income. Others are flat, such as *property taxes*, which are paid on the value of an investment. For instance, a rate may be fixed at 5% of assessed property values. There are many other tax provisions that are not discussed in this book—for example, surcharges, investment tax credits, capital gains and losses, carry amount forward and backward of losses, and many others. Details on these can be explained by other professionals and textbooks. But remember, they are important.

Prominent income taxes for the evaluation of engineering projects are given as federal and state. The range of federal taxes rises to 38%, and for state taxes the rate is from 0 to 5 or 6% and slightly higher in some states.

In addition, assume that the common case is found, where taxable income is settled for federal and state similarly, and federal taxable income has state taxes deducted from it, but state taxes are unaffected by federal taxes. Furthermore, repeating Equation (9.2) an *effective income tax rate* combines both, as follows:

$$t \cong \text{federal rate} + \text{state rate} - (\text{federal rate})(\text{state rate}) \qquad (10.14)$$

where $t \cong$ effective corporate income tax rate on increments of taxable income of importance to engineering economy studies, decimal

federal rate $\cong$ approximation of corporate tax rate, decimal

state rate $\cong$ approximation of corporate tax rate, decimal

For example, let the federal and state tax rates be 34% and 6%, then $t \cong 38\%$.

let taxes $\cong$ amount of federal and state consolidated taxes, dollars (10.15)

taxes $\cong t \times$ (taxable income), dollars

taxable income = amount used in project analysis due to project investment, dollars

$\cong$ before tax cash flow $-$ depreciation charges $-$ other appropriate charges, dollars

other appropriate charges $\cong$ interest for project loan, etc., dollars

Special tax provisions and charges for depreciation and interest are applied to the before-tax cash flow for the year to determine an after-tax cash flow. The distinction of whether or not the firm uses accrual or cash accounting makes a difference in the approximations. The cash flow may have no similarity to taxable income if the firm is using the accrual method of accounting. Professional advice may be helpful for real situations.

A tabular format is convenient for these calculations, and economic comparisons are thus displayed for understanding. These procedures are followed in several later examples.

$$\text{After tax MARR} \cong (\text{before tax MARR})(1 - t) \tag{10.16}$$

For instance, if the before-tax MARR is 20%, and the effective tax rate is 38%, then the approximate after-tax is roughly 12.4%. Thus the hurdle rate for projects declines when taxes are added as an expense to gross profits before taxes.

"Cash flow" needs to be understood in the context of this chapter. A broadly used term, *cash flowback*, is flow back of money into the firm from net profits plus depreciation. It is important to realize that this flowback is the same as the difference in disbursements between alternatives—namely, the savings in annual disbursements because savings minus depreciation equals profit. The major exception is the situation where the company is operating at a loss and thus is unable to squeeze depreciation from its expenses.

A company invests in projects, and significant money is risked in the business. These investments returnable to the firm as depreciation can cause dramatic changes in the profitability of the venture, and earlier return of the investment through accelerated methods are desirable. There are several methods of depreciation, and this chapter considers only two: straight-line and modified accelerated capital recovery systems (MACRS). The total amount of the depreciation monies are approximately the same, but the timing of the money can vastly affect the time value of money opportunities. MACRS deploys the return money more quickly than straight-line method. The effects of present worth, annual worth, etc., are influenced by the timing of the cash flows. These methods result in an identical conclusion regarding the investment decision provided that accurate principles are followed. For purpose of tabulations that follow, we define

$$\begin{aligned} \text{after tax cash flow} &\cong \text{before tax cash flow} - \text{taxes} - \text{loan cash flow} \\ \text{where after tax cash flow} &\cong \text{computation for project, dollars} \\ \text{Loan cash flow} &= \text{capital recovery of loan, dollars} \end{aligned} \tag{10.17}$$

Before- and after-tax replacement evaluations are conducted in the same fashion once the tax effects have been imposed on the cash flow patterns. Significantly, taxes have a part in the evaluation of engineered projects. There is variety to the way the economics of engineering design projects are handled. Here are considerations for after-tax analysis:

- Costs of investment, operation, revenue, and savings of designs
- Depreciation
- Taxable income
- Cash flow effects
- Equivalent annual cost, present worth for equal service life or rate of return calculations
- Decision process
- Consideration of noneconomic/engineering factors in decision making

Now consider the following examples having tax influences.

Example: Defender and Challenger with MARR and Tax Consequences for Remediation Engineering Project

A firm involved with remediation engineering design and recovery of EPA pollution supersites is confronted with a radioactive waste dump project. The project requires equipment for the next three years, and after that the site will meet minimum standards and the project will conclude. Present equipment has only three years remaining. Rebuilding of the equipment is expected to cost \$11 million, which will qualify the equipment to meet the project requirements and simultaneously reduce annual remediation operating costs to \$14 million. Accounting records show a remaining undepreciated value of \$12 million. The old processing equipment can be sold currently on the market for \$3 million. At the end of the project, the equipment will be unable to meet rigid EPA standards and will be sold for secondary service for \$2 million. The existing equipment uses straight-line depreciation methods.

New equipment is available on the market and the best quotation asks \$28 million. The new processes and equipment will give an annual operating cost of \$7 million. The depreciation method for the challenger is Modified Accelerated Capital Recovery System (MACRS.)

The firm's all-inclusive tax rate is 35%. Furthermore, the after-tax return on investment is 10%. Which system is preferred? Apply the equivalent annual cost method.

The defender's capital cost for the replacement study is its current market value, \$3 million, and this amount is considered to be outlay at time zero for the comparison shown in Table 10.5. But during the next three years it will have a straight-line depreciation charges of $(12 - 2)/3 = \$3.3$ million. Operating costs of the refurbished equipment during the first year include the rebuilding of the equipment of \$11 million and the remediation cost of \$14 million for \$25 million, which is shown as a negative number in Table 10.5 for end of year 1. There is no income attributed to the remediation, since the defender and the challenger have the same income from the contract terms for the project.

Because no income is assigned to the system's performance, all before-tax cash flows are negative, except the salvage value of the defender. IRS code for MACRS recovery for the challenger does not allow a salvage value, and in this practice salvage is 0, but if salvage value is received at the end of the useful life, it is taxable as ordinary income to the firm. The annual recovery rates for the MACRS schedule are shown in Chapter 4, and the student will want to review that material.

The market value is considered to be the mythical "purchase price" of an asset already owned, and it represents the best current estimate of ownership cost. A book-value loss, such as the sunk cost of $\$12$ million $-$ $\$3$ million $=$ $\$9$ million has no bearing on the current situation, and previous instituted depreciation charges remain in effect. It should be noted that overhaul and repair expenses are treated as current expenditures deductible from current income. They are not capital costs of ownership. Overhaul and repair expenses can affect capital cost only by increasing the salvage value as a result of more conscientious maintenance.

The reverse of sunk cost occurs when the amount received for an asset at the end of its useful life is greater than the estimated salvage value. The sale of an asset below its book value is treated as an ordinary loss. When the selling price is greater than the book value but less than the original purchase, the difference is taxed as ordinary income. The portion of the selling price that is above the original capital cost is considered to be a capital gain and is taxed according to capital gain tax policies, which are not discussed in this text. The loss or gain is declared in the year in the which the transaction takes place.

A summary income statement for year 2 shows the increase in operating cash flow on an after-tax basis, and is given as follows:

Partial Income Statement at End of Year 2, $$10^6$	Defender	Challenger
1. Savings or revenue due to equipment	$0	$0
2. Operating costs	14	7
3. Before tax cash flow = savings or revenue − operating costs, 1 − 2	−$14	−$7
4. Depreciation charges	3.3	12.6
5. Deductible charges = net increase in profits before tax, 3 − 4	−$17.3	−$19.6
6. Tax savings = tax on increase in profits before tax at 35%, 0.35 × 5	−6.1	−6.9
7. Increase in profit after tax due to investment in equipment, 5 − 6	−$11.2	−$12.7
8. Increase in operating cash flow = after-tax cash flow, 3 − 6	−$7.9	−$0.1

Line 8 is an entry for Table 10.5 for the after-tax cash flow column for end of year 2.

TABLE 10.5 Comparison of After-Tax Costs for a Defender and Challenger for an Effective Tax Rate of 35m%

End of Year	Before Tax Cash Flow (1)	Depreciation Rate (2)	Depreciation Charges (3)	Deductible Charges (4) = (1) − (3)	Tax Savings at 35% (5) = (4) × 0.35	After Tax Cash Flow (6) = (1) − (5)
			Defender ($$10^6$)			
0	−$3					−$3
1	−25	0.33	$3.3	−$28.3	−$9.9	−15.1
2	−14	0.33	3.3	−17.3	−6.1	−7.9
3	−14	0.34	3.4	−17.4	−6.1	−7.9
3	2					2
			Challenger ($$10^6$)			
0	−$28					−$28
1	−7	0.33	9.2	−$16.2	−$5.7	−1.3
2	−7	0.45	12.6	−19.6	−6.9	−0.1
3	−7	0.22	6.2	−13.2	−4.6	−2.4

The after-tax cash flows are used in the equations to the EAC for the defender and the challenger.

$$\text{EAC}(D) = [-3 - 15.1(P/F\ 10\%,1) - 7.9(P/F\ 10\%,2) - 7.9(P/F\ 10\%,3) + 2(P/F\ 10\%,3)](A/P\ 10\%,3)$$

$$\text{EAC}(D) = [-3 - 15.1(0.9091) - 7.9(0.8265) - 7.9(0.7513) + 2(0.7513)](0.4021)$$

$$= -\$11.1 \text{ million}$$

$$\text{EAC}(C) = [-28 - 1.3(P/F\ 10\%,1) - 0.1(P/F\ 10\%,2) - 2.4(P/F\ 10\%,3)](A/P\ 10\%,3)$$

$$\text{EAC}(C) = [-28 - 1.3(0.9091) - 0.1(0.8265) - 2.4(0.7513)](0.4021) = -\$12.5 \text{ million}$$

The challenger is preferred, as the negative costs represent disbursements that are treated as positive costs, so the challenger is preferred over the defender.

The defender and the challenger have differing depreciation methods. The straight-line method has the disadvantage that the capital money is returned later to the investment cash flow, thus a favorable advantage for the MACRS and the challenger.

Example: After-Tax Cash Flow Present-Worth Analysis for Frosty Steamy Mug

A start-up design-manufacturing company has applied for patent for the product "Frosty Steamy Mug." This student design uses the Peltier principle of physics, where two dissimilar metals joined similar to a thermocouple can behave both as a heat sink to heat or to cool in way that a drinking mug with hot coffee or ice tea can be heated or cooled. Batteries are the source of power, which are hidden in the mug's handle. The student's plan is evolving into a consumer product that will be on the market within the first year, as the design is complete, that being achieved in an inventor's class. But purchased equipment and space rental for the processes are necessary. Especially tricky is the process to seal an outer stainless steel jacket and plastic inter-liner with a vacuum of 10^{-4} torr within the 8 mm space between the two materials.

Venture capitalists, who fund start-up companies for a later and a big share of the business, are willing to grubstake the students. The students find it necessary to develop a business plan, of course, and secure some of their own financing, in addition to the loan by the VCs. In effect, the VCs in first-stage financing behave as bankers. Already the students have learned that banks are unwilling to lend money because of the risk of this unknown principle and a product where the market is undeveloped.

VC's insist on after-tax rate of return of 20% as the MARR to prove the investment plan. The VCs insist as a condition of the loan that they be presented with an analysis of the investment ROI with the VCs as the owners of the capital. Thus the students need to explain their plan with a present worth of the cash flow of income and expenses.

The student team has developed a business plan indicating that the team members are able to secure $400,000 from family and friends. Furthermore, a loan of $1.8 million by the VCs is necessary, who require an interest rate of 5 percent, a generous rate from the viewpoint of the students, but the VCs want a successful business, and they will get their "pound of flesh" as the business progresses to issuing shares, and becomes stable, that probably happening after year 5. The plan specifies a depreciable amount of $2.2 million is necessary. Working capital for labor, materials, indirect costs and salaries is subsumed into a before-tax cash flow. Income from sales of the mug contribute to a positive cash flow early in the financing plan. Follow the development in Table 10.6.

TABLE 10.6 After-Tax Cash Flow for Frosty Steamy Mug

End of Year	Before-Tax Cash Flow	MACRS Rate	Depreciation Charges	Interest on Loan	Taxable Income	Tax Rate	Taxes	Loan Cash Flow	After-Tax Cash Flow
0	−$2,200,000							$1,800,000	−$400,000
1	600,000	0.20	$440,000	$90,000	$70,000	0.15	$10,500	−415,800	173,700
2	800,000	0.32	704,000	73,710	22,290	0.15	3,344	−415,800	380,857
3	1,200,000	0.24	528,000	56,606	615,395	0.34	209,234	−415,800	574,966
4	1,200,000	0.16	352,000	38,646	809,354	0.34	275,180	−415,800	509,020
5	1,200,000	0.08	176,000	19,788	1,004,212	0.40	401,685	−415,800	382,515

Loan repayments can take several forms. The equal-annual payment plan is a common type of business loan. Annual payments, which include the interest charges at 5 percent on the unpaid balance, are

$$\text{Annual loan repayment} = 1{,}800{,}000\ (A/P\ 5\%,5) = 1{,}800{,}000\ (0.2310) = \$415{,}800$$

which is the loan cash flow repaid to the VCs.

Annual interest on the unpaid balance for year 1, 2, and 3 is

$$\text{Year-1 interest} = 1{,}800{,}000(0.05) = \$90{,}000$$

$$\text{Year-2 interest} = [1{,}800{,}000 - (415{,}800 - 90{,}000)]0.05 = \$73{,}710$$

$$\text{Year-3 interest} = [1{,}800{,}000 - 325{,}800 - (415{,}800 - 73{,}710)]0.05 = \$56{,}606$$

Then taxable income equals before-tax cash flow minus depreciation charges minus interest on loan. Year-1 taxable income = 600,000 − 440,000 − 90,000 = \$70,000. These values are entered in Table 10.6.

We find the taxes equal tax rate that depends on the taxable income multiplied by taxable income. This is the after-tax cash flow when a loan is repaid with uniform payments. In Table 10.6 after-tax cash flow equals before-tax cash flow minus taxes minus loan cash flow. The loan cash flow is to the VCs and from the students' viewpoint, is negative. For example,

$$\text{Year-1 after-tax cash flow} = 600{,}000 - 10{,}500 - 415{,}800 = \$173{,}700$$

$$\text{Year-2 after-tax cash flow} = 800{,}000 - 3{,}344 - 415{,}800 = \$380{,}857$$

With the after-tax cash flows determined, the next step is to find the present worth sum, using the interest rate of 20%, which is the VCs required rate.

$$\begin{aligned} PW = & -400{,}000 + 173{,}700(P/F\ 20\%,1) + 380{,}857(P/F\ 20\%,2) + 574{,}966(P/F\ 20\%,3) \\ & + 509{,}020(P/F\ 20\%,4) + 382{,}515(P/F\ 20\%,5) \end{aligned}$$

$$\begin{aligned} PW = & -400{,}000 + 173{,}700(0.8333) + 380{,}857(0.6945) + 574{,}966(0.5787) \\ & + 509{,}020(0.4823) + 382{,}515(0.4019) = \$741{,}215 \end{aligned}$$

This present worth amount of \$741,215 is well above zero, and is a strong recommendation to the VCs.

SUMMARY

This chapter introduces the principles of engineering economy. Starting with average annual rate-of-return and payback methods, the overarching time-value-of-money method is described and various formulas are presented along with tables. Present worth, future worth, annual cost, and rate of return are prominent quantities that have important meaning in the analysis of engineering projects. Engineering economics provides a rigorous methodology for comparing investment alternatives.

Practical examples of selection of alternatives, replacement choices for equipment, and taxation effects on the project are discussed. The inputs of the engineering-economy analysis are estimated, thus connecting this chapter to earlier presentations.

The student will want to remember these principles for analysis of investment opportunities.

QUESTIONS FOR DISCUSSION

1. Define the following terms:
   - Annual worth
   - Average rate of return
   - Capitalistic efficiency
   - Capital recovery
   - Compound interest
   - Defender
   - EAC
   - Effective income tax rate
   - Engineering economy conventions
   - Equipment life
   - ERR
   - IRR
   - MARR
   - MACRS
   - Net present worth
   - Owner viewpoint
   - Payback
   - Replacement principles
   - Simple interest
   - Sunk cost
   - Taxable income
   - Time value of money
   - Unequal life alternatives
2. Discuss the importance of engineering economy and give examples where it is best applied.
3. Compare the advantages and disadvantages of the payback method.
4. Give other measures of return. For instance, when a shareholder evaluates a stock, return = (dividend per year)/(stock price). Consider return on savings account, price appreciation, real estate, and cost savings for an engineering improvement.
5. Why are engineering investments handled by engineering-economy methods? Describe those situations where this type of analysis is overlooked.
6. What does "discounting" imply?
7. List the pros and cons of the metrics "net present worth" and "internal rate of return" for the evaluation of an engineering project analysis.
8. List the usual assumptions required for engineering economy project analysis.
9. Intangible factors often out weigh the economic factors. Describe situations where this is found.
10. How are unequal-life alternatives judged?
11. Sunk costs are retractable or irretractable. Discuss.
12. What is the effect of taxes on the evaluation of projects having revenue projections?

PROBLEMS

10.1 (a) New manufacturing equipment costs $225,000, salvage value is $25,000, and an average annual earning of $20,000 after taxes is expected. Find the average annual rate of return. If earnings are doubled, then what is the rate?

(b) Consider part (a). The investment has an economic life of 10 years. Straight-line depreciation for 8 years is used. Find the percent return.

10.2 Investment for new equipment is $100,000, and the salvage value is $10,000 8 years hence. Average yearly earnings from this equipment are $15,000 after taxes. What is the non–time-value-of-money return? When is the payback?

10.3 (a) What is the amount of interest at the end of 2 years on $450 principal for a simple interest rate of 10% per year?
(b) If $1600 earns $48 in 9 months, what is the nominal annual rate of interest?
(c) An investment of $50,000 is proposed at an interest rate of 8%. What is the future amount in 10 years? Consider both the simple and compound cases.
(d) What is the present worth of $1000 for 6 years hence if money is compounded 10% annually?
(e) What is the compound amount of $3000 for 15 years with interest at 7.25%?
(f) Find the annual equivalent value of $1050 for the next 3 years with an interest rate of 10%.
(g) How many years will it take for an investment to triple itself if interest is 5%.
(h) An interest amount of $500 is earned from an investment of $7500. What is the interest for 1 year? For 2 years?

10.4 (a) Find the principal if the interest amount at the end of 2.5 years is $450 for a simple interest rate of 10% per year.
(b) A loan of $5000 earns $750 interest in 1.5 years. Find the nominal annual rate of interest.
(c) In 1626 Native Americans bartered Manhattan Island for $24 worth of trade goods. Had they been able to deposit $24 into a savings account paying 6% interest per year, how much would they have in year 2000? At 7% interest?
(d) What payment is acceptable now in place of future payments of $1000 at the end of 5, 10, and 15 years if the interest rate is 5%?
(e) What is the compound amount of $500 for 25 years with interest at 10%?
(f) Calculate the annual equivalent value of $1000 for the next 4 years with an interest rate of 10%.
(g) How many years will it take for an investment to double itself if the interest rate is 10%?
(h) An investment of $10,000 earns an interest amount of $750. Find the rate of nominal interest if the amount is earned over 1, 2, or 3 years.

10.5 Two investment opportunities are proposed:

	Enterprise A	Enterprise B
Total investment	$60,000	$60,000
Revenue (after tax)		
Year 1	20,000	30,000
Year 2	20,000	30,000
Year 3	20,000	30,000
Year 4	20,000	
Year 5	20,000	

Find simple payback. For an interest rate of 10%, which Enterprise has the least period of time before investment recovery is complete? (*Hint*: Use the cumulative amount of present worth to determine the investment payback number.)

10.6 Two bids are compared for performing the same work:

	Bid A	Bid B
Total bid	$100,000	$150,000
Revenue (after tax)		
Year 1	25,000	25,000
Year 2	30,000	40,000
Year 3	35,000	55,000
Year 4	30,000	40,000

(a) On the basis of payback, which bid is preferred?
(b) For an interest rate of 5%, which bid has the least period of time before the bid value is returned? (*Hint*: Use the $(P/F, i\%, n)$ factor to find the cumulative amount of present worth to determine a payback number. Apply Table 10.3.)
(c) Discuss the pros and cons of the payback measure of evaluating an investment option.

10.7 A prospective venture is described by the following receipts and disbursements:

Year End	Receipts	Costs
0	$ 0	$800
1	200	0
2	1,000	200
3	600	100

For $i = 15\%$, find the desirability of the venture on the basis of net present worth.

10.8 An estimate has the following cost and revenue cash flows. The cash flows are assumed to occur at the end of the year.

Year	Cost	Revenue
0	$800	$ 0
1		450
2		425
3		400

(a) If interest is 10%, find the net present worth, net future worth, and net annual equivalent worth.
(b) Find the rate of return. (*Hint*: The guessing range is 25 to 30%).
(c) Present a summary of the four methods.

10.9 The cash flow of cost and revenues for a project are given as follows:

Year	Cost	Revenue
0	$1025	$ 0
1	0	450
2	0	425
3	0	400

(a) Calculate the net present worth, net future worth, and net equivalent annual worth if the interest rate is 11%.
(b) Repeat (a) for an interest rate of 9%.

10.10 Cash flows of two competing choices, A and B, are given below. Project life is over after 5 years. Use an interest rate of 20%. Present a summary of the four methods of engineering economy evaluation. Is A or B preferred? Discuss the advantages of the time-value- of-money concepts for a situation like this.

	Equipment A	Equipment B
Investment	$60,000	$60,000
Annual after tax earnings		
Year		
1	20,000	30,000
2	20,000	30,000
3	20,000	30,000
4	20,000	
5	20,000	
Total	$100,000	$90,000

10.11 (a) Evaluate the cash flow diagram given by Figure P10.11. Determine a present sum, equivalent annual payment, and a future sum. Use an interest rate of $i = 10\%$. (*Hint*: Arrows represent end-of-the-year cash flows. Those arrows pointing down represent costs and those pointing up are revenues.)
(b) Repeat (a) with an interest rate of $i = 20\%$.

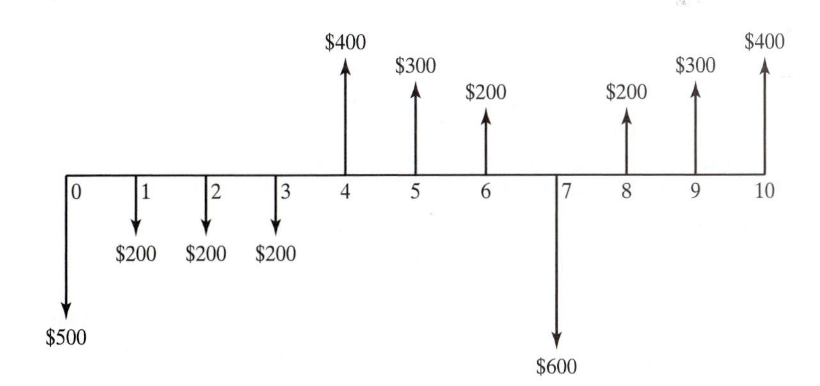

Figure P10.11

10.12 (a) Evaluate the cash flow diagram, Figure P10.12. Find present value, annual value, and a future value. Use an interest rate of $i = 10\%$. (*Hint*: Arrows are end-of-the-year cash flows. Arrows pointing down are costs and those pointing up arrows are revenues. The present value is timed to occur at $n = 0$, and the future value is timed to occur at $n = 10$.)
(b) Repeat (a) with $i = 25\%$. Use a spreadsheet to find your solutions.

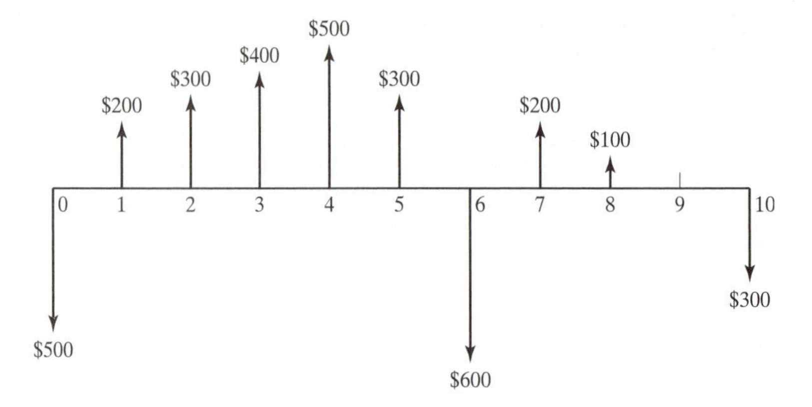

Figure P10.12

10.13 The "Rule of 72" can be used as an approximation to find the interest or period of time, given the other quantity. The formula is given as $ni \cong 72$. If \$1 is invested for 10 years, what compound rate is necessary for the money to double? Find the exact answer. How long does it take for money to double at 12%?

10.14 The interest rate is modeled by several equations. Consider the following one:

$$i_{eff} = \left(1 + \frac{r}{m}\right)^m - 1$$

where i_{eff} = effective interest, and sometimes called the Annual Percentage Rate, APR, decimal

r = nominal interest rate per year, decimal

m = compounding periods per year, number

If the nominal interest rate per year is 10%, and it is compounded monthly, or 12 periods per year, find the effective interest.

10.15 Suppose a gasoline credit card charges a rate of 1% per month upon the unpaid balance. Find the nominal rate and the APR.

10.16 Banks A and B advertise differing interest rates. Bank A pays a nominal rate of 5% per year compounded quarterly, and Bank B pays depositors $4\frac{3}{4}$% per year compounded daily. Which bank has the better rate? (*Hint*: Compare the rates using effective interest.)

10.17 Continuous compounding of interest is found in finance, banking, and large money transactions, though it is not normally used for engineering analysis, however. For continuous compounding,

$$i_{eff} = e^r - 1$$

where r = nominal rate per year, decimal

Suppose that \$1000 is invested into a continuously compounded account of 8% nominal interest for 5 years. What is the future amount? (*Hint*: Find the effective interest rate for continuous compounding, then find the amount.) Repeat for annual compounding.

10.18 Find the best loan rate between the following two choices: Loan A is quoted at the rate of 9%, compounded monthly, and second quotation gives 8.8% compounded continuously.

Both the principle and the interest are paid back to the lender on the first-year anniversary date of the loan. (*Hint*: Consider yourself as the borrower.)

10.19 Suppose that you will be buying a new house that sells for $230,000. Currently, you have a smaller house that will provide $30,000 for the down payment for the new house. The financing is a 30-year mortgage with an interest rate of 10%.

(a) Find the approximate monthly payment to buy the house using the Tables where $n = 30$ years. (*Hint*: Solve for the annual amount A, and divide by 12, this gives the approximate monthly payment.)

(b) Find the exact amount of the monthly payment. An exact solution requires revising Equation 10.10, which gives the modified equation

$$\frac{A}{m} = P\left[\frac{\frac{i}{m}\left(1 + \frac{i}{m}\right)^{nm}}{\left(1 + \frac{i}{m}\right)^{nm} - 1}\right]$$

where $m = 12$ months per year

(c) Calculate the total interest paid.

10.20 An engineer is considering an individual retirement account (IRA). Our engineer plans on depositing $1500 annually into the IRA, and a mutual fund is selected which claims to return 10% per year. Assume that the engineer's earnings are such that there are no current tax disadvantages or other penalties, and the federal tax code will remain constant on this provision over this period of time. Furthermore, the amount will be untouched until retirement 35 years from now. What amount can be expected at retirement?

10.21 Designers are evaluating two systems for heating and cooling the air for a new building. The life of the building is 40 years, and the alternatives are expected to have a 20-year life. Evaluate the alternatives and find the preferred design using an interest rate of 10%. Employ the annual cost method.

	Natural-Gas Engine Powered Heat Pump	Refrigeration and Domestic Heating
Investment	$900,000	$600,000
Annual energy costs	$15,000	$35,000
Annual maintenance costs	$22,000	$30,000
Incremental annual taxes	$4,000	

10.22 A company is considering two locations in addition to its currently operating plant location. Advantages are to be gained by relocation, but is it worth it? This firm uses a 20% rate of return before taxes, and the economic life for the plant-relocation opportunity is 25 years. Salvage value on the building is 20%, while there is 100% salvage on the land. Use a present worth analysis. Estimates are made for factors that vary between the locations. All other factors are considered neutral to the analysis. Determine the most favorable location.

	Present Location	Location C	Location W
Land cost	$5,000,000	$2,500,000	$7,500,000
Factory cost	5,000,000	35,000,000	37,500,000
Property taxes, annual	2,500,000	1,000,000	1,300,000
Raw material transportation, annual	2,000,000	500,000	1,400,000
Utilities, annual	15,100,000	15,500,000	18,500,000
Labor, annual	22,500,000	16,000,000	23,000,000
Product distribution, annual	1,000,000	1,500,000	1,400,000

10.23 Three mutually exclusive investments are suggested for an engineering design. These opportunities have initial capital investments that are producing positive cash flows from increased revenue. The firm's ROI is 20% per year, which is the desired rate before income taxes. The alternatives have identical useful lives of 10 years.

	Alternative Design		
	A	B	C
Investment (first) cost	$370,000	$920,000	$680,000
Net annual receipts less expenses	$169,000	$267,000	$233,500

(a) Using the present worth method, find the order of preference of the investments. (*Hint*: Consider the investment as a negative cash flow while the net revenue is positive.)
(b) Using the annual worth method, find the order of preference of the investments.
(c) Using the future worth method, find the order of the preference of the investments.

10.24 A product has a certain life of 5 years, and a heavy stamping and forming production line is necessary for this opportunity. There are four basic types of frame design for stamping and forming, which are gap, arch, straight side, and horn. These four types have individual requirements for mechanization and part handling, labor costs, power, property taxes and insurance, and maintenance costs that differ depending on the fundamental design. Though the product design and quantity are identical, the costs of press investment and operation are different.

The firm foresees no salvage value at the end of the product cycle. The firm desires a 10% rate of return on the investment before taxes. Use the three methods of present worth, annual cost, and future worth to choose the least costly press. Indicate the preference order. Consider the following facts.

	Frame Design for Press			
	Gap	Arch	Straight Side	Horn
Investment, installed	$600,000	$760,000	1,240,000	$1,300,000
Useful life, years	5	5	5	5
Annual operation and maintenance cost				
Power	68,000	68,000	120,000	126,000
Labor	660,000	600,000	420,000	370,000
Maintenance	40,000	45,000	65,000	50,000
Property taxes, insurance	12,000	15,200	24,800	26,000

10.25 Two business opportunities are under scrutiny for a "due diligence" investigation by venture capitalists. The purpose of the analysis is to find the best choice between A and B. Revenues and costs are known, and the alternatives have differing lifetimes. For an interest of 10% before taxes, which investment is preferred? The estimates are as follows:

Alternative	A	B
Investment cost	$350,000	$500,000
Annual revenue	190,000	250,000
Annual cost	64,500	138,300
Salvage value at end of useful life	0	0
Useful life, year	4	8

(a) Use the present worth method to make the preferred choice. (*Hint*: Use the repeatability assumption for the shorter-lived alternative.)
(b) Use the annual worth method to make the preferred choice. (*Hint*: Have the annual worths compared over their useful life.)

10.26 A venture capital mutual fund is evaluating the options of an electronics manufacturer who is considering either contract assembly abroad and out-sourcing or home manufacture in the United States. This decision is rooted in the economic comparison of the two choices. The mutual fund associate is presented with the information that the company is evaluating.

Product revenues for identical production and sales volume are assumed to be equal though annual expenses are different. The annual expenses are sufficient to manufacture the product to identical specifications, quality, and delivery location. The rate of investment that the mutual fund is seeking is 20% before taxes. Values are $\$10^4$, but that is not indicated in the table below.

	Contract Manufacture	Home Manufacture
Investment of plant	$12,0000	$30,000
Salvage value of plant at end of plant life	0	10,000
Annual expenses	2,200	1,000
Useful life, year	10	15
MARR before taxes	20%	20%

Use the repeatability assumption to determine which investment plan is preferred. (*Hint*: Repeated life for a common multiple for the two alternatives is 30 years.) Recommend the preferred location for manufacture of the product. Use the present worth and annual cost methods.

10.27 A 3-year old induction furnace is too small for future heat-treating production requirements. The company has been forced to subcontract half of the total heat treating requirements during the past year. A new induction furnace, which duplicates the present furnace, can be purchased and installed for $600,000. The net realizable value of the present furnace is $400,000. Both furnaces are expected to have a 6-year economic life from this date.

At the conclusion of this life, the salvage value of the new furnace is expected to be $90,000 and that of the old equipment is $80,000. Operating costs for either machine running at 100% capacity are $400,000 annually. Consider this problem as replacement of equipment having inadequate capacity.

A single furnace can be obtained to handle the load, and it will deliver 125% of the joint output of the two furnaces. Its installed cost is $1,100,000 with a salvage value of $165,000 at the end of the predicted economic life of 6 years. Annual operating costs are $750,000 when it is producing at rated capacity, and $600,000 when equaling the rated output of the two smaller furnaces.

The minimum required rate of return is 10%. For this before-tax analysis, construct three cash flow diagrams for the cases of present unit, added unit, and large furnace. Find the annual cost of two furnace systems with identical output. (*Hint*: If operating costs do not provide identical output, one alternative will be burdened with costs that are not charged to other units.)

Think about this question for discussion: What about using the salvage value returned from the present small machine as salvage value for the large machine?

10.28 An old process is valued presently at $8,000,000 and is expected to have a life of 3 years, and its salvage value is zero while annual operating costs are $16,000,000 a year. The newer process will cost $40,000,000 and will have an 8-year economic life with zero salvage and operating costs of $11,000,000 yearly. The minimum attractive rate of return is 10%.

Sketch the cash diagram. Calculate the annual cost of the alternatives with the 3- and 8-year lives and find the numeric advantage of the lower-cost alternative. (*Hint*: The annual cost method for the new alternative is "recovered" over 8 years. Therefore, the period of comparison is limited to 3 years. This method disregards the future events and their consequences beyond the life of the shorter-lived process.)

Using present worth, find the economic value of the two alternatives for a 3-year period. (*Hint*: Compress the 8-year alternative of annual costs to a 3-year life. Also for the shorter lived process, find the present value of the alternative.) Does the choice of the preferred method change between the annual-cost or present-worth methods?

10.29 A small manufacturing company is facing the following situation. A process is needed for just 3 years. It has a initial cost of $920,000 and is expected to have a salvage value of $200,000. The process is expected to generate annual income of $500,000. A loan of $400,000 at 10 percent interest is necessary for the purchase. Terms for the loan are repayment in three equal installments, which includes interest. The firm uses straight-line depreciation, and its effective income tax rate is 35%. What is the present worth of the after-tax cash flow for the process when the minimum after-tax rate of return for ventures is 20%? (*Hint*: Consider the problem as an after-tax present worth with interest and depreciation.) What is your recommendation for the purchase of the system?

10.30 A product requires a machine tool for the next 4 years and after that the need no longer exists. But the present defender has only 3 years remaining on its original life. The equipment can be refurbished for $5000, which will extend its life and reduce operating costs to $16,000 per year. Accounting records indicate a book value of $9000. The old machine can be sold for $8000 now. At the conclusion of life, there will be no salvage value.

A challenger is available and its price is $36,000. The more advanced equipment has operating costs of $12,000 per year and salvage value of $6000 at the conclusion of a 4-year life.

The firm's all-inclusive tax rate is 40%, and the company requires an after-tax rate of 10% as the minimum hurdle for investments. Analysis for depreciation is the simplified straight-line method.

Which configuration of equipment is desired? Use the equivalent annual-cost method for this after-tax comparison. (*Hint*: The market price is considered to the purchase price of any existing asset. Overhaul or repair expenses are treated as current expenditures. Revenue is the same between the defender and the challenger.)

CHALLENGE PROBLEMS

10.31 A company is planning the design and production of a product with the following estimates: total direct labor is 12.500 hr/100 units; average gross hourly cost is $25/hr; material costs are $17/unit; general overhead, which includes salaries, office expenses, and utilities, is based on direct labor cost and is 150%.

Equipment that is necessary to produce the design will cost $400,000 and is expected to have an equipment life of 10 years. The company uses a return of investment of 20%. Whenever the equipment is replaced, residual value is expected to be 10%.

Activity-based costs for product development are $85,000 and is a first-year cost only. Marketing indicates that the product will have an economic life of 8 years, and the annual sales are 35,000 units.

Find the full cost of the product. (*Hint*: Assume that the return-of-investment contribution is a cost of the product. Treat the activity-based costs as an annual cost with a ROI of 20%. Have market life override the equipment life.) Determine the unit product cost on the basis of annual costs.

10.32 Two alternate sand-handling belt materials are candidates for a gray-iron foundry. Estimates are given as follows:

Design	First Cost, $	Life, Years
Steel grate	80,000	10
Reinforced rubber	100,000	15

Any salvage value of the replaced sand-handling material is offset by its removal cost at the end of its life. Maintenance costs are equal for both designs.

(a) Sketch the cash flow diagrams. Find the preferred design for an interest rate of 10% for a foundry life of 30 years. Find the annual cost of both designs. Use functional notation to solve this problem. (*Hint*: In the sketch have the project estimate cost arrows point downward and the equipment annual cash flows upward. Assume two replacements of the steel grate and one replacement of the reinforced rubber material. There is no replacement at the thirtieth year period. Why? Use the 10% table in the appendix.)

(b) Repeat part (a) for 20%.

10.33 Manufacture of a printed circuit board product is to last 2 years. The work requires moving a printed circuit board mounted with electronic components in an assembly-line fashion. Two solutions are proposed: A fixed-belt conveyor costing $75,000 is estimated to have an annual operating cost of \$20,000 and \$15,000 salvage value at the end of the second year. An alternate choice is mobile equipment costing \$30,000 and its estimated operating cost is \$70,000 annually. Salvage value of the mobile equipment is \$10,000 at the end of 2 years. These alternatives have a longer physical life than the project life.

(a) Sketch the cash flow diagrams. Find the preferred method for an interest rate of 10%. Find the annual cost of both methods. (*Hint*: Use functional notation and the 10% table in the appendix to solve this problem. Initially use the present worth method to find the cost. The operating costs are applied at $n = 0$ and 1 end-of-year periods.)
(b) Repeat part (a) for 20%. Discuss the effects of raising the interest rate.

10.34 An architect is suggesting either a brick or wood exterior surface for your residence. If the appearance is neutral to the choice between the alternatives, select the preferred design. Evaluate the proposals at 10% interest using the annual cost model. (*Hint*: For the annual cost model, it is not necessary to consider replacement for the shorter-lived material.) Estimates are given in the following table:

Alternative	Brick	Wood
Initial cost of residential surface	\$20,000	\$7,000
Expected life, year	80	40
Annual maintenance	\$50	\$100
Painting maintenance		\$3000
Painting cycle, year		10

PRACTICAL APPLICATION

There are many opportunities for a practical experience in engineering economy using the objectives of this chapter. Under your instructor's guidance, form a team to conduct this application. Your assignment is to call an engineering firm that conducts studies on projects, and interview them. Your instructor will give additional and refining ideas for the report.

Expand your understanding of project analysis using time-value-of-money concepts as studied in this chapter. Prepare a list of questions for the interview. Consider the following: Are they able to provide you a report on a typical study? What are the chief methods of analysis, and where do they have the biggest obstacles? What confirming information convinces you of the accuracy or appropriateness of the conclusions? Do they recommend a choice among alternatives? Conclude your practical application by word processing a report on your findings.

Send the engineers a "thank you" and include your report.

CASE STUDY: SEASONAL PRODUCTION

Management is aware of the economic advantages of equipment size and capacity. A particular question naturally arises: What are the cost advantages to be gained by change from one equipment capacity to another? This type of problem may arise from seasonal fluctuations in consumer demand. The advantage of a smaller plant or equipment capacity is low first cost and investment charges, and the advantage of larger capacity is the reduction in finished goods inventory and storage costs and the avoidance of potential losses from fickleness of consumer preferences or competitive design which would reduce the value of stored product. Remember that increases in productive capacity do not result necessarily in increased sales, but only in reduction of the average and maximum inventory.

Consider a seasonal product of "tire chains," which sells briskly during periods of snow and natural hazard, but not during other periods of the year. A manufacturer produces and sells 240,000 sets a year to dealers, and sales occur over a relatively short time season as follows:

September	30,000
October	50,000
November	80,000
December	40,000
January	40,000

The process for producing tire chains uses automatic forging equipment that transforms hot coiled AISI 1035 steel into links and then chains them into the weave and warp of the design. Capacity of the present alternate is 20,000 pairs a month. The current automatic hot forger machine originally cost $410,000, but now has a net realizable value of $300,000.

Additional equipment can be purchased, along with forging dies, for $340,000, and that will bring the plant production capacity to 40,000 pairs per month. The equipment and tooling investment for 40,000 pairs is $640,000.

The plant carries an inventory during various periods of the year in order to meet customer demand. It is company policy to maintain a base inventory of 5000 pairs at all times. A rule to find the end-of-period inventory is

initial starting inventory at start of month
− sales in month + production during month.

Thus the annual cost of storage of chains is $0.15 per set per year, which is based on the maximum number of pairs in storage, as space must be provided for the maximum number.

The total cost of making a chain is $9 per pair, and the variable cost is $5. Insurance and taxes are 5% of the average number of chain-pairs in storage.

The minimum required rate of return is 20% before taxes. The equipment has an economic life of 10 years and has essentially zero salvage value. Use the annual cost method.

Find the best choice for capacity, either 20,000 or 40,000 chain pairs per month. Base the decision on total annual cost including cost of storage, insurance and taxes, interest on inventory, and capital recovery on investment. *Hint*: Find the maximum and average number of stored product in inventory under the two production rates. The rate of return affects the capital cost and average number of inventory in stock. Consider the following column headings for a spreadsheet analysis:

Sales (pairs)	Month	Production Capacity (pairs per month)	
		20,000	40,000
		Equipment Investment	
		$300,000	$640,000
		End-of-Month Inventory After Sales	
40,000	January	0	0
0	February	20,000	0
	March		

Chapter 11

The Enterprise, Entrepreneurship, and Imaginamachina

This chapter introduces concepts about creativity and the nuts and bolts for starting a business or augmenting the opportunities of an existing business. While the slogans "Creativity is contagious, pass it on," or "The greatest threat to organizations today is not the lack of resources, but the lack of imagination," are attractive, they do not teach students to think like entrepreneurs. Nowadays, classes attempt to enlighten students in thinking ingeniously and to exploit those engineering images.

A humbling statistic comes from one study suggesting that fewer than one in 100 brand new products make money. Furthermore, according to the U.S. patent office only two in 100 patented products are licensed to manufacturers who can then sell them.[1] These statistics are for endeavors involving only a few individuals or the garage tinker with design and hardware. They hide the realization that intrapreneurs are as likely to be found in the biggest of corporations. These experiences point out that business competition is intense, but success is improved whenever advanced technical skills are found alongside shrewd business planning.

The focus of this chapter will be on cost analysis and estimates, not the technical feasibility of the design, equipment, plant, and software. This chapter also provides some thoughts about economic evaluation in the context of the enterprise and entrepreneurship. Topics include designing for profit, preparation of the enterprise plan, and financing with stocks, bonds, or credit.

Figure 11.1 presents our global view of the enterprise and entrepreneurship. With imaginamachina, the entrepreneur begins the process of evaluating the market

[1]Patent and trademark information is available at www.uspto.gov for the United States, www.wipo.int for International, and www.patent.gov.uk for the United Kingdom.

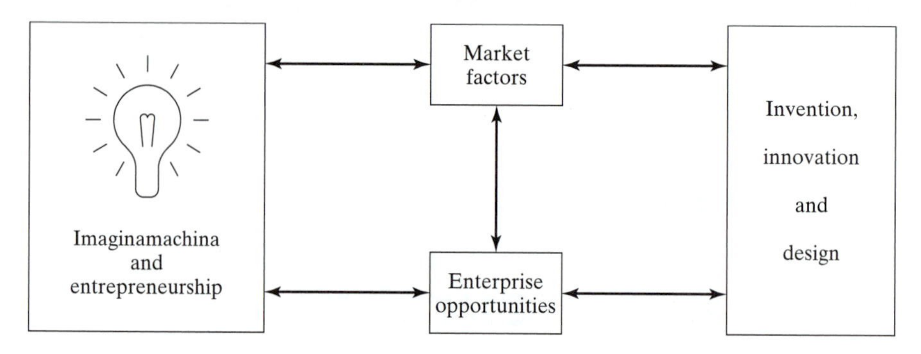

Figure 11.1 Overview of the enterprise, market factors, invention and innovation, entrepreneurship and imaginamachina.

and invention. Ideas do not result from serendipity but are instead the forced consequence of searching and design or even a result of someone's frustrations. In the Chinese language, there is a character that can be translated as both threat and opportunity. The creation of ideas can involve such extremes. Thomas Edison expressed the process of idea generation with both wit and insight: "Genius is 1% inspiration and 99% perspiration."

During these primitive stages, product and process designs and business planning take shape. This planning is preliminary, as we shall see, which leads to its grooming and improvement. That is what this chapter is about.

11.1 THE ENTERPRISE

Manufacturing business is notable for its *free enterprise*, a trite phrase and generally its meaning is obscure. The original interpretation implied that a firm is free to make product or not. How to design, produce, price, and distribute the product is their free choice. Furthermore, a customer has similar freedoms to conduct business with a business. The customer is not compelled to restrict the choices of whom, when, or how to transact business or to buy product. For the business or customer, the earning of profit or competent stewardship of resources is a primary goal for survival and self-benefit. This is our description of the free enterprise system.

The sectors of the free enterprise system in North America include manufacturing, construction, and the information/service industries. These businesses are profit motivated; they pay taxes and are generally not subsidized, federally, which lessens the competitive drive of self-interest. The players in the enterprise include owners and shareholders, designers, manufacturers, customers, and employees.

Manufacturing can be defined as the act of "making things." There is in this definition both process and product. The word "making" is the process, while "things" is the product. Size, location, process, and degree of technical sophistication do not alter the notion of the manufacturing enterprise.

Figure 11.2 shows the importance of market factors on product pricing strategies and places the product launch function within the enterprise. But if the enterprise is to endure and prosper, it needs synergy with the entrepreneurs who provide the imaginamachina. It is the actions by the individuals and teams that generate technical leadership. Maintaining the status quo should never be a policy since cost entropy

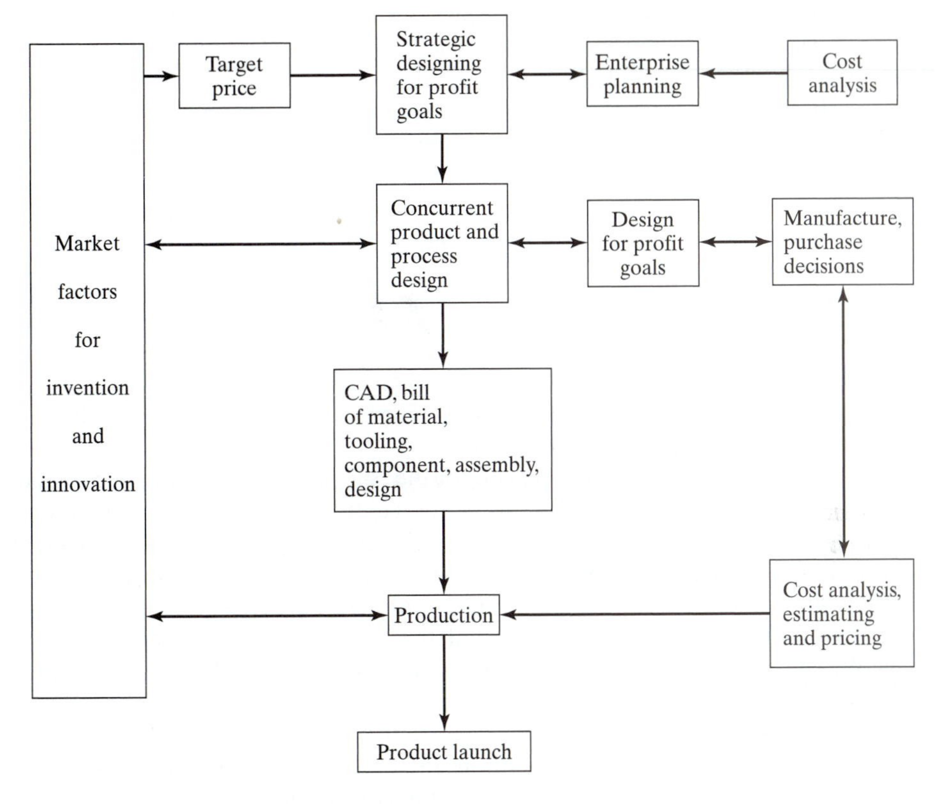

Figure 11.2 Overview of launching a product.

and competitive market forces require improvement even to maintain existing market share. Engineering must continually update, create new products, and reduce costs. Standing still is not an option.

11.2 ENTREPRENEURSHIP

Technology was the catalyst for the cauldron of invention during the twentieth century. The impacts of technology on lifestyle, employment, education, and war (regrettably) were significant. But the concepts that drove technology for that century need to be reexamined for the twenty-first century.

Is the maxim "necessity is the mother of invention" still true today? We invent new products not because they are found necessary; products are not deemed important because of consumer demand. Much of what is new is a recombination of existing technologies, a process described as *technology convergence*. For example, the VCR and electronic mail were not created in response to consumer demand, nor were any new scientific breakthroughs used. Rather, there were existing technologies that could be recombined. Creative ideas and a kaleidoscope of dynamics led to these products. It is later the job of marketing to create demand for such inventions, and money at risk is wanted to build the factories and infrastructure to bring the products to the customers.

Think about another maxim: Technologies mature, but technology does not. A new technology will emerge (abacus, slide rule, comptometer, mechanical, electro-mechanical, electrical, and electronic calculator, and the computer) and have several stages of development, including innovation, incubation, growth, standardization and the addition of features before reaching maturity. But maturity is foreordained, and what was experimental and new becomes the next commodity. The result is a spectacular sequence of growth, stabilization, and decline in industries, unless they are able to reinvent themselves, and once again finding the entrepreneurs able to furnish imaginamachina and inventions.

11.2.1 Inventors and Innovators

There are many entrepreneurs and their names are prominently etched in our technological history. Thomas Edison and Henry Ford were early pioneers of the industrial twentieth century. Can you name others?

Figure 11.2 may be inappropriate for the sole inventor, who, in a sense, does it all. The inventor, an individual driven to be the creator of an idea, discovers the process, product, or contrivance, and energizes the total scheme. In a way, the steps of Figure 11.2 are a part of the sole inventor's thinking, though not necessarily in the precise sequence that is shown. We do not attempt to fingerprint the characteristics of an inventor.

In addition, the entrepreneur and the enterprise are synchronous parts of a whole. The *entrepreneur*, for our purpose, is defined as a gifted individual, team, or collaboration of participating enterprises, promoting progress and profit for capitalists who invest their money for business opportunity. Entrepreneurship, when speaking of individuals, is characterized by invention, and innovation. There must be a strategy that values profit and growth where investment may be required.

To have a chance of success, entrepreneurs need a clear, long-term vision of the market niche that they are pursuing, and a short-term willingness to revise the vision quickly. Nearly all start-up companies are backed by venture capitalists and entrepreneurs who display their alertness by altering course in midstream. Understanding the sales cycle and decision-making process are essential to this action.

Individual entrepreneurs must be high in self-confidence but low enough in ego in order to surround themselves with high-powered talent, and even to step aside if it is in the best interest of the enterprise. Successful leaders are people magnets, who can attract critically needed talent. Knowing the shakers and movers is critical, say the money lenders. Connections are important.

Another key trait is agility. No matter how good an idea may be, or how much research is done, entrepreneurs cannot become too enamored of their ideas, inventions or business plans, which in a competitive market can change at any moment.

Some venture capitalists prefer leaders having technical skills along with a knack for sales and marketing. Some investors look for those who are accelerators having skills and knowledge that get a product or service out at warp speed. Forging strategic partnerships and creating realistic revenue models are important. A trait of being profitable is important, many venture capitalists say.

For the cost analysis and estimating step in entrepreneurship, it starts with the method of "designing for profit." It is in that direction that we now turn.

PICTURE LESSON Household Appliances

Wisconsin Historical Society.

Household appliances dramatically changed the twentieth century lifestyle. The engineer's role in transforming the domestic environment throughout the century has been enormous. It began with electrification, which brought light and power into homes. The household appliances that followed in the first half of the century depended on two basic engineering innovations: resistance heating and small, efficient motors. Engineering originality produced a variety of devices, including electric ranges, vacuum cleaners, dishwashers, and dryers.

In the second half of the century new technologies like the magnetron and microprocessor transformed the household environment yet again, spawning new appliances with sensors, timers, and programming devices. Always, design innovations focused on making appliances lighter, smaller, more energy-efficient and useful.

Until these products arrived, most women organized household work by day. Families were generally large, and chores took a long time. One day was set aside for laundry. The tools at hand were wash boards and tubs, boilers, and clotheslines.

Another day was for ironing and sewing. Ironing involved heating heavy flat irons on the stove, keeping two or three going at the same time so there was always a hot one, and keeping the fire stoked with coal or wood. Sewing involved making patterns for dresses and shirts, then cutting cloth and stitching it by hand.

Another day was for cleaning—sweeping with brooms, scrubbing and waxing floors by hand, taking rugs to the clothesline to beat the dust and dirt away. Finally, one day was set aside for baking, or canning and preserving. The 1890 photograph shows a domestic staff for a well-to-do family, consisting of laundresses, cooks, parlor maids, and scullery girls with their work utensils.

The first U.S. patent for an electric vacuum cleaner occurred in 1908, and the first Hoovers weighed 40 pounds, most of it the weight of motor. The electric toaster was a small but important device to many. Numerous attempts to find the proper heating element were tried by dozens of engineers and inventors, including Thomas Edison. In 1905 an engineer solved the problem with a patent on Nichrome, an alloy of nickel and chromium.

A major event that contributed to the success of electrical appliances was the standardization of electric outlets and plugs in the 1920s. Early versions of many appliances had wiring with plugs that screwed into an overhead light or a wall sconce—wall outlets did not appear until later.

Household appliances give us more free time, and their related industries contribute significantly to our economy. Their impact on life in the 20th century has been immense.

11.3 DESIGNING FOR PROFIT

For the most part, engineers "build up" an estimate starting with operation estimates and concluding with product cost and price. This activity begins with minimal information if preliminary estimates are made. If bills of materials and designs are complete, then detail estimates are prepared. Most detail estimating is "bottoms up."

Occasionally a reverse direction—"top down"—is preferred. This is the frequent case for entrepreneurs who have an idea and need price and cost analysis. A procedure begins with a market price, which is used as the target price.[2] The essentials of the technique has the engineer calculating downward from a market price to find the cost for design assemblies and components. This practice is called *designing for profit.* Other names are design to cost, or target costing and have similar philosophies. Much of the these costs are rough-order-of-magnitude numbers. Those targets become goals for engineering, sales, procurement, and production. If the cost goals are realized, then a profit is ensured because the procedure began at the top, or a price that ensures market success, so it is reasoned.

Design for profit involves the early estimating of the product before design. Of course, there may be sketches, preliminary component and assembly designs, and specifications. A prototype product may be at hand, and development may have progressed to that point. Perhaps, the competitor's actual hardware may be available for estimating. At times a competitor's product is *reverse engineered*, which implies that the product is demanufactured from the final assembly to its components where materials and operations are estimated.

What is meant by "early" cost estimating? Figure 11.2 shows that immediately after the strategic designing for profit goals are determined, cost analysis and estimating continues. Indeed, the activity can be thought of as continuous throughout the various stages.

[2]Marketing information and sources for the American Marketing Association are found at www.ama.org and for the National Retail Federation at www.nrf.com.

The design is again estimated after the drawings become available, and these estimates are compared to the design-for-profit goal. The essential purpose of a design-for-profit program is to ensure that the product meets a market price to allow product competition.

Design-for-profit programs supply a target for direct labor and direct material for significant hardware. Those designs are costed by opinion, comparison, and other preliminary methods. Remember, design for profit is before a bill of material and final designs are available. This activity strongly assumes that design and manufacturing engineering play a prominent role in the actual cost of hardware and software. Thus the designer knows that he or she controls the design and is encouraged to meet the target.

The procedure of designing for profit starts at the top with *market price*. Retail markup is removed, leaving the manufacturing cost. Profit and other overhead items are removed, many times by using ratios. Eventually, the cost is stripped down to the product parts and subassemblies. It is at this point that the design-to-cost targets are applied.

Example: Human Hand Prosthesis Product

A prototype is available, for a human hand prosthesis product, as shown by the isometric assembly in Figure 11.3, where the plastic hand cover is removed. Other important engineering details are not yet finalized and changes are possible.

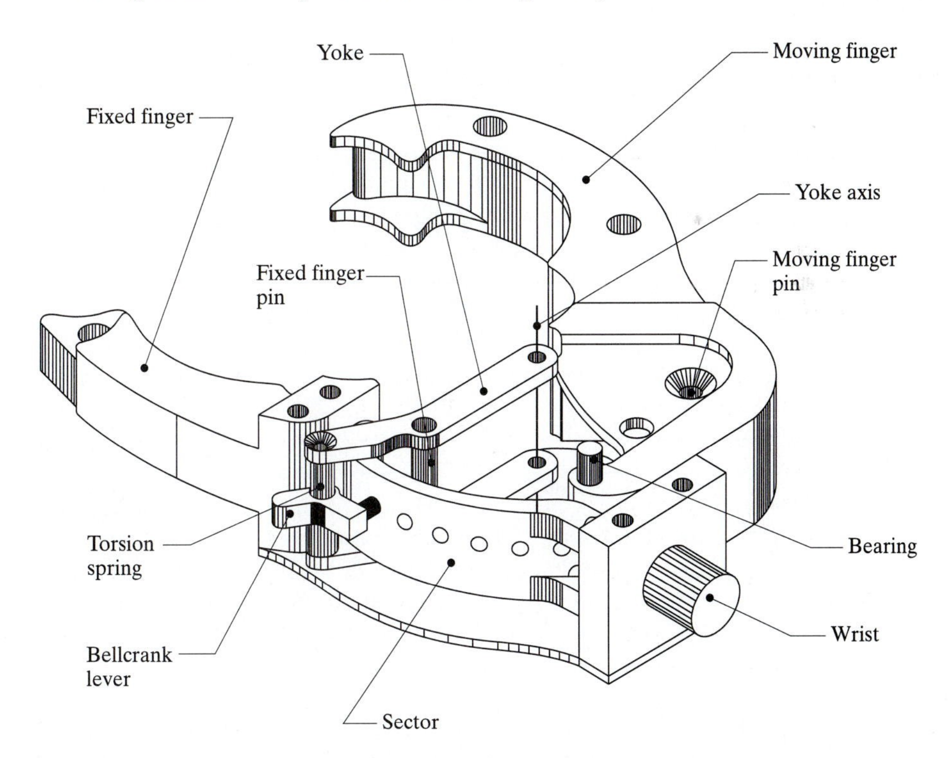

Figure 11.3 Isometric view of Vector Grip human hand prosthesis with sized-replica of soft plastic hand removed.

As the design and cost analysis progress, it may become apparent that design-to-profit goals are unobtainable. It may be prudent to forget about the development of the product and kill the product. It is better to make this kill strategy early, some companies say. If the product is to compete with another product, and if there are no superior features, the early dropping of the product can be cost effective overall. On the other hand, if as the steps shown in Figure 11.2 mature, early cost reduction efforts are mandatory in order for the product development to continue. These are serious management decisions.

Example: Design-for-Profit Chart for a Body-Powered Upper Limb Prosthesis Product

Figure 11.4 presents a design-for-profit chart for a body-powered upper limb prosthesis product. Market investigation of retail clinics in the orthotics and prosthesis business

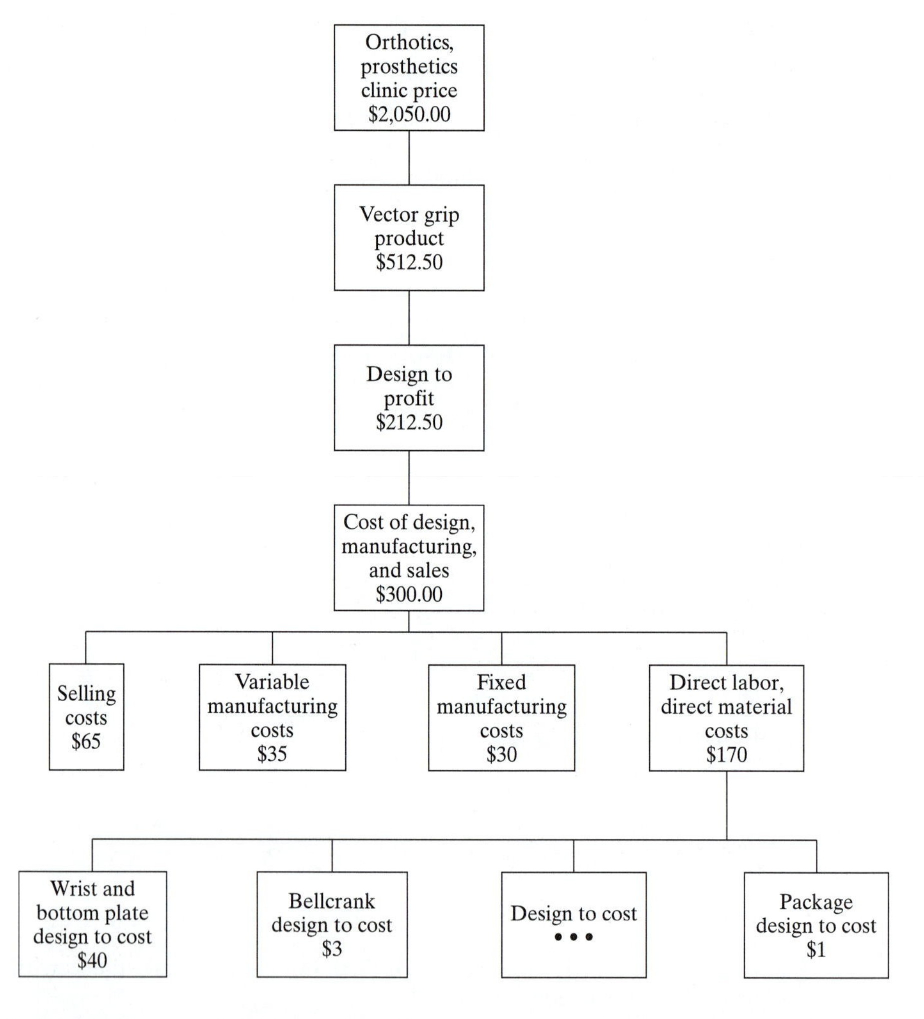

Figure 11.4 Design for profit for Vector Grip prosthetic product.

shows that a new product, called a Vector Grip prehensor, is likely to sell at about $2050 per unit because of new and innovative technical features. These retail business provide much individual fitting and service of the prosthesis for the customer, and thus a markup of 75% on the sale value gives an over-the-counter price of $2050. Working backward, a discount of 75% results in a selling price by the manufacturer of $512.50—or $2050 = $C_i/(1 - 0.75)$, Equation (8.15). The $512.50 sale price to the distributor is mandatory for the entrepreneur company to be competitive in the marketplace.

This new business in turn believes that a profit of $212.50 is necessary, leaving a cost of design, manufacturing, and sales of $300 per unit. Other prominent business costs of sales, design, and direct labor and material are set. The expenses of $65, $35, $30, and $170 add upward to $300. Notice that direct material and direct labor are goaled at $170 per product unit.

It is at the block of direct labor and material that the lower blocks become important to design and manufacturing. These lower blocks partition the hardware of the product. They are tangible and can be analyzed with careful procedures. The subassembly "wrist and bottom plate" are given a $40 *design-to-cost goal.* It is this designation that gives the name for the practice. Thus the engineering team is to design and manufacture the wrist and bottom plate to this $40 goal.

Naturally, it is important that the total product cost goal of $170 of direct labor and material be distributed fairly to subassemblies and components and that no subassembly or component be favored at the expense of another. As the design effort progresses, the usual pattern of preliminary and detail cost estimates are undertaken. Obviously, it is important that the final price harmonize with design-for-profit goals.

The design-for-profit values are important as the entrepreneur needs a product price for the business plan and for investor information.[3] The business plan applies the design to profit values, and reliable information is important.

11.4 DESIGNING FOR MANUFACTURE

The prototype design for the wrist and bottom plate is shown as Figure 11.5. It is said that the first design is usually not optimal, and "try and try again" is a slogan for improvement. Designing for manufacture are efforts that use engineering skills to reduce cost. A variety of practices are considered, such as opening of tolerances, combining or eliminating parts, selecting material and processing, choosing lot quantity, and specifying tools. Indeed, a check-off guide for cost reduction should be used to confirm that nothing is overlooked.

Example: Two Alternative Designs for Vector Grip

A prototype design is shown in Figure 11.5 (a). Improved design combines two parts into one, and adds location and pin holes for other functions, as shown in Figure 11.5 (b).

[3]The following two Websites provide resources for the budding entrepreneurs: www.sba.gov and www.entrepreneur.com

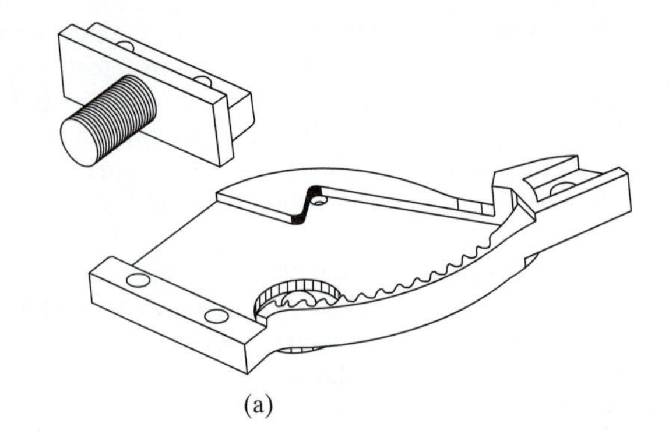

(a)

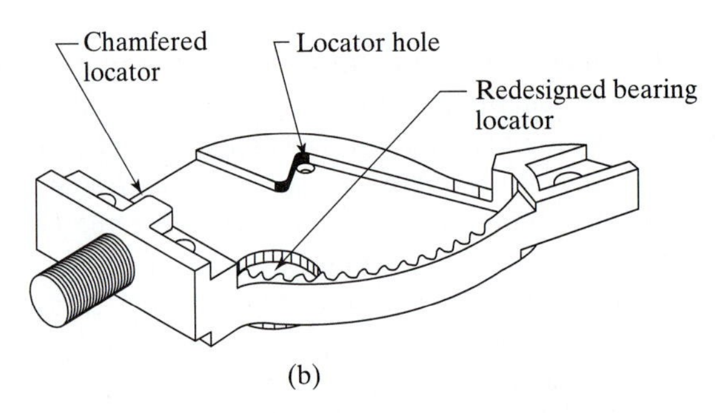

(b)

Figure 11.5 Wrist and bottom plate of vector grip: (a) prototype design, and (b) redesign to reduce complexity and number of parts.

Table 11.1 summarizes these evaluations. Significant differences in the basic manufacturing processes are proposed. For these cases different designs are necessary.

The lowest unit cost shown is $45.93, but the titanium process is selected for its superior engineering performance. The titanium part is the lightest and strongest of the designs, and the teeth of the ratchet, which are integral to the final casting, do not require machining. The titanium casting design, while increasing cost over a single aluminum casting, gives improved engineering properties. The wrist and bottom plate is the most costly of the components, At this time, its cost of $54.66 does not meet the design-for-profit goal of $40 shown in Figure 11.4.

The consideration of material, process choice, tolerance, and lot quantity is apparent in Table 11.1. Those properties are a part of the important knowledge base that engineering brings to the analysis. Table 11.2 illustrates a small sampling of the knowledge base for design and manufacturing.

The design-for-manufacturability analysis advises on where the components are obtained. Are parts and assemblies manufactured in-house, or are they purchased from outside suppliers? From a cost standpoint, this analysis is important to the enterprise because of its influence on business planning. It is this topic that we now study. At this stage of analysis, final drawings and specifications are available.

TABLE 11.1 Estimated Unit Cost of Wrist and Bottom Plate Under Varying Production Quantities and Methods

Lot Quantity	Machining Separate Parts from Solid Aluminum Stock	Die Casting Aluminum Parts and Machining	Die Casting as Single Aluminum Part and Machining	Single Part Titanium Casting, Opening of Tolerances, No Machining
10	$212.62	$463.79	Impractical	Impractical
100	93.17	59.63	$57.21	$173.73
500	91.32	57.42	45.93	54.66

TABLE 11.2 Illustrative Technical Properties of Aluminum Casting Versus Titanium Casting for Wrist and Bottom Plate

	Die Casting Aluminum Alloy and Machining	Investment Casting of Titanium Alloy
Surface finish	64–125 μ in.	25–64 μ in
Tolerance range	±0.002 in.	±0.005 in.
Limitations	Part is difficult to fixture, and cost does not decrease significantly with larger lot sizes	Expensive pattern tooling
Advantages	Good for small lot sizes	Good dimensional accuracy and cost dramatically decrease for increasing lot quantities. Reduces machining.
Reliability	1,500,000 cycles	3,250,000 cycles
Hardness	Soft	Harder penetration in impact

11.5 SELF-MANUFACTURE OR PURCHASE?

Estimated unit costs are used for many purposes, including pricing and the *make-versus-buy analysis* that we now consider. Operation estimating and product pricing were studied in Chapters 7 and 8, and the student will want to scan those chapters. Those principles are necessary for manufacturing or purchase selection.

In this analysis engineering determines if it is cheaper to self-manufacture the product (make) or to purchase it from an external source (buy). The engineer compares the price as promised by the vendor's quotation to a calculated make value. If a purchased part is cheaper, then the firm will buy from a supplier. If the calculated make cost is less, then the company will manufacture the product. The objective is to select the cheapest source.

So, how does outsourcing save money for the firm that has the choice of "make inside" or "outsource" to a specialty firm? It is argued that contract suppliers have "leverage." The suppliers will have an assembly line, for example, that is suitable for multiple customers. The leverage is that the supplier is able to purchase more specific and expensive equipment than a company can. This capital advantage and lower overhead costs are in favor of the supplier.

There are tactical issues to the overall analysis. Capability and quality, schedule to complete and deliver, and future activity are factors beside cost that influence the subcontract decision. If those factors are more or less equal between the supplier and the manufacturer, it is to the firm's economic advantage to base the decision on the results of a make-versus-buy analysis.

Suppliers are given the identical design, order requirements, and specifications that the manufacturer evaluates. Of course, the "buy side" of the comparison—i.e., the quotation as contractually promised—is furnished by the supplier.

The manufacturer prepares an estimate using the plant's production resources for this side-by-side comparison. A summary of the cost elements included in the "make side" estimate is shown in Table 11.3. Selection of the cost element depends on the level of the plant utilization.

"Plant utilization" refers to whether or not the plant is operating at less than 100%, or greater than 100% of fixed capacity. A plant operating below 100% fixed capacity will have fewer employees than normal. Idle equipment is apparent and sales are down. If the plant is operating greater than 100% of fixed capacity, there is overtime, new equipment is required to meet schedule and demand, and hiring of new employees may be ongoing. Plant utilization affects the selection of the cost elements for the make side of the comparison.

Direct labor, direct material, and variable overhead are always included in the make side calculation despite the level of plant utilization.

The cost element *fixed overhead* as an inclusion if the make-side calculation depends on the level of plant activity. Fixed overhead costs are interest, depreciation, and those account items for plant operation that are independent of plant activity. If the plant is operating at less than 100% capacity, then fixed costs are not usually included in the make side of the analysis. The fixed costs must be paid, but not necessarily from the production of the make part under consideration. If the plant load is less than 100%, other income of the business, it is assumed, must pay those fixed costs. There are serious consequences to the enterprise if interest and other fixed obligations are not paid promptly.

TABLE 11.3 Summary of the Cost Elements to Include in the "Make" Side of the Manufacture or Purchase Cost-Estimating Analysis

	Plant Utilization	
Cost elements	Less than 100%	Greater than 100%
Direct labor	Include	Include
Direct material	Include	Include
Variable overhead	Include	Include
Fixed overhead	Omit	Include
Marginal cost	Include	Include
Sunk cost	Omit	Omit
Profit	Omit	Management choice

When the fixed costs are not included, the make cost is less, allowing the factory to be favored when compared to purchased parts. A company will adopt this policy of not including fixed costs because it wants to keep its employees and plant operating.

Sunk cost for the analysis of make-versus-buy are those costs where the money is already spent, and it does not influence future decisions in a meaningful way. If the product is already designed, then the cost of engineering is sunk and immaterial to the analysis of the make-versus-buy, as those engineering monies have been spent. If the manufacturer does not include the design costs, neither should the supplier. Of course, the manufacturer will include the design costs for the pricing analysis, but they are ignored for this special make-versus-buy study.

The *marginal cost* element is a calculation that results from the decision to make or buy. Consider an example where plant activity is already less than 100% capacity. If the decision to buy results in employee terminations, then the unit cost approximation to reflect increased unemployment insurance is added because of the decision. A plant operating at more than 100% capacity will have overtime and rushed work and may be hiring unskilled employees. If the plant capacity is greater than 100% and the "make" decision requires overtime, new equipment, new plant space, or additional shifts, then a marginal cost addition to the make side is made that reflects this increased cost.

Example: Make Versus Buy for Wrist and Bottom Plate

The wrist and bottom plate, as first listed in the design for profit of Figure 11.4, is designed and drawings and a bill of material are available. We wish to determine if the wrist and bottom plate is to be made in-house, or purchased externally from a supplier. A make-versus-buy analysis of that product is necessary, and there is additional information at this point in time. Quantities of 100 and 500 are chosen for lot releases. Preliminary estimates were used for the design for profit of Figure 11.4. The wrist and bottom plate, which is cast of titanium, has an annual marketing requirement of 3200 units. Table 11.4 presents an analysis of the make-versus-buy of the bottom plate.

The information direct labor, direct material, etc., for the make-versus-buy analysis are removed from the estimates. The engineering costs are sunk, and have been spent,

TABLE 11.4 Make-Versus-Buy Unit Costs for the Cast Titanium Wrist and Bottom Plate of the Vector Grip Prosthesis

	Self-Production		Supplier	
Lot quantity	100	500	100	500
Cost Element				
Direct labor	$89.55	$16.56		
Direct material	34.92	23.72		
Variable overhead	5.72	2.68		
Fixed overhead	8.84	4.76		
Engineering (sunk)	0	0		
Tooling	34.70	6.94		
Profit	0	0		
Total unit cost	$173.73	$54.66	$103.73	$44.66

as the drawings and specifications are completed. They are not included for either the manufacturer or the supplier for this analysis. Nor is any profit included for the manufacturer in this comparison.

A supplier submits a quotation of $103.73 and $44.66 per unit for lots of 100 and 500 units, which includes the supplier's profit. The supplier is not required to give any details for the quotation. As the supplier's price is less than the in-house value of production, the supplier is chosen. The $44.66 cost is closer to the design-to-cost goal of $40, though it remains above. Continuing improvement is a possible option. Selection of final lot size for purchase is made later.

Comparing in-house production costs to outside supplier prices, and the refinement of the cost of the prehensor is an ongoing procedure. Eventually, the full cost and price are found; reliability of the values are confirmed, and the planning proceeds. There are numerous routes for enterprise planning for a new product. The method we are discussing now proceeds when the preproduction analysis is essentially complete.

The custom of the "self-manufacture or purchase" comparison in recent decades has taken on a newer meaning than historically understood. In twentieth-century practice, it was not uncommon that the large enterprise buyer was in a position to either make or buy the product. What was true in that century is probably not true in the twenty-first century. The management policy is first "design and make anywhere," and contract manufacturing is more the norm than the exception.

11.6 ENTERPRISE PLANNING

The many methods of enterprise planning can be confusing, but the first objective is that planning, both technical and cost, be done. Enterprise plans are sometimes called *prospectus*, from the Latin word meanings "distant view." A prospectus is important for the enterprise and investors. In addition, the intent is to provide an overview of information for investors, bankers, and interested individuals. The plan is going to indicate how much money is necessary for the venture, if it should be debt or equity, and when it is needed to accomplish the tasks. Essential parts of an enterprise plan, as we see it, include the following:

- Executive summary
- Objectives and potential
- Design, manufacturing, and technology
- Keys to success
- Market analysis
- Income and cost analysis
- Assessment and due diligence

Enterprise plans can be prepared for various internal needs or for external purposes, such as to raise money. While giving information about the business investment, and if the plan continues, it may suggest and even pitch the final decision and implementation stage, or leave that action to the investor.

A much reduced plan is given for a student startup business called Premier Prosthetics Inc. Enterprise plans can fill many notebooks and range to those that are simple, such as the one illustrated in the following example.[4]

Example: The Enterprise Plan for Premier Prosthetics Inc.

Executive Summary

Premier Prosthetics Inc. designs, manufactures, and sells personal devices for the prosthetics market. PPI has developed advanced products for upper limb amputees. These products will revolutionize and significantly improve the ease of human hand manipulation for the potential 175,000 customers in the United States and over 42 million amputees worldwide who employ prosthetic hand appliances. The lifelike products will be sold to a network of over 6500 prosthetic and orthotics clinics in the U.S. domestic market.

For PPI to meet these objectives, it intends to raise over $820,000 in an initial public offering of common stock within the next year to allow the mobilization of a fabrication plant dedicated to the production of a voluntary opening prosthetic prehensor, called a Vector Grip, and the continuation of product development.

Developed in a university laboratory, a prototype of the Vector Grip is available where the claims of superior technical features can be substantiated through comparison and human testing. The prototype unit is shown in Figure 11.6.

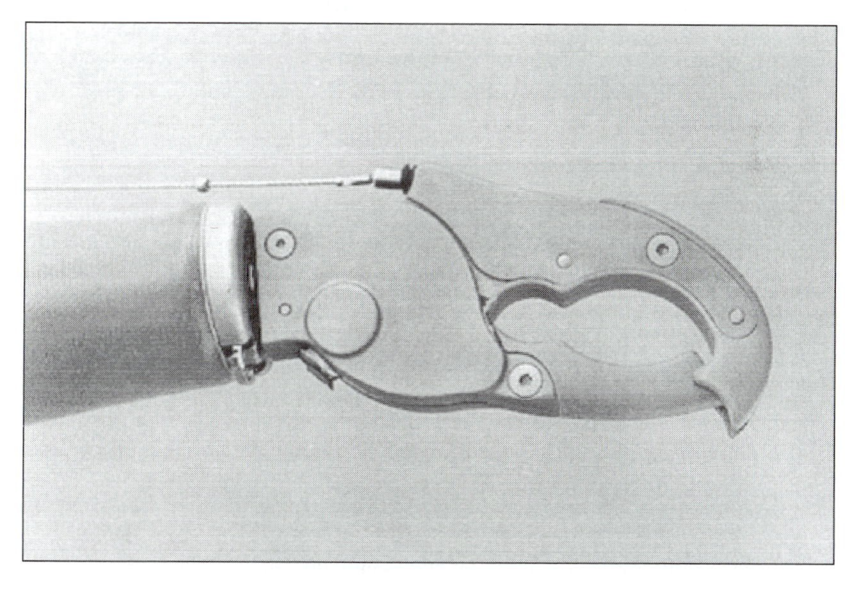

Figure 11.6 Prototype Vector Grip prehensor.

[4]Many companies put their business plan on the Web. Check some out using various search engines, or look at the *Wall Street Journal* at www.wsj.com.

Engineering inventors Professor Lawrence Carlson and student Robert James Young are the founders of PPI. They have formulated the following objectives.

Objectives and Potential

Our most important business objectives are a first-year revenue of nearly $2 million and an after-tax yield of better than 29% on the investment of $820,000. We are positing a doubling of revenue within 10 years.

PPI offers customers a reliable, high-quality, world-class alternative to the conventional Dorrance hook. A disadvantage of commercially available voluntary opening (VO) prehensors is that the gripping force is constant. Therefore, an amputee must often settle for one force that can perform a limited range of functions. A VO prehensor, which is able to easily adjust the gripping force over a wide range of forces, is of great benefit. The human hand can perform a wide range of motions and forces, which the average person takes for granted. The ability to pick up a Styrofoam cup and seconds later pick up a heavy bag is possible because of the range of forces that the hand can generate. An upper limb amputee cannot easily perform these functions.

PPI will establish manufacturing and product sale for the U.S. market using an established manufacturer's representative chain and electronic commerce to expedite customer service. Each retailer will have product specifications, pricing, availability, order status, and delivery dates in real time in an electronic commerce venue.

Investors and clients must know that working with PPI is a more professional and less risky way to develop new technological products as compared to traditional design and production/logistical methods.

PPI must also maintain a strong financial balance, charge a high price for its products and services, and deliver an even higher value to its clients.

PPI is an excellent place to work and has a professional environment that is challenging, rewarding, creative, and respectful of ideas and individuals.

PPI provides excellent value to its customers and a fair reward to its owners and employees.

Design, Manufacturing, and Technology

Design and research are accelerating in these areas:

1. Production and design engineering for the components and assembly of the vector prehensor prosthesis are underway. Negotiation for rental of office and plant space, purchase of equipment, and the employment of skilled machinists and assemblers are beginning.
2. Prosthetics covers, which greatly enhance the physical appearance of the vector prehensor appliance, are under development. Older limbs are rigid and hard, are often made of metal, and do not have a human appearance. It is intended that our products will have plastic hand covers that will dramatically change the appearance of the Vector Grip prosthetic.

Keys to Success

Success in product engineering and manufacturing rests on these advantages:

1. First in its class, the Vector Grip prehensor prosthesis. Protected by U.S. Patent 5,800,571, Figure 11.7 shows a patent drawing with the fixed finger or thumb in the open position with the lock enabled and unlocked, and the lever uncocked.
2. Excellence in fulfilling the business order, trustworthy expertise and information.
3. Leveraging from a single pool of engineering and computing expertise into multiple revenue generation opportunities, retainer and project consulting, market research,

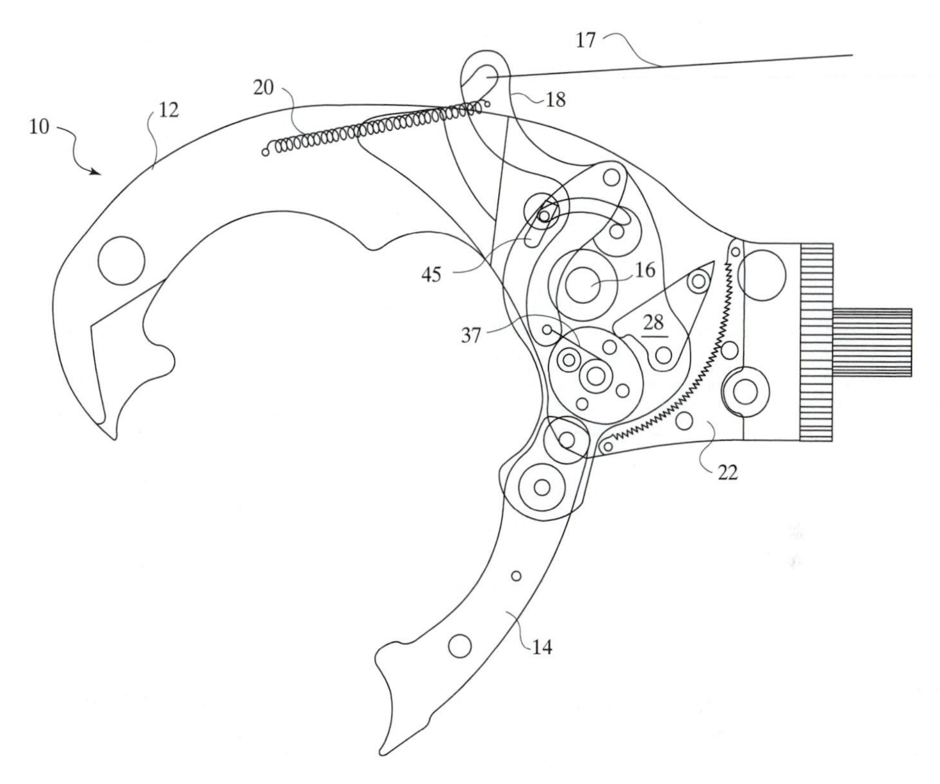

Figure 11.7 Patent drawing for voluntary closing prosthetic prehensor.

and published reports assisting customers wanting advanced medical and engineering appliances.

Market Analysis

A market survey indicates that PPI can anticipate annual sales of 3200 prehensor systems priced at $512.50 each. This penetration is 2% of the existing market. The price of $512.50 to the retailer is about 50% above the competitive Dorrance hook. Advanced features of the PPI hand should allow this premium pricing.

The 95% range of spending for upper limb amputees is about $130 to $410 per year for products and maintenance of the prosthetic, according to surveys conducted by Market Research International.

Income and Cost Analysis

Cost analysis indicates that production costs are proportionate to competitive products, and for some components there is an advantage to PPI. Fixed and variable expenses are shown in Table 11.5. The data are either fixed or variable, except for utilities, which is a mixed account, or both. The variable costs are driven by the number of units produced annually.

Significant sales effort is projected for the product. Fixed sales cost is determined to be $25,600 and variable expenses are $54 per unit for the marketing campaign. (The calculation of income taxes is studied in Chapter 9, and the student will want to review that material.)

TABLE 11.5 Annual Cost Data for Vector Grip Prosthetic

Expense	Fixed Cost	100% Plant Capacity Cost
Rent	$15,000	
Utilities: heat, power and communications	16,000	$116,512
Insurance and protection	3,500	
Supervision salary	36,140	
Annual depreciation	32,800	
Maintenance: 3200 units at $19.42 per unit		62,144
Miscellaneous: 3200 units at $16.99 per unit		54,368
Direct labor: 3200 units at $97.10 per unit		310,720
Direct material: 3200 units at $72.83 per unit		233,056
Total	$103,440	$776,800

Table 11.6 presents the cost analysis for the vector prehensor product. While 100% plant capacity is planned for production of 3200 units, other capacity levels are shown. Both total and unit cost and income values are provided. Income taxes, which are planned at current rates, do not inhibit strong earning rates. Furthermore, analysis in terms of plant capacity demonstrates that strong profitability begins early in production and the after-tax yield on investment at full production of 3200 units is 29%.

The operating cost and income charts for the Vector Grip product are given in Figure 11.8 for the first year. The charts are found using Table 11.6. Notice the vertical step in unit income taxes, which is a consequence of increased income results in a higher IRS

TABLE 11.6 Cost Analysis for Vector Grip Prosthetic Product

1. Plant capacity	0%	25%	50%	75%	100%
2. Units made and sold	0	800	1600	2400	3200
3. Income (2 × $512.50)	0	$410,000	$820,000	$1,230,000	$1,640,000
4. Total fixed manufacturing cost	$103,440	$103,440	$103,440	$103,440	$103,440
5. Total variable manufacturing cost (1 × 776,800)	0	$194,200	$388,400	$582,600	$776,800
6. Total sales costs	$25,600	$68,800	$112,000	$155,200	$198,400
7. Total manufacturing and sales costs (4 + 5 + 6)	$129,040	$366,440	$603,840	$841,240	$1,078,640
8. Unit operating cost (7/2)		$458.05	$377.40	$350.52	$337.08
9. Net profit before income tax	$(129,040)	$43,560	$216,160	$388,760	$561,360
10. Yield on investment before income tax (9/820,000)	−15.7%	5.3%	26.4%	47.4%	68.5%
11. Income tax	0	$4,356	$32,424	$216,060	$237,560
12. Net profit after income tax (9 − 11)	$(129,040)	$39,204	$183,736	$216,060	$237,560
13. Yield on investment after income tax (12/820,000)	−15.7%	4.8%	22.4%	26.3%	29.0%
14. Total operating cost including income tax (7 + 11)	$129,040	$370,796	$636,264	$1,013,940	$1,402,440
15. Unit operating cost including income tax (14/2)		$463.50	$397.67	$422.48	$438.26

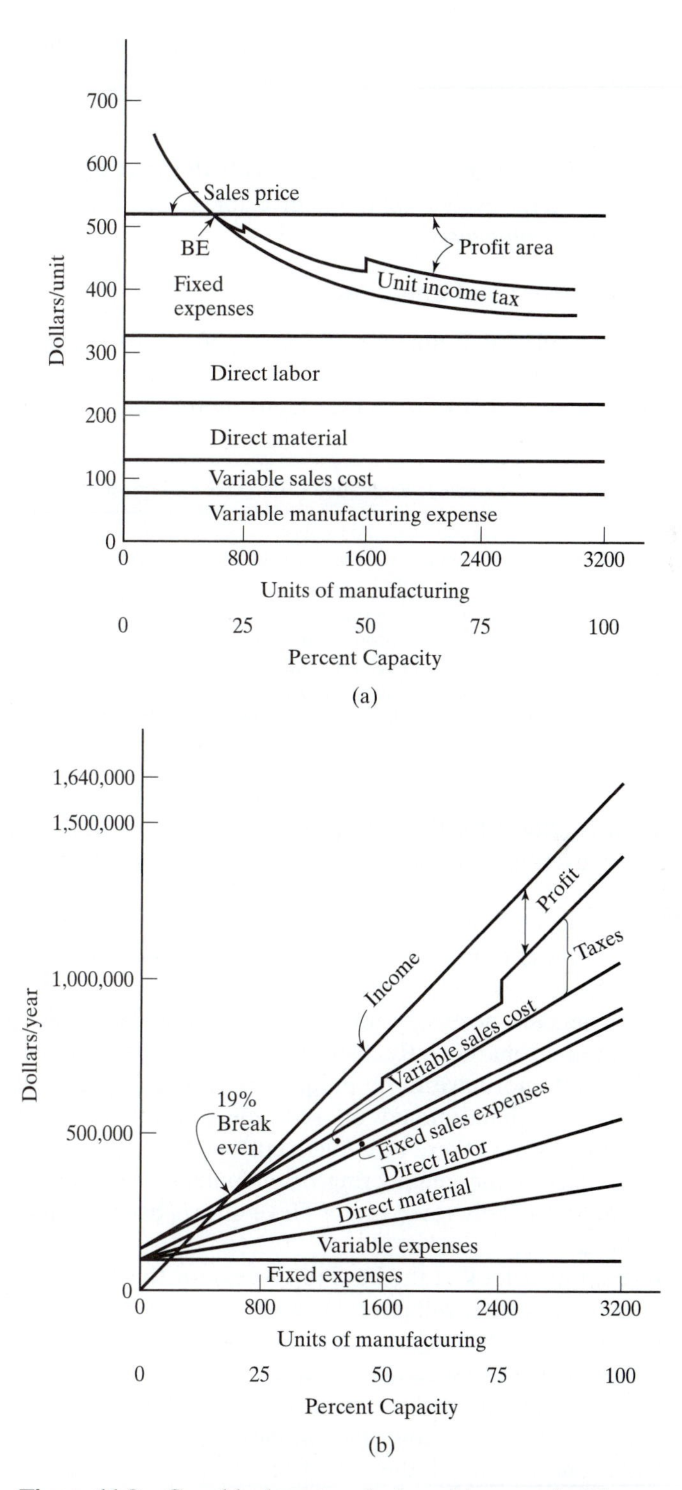

Figure 11.8 Graphical cost analysis and income for Vector Grip product: (a) combined unit operating cost and income chart, and (b) total manufacturing, sales and income chart.

rate for the enterprise. Conservative projections suggest that break-even is at 19% operation of plant capacity. Profit is expected above 700 units.

An enterprise plan may conclude with the cost analysis and leave questions about any systematic evaluation to the investor. The investor's evaluation is sometimes referred to as *due diligence*.

11.6.1. Assessment and Due Diligence

The enterprise plan is frequently the point of the first impression for an investor, banker or stock fund. Its importance cannot be understated. The plan is analyzed in many ways. Analytical criteria include many metrics, and a few are given as follows:

Economic Evaluation Criteria for Enterprise Plans
Affordability and total cost
Percent profit on investment
Percent profit on sales
Annual cost of production of the product
Unit cost to produce and sell
Unit cost to produce and sell, including profit and taxes
Payout period and return on investment
Amount and time strategy for liquidation
Other "yield" calculations on investment

But it is critical for the entrepreneur to have asked many questions during the preparation, even though the answers may be left unwritten. The entrepreneur needs to raise self-questions that are difficult, because hard questions cause superior thinking and improve the chances for success, both for the enterprise and the investor. The enterprise must be so thoroughly studied, you can answer any question that an investor can ask.

Customers are necessary for a business. Does the mission statement clearly define the potential market? Realistically now, are there enough people wanting the product? You do not have to be unique, but you do have to offer people something they will want to spend money on. Ask yourself how many of those customers will spend that money. Will there be enough customers?

What stage is the engineering development? Are the manufacturing costs too high in relation to likely retail costs? Have efforts been given to design and manufacturing value analysis?

Take another look at the keys to success for this business. Realistically, how can you start a business that will have those key elements? Can you get the resources? Do you have a team that meets your success criteria, or can you gather that team? Can you get together the keys to success? Yes, no?

Now look at your break-even analysis again. It tells you how many units you need to meet your costs, as well how much money you will lose if sales do not meet expectations. Is this business worth the risk? Can you afford to start it and lose your start-up costs if you are wrong?

What is the fatal flaw in the planning? Are there torpedoes that can sink the plan? Can these flaws be fixed, and is there an exit scheme that will recoup investment?

Whether the enterprise is a new start-up firm or is an established one, there are requirements for money to carry on the vision of the entrepreneur. An existing company, admittedly, will have ongoing funds from the corporation, such as the surplus generated by profits, and may have an easier opportunity for raising capital, so it would seem. But there are limitations even in those circumstances. Financing can take many forms, as we now see.

11.7 FINANCING AND ENTERPRISE FUNDING (OPTIONAL)

A start-up company is challenged to provide funds for the development of the idea. A great variety of resources are available. We now study the topic of raising money for the vision of the entrepreneur.

11.7.1 Raising Money

Getting funding for a start-up company or the development of an idea in an existing company is paramount to the enterprise. Market potential, design/technology/patents, and business strategies are part of the endeavor to raise money.

For the smallest enterprise, finding money usually begins with family, friends, credit cards, second mortgages, and personal collateral for small business loans. Entrepreneurs sometimes get funding from angels, those wealthy individuals that back start-ups. Also, a secretive bunch (on the Internet they may be listed by code name) and others will band together and screen potential investments. Some of these investors can be demanding, as they invest nominal amounts but claim ownership of 51% or more of the shares of stock. Other investors are openly solicitous and can be tracked through magazines, Websites, and regional trade groups.[5] Even syndicated TV shows exist for the money seekers. Then start-up companies will advertise for financial backing. Venture mutual stock funds may rely on the recommendations from attorneys and accountants who do incorporation and public-offering work.

For the larger enterprise, there is much exposure in the financial markets with large investment banks, mutual funds, and investor groups able to provide venture capital. Private venture funds, with limited funds of their own, usually gather their money from investors before they buy stakes in new ventures. Still there are funds that have sufficient capital of their own to invest, and later reoffer the investments to their own clients.

The first-round venture funds may be followed with a second round of investment, where the track record of the first offering is carefully examined.

Whether the company is existing and ongoing or a start-up venture, the entrepreneur needs to know techniques for capitalizing and financing. In simple terms, this implies raising the money to carry on the proposed venture to an eventual success. Those opportunities include the following broad categories:

- Stock
- Bonds

[5]Magazines include *Red Herring* and *Upside*; see also the National Venture Capital Association at www.VentureOne.com.

- Debt
- Alliances

If the entrepreneur is an individual, then other possibilities, such as having partners may be necessary if the money requirements exceed that which the individual is able to have on hand. If the entrepreneur is within a corporation, and the design exceeds cash/receivables/and other liquid accounts on hand, raising money using stock is a viable opportunity.

Raising money by stock is a popular option. There are two classes of stock: (1) common and (2) preferred. We now study the important lessons of stock, or ownership of a company.

11.7.2 Stock

Investing means putting your money to work to make it grow. The "work" happens because stocks are *ownership shares* of a company, and investors buy stock to participate in the growth of a company. If the company's profits grow over time, its stock price tends to increase. As profits materialize, some companies may pay out a portion of earnings by issuing periodic dividend checks to shareholders. The company will report its results to the owners each period or (at least at the end of the fiscal year). Owners are invited to the company's annual meeting and are granted voting rights on matters pertaining to the governance of the company, composition of the board of directors, selection of the auditor and other issues.

Investors benefit from owning stocks in two ways: (1) through capital appreciation (when a stock's share price rises) and (2) through dividends. The downside of stock ownership is that there is no guarantee that the shares will increase in value. There is the possibility they will decline—or even plunge to zero value. If a company has a bad financial period, it might eliminate its dividend, if it ever paid one. Common stockholders are generally last to receive anything if the company declares bankruptcy and ceases operation.

Indeed, the choices of investing are mind-boggling for the investor: small stocks, big stocks, mutual funds, index funds, foreign stocks, junk bonds, Treasury bills, real estate, etc. There are over 10,000 U.S. stocks available to investors (excluding over-the-counter regional stocks which may number over 50,000, and foreign stocks, as well).

Equity or *stock financing* techniques are appropriate strategies to raise capital. To make equity financing cost effective, the firm must be able to get a fair market for its shares, because in exchange for financing from the sales of shares, the existing owners and management must give up some percent of ownership and control. The best possible share price requires convincing potential investors that the company is valued more than the competition and realistic earning projections reflect growth as a consequence of the infused capital.

Companies issue stock for a number of reasons, but the primary one is they need money for expansion. True, there are bank loans. But some companies grow so fast that they are unable to spare the cash to make the required interest payments, ignoring for the moment the principal repayment. Others may not have much for collateral, or are unable to find a lender. Still others might find investors who are anxious to own shares in their business and pay a price for the privilege. For example, investors wanted to own shares in new Internet companies—which sometimes had very little hard assets—even before they had earned a profit.

While it is true many entrepreneurs are anxious to have cash turn into a big cash payday, not all want to be publicly held. Being a public company means filing quarterly reports with the Securities and Exchange Commission (SEC),[6] providing financial information to outsiders on Wall Street and dealing with shareholders who naturally want quarterly earnings updates and annual reports, not to mention explanations when things go wrong.

If the start-up company has listed stock, then a downturn in the market can be a boon for investors. For many investors, a downturn simply means the possibility of cheaper prices for shares in the companies they are looking to invest. If the prices become inflated, especially in the later rounds of private funding, there is increased risk for ventures capitalists, especially if these stocks are unable to generate colossal returns.

A solid company with highly valued assets can sell higher-priced shares in both private and public markets as the shares represent ownership of something of value. The higher the share price relative to their value, the lower the cost of giving up equity in the company. In a similar way, an early stage company with a low current valuation and potential earnings growth can raise the price of its shares because future growth justifies a higher share price.

Capital can be raised from the public market, either through an underwriter or through self-underwriting by the company. The underwriter is the securities firm that helps bring a company public. When a company wishes to sell stock to the public, it contracts with an investment banker for advice. The investment banker helps the company determine the number of shares to be sold and the market price per share. The underwriter sets the IPO (initial public offering) price and determines how many shares other investment firms in the underwriting group get to sell to investors.

For example, say a company wants to issue $25 million of stock and decides that $5 per share is the best price to the public. Five million shares would be sold. The company sells its shares to its investment bankers for $25 million, less the fee it pays for the service. After all the investment banking firm is putting its capital at risk for the duration of the stock offering—usually a few hours to days. If the banker's fee is 1%, it would earn $0.25 million for its efforts. So the company would raise $24.75 million from its equity issue, net of paying the investment banker.

After the stock issue is executed and its shares are freed from the syndicate to trade, investors may sell their shares or hold them as they wish. The price at this point is no longer determined by the investment banker, but rather is now determined by the market. If the company performs well and investors think that is likely to continue, the stock price will probably increase. If the company hits a rough period in its business, the price is likely to fall.

The *market capitalization* of a company equals its per-share price multiplied by the number of shares outstanding. Stocks in companies that pay dividends tend to be less volatile—meaning they tend to move up and down less often—than stocks with no

[6]Information about the SEC can be found at www.SEC.gov. Information about exchanges can be found at www.NASDAQ.com, www.NYSE.com, and www.AMEX.com.

dividend. Dividends are partially taxable when the investor receives them as Congress considers allowances for dividends. A major complaint among some business economists is that corporate dividends are in effect taxed twice, once when the corporation pays taxes on its profit before paying out its dividends, and again when the investor includes them in their income and then pays taxes.

Some companies choose to undergo self-underwriting—the process where a private company, usually small, goes public without the assistance of a brokerage firm. For the most part, rules of public disclosure remain the same, and in some states the firm may be prohibited from issuing IPOs. Self-underwriting may raise debt capital, in addition to equity, from private investors.

An IPO is the process where a privately held company sells previously issued or newly created shares of the stock to the public for the first time, discloses operations to the public, and registers with and operates within the provisions of the SEC, and other agencies and state regulatory bodies. Following completion of the IPO, the company will be a registered public entity, subject to federal and state laws. The company is eligible for listings on various exchanges, such as the New York Stock Exchange (NYSE), American Stock Exchange (AMEX) or the National Association of Securities Dealers Automated Quotations (NASDAQ) system. Publicly held shares of stock can be freely bought and sold without restriction among public shareholders. Disclosure of information about the company is required by the SEC and shareholders on a regular basis during the life of the corporation.

There are advantages and disadvantages to IPOs. The primary benefit is that large or small amounts can be raised and these funds can be used for almost anything that the company desires, such as paying off long-term debt. The company has a broader equity base and larger net asset value. It may become easier to raise future equity capital and borrow additional funds because of improved debt-to-equity ratios. But going public to raise capital is one of the more expensive ways, especially for smaller companies. Sensitive and proprietary information may be disclosed, and going public may preclude the owners from raising capital by other means.

The measure of importance to investors and shareholders is *total return*. Sometimes yield is the term used, and yield is calculated in several ways. Total return is the yardstick that puts shareholders first. Total return to shareholders includes changes in share prices and reinvestment of any dividends, rights and warrants offerings, and cash equivalents, such as stock received in spinoffs. Returns are adjusted for stock splits, stock dividends, and recapitializations. There are other measures, such as price of share to earnings ratios—a company's stock price divided by its per-share earnings.

Preferred stock is a type of security, along with bonds and common stock, that is used to raise capital. It is a class of capital stock, junior to debt obligations, which are always paid first in the event of asset liquidation. Preferred stock can be issued by private or public companies. It pays dividends at a specific rate and has preference over common stocks in the payment of dividends. Like common stock, some preferred shares carry voting rights in the company and may have special privileges. In addition to paying dividends, preferred issuers can contract to buy back an amount of preferred stock annually. As the number of preferred stock dwindles, the value of outstanding shares may increase.

Warrants entitle a holder to buy an amount of the company's common or preferred stock at a preset price before an expiration date. The common share price of large mature companies often rises above the warrant price before the warrants are exercised. This means that a shareholder who has a warrant for more common stock can buy a $10 share for $5, for example. The common share price of smaller, emerging companies often does not rise above the warrant price before the warrants are exercised. A shareholder who has a warrant for more common stock may be faced with buying a share that currently sells for $10 in the public market for the higher warrant price of $20. When the warrant price is higher than the stocks market price, most warrants are not typically exercised. A warrant will be worth its intrinsic value because it represents the discount a holder can buy stock below the market price. A warrant can function as a promotional incentive for small companies needing a boost to sell the security. There are other advantages and disadvantages for warrants not discussed here.

The enterprise may need to raise capital by debt, a meaningful method. All important companies employ debt for a variety of reasons. We first study the nature of debt financing.

11.7.3 Debt

Debt financing techniques can be effective strategies for raising capital. The optimal use of and benefits from a debt issue needs to be reckoned in terms of the company's capital structure, business activity, debt ratios, and the market for borrowing. In addition, there are good and bad points for debt, and the service of interest of the debt can strap the future of the company. Pivotal factors in debt financing are a favorable record of stability in revenues and earnings and a cash flow to cover interest payments and principal. For a stable company and a strong cash balance with favorable overall debt ratios, debt can be used to effectively provide operating and investment money. In contrast, if the enterprise is unable to show stability and cash flow from revenues, the interest is higher and the protective covenants are more restrictive. A company will have a credit rating that influences the interest rate, collateral, payment stream, and other terms of the borrowing.

If the entrepreneur is an individual, the borrowing of money can range from family, friends, and banks or other institutions. The borrowing may conclude with a handshake, IOU, or a contract with terms and collateral and in the event of loan failure, the legal obligations of the borrower to the lender.

Corporate borrowing from a financial institution or investor provides the use of an amount of capital to be repaid within a certain period of time along with the interest in the form of cash and or stock, warrants, royalties or licensing fees. If the company is liquidated before the loan is repaid, lenders are paid off before stockholders receive distributions. Investors with such debt instruments as commercial paper and bonds are considered lenders with the same rights and powers as a bank.

There are various ways to borrow. A long-term loan has a time period in excess of a year. Repayment is made to a set interest rate tied to the prime or *U.S. Treasury bill rate*. Long-term loans are obtained from banks, special or institutional investment funds, or government agencies like the Small Business Administration.

Loans may be short-term, perhaps up to two years, which is a tailored loan maturing and paid back at a higher rate than the long-term note. Once again the interest rate may be tied to the prime rate or the U.S. Treasury bill rate. Sometimes called bridge financing, it is money that supports a company in initial product sales through going public or acquisition, for example.

A line of credit is the bank's commitment for the availability of a negotiable amount of financing contingent on the company's or management's credit worthiness for varying periods of renewable time, usually one year. It is common for banks to require that at least 10 percent of the line of credit value is retained in a bank account as a compensating balance. A collateral line of credit can be initial financing for a start-up company or supplementary capital for a later-stage company.

A letter of credit is a specialized way to borrow against an account receivable (money owed by customer). The customer will be paying off the account sometime in the near future.

A leveraged buyout is the purchase of a company's assets using borrowed money. This buyout of the company's outstanding stock uses as little of its own capital as possible. Another approach is the purchase of a company with borrowed capital, thus the "levering." There are tax-free buyouts too. A buyer can borrow using what it owns as collateral. A publicly held company can become privately owned, and vice versa. The buyer can increase its stake in the firm, which may result in improved productivity and better returns to the equity owners and investors. On the other hand, a leverage buyout can fail if the debt-to-equity ratio creates high interest payments that add to the company's burden, and endanger the company's credit rating.

11.7.4 Bonds

When a company raises capital to build a new plant, improve its product or production, or acquire another business, it may issue bonds instead of selling stock to the public. Stock is considered the most expensive form of capital, and the most permanent. Once stock is outstanding, it remains in the marketplace until the company buys it back. Although buybacks are common, they are something most companies are not always anxious to do. Also, each new share of stock dilutes ownership in the corporation, which means the dividends may be spread over more shares. That can negatively affect the value of the shares as perceived by the market.

In contrast, bonds are not forever. Maybe a company wants to build a new manufacturing plant with an estimated useful life of 30 years. The company floats a bond issue for the amount it needs to pay for the plant, say, $20 million. That way the company has match-funded an asset—the plant—with a corresponding liability—the bond issue.

Bonds are debt, or IOUs that promise repayment by the corporation that is borrowing the money from investors. Sometimes, bonds are called "fixed-income securities" because they typically give a set interest rate ("coupon") that is fixed for the life of the bond ("maturity").

Buying bonds does not make the investor a part owner of a company. A bondholder is a creditor to the company. The investor gives the issuer cash in return for the pledge of an income stream (the coupon interest payments) for a fixed period of time, and the guarantee that your principal will be returned at maturity. Thus bond investors often desire a steady cash flow from the interest payment as compared to loans where payment includes both interest and principal.

A company that wants to raise debt capital will sell an issue of bonds to an investment bank. It is the investment bank that gives the firm the money for the bonds, less the cost of issuance. The bonds are priced in relation to the 30-year U.S. Treasury bond benchmark. Depending on the credit quality of the issuing corporation, the new bonds will be offered at a price that will yield Treasuries plus a certain margin.[7] This spread over Treasuries is described in terms of basis points, one-hundredths of a percentage point. One percentage point thus equals 100 basis points. The premium over Treasuries is the least for AAA-rated corporate bonds, which have the highest quality possible. It is the widest for the least credit-worthy issuers. These latter bonds are known as "high-yield" or "junk" bonds.

For a "spread," or fee per bond, investment banking companies will syndicate bonds to other firms, who then sell the bonds to institutional investors. Bonds are rated, or evaluated, according to their credit quality, by rating agencies.

Face value is the value of a bond, note, or other security that is shown on the instrument. The stated or face value of the bond is also known as its par (or maturity) value. It represents how much the firm promises to pay the bondholder at the maturity date. Average face value for a corporate bond is $1000. In the past, a bondholder submitted a coupon or ticket to receive interest payment. This has given rise to the term *coupon rate,* which is the nominal rate on which interest payments are made. Coupon books are seldom used with modern bonds, but the term survives. The coupon rate determines the magnitude of the interest payments during the life of the bond.

A bond's price is stated as its "par" or "face" value, which most often is $1000. In most cases par means 100% of the face value. Par represents in percentage terms the dollar value of the bond at the time it was originally issued into the primary market for bonds.

The enterprise pays interest to bondholders, which is unlike the dividends it pays on stock. Furthermore, the company deducts the interest on bonds from its income as an expense of doing business, which is similar to an individual taxpayer who deducts home mortgage interest. After-tax cost of the plant and adding new productive capability to the company can be attractive, if bonds are used to raise money.

Suppose a corporation issues a bond at 100% of par, or $1000 face value, and a 6.00% coupon for 30 years. The buyer of this bond is entitled to the stream of interest payments at 6.00% for 30 years hence, and will be repaid $1000 at maturity. In this case, the stated coupon and the yield are the same, 6.00%.

Sometimes bonds are issued at a slight discount from par. Say that the 6% bond in the example above was issued at a discounted par value of 99% of $1000, or a dollar price of $990 ($0.99 \times \$1000$). Remember, the coupon stays at 6%. Because the price of the bond is lower than par (100%), mathematically, the yield must be higher than the stated coupon rate. Again, to get the yield, we divide coupon rate by price. In this case, it is 6 divided by 99%, so our yield moves up to 6.06%.

When the price of the bond is lower, the yield is higher. Prices and yields move in the opposite direction from one another. This is the one main rule to remember about bonds: bond prices and bond yields move in opposite directions.

A private placement of bonds is a direct sale by the company to life insurance companies, pension funds, bond funds, and similar financial institutions, or to individuals. A private bond placement is also known as corporate debt financing. In general, privately placed bonds are unsecured senior notes with fixed interest rates and maturities

[7]Credit quality is found on www.moodys.com and www.standardandpoors.com.

in the 12- to 18-year range. A bond then is a discounted or interest-bearing corporate or government security that requires the issuer to pay bondholders a specified amount of money at intervals as interest, and to repay the principal amount of the loan at maturity.

Although the terms "bond" and "debenture" are sometimes used interchangeably, technically speaking a bond is backed (or "secured") by an asset such as a manufacturing plant, while a debenture is an obligation backed by the issuer's general credit.

A new bond is sold in the primary market, with the proceeds going to the issuing unit. At the time the bonds are initially sold, they are priced to sell close to their par value. Outstanding bonds, however, refer to bonds that have been previously issued and are outstanding. They may be bought or sold in a secondary market, and their price may be close or far away from the par value.

Calculations for bonds are closely related to topics discussed in Chapter 10, and the student will want to review that material. The value of a bond at the present time is approximated by the following formula.

$$P_0 = I(P/A, i_b, n) + M(P/F, i_b, n) \tag{11.1}$$

where P_0 = present value of bond, dollars

I = annual interest payments received by bondholder, calculated as bond value multipled by coupon rate, dollars

M = maturity value of bond stated on certificate, dollars

i_b = interest rate required by bondholder, decimal

n = annual compounding periods to maturity, number

$(P/A, i_b, n)$ = present value given annuity, factor from interest tables

$(P/F, i_b, n)$ = present value given future value, factor from interest tables

The market price of a bond P_0 is equal to the present value of the series of interest payments to be received over the bond's life, plus the maturity value of $1000, all discounted at a required rate of return for the bond. The interest rate, i_b, varies with each investor. Chapter 10 describes the individuality of this rate, which is known as minimum attractive rate of return (MARR).

The fee as collected by an underwriter, if one is used, for a bond issue is about 2 percent of the amount borrowed. Most national or local brokerage firms place and sell private bond issues.

The repayment schedule is often set so that payments by the firm do not begin until a few years have elapsed, and this nonpayment period is called the *blind spot.*

Bonds are traded regularly through financial markets. Depending on the prevailing interest rates at a moment in time, a bond may sell for a price that is more than, less than, or equal to its face value. When the owner of a bond, which has been purchased previously, seeks to sell it before maturity, the original purchase price and premiums already received have no bearing on the pending sale price. Only future cash flows have consequences on the pending market price of the bond.

The current market rate of interest strongly affects bond prices. Higher market rates tend to lower bond prices by decreasing the present worth of payments promised by the bond. The key point is that bond fund prices and interest rates move in opposite directions. Why? Assume that an investor buys a 20-year $1000 U.S. Treasury bond with a 5% yield, and interest payments total $50 a year. If interest rates immediately rise to 6%, another investor could buy a $1000 Treasury bond and get $60 a year in interest, so no one would be anxious to buy the older bond for the $1000 paying 5%, and the older bond would decline in price, say $884. Yet if interest rates fell and new Treasury bonds having similar interest rates and maturities were offered with 4% yield ($40 a year in interest), the investor is able to sell the original 5% bond for more than the original purchase price, or in this case $1137.

There are disadvantages to the raising of funds by using the bond market. High interest rates can make a bond issue expensive for the company over the long term. If the company has a weak credit rating, the company may give up control of the company to get debt financing. It is necessary that long- and short-term debt ratios meet industry averages to structure a bond issue with terms that are favorable to the company.

Example: $1000 bond with 10% Rate

To see how this works, consider a $1000 bond that has a 10 percent coupon rate and a 25 year maturity. If this bond has a required rate of return of 10% and pays interest annually, its value is $1000. In this example the bond has a current market value of $1000, which is equal to its par value.

$$
\begin{aligned}
P_0 &= 100(P/A,i_b,n) + 1000(P/F,i_b n) \\
&= 100(P/A,0.10,25) + 1000(P/F,0.10,25) \\
&= 100(9.0771) + 1000(0.0923) = \$1000
\end{aligned}
$$

If the investor's required rate of return i_b is not identical to the bond's coupon rate, other bond values will be found. For example, if the coupon rate is 5% while the required rate of return i_b is 10%, the bond value is $546 in 25 years. The coupon rate of 5% will result in $50 annual interest to the bondholder. The i_b is the required rate of return by the bondholder, which varies with each bondholder. Or if the coupon rate is 10% and the required rate of return is 5%, the bond value is $1705. It needs to be pointed out that the amount of interest can be paid more than once per year, in which case, Equation 11.1 is modified.

An industrial development bond is issued by a political entity, such as state, county, or municipal government in order to attract a company to the area by financing capital expenditures, or to encourage the expansion of a local company.

Example: Costs of Raising Capital

Methods of raising capital can be roughly compared. Table 11.7 presents the *cost of money* for raising $8,000,000. Rates, dividends, interest, and after-tax costs, which are assumed for the example, have an influence on the costs of raising capital.

TABLE 11.7 The Cost of Raising Money for Three Methods

Security Type	Rate	Price	Amount of Financing First Year	Interest or Dividend Cost to Raise Capital	
				Before Tax	After 50% Tax Rate
30-year bond (8,000 certificates)	7%	$1000	$8,000,000	$560,000	$280,000
Preferred stock (80,000 shares)	7%	100	8,000,000	560,000	560,000
Common stock (400,000 shares)	$1.20 dividend	20	8,000,000		480,000

11.7.5 Alliances

Alliances or partnerships can be structured in a variety of ways. Variations on the joint venture theme provide capital, technological assistance, facilities, or distribution opportunities. Limited partnerships can be hybridized to provide income for the investors for a research and development program. A variation would be licensing agreement, in which the right to make, use, or sell a product is granted to another party for a specific time and under special terms in exchange for compensation in the form of fees, royalty payments, or a percentage of income.

A *joint venture* is a partnership between two or more parties to research, develop, produce, market or distribute a product for profit by forming a separate project that is owned, operated, and controlled by a group of investors. A joint venture can be organized as a corporation, partnership or undivided interest. A joint venture can include, but is not limited to a licensing agreement or an effective research and development partnership with the joint venture party.

Venture capital is early-stage risk financing offered by private individuals or funds, publicly held funds, or subsidiaries of banks or corporations. Venture capitalist firms exist to lend money to start-up companies, for example.

A *limited partnership* is a for-profit business entity composed of a small number of general partners and a larger number of limited partners. The general partners manage the day-to-day activities of engineering, marketing and so on. The limited partners finance the project. An important reason for a limited partnership is the flexibility and tax benefits. A form is the research and development limited partnership between a sponsoring company with technology that needs to be funded and a group of partners who function like stockholders in a corporation.

An offshoot of partnerships is barter. Traditionally, there are two types of barter exchanges: (1) retail and (2) corporate. In retail barter exchanges, moms and pops agree to swap goods and services, trading for example pet products for legal consultation. In corporate exchanges, companies usually purchase advertising in exchange for inventory or real estate, sometimes together with some cash. An inventor may find partners, who for a share of sales, a percentage of royalties and/or a cut of licensing fees, contribute financing to the venture.

There are other ways to raise capital for the new or expanding enterprise. They include industrial development bonds, leverage buyouts, franchising, licensing, divestiture, and employee stock ownership plans. Those plans are beyond the objectives of this book.

11.7.6 Offering

Start-up companies adopt four steps to secure venture capital, and collectively they are called the *offering*:

- Enterprise plan
- Presentation to interested parties
- Assessment and due diligence by parties
- Negotiation and closing between entrepreneur and investors

The enterprise plan provides potential investors detailed information on the market, company, officers, design/production, and technology. The enterprise plan gives an overview, as shown by the executive summary, and discusses the financial needs for a three-year period (or longer). A very brief plan was presented earlier in this chapter. A plan for an investor examination will provide other financial metrics for the future of a business. Typically, the plan can fill many notebooks and files.

The enterprise plan is formally presented to potential investors either at the company's or the investor's offices. Questions and answers, impressions, engineering demonstration of the prototype or other product information, and fact giving are the goals for the presentation.

The investors are advised about "due diligence" for the opportunity. The investors are interested in their rate of return,[8] and they look at the capital growth and dividends that the opportunity will present. After investors have had time to assess the company's materials, interested investors meet with company management and the technical staff to learn more.

If these stages proceed, the enterprise negotiates further with expected valuation offers. Those investors within the boundaries of a company's range of advantage are invited to submit a financing offer—or a term sheet—to the company. The company may select a lead investor and then determine how many investors or investing companies will be included and the financing amount from each investor. While the lead investor determines valuation for all others, each investor undertakes its own due diligence, legal review, and negotiation of the terms. The company will formally accept those investor deals that are for its best benefit.

[8]A study reported in the *Wall Street Journal* indicated that the internal rate of return for venture funds that invested in new business was an average of 20%, according to Cambridge Associates.

11.8 LEGAL REMINDERS (OPTIONAL)

The legal organization of an enterprise can be formed in three ways, each with its advantages and disadvantages:

- Sole proprietorship
- Partnership
- Corporation

Sole proprietorship and partnerships are subject to unlimited personal liability when it comes to business debt. Creditors of the business can hold the owners of the business personally liable for debt and can move to seize the proprietor's or partner's home, savings, or other personal assets. If a sole proprietor or partner dies, the business ends or it may become involved in various legal entanglements. With sole proprietorships and partnerships, investors are harder to attract because of the personal liability issue. For example, if investors in a sole proprietorship (or some forms or partnerships) want a share of the business for their capital contribution, they could become subject to a demand on their personal assets from creditors if the business becomes insolvent.

The shareholder of a corporation has only the money placed into the company to lose, and usually no more. A corporation has the most enduring legal business structure. Since a corporation has a life of its own, it may continue on regardless of what may happen to its individual officers, managers, or shareholders. In addition, ownership of the business may be transferred, without disrupting operations, through the sale of stock. Capital can be more easily raised with a corporation. This may be accomplished through the sale of stock or other equity interests. There is more legal formality under a corporation banner with attendant state and federal rules and regulations.

With partnerships, each individual general partner may bind the business to arrangements that may result in serious financial difficulty. A corporation's shareholders cannot legally commit the company by their acts simply because they have invested in it.

The general corporation is the most common corporate structure. The corporation is a separate legal entity that is owned by stockholders. A general corporation may have an unlimited number of stockholders that, due to the separate legal nature of the corporation, are protected from the creditors of the business. A stockholder's personal liability is usually limited to the amount of investment in the corporation and no more.

Businesses must comply with federal, state and local laws and regulations.[9] This is not a small matter. But the business needs to be informed about the legal requirements affecting the business. This is a vast and costly requirement that calls for attorneys and accounting specialists. Compliance may mean having work certificates or a license and registration from the state (a business's name), incorporation papers, sales tax number, separate business bank account, patents, trademark, and copyright permission. The list is almost endless, but it is essential for the conduct for the business of the enterprise.

[9]The U.S. Business Advisor site at www.business.gov identifies regulations and gives links to the Internal Revenue Service, Social Security Administration, and the Occupational Safety and Health Administration as well as numerous other federal agencies.

The business is responsible for withholding income and social security taxes and complying with laws covering employee health and safety, and minimum wage. Start-up businesses need to be aware of the city's zoning regulations. There are restrictions on certain products, such as drugs, medical products, toys, food, drink, or clothing, for example.

11.9 ETHICS AND ENGINEERING

It is fitting that a book on cost analysis and estimating have a discussion of ethics in the chapter on enterprise. It is no less important than the very first sections of the book, which deal with the necessity of profit and wise stewardship. Consider the *American Heritage Dictionary's* definition of the word *ethics*: (1) a principle of good conduct; (2) a system of moral principles or values.

This book focuses on an important role performed by the engineer, but we have not defined engineering until this point:

> Engineering is an important and learned profession. As members of this profession, engineers are expected to exhibit the highest standards of honesty and integrity. Engineering has a direct and vital impact on the quality of life for all people. Accordingly, the services provided by the engineers require honesty, impartiality, fairness and equity, and must be dedicated to the protection of the public heath, safety, and welfare. Engineers must perform under a standard of professional behavior that requires adherence to the highest principles of ethical conduct.[10]

The task facing the engineer is providing a measure of the economic want for the design. During the planning period, it is not uncommon that politics or business pressures are applied on the people who are estimating. It is natural that a design engineer will believe that the new design is cheaper, or that the sales staff will promote an opportunity that gives encouragement to marketing. Though those motives are understandable, engineers need to maintain objectivity in fact finding and cost analysis. Subjectivity that may inappropriately influence estimating out-of-pocket future costs seems to be unprofessional.

The estimate deals with elements of material, labor, and money. Because competition is the nature of business, there are occasions in which the propriety of some trade practices is questionable. Revealing confidential quotations to subcontractors or vendors with the hope that a new bidder will submit an even lower bid is improper. "Bid shopping" and "bid peddling" are the terms applied to those practices. On the other hand, some contracts are required to be public knowledge, and on those occasions integrity would require that the same value be disclosed equally to all candidate bidders.

Firms known to be unqualified to perform work or supply the product should not be invited to bid. Unless it is understood as a clause in the contract or is mandated by public law, the price and cost estimates of one competitor should not be made known to another competitor.

Acts of collusion or conspiracy with the implied or express purpose of defrauding clients, suppliers, or subcontractors and business practices that are not fair and

[10]National Society of Professional Engineers; additional comments can be found at www.nspe.org.

honest are condemned and may be illegal and unconscionable to professional engineering practice. The National Society of Professional Engineers have comments on those practices.

SUMMARY

This chapter has summarized the opportunities for the enterprise when coupled with the entrepreneur. It discusses the steps of product and idea development and carrying that on to the enterprise plan. Calculation of break-even, make-versus-buy, investor's yield, values of stocks and bonds and their influence on the enterprise were introduced.

Have you wondered about the meaning of the word *imaginamachina*?[11] The word means ideas and imagination coupled to the engineering thinking process, which is essential to the enterprise and entrepreneurship.

QUESTIONS FOR DISCUSSION

1. Define the following terms:

Alliances for the enterprise	Imaginamachina
Code of ethics	IPO
Common stock	Joint ventures
Corporation	Make versus buy
Cost of money	Offering book
Coupon rate of bond	Par value of bond
Debenture	Preferred stock
Designing for manufacture	Prospectus
Due diligence	Sole proprietorship
Free enterprise	Total return
High yield bonds	Warrant

2. What distinguishes concurrent design and manufacturing?
3. Is the slogan "Necessity is the mother of invention" true any longer?
4. Discuss the importance of designing for profit in product development. What happens when there is distortion in these goals?
5. If the plant theoretical capacity is below 100%, what adjustments are made to the calculation of the "make" cost in a manufacture or purchase comparison?
6. Why is business planning so vital to the start-up enterprises? List these points.
7. Describe several ways that money is raised to infuse cash into the new enterprise.
8. Investors benefit from owning stocks in two ways. Discuss the financial importance from the investor's viewpoint.
9. Compare the risks and benefits of raising money by bonds or stocks.
10. The offering book is important to the enterprise and the investor. Write out the steps for its preparation.

[11]This original word is the name of a University of Colorado engineering sculpture and was originated by David M. Griggs, an artist from Denver, Colorado, who visualized a combining of "imagination," "magic," "animation," and "machina" seen in the word. The Latin word *machina* means to devise, plan, and plot artfully, an appropriate engineering objective. We use *imaginamachina* for the first time in this book.

11. Compare the advantages and disadvantages of the corporation, sole proprietorship, and partnership.

12. Examine the Website www.SEC.gov and list the precautions given to investors.

PROBLEMS

11.1 Conventional wisdom in design for manufacture encourages the rule "reduce the number of parts into the fewest number." Consider the two views views shown in Figure P. 11.1.

Some properties of the material and orders of magnitude for cost are given as shown below.

	Two Parts		One Part
	Locator	Plate	Locator and Plate Combined
Primary manufacturing process	Investment cast	Stamped	Investment cast
Material	Titanium (T16A14V)	Stainless steel, (17-4 PH)	Aluminum
Cost range of magnitude	\$50/lb	\$2/lb	\$3/lb
Density, lb/in.^{3}	281.4	495	169
Young's modulus of stiffness, 10^6 psi	10–11	28–30	15–17

Discuss keeping the design as two separate parts or combining the two parts into a single component. What effect does quantity have on your thinking? (*Hint*: If there are two parts, fastening and special holes for that manufacturing operation are necessary. If there is one part, it is a common material.)

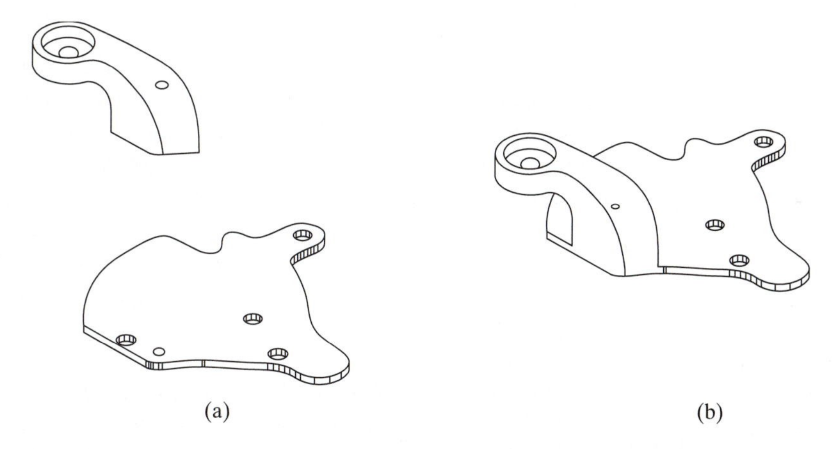

Figure P11.1 (a) Locator arm and plate, and (b) combined locator-plate.

What about machining of the separate parts or common part using a computer numerical controlled milling machine? Discuss selection of the two designs for a least-cost criterion. Which design do you choose? Why?

11.2 A company has 14.2 million shares outstanding. It pays \$0.32 as an annual dividend. The company reports annual profits of \$20.67 million as its stock closes at \$40. Find the earnings per share, price to earnings ratio, and dividend yield.

11.3 The enterprise is conducting a make-versus-buy analysis where the significant cost elements for the manufacturer are as follows:

Cost Element	Unit Value
Direct labor	\$34.10
Direct material	\$8.40
Variable overhead ratio of labor and material	100%
Additional fixed costs	\$5.00
Profit	\$2.50

What is the price of the product? If the plant is operating at full capacity, and a supplier has submitted a bid of \$87.50, is the decision to make the product or buy it from a supplier?

11.4 A design and manufacturing company is anticipating a new product where 3200 annual units is the 100% plant capacity. Cost facts are given as follows.

Cost Element	Amount
Investment cost of project	\$4,100,000
Annual income	\$8,200,000
Operating cost	\$5,384,200
Income tax, all kinds	\$1,319,150

Find the profit before income taxes, yield before income taxes, profit after taxes, yield after taxes, unit operating cost before income taxes, and unit operating cost after taxes.

11.5 Engineering students are planning for the production and sale of a bicycle accessory, called "Warm Cup Holder." While cold water bottles for bikers are common, cups and bottles for hot beverage are not. The design features a quick release mechanism for safe handling during biking and is stronger than the common bicycle water-bottle receptacle. Sale of the garage-produced product, with ample subcontract purchase of plastic molded parts, is to over-the-counter sales, wholesalers, and from their own Website with UPS delivery of product. A prototype product and completed CAD designs are available. The students have estimated the following information:

Cost Element	Amount
Annual units made and sold, number	16,500
Average price for unit sold	\$21.75
Full variable cost of manufacturing, sales, and delivery, dollars per unit	\$6.15
Fixed cost, dollars	\$35,000
Income tax rate	18%
Investment, dollars	\$320,000

Find the profit before income taxes, yield before income taxes, profit after taxes, yield after taxes, unit operating cost before income taxes, and unit operating cost after taxes. (*Hint*: Unit operating cost includes direct labor, direct material, overhead or full variable cost, and fixed cost prorated over number of units.)

11.6 An engineering student team has designed and built a prototype "winch" product for raising and lowering a maximum of 200 pounds of supplies, food, climbing gear, water, etc. over near-vertical mountain walls. The product is intended for recreational and professional mountain-climbing enthusiasts. The product is composed of aluminum pulleys, gears, stop clamps, rope guides, housing, etc. The bill of material is composed of 18 items. Manufacturing information is estimated as follows:

Cost Element	Amount
Direct labor cost, unit	$53.50
Direct material cost, unit	82.15
Variable production cost, unit	11.05
Variable sales cost, unit	15.25
Annual fixed production and sales cost	42,500

Market analysis indicates that this innovative technical product will command price and volume as follows:

	Price	First-Year Units
Wholesale distributor	$285*	325
Internet	$470	200

*Suggested over-the-counter retail price of $480

Advice from investors suggests the following financial obligation for the business:

Yield after income tax	75%
Effective tax rate	15.5%

Find the allowable investment to meet yield requirements. Discuss the financial requirement of yield in respect to the opportunity.

11.7 A student design/manufacturing team is presenting its production costs to the class jury. The product, an electronic relay that controls the on/off function of a portable radio where the listener/wearer is unable to control the unit through manual control, is voice activated by commands "on" and "off." Their summary is given as follows:

Expense	Fixed Cost	Variable Cost per Unit
Annual plant cost	$62,000	
Direct labor		$5.25
Direct material		$4.75
Variable production support		$2.85

The team believes that 100 percent theoretical capacity of their world-class plant is an annual 12,000 units. Plot the total and unit manufacturing cost of the components. Significant production points are 4000 and 8000 units in addition to the maximum capacity of the plant. Let the x-axis be "Units of production" and "Percent theoretical capacity." (*Hint*: Note that the problem is dealing only with costs and is overlooking income.)

11.8 An investment is $1 million. The expected profit rate is 12%. The effective federal and state tax rate is 40%. What is the pretax earning rate required for the project? Find the expected profit plus income tax. What is the expected profit after income tax? (*Hint*: Income tax is a tax imposed on net incomes by the federal and the state government. It is necessary to consider taxes when considering the earning power of a project.)

11.9 The present value of bonds paying semiannual interest is approximated using the following equation:

$$P_0 = \frac{I}{2}(P/A, \frac{i_b}{2}, 2n) + M(P/F, \frac{i_b}{2}, 2n)$$

where P_0 = value of bond at present time, dollars
I = annual interest, dollars
M = maturity value of bond, dollars
i_b = interest rate desired by bondholder, decimal
n = annual periods, number

A $1000 bond has semiannual payments and the coupon rate is 10%. If the required rate of return is 20%, find the value of the bond for a 25-year period and for a 15-year period.

11.10 What is the current market price of a $1000 par 5% coupon rate bond if interest is paid annually or semiannually, and there are 12 years to maturity. The required rate of return is 10%.

11.11 A company sold an issue of 10% bonds 6 years ago. Each bond has a face value of $1000 at maturity, and is due in 14 years. It pays interest twice a year. Even as interest rates are climbing, the bond can be sold on the market for $1100. If buyers expect their money to earn 10% compounded semiannually, and they must pay a brokerage charge of $20 to purchase each bond, is the proposed selling price reasonable?

11.12 Find the current market value of a 20-year, 10% coupon rate bond with a par value of $1000, if interest is paid annually for current market rates of 15% or 5%. What are the current market prices if everything is the same except that the bond has only 10 years to maturity. Discuss the relative influence of changing market interest rates on the market prices of short-term versus long-term bonds.

11.13 Find the stock and bond value for a company having outstanding bonds and stocks. The outstanding bonds amount to $3,000,000 and the issued par value is $1000. Existing market value for the bonds is $1030. There are 6000 shares of stock, and its current market value is $160 each.

11.14 Construct an annual income statement for the company Enterprise Ltd. The all-inclusive federal and state income tax rate is 30%. What is the final net income after taxes? What is the addition or subtraction for the year profit or loss to retained earnings for the balance sheet?

Stocks and bonds have been issued for Enterprise including bond interest, $5\frac{3}{4}$% coupon rate for 900 certificates, par value of $1000; preferred stock, 7% rate, 3000 shares at $100 each share; and common stock, $1 dividend for 14,000 shares currently valued at $50 each.

Closed Account	Amount
Utilities	$60,000
Interest from owned securities	95,000
Fees for consulting and engineering	85,000
Salaries	1,250,000
Sales from products	6,250,000
Equipment lease	72,000
Supplies	125,000
Manufacturing costs not included elsewhere to produce product	2,100,000
Patent royalty from designs	150,000
Rent	48,000
Travel and hotel	50,000

CHALLENGE PROBLEMS

11.15 There are many methods for raising capital for an invention. An entrepreneur will find it necessary to prepare a balance sheet as well as a profit and loss statement for different circumstances of the business opportunity. The following parts are not complete in all details, but the student needs to pay attention to the financial sheets for their differences and similarities.

(a) An entrepreneur has determined a balance sheet and profit and loss statement for an investment, which are as follows:

Balance Sheet Statement

Fixed assets, plant	$1,660,000	Owner's equity	$1,660,000

Profit and Loss Statement

Income from engineering design, plant	$240,000
Operating cost	110,840
Gross profit	129,160
Income tax of individual entrepreneur	37,940
Net profit after income tax	91,220

Find the entrepreneur yield. (*Hint:* For this case assume that the entrepreneur has the full amount of money for the project. Though not a part of this problem, the income tax is determined as for an individual.)

(b) Assume that the entrepreneur has in hand $1,460,000, and she raises another $200,000 in an unsecured loan from a friend for 6% yearly. Balance sheet and income statement are as follows:

Balance Sheet Statement

Fixed assets, plant	$1,660,000	Owner's equity	$1,460,000
		Loan from friend	200,000
	$1,660,000		$1,660,000

Profit and Loss Statement

Income from engineering design, plant	$240,000
Operating cost	110,840
Gross profit	129,160
Interest on $200,000 loan at 6%	12,000
Net profit subject to income tax	$117,160
Income tax	32,920
Net profit after income tax	$84,240

Find the entrepreneur's yield. (*Hint:* The analysis does not provide for paying back the loan. If the investor desires to repay the loan, it must be done out of the profit. If the payment of the loan is made, and is considered a financial expense, the effect is to increase the investor's equity in the project, and an unwarranted increase in the yield.)

(c) The entrepreneur chooses to raise funds via bonds as secured funds where the plant and design provides the mortgage collateral. The investor will invest $260,000 into the enterprise, leaving the remainder as units of $500 per bond at a rate of 5 percent. Balance sheet and income statement are as follows:

Balance Sheet Statement

Fixed assets, plant	$1,660,000	Owner's equity	$260,000
		Funded debt, first-mortgage bond issue	1,400,000
	$1,660,000		$1,660,000

Profit and Loss Statement

Income from engineering design, plant	$240,000
Operating cost	110,840
Gross profit	129,160
Interest on bond issue, $1,400,000 at 5%	70,000
Net profit subject to income tax	59,160
Income tax	13,380
Net profit after income tax	$45,780

Find the entrepreneur's yield. (*Hint:* Taxes for the investor are calculated as an individual, and are not shown.)

(d) The enterprise is incorporated with capital raised as common and preferred stock. Balance sheet and income statement are as follows:

Balance Sheet Statement

Fixed assets, plant	$1,660,000	Stock, common, 12,600 shares at $100	$1,260,000
		Preferred stock, (7%, 400 shares at $1000	400,000
	$1,660,000		$1,660,000

Profit and Loss Statement

Income from engineering design, plant	$240,000
Operating cost	110,840
Gross profit subject to income tax	$129,160
Income tax at 25%	32,290
Net profit after income tax	96,870
Dividend on preferred stock, $400,000 at 7%	28,000
Net profit on common stock	$68,870

What is the yield on the preferred stock? Find the yield on common stock.

11.16 Product PN 8871 is to be analyzed for a make-versus-buy decision. The labor estimate is as follows:

Operation	Unit Labor Cost
1	$0.0006
2	0.0130
3	0.0130
4	0.0007
5	0.0068

Direct material and material overhead is $0.0084 per unit. Variable overhead is 75% of direct labor, and fixed cost is $0.05 per unit. This company believes that the part deserves a profit of $0.025. (*Hint*: Each of the subparts of the problem are independent. There are no quantity considerations.)

(a) Find the total cost and price of the component as if there is no make-versus-buy comparison.
(b) If the company's plant capacity is underutilized, is the decision make-or-buy if a vendor's price is $0.075?
(c) For 100% plant utilization, what is the decision for a supplier's $0.118 price?
(d) If the plant chooses to make the article while at undercapacity, it will incur a 15% increase in direct wages owing to marginal costs of inefficient production. What is the decision for a supplier's $0.105 price?
(e) Evaluate the choices for a nonrecurring initial fixed price of tooling designed, manufactured, and paid for. Those initial costs for 2500 units were $25. What are the nonquantitative considerations of this sunk cost? Also, perform the analysis as if the $25 has not been spent.

11.17 A company is considering the raising of capital by one of three means: 30-year bonds, preferred stock or common stock. It will select the best financing method on the basis of minimum cost after taxes.

The $1000 bonds have a 6% annual coupon rate. The $100 preferred stock has a guaranteed dividend rate of 6%. Common stock is anticipated to have a $50 stock value and dividends are expected to be $1.50 per share. All values are annual. The desired amount of financing for the first year is $15,000,000. Assume that the company is raising the money itself, without the assistance of an underwriting firm, thus avoiding that cost of raising the money. Furthermore, the three methods are assumed to be substantially equal in the cost of financing. The all-inclusive federal and state tax rate is 40%.

Find the preferred financing method.

The market will probably demand a higher rate of return for the bonds than for the preferred stock, since bonds do not tend to have much upside potential, whereas preferred stock has a potential to increase in value.

11.18 Construct a balance sheet for the company Enterprise Ltd., third quarter. The closed ledger accounts, bonds, and stock information are given as follows:

Retained earnings, end of second quarter with profit included	$1,020,000
Finished goods inventory	250,000
Total fixed assets	2,150,000
Preferred stock, 7%, 3000 shares at $100 each	
Short-term borrowing, leases and rentals	80,000
Cash on hand	300,000
Bonds, 5 $^{2}/_{4}$% coupon rate, 900 certificates at $1000 each	
In-process materials	100,000
Common stock, 14,000 shares at $50 each	
Purchased materials inventory	200,000

(*Hint*: Set up a balance sheet having the appropriate formal heading—assets, liabilities, and net worth; see Chapter 4.)

11.19 A student team submits a bookstand prototype and a CAD dithered rendition to the buyer of a commercial mail catalog that specializes in "tools for the serious reader" for purchase consideration. The bookstand, called "Futura," will handle small and large books, measures 5 × 6 × 7 in., weighs 2 lb, folds down for storage, is made of formed and brushed stainless steel and has protective felt feet. The catalog buyer responds, saying that the supplier also delivers the product in a paper box having the appearance of gift wrapping. The catalog will list the product for sale at $29.90, and they are prepared to offer $14.95 a unit. The buyer indicates a probable order of 20,000 units yearly for the foreseeable future.

The team finds that the plant requires equipment such as a shear (sizing the blanks), press brake (forming), flat sander (wire brushing the finish), and a 20 ft conveyor and benches for assembly. Tooling is necessary for piercing, blanking, and forming. The student team is renting a 2000 ft^2 factory, hiring two machinists and is managing service, sales, book keeping, billing, design, and supervision. Estimates are found as follows:

Expense	Fixed Cost	Variable Cost per Unit
Annual plant cost including power, heat, insurance, depreciation, maintenance, working capital, etc.	$75,000	
Direct labor		$1.25
Direct material		1.75
Variable management support by student team		1.00

The team borrows $195,000 from relatives and friends. A certified public accountant indicates that the combined federal and state income tax rate will be 17.5% for their likely profit range of business. Plot plant capacity versus total operating costs and unit operating costs and find the break-even point of operation graphically. What is the yield on investment before and after income taxes?

(*Hint*: Use plant capacity of 0, 25, 50, 75, and 100% where maximum capacity is 20,000 annual units. Two plots are required.)

Would you invest in this opportunity? Why or why not?

11.20 A student team is developing a design concept to ease the movement of a wheelchair. Now the occupant moves the chair by hand grabbing an outer soft-rubber wheel, rolling the wheel forward about 15 to 25°, releasing the wheel and quickly repeating the motion sequence.

The concept introduces a crank motion with a kinematic mechanism where mechanical leverage is improved, the chair is able to move forward or backward more effortlessly

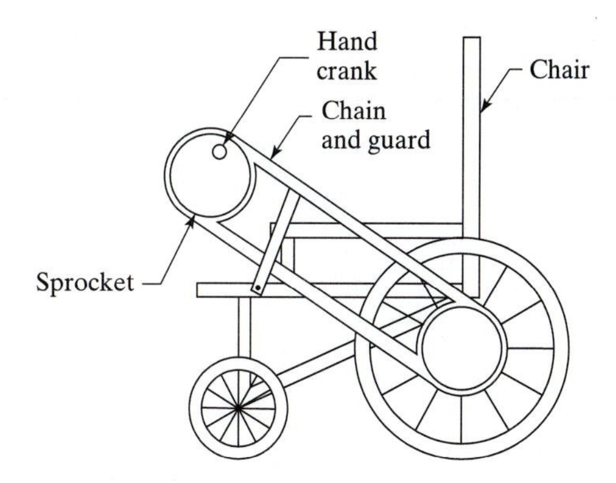

Figure P11.20

and is ergonometrically complaint, according to the team. While the human work to move the chair depends on a number of physical conditions, exercise is also enhanced.

A preliminary design visualizes two hand cranks mounted on the arms of the chair and connected by chains and sprocket-ratchets to the rear wheels. Major assemblies include front and rear wheel and frame attachment, hand crank, frame and chain guard for each side. Figure P11.20 illustrates the concept.

The product is visualized as an after-market sale. What advice can you give to the engineering team as they consider final design and specifications of the product. (*Hint:* Consider design synergy between bicycles and this application. You may wish to survey wheelchair college students to help sharpen your recommendations.)

PRACTICAL APPLICATION

Form a team for this application. Your instructor will define the limits and operation of the team for a design-for-profit assignment. Select a simple product and with a team-conference approach construct a tree that is representative of the product. (*Hint*: Visualize the assembly process first, finding any major or minor assemblies before those assemblies are reduced to components.)

The product can be a real object, or it may be from a design that you are considering in another course. Notice the subassemblies through reverse engineering. Take the product to the component level, suggesting a part number for each part, then estimate a cost for each subassembly level and component. What is the price of the product? Use the market price, if available, and reduce it to an appropriate cost for production. Show the tree as the result of the assignment. (*Hint*: Find a simple product like a fingernail clipper for example.)

CASE STUDY: ROUND PLATE INC.

Round Plate Inc. makes and sells round plates. This company has standardized its product lines into the two families shown in Figure C11.1. These plates are used in many applications, such as bearing plates, post rests, flanges, bases, and any engineering design that starts with 10- and 12-inch. round bar stock and ends with similar shapes. Manufacturing operations are limited to turning, drilling, tapping, reaming, step milling, and grinding.

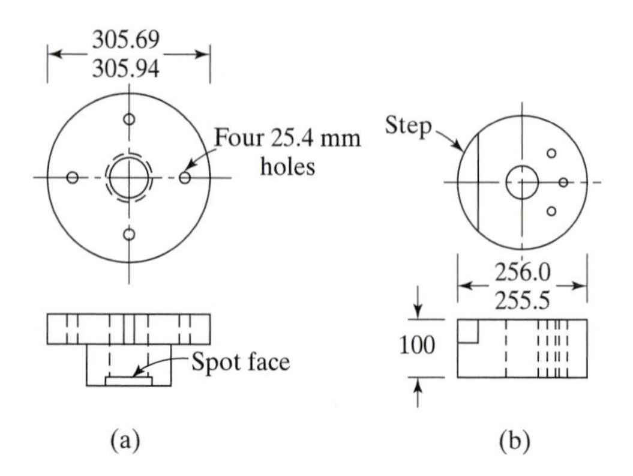

Figure C11.1

Edward Lyell, owner of the shop, tells any potential investor, "I can win any bid starting with standard bar stock and concluding with no more features than shown on the sketches." He has been talking to you about investing in his company.

A registered and legal financial prospectus about Round Plate says that future capitalization with infused shareholders' investment will result in a modern factory of the future equipped for vertical market penetration in the round plate business. This factory will feature standard pallets, fixtures, grippers, and robots for the loading and automatic machining of the bar stock. Round Plate's operation is a garage-style cell.

The prospectus gives the number of outstanding shares, income and balance sheet for the past year, strategy for a 25% control of the market, and pledges for an after-tax yield of 47% on the investment of $1,250,000. The prospectus claims that Round Plate will employ only the president, maintenance operator, tool room grinder, computer programmer, manufacturing engineer, robot maintainer and teacher, accountant and bill collector, and finally a material handler for shipping product, raw stock receiving, and chip and waste removal. All work is computer numerical controlled and does not require any direct labor for the operations.

"We will operate the night shift without anybody," boasts Lyell. "Even the lights are out," he says. "This is a good time to buy shares." He looks at you expectantly.

You pause and then reply, "Well, everything sounds pretty good, and I do like your ideas. But I don't buy into businesses without first sleeping on it. I'll see you tomorrow and we will go over some details."

That night you begin to ponder the offer. You mutter, "What are the critical engineering and manufacturing requirements for this business? What is necessary for success in this enterprise?

Continue your questioning attitude and develop other questions for Mr. Lyell. Develop your questions along these general groupings: (1) equipment and ancillary support for tooling, material movements, controls, and machine vision; (2) market strategy and factors for the round plate business; and (3) enterprise plan. Conclude by preparing a written page on "due diligence" that you as an investor need to use for assessment.

APPENDICES

APPENDIX 1 Values of the Standard Normal Distribution Function

APPENDIX 2 Values of the Student *t* Distribution

APPENDIX 3 10% Interest Factors for Annual Compounding Interest

APPENDIX 4 20% Interest Factors for Annual Compounding Interest

APPENDIX 5 Values of Learning Theory

APPENDIX 1 Values of the Standard Normal Distribution Function

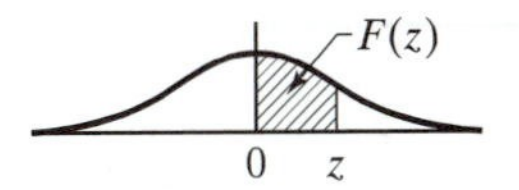

Areas under the Normal Curve

$$F(z) = \int_0^z \frac{1}{\sqrt{2\pi}} e^{-z^2/2}\, dz$$

z	*0.00*	*0.01*	*0.02*	*0.03*	*0.04*	*0.05*	*0.06*	*0.07*	*0.08*	*0.09*
0.0	0.0000	0.0040	0.0080	0.0120	0.0159	0.0199	0.0239	0.0279	0.0319	0.0359
0.1	0.0398	0.0438	0.0478	0.0517	0.0557	0.0596	0.0636	0.0675	0.0714	0.0753
0.2	0.0793	0.0832	0.0871	0.0910	0.0948	0.0987	0.1026	0.1064	0.1103	0.1141
0.3	0.1179	0.1217	0.1255	0.1293	0.1331	0.1368	0.1406	0.1443	0.1480	0.1517
0.4	0.1554	0.1591	0.1628	0.1664	0.1700	0.1736	0.1772	0.1808	0.1844	0.1879
0.5	0.1915	0.1950	0.1985	0.2019	0.2054	0.2088	0.2123	0.2157	0.2190	0.2224
0.6	0.2257	0.2291	0.2324	0.2357	0.2389	0.2422	0.2454	0.2486	0.2518	0.2549
0.7	0.2580	0.2611	0.2642	0.2673	0.2704	0.2734	0.2764	0.2794	0.2823	0.2852
0.8	0.2881	0.2910	0.2939	0.2967	0.2995	0.3023	0.3051	0.3078	0.3106	0.3133
0.9	0.3159	0.3186	0.3212	0.3238	0.3264	0.3289	0.3315	0.3340	0.3365	0.3389
1.0	0.3413	0.3438	0.3461	0.3485	0.3508	0.3531	0.3554	0.3577	0.3599	0.3621
1.1	0.3643	0.3665	0.3686	0.3708	0.3729	0.3749	0.3770	0.3790	0.3810	0.3830
1.2	0.3849	0.3869	0.3888	0.3907	0.3925	0.3944	0.3962	0.3980	0.3997	0.4015
1.3	0.4032	0.4049	0.4066	0.4082	0.4099	0.4115	0.4131	0.4147	0.4162	0.4177
1.4	0.4192	0.4207	0.4222	0.4236	0.4251	0.4265	0.4279	0.4292	0.4306	0.4319
1.5	0.4332	0.4345	0.4357	0.4370	0.4382	0.4394	0.4406	0.4418	0.4430	0.4441
1.6	0.4452	0.4463	0.4474	0.4485	0.4495	0.4505	0.4515	0.4525	0.4535	0.4545
1.7	0.4554	0.4564	0.4573	0.4582	0.4591	0.4599	0.4608	0.4616	0.4625	0.4633
1.8	0.4641	0.4649	0.4656	0.4664	0.4671	0.4678	0.4686	0.4693	0.4699	0.4706
1.9	0.4713	0.4719	0.4726	0.4732	0.4738	0.4744	0.4750	0.4756	0.4762	0.4767
2.0	0.4772	0.4778	0.4783	0.4788	0.4793	0.4798	0.4803	0.4808	0.4812	0.4817
2.1	0.4821	0.4826	0.4830	0.4834	0.4838	0.4842	0.4846	0.4850	0.4854	0.4857
2.2	0.4861	0.4865	0.4868	0.4871	0.4875	0.4878	0.4881	0.4884	0.4887	0.4890
2.3	0.4893	0.4896	0.4898	0.4901	0.4904	0.4906	0.4909	0.4911	0.4913	0.4916
2.4	0.4918	0.4920	0.4922	0.4925	0.4727	0.4929	0.4931	0.4932	0.4934	0.4936
2.5	0.4938	0.4940	0.4941	0.4943	0.4945	0.4946	0.4948	0.4949	0.4951	0.4952
2.6	0.4953	0.4955	0.4956	0.4957	0.4959	0.4960	0.4961	0.4962	0.4963	0.4964
2.7	0.4965	0.4966	0.4967	0.4968	0.4969	0.4970	0.4971	0.4972	0.4973	0.4974
2.8	0.4974	0.4975	0.4976	0.4977	0.4977	0.4978	0.4979	0.4980	0.4980	0.4981
2.9	0.4981	0.4982	0.4983	0.4983	0.4984	0.4984	0.4985	0.4985	0.4986	0.4986
3.0	0.4987	0.4987	0.4987	0.4988	0.4988	0.4989	0.4989	0.4989	0.4990	0.4990
3.1	0.4990	0.4991	0.4991	0.4991	0.4992	0.4992	0.4992	0.4992	0.4993	0.4993

*This table gives the probability of a random value of a normal variate falling in the range $z = 0$ to $z = z$ (in the shaded area in figure). The probability of the same variate having a deviation greater than z is given by 0.5 − probability from the table for the given z. The table refers to a single tail of the distribution; therefore the probability of a variate falling in the range is $\pm z = 2 \times$ probability from the table for the given z. The probability of a variate falling outside the range $\pm z$ is $1 - 2 \times$ probability from the table for the given z.

Source: From Phillip F. Ostwald, *Engineering Cost Estimating* (Englewood Cliffs, NJ: Prentice Hall, 1992). The values in this table were obtained by permission of author and publishers from C. E. Weatherburn, *Mathematical Statistics*. (London: Cambridge University Press, 1946).

APPENDIX 2 Values of the Student *t* Distribution

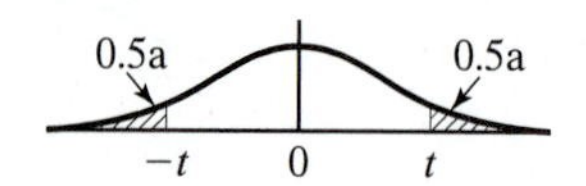

Degrees of Freedom ν	Probability α 0.10	0.05	0.01	0.001
1	6.314	12.706	63.657	636.619
2	2.920	4.303	9.925	31.598
3	2.353	3.182	5.841	12.941
4	2.132	2.776	4.604	8.610
5	2.015	2.571	4.032	6.859
6	1.943	2.447	3.707	5.959
7	1.895	2.365	3.499	5.405
8	1.860	2.306	3.355	5.041
9	1.833	2.262	3.250	4.781
10	1.812	2.228	3.169	4.587
11	1.796	2.201	3.106	4.437
12	1.782	2.179	3.055	4.318
13	1.771	2.160	3.012	4.221
14	1.761	2.145	2.977	4.140
15	1.753	2.131	2.947	4.073
16	1.746	2.120	2.921	4.015
17	1.740	2.110	2.898	3.965
18	1.734	2.101	2.878	3.922
19	1.729	2.093	2.861	3.883
20	1.725	2.086	2.845	3.850
21	1.721	2.080	2.831	3.819
22	1.717	2.074	2.819	3.792
23	1.714	2.069	2.807	3.767
24	1.711	2.064	2.797	3.745
25	1.708	2.060	2.787	3.725
26	1.706	2.056	2.779	3.707
27	1.703	2.052	2.771	3.690
28	1.701	2.048	2.763	3.674
29	1.699	2.045	2.756	3.659
30	1.697	2.042	2.750	3.646
40	1.684	2.021	2.704	3.551
60	1.671	2.000	2.660	3.460
120	1.658	1.980	2.617	3.373
∞	1.645	1.960	2.576	3.291

* This table gives the values of t corresponding to various values of the probability α (level of significance) of a random variable falling inside the shaded area in the figure, for a given number of degrees of freedom ν available for the estimation of error. For a one-sided test the confidence limits are obtained for $\alpha/2$.

Source: From Phillip F. Ostwald, *Engineering Cost Estimating* (Englewood Cliffs, NJ: Prentice Hall, 1992). This table is taken from Fisher and Yates, *Statistical Tables for Biological, Agricultural, and Medical Research*, Edinburgh: Oliver & Boyd Ltd., 1963, table 3).

APPENDIX 3 10% Interest Factors for Annual Compounding Interest

n	Single Payment		Equal Payment Series			
	Compound-Amount Factor	Present-Worth Factor	Compound-Amount Factor	Sinking-Fund Factor	Present-Worth Factor	Capital-Recovery Factor
	To Find F Given P F/P i, n	To Find P Given F P/F i, n	To Find F Given A F/A i, n	To Find A Given F A/F i, n	To Find P Given A P/A i, n	To Find A Given P A/P i, n
1	1.100	0.9091	1.000	1.0000	0.9091	1.1000
2	1.210	0.8265	2.100	0.4762	1.7355	0.5762
3	1.331	0.7513	3.310	0.3021	2.4869	0.4021
4	1.464	0.6830	4.641	0.2155	3.1699	0.3155
5	1.611	0.6209	6.105	0.1638	3.7908	0.2638
6	1.772	0.5645	7.716	0.1296	4.3553	0.2296
7	1.949	0.5132	9.487	0.1054	4.8684	0.2054
8	2.144	0.4665	11.436	0.0875	5.3349	0.1875
9	2.358	0.4241	13.579	0.0737	5.7590	0.1737
10	2.594	0.3856	15.937	0.0628	6.1446	0.1628
11	2.853	0.3505	18.531	0.0540	6.4951	0.1540
12	3.138	0.3186	21.384	0.0468	6.8137	0.1468
13	3.452	0.2897	24.523	0.0408	7.1034	0.1408
14	3.798	0.2633	27.975	0.0358	7.3667	0.1358
15	4.177	0.2394	31.772	0.0315	7.6061	0.1315
16	4.595	0.2176	35.950	0.0278	7.8237	0.1278
17	5.054	0.1979	40.545	0.0247	8.0216	0.1247
18	5.560	0.1799	45.599	0.0219	8.2014	0.1219
19	6.116	0.1635	51.159	0.0196	8.3649	0.1196
20	6.728	0.1487	57.275	0.0175	8.5136	0.1175
21	7.400	0.1351	64.003	0.0156	8.6487	0.1156
22	8.140	0.1229	71.403	0.0140	8.7716	0.1140
23	8.954	0.1117	79.543	0.0126	8.8832	0.1126
24	9.850	0.1015	88.497	0.0113	8.9848	0.1113
25	10.835	0.0923	98.347	0.0102	9.0771	0.1102
26	11.918	0.0839	109.182	0.0092	9.1610	0.1092
27	13.110	0.0763	121.100	0.0083	9.2372	0.1083
28	14.421	0.0694	134.210	0.0075	9.3066	0.1075
29	15.863	0.0630	148.631	0.0067	9.3696	0.1067
30	17.449	0.0573	164.494	0.0061	9.4269	0.1061
31	19.194	0.0521	181.943	0.0055	9.4790	0.1055
32	21.114	0.0474	201.138	0.0050	9.5264	0.1050
33	23.225	0.0431	222.252	0.0045	9.5694	0.1045
34	25.548	0.0392	245.477	0.0041	9.6086	0.1041
35	28.102	0.0356	271.024	0.0037	9.6442	0.1037
40	45.259	0.0221	442.593	0.0023	9.7791	0.1023
45	72.890	0.0137	718.905	0.0014	9.8628	0.1014
50	117.391	0.0086	1163.909	0.0009	9.9148	0.1009

APPENDIX 4 20% Interest Factors for Annual Compounding Interest

n	Single Payment		Equal Payment Series			
	Compound-Amount Factor	Present-Worth Factor	Compound-Amount Factor	Sinking-Fund Factor	Present-Worth Factor	Capital-Recovery Factor
	To Find F Given P F/P i, n	To Find P Given F P/F i, n	To Find F Given A F/A i, n	To Find A Given F A/F i, n	To Find P Given A P/A i, n	To Find A Given P A/P i, n
1	1.200	0.8333	1.000	1.0000	0.8333	1.2000
2	1.440	0.6945	2.200	0.4546	1.5278	0.6546
3	1.728	0.5787	3.640	0.2747	2.1065	0.4747
4	2.074	0.4823	5.368	0.1863	2.5887	0.3863
5	2.488	0.4019	7.442	0.1344	2.9906	0.3344
6	2.986	0.3349	9.930	0.1007	3.3255	0.3007
7	3.583	0.2791	12.916	0.0774	3.6046	0.2774
8	4.300	0.2326	16.499	0.0606	3.8372	0.2606
9	5.160	0.1938	20.799	0.0481	4.0310	0.2481
10	6.192	0.1615	25.959	0.0385	4.1925	0.2385
11	7.430	0.1346	32.150	0.0311	4.3271	0.2311
12	8.916	0.1122	39.581	0.0253	4.4392	0.2253
13	10.699	0.0935	48.497	0.0206	4.5327	0.2206
14	12.839	0.0779	59.196	0.0169	4.6106	0.2169
15	15.407	0.0649	72.035	0.0139	4.6755	0.2139
16	18.488	0.0541	87.442	0.0114	4.7296	0.2114
17	22.186	0.0451	105.931	0.0095	4.7746	0.2095
18	26.623	0.0376	128.117	0.0078	4.8122	0.2078
19	31.948	0.0313	154.740	0.0065	4.8435	0.2065
20	38.338	0.0261	186.688	0.0054	4.8696	0.2054
21	46.005	0.0217	225.026	0.0045	4.8913	0.2045
22	55.206	0.0181	271.031	0.0037	4.9094	0.2037
23	66.247	0.0151	326.237	0.0031	4.9245	0.2031
24	79.497	0.0126	392.484	0.0026	4.9371	0.2026
25	95.396	0.0105	471.981	0.0021	4.9476	0.2021
26	114.475	0.0087	567.377	0.0018	4.9563	0.2018
27	137.371	0.0073	681.853	0.0015	4.9636	0.2015
28	164.845	0.0061	819.223	0.0012	4.9697	0.2012
29	197.814	0.0051	984.068	0.0010	4.9747	0.2010
30	237.376	0.0042	1181.882	0.0009	4.9789	0.2009
31	284.852	0.0035	1419.258	0.0007	4.9825	0.2007
32	341.822	0.0029	1704.109	0.0006	4.9854	0.2006
33	410.186	0.0024	2045.931	0.0005	4.9878	0.2005
34	492.224	0.0020	2456.118	0.0004	4.9899	0.2004
35	590.668	0.0017	2948.341	0.0003	4.9915	0.2003
40	1469.772	0.0007	7343.858	0.0002	4.9966	0.2001
45	3657.262	0.0003	18281.310	0.0001	4.9986	0.2001
50	9100.438	0.0001	45497.191	0.0000	4.9995	0.2000

APPENDIX 5 Values of Learning Theory

N	$\phi = 80\%$			$\phi = 90\%$		
	T_u	T_c	T_a	T_u	T_c	T_a
1	1.0000	1.0000	1.0000	1.0000	1.0000	1.0000
2	0.8000	1.8000	0.9000	0.9000	1.9000	0.9500
3	0.7021	2.5021	0.8340	0.8462	2.7462	0.9154
4	0.6400	3.1421	0.7855	0.8100	3.5562	0.8891
5	0.5956	3.7377	0.7475	0.7830	4.3392	0.8678
6	0.5617	4.2994	0.7166	0.7616	5.1008	0.8501
7	0.5345	4.8339	0.6906	0.7439	5.8447	0.8350
8	0.5120	5.3459	0.6682	0.7290	6.5737	0.8217
9	0.4929	5.8389	0.6488	0.7161	7.2898	0.8100
10	0.4765	6.3154	0.6315	0.7047	7.9945	0.7994
11	0.4621	6.7775	0.6161	0.6946	8.6890	0.7899
12	0.4493	7.2268	0.6022	0.6854	9.3745	0.7812
13	0.4379	7.6647	0.5896	0.6771	10.0516	0.7732
14	0.4276	8.0923	0.5780	0.6696	10.7211	0.7658
15	0.4182	8.5105	0.5674	0.6626	11.3837	0.7589
16	0.4096	8.9201	0.5575	0.6561	12.0398	0.7525
17	0.4017	9.3218	0.5483	0.6501	12.6899	0.7465
18	0.3944	9.7162	0.5398	0.6445	13.3344	0.7408
19	0.3876	10.1037	0.5318	0.6392	13.9735	0.7354
20	0.3812	10.4849	0.5242	0.6342	14.6078	0.7304
21	0.3753	10.8602	0.5172	0.6295	15.2373	0.7256
22	0.3697	11.2299	0.5104	0.6251	15.8624	0.7210
23	0.3644	11.5943	0.5041	0.6209	16.4833	0.7167
24	0.3595	11.9538	0.4981	0.6169	17.1002	0.7125
25	0.3548	12.3086	0.4923	0.6131	17.7132	0.7085
30	0.3346	14.0199	0.4673	0.5963	20.7269	0.6909
40	0.3050	17.1935	0.4298	0.5708	26.5427	0.6636
50	0.2838	20.1217	0.4024	0.5518	32.1420	0.6438
100	0.2271	32.6508	0.3265	0.4966	58.1410	0.5814
500	0.1352	98.8472	0.1977	0.3888	228.7851	0.4576

Selected Answers

CHAPTER 1

1.1 (a) 1.58 m^2, 0.22 m^2, 0.29 m^2, 464.5 m^2.
1.3 (a) 5.5 m, 0.61 m.
1.5 (a) 240.29 kg/m^3; (b) 11.24 lbm/ft^3.
1.7 (a) 0.01 m^3, 3.54 m^3, 19.82 m^3; (d) 0.015 $in.^3$, 0.00488 $in.^3$, 0.0915 $in.^3$.
1.9 (a) € 18.781.
1.11 (a) $100; (b) 113.036 ¥/$.
1.13 (a) 5,093,338 renolas, $356,177.
1.15 $10,014,675; (b) $7,300.

CHAPTER 2

2.1 (a) 7200, 5400; (b) 116.1, 104.5.
2.3 (a) 0.633 min.; (b) 94.8; (c) 1.05 hr; (d) $0.156.
2.5 (a) 2.9, 3.41, 5.68, 17.6; (b) $0.87.
2.7 (a) 1.060, 1.250, 20.8; (b) $18.89.
2.9 (a) 2.27; (b) 3, $156,000.
2.13 (a) 4610; (b) 18,440.
2.15 (b) $394.13; (c) $432.69.
2.17 $25.63.
2.19 $38,773.
2.21 $17.00/hr.
2.23 (a) $0.33; (b) $46.67, 133%.
2.25 $49,350.

2.27 Part 1 (a) \$0.59; (b) \$0.79; (c) \$0.70; Part 2 (a) \$0.59; (b) \$0.38; (c) \$0.47.

2.29 (a) 25%; (b) \$4.35.

2.31 (a) 3.98; (b) \$712.02, \$8544.24.

2.33 (a) 2860; (b) 39.5.

CHAPTER 3

3.1 (a) 4; (b) \$329.11; (c) \$158.22, \$15.82; (d) 88.0%.

3.3 (a) 0.0685 in.^{3}, 0.323 in.^{3}; (b) 20.4%; (c) \$0.084.

3.5 (a) 17.745 in.^{3}; (b) \$0.55.

3.7 (a) 0.297 in.^{3}; (c) 0.3997 in.^{3}; (e) \$0.042; (g) \$0.037.

3.9 (a) \$7.00; (b) \$12.00; (c) \$12.00; (d) \$12.95; (e) \$9.98; (f) \$11.98.

3.11 (a) \$4.40, \$0.27, \$4.67; (b) \$4.50, no.

3.13 \$3.01.

3.15 (f) \$9.78.

3.17 \$0.71.

3.19 33 bars, \$3168.65 material, \$3256.43 labor and material.

3.21 66 pieces/sheet, 95.5%, \$1.282/piece.

CHAPTER 4

4.1 Assets = \$15,100, liabilities + N.W. = \$15,100.

4.3 P&L, profit = \$1570; Balance Sheet, assets = \$7210, liabilities + N.W. = \$7210.

4.5 Retained earnings = \$325,000.

4.7 Assets = \$485,000, liabilities + N.W. = \$485,000.

4.9 Income, profit = \$57; Balance Sheet, assets = \$139, liabilities + N.W. = \$139.

4.11 Income, profit = \$105,000; balance sheet, assets = \$540,000, liabilities + N.W. = \$540,000.

4.13 Depreciation = \$0.35/mile; \$10,869/year; \$1.09/mile.

4.15 (a) \$18,920; (b) \$12,000.

4.17 First year depreciation for accelerated \$50,000; straight-line \$41,667; production units \$41,667.

4.21 Total variance = \$70 (unfavorable).

4.23 Book value in year 10: straight-line \$292,000, MACRS \$0; production \$292,000.

4.25 Variance: hours \$211.69 (unfavorable); hourly rate \$6.75 (unfavorable); net \$218.44 (unfavorable); 96% productivity.

4.27 Productive hour costs: Fab \$74.41; Assm \$82.81; Test \$84.54.

4.29 Unit cost \$34.15; lot cost \$41,496.26.

CHAPTER 5

5.3 (a) mean 4.9, median 5, mode 5, range 8, Var. 5.664, Std. Dev. 2.38.

5.5 (c) years 0–9: $y = 92.3 + 7.785x$ and for year 10, $y = 170.2$; years 4–9 only: $y = 68.106 + 11.291x$ and for year 10, $y = 181$.

5.7 $T = 3206(N)^{-0.709}$, for $N = 60, T = 175.9$.

5.9 (a) $t_\alpha = 1.812$; (b) for $v = n - 2, t_\alpha = 2.101$; (c) $t_\alpha = 1.98, Z = 1.96$; (d) $\alpha = 0.016, n = 17$.

5.11 (b) $y = 101.4 + 4.736x$; (c) for year 7, $y = 134.5$ for an error of +1.43%, range = (130.03, 138.96); (d) \$29.35–\$31.36.

5.13 (a) $y = -0.116 - 0.0608x$, for 15 spots, $y = 0.752 - 0.841$.

5.15 (b) $T = 33.287(0.935)^x$; (c) for year 5, $T = 23.787$ hours.

5.17 Correlation = −0.127.

5.19 (a) exponential, $y = 4.00867(1.00022)^x$, polynomial, $y = 5.84 - 8.65 \times 10^{-4}x + 4.47 \times 10^{-7}x^2$; (b) exponential, $\Sigma\varepsilon^2 = 2.7856$, polynomial, $\Sigma\varepsilon^2 = 1.6974$ the better choice.

5.21 Expected cost in period 11: by moving average, \$67.90; by smoothing, \$68.57.

5.25 $I_3 = 1648, I_6 = 4264.5, C_6 = \$329.80$.

5.29 (a) $y = 11.81 + 0.1733x_1 - 1.3167x_2$; (b) \$9.35/unit.

5.31 France \$2.8501, United States \$2.6868 choose; United States.

CHAPTER 6

6.3 For weight, $y = 0.0483 + 0.235x_1$; for girth, $y = 0.3575 + 0.3873x_2$; for fixture $y = 6.0634 - 1.0341x_3$; weight has the best correlation.

x_1	5.0	6.3	7.9	10.0	12.5	15.7	19.7
y	1.22	1.53	1.91	2.38	2.98	3.73	4.66

6.5 (a) 5971 hr; (b) for 4th unit, 13,624 hr, for 6th unit, 12,558 hr; (c) $K = 7179, T_{11} = 6011$; (d) for 92%, $T_1 = 2 \times 10^6, T_2 = 1.84 \times 10^6, T_3 = 1.69 \times 10^6$; (e) $T_{25} = 191, T_{50} = 176$; (f) $T_u = \$131{,}915$; (g) $\Phi = 85\%$.

6.7

N	T_u	T_c	T_a
5	6857	40,311	8062
6	6570	46,881	7813

N	T_u	T_c	T_a
5	6937	36,434	7287
6	6844	43,278	7213

6.9 Design A (preferred) \$1,669,100; Design B \$1,862,000.

6.11 \$27.73 million.

6.13 $m = 0.28$, \$149.

6.15 (a) Expected profit from higher bid \$1.8 million, from lower bid \$2.4 million (preferred); (b) \$50; (c) \$2950; (d) 40% greater risk; (e) \$87.25.

6.17 (a) in year 8

6.19 (a) in millions, total cost \$28.75, total variance 3.007; (b) 94.41% probability of exceeding.

6.21 5.3.

6.23 2.6.

6.25 Task 1:

Robot Distance	8.0	10.0	12.6	15.8	19.8
Minutes	0.027	0.033	0.041	0.052	0.065

Tasks 2 plus 3: constant 0.069 min.

6.27 (a) 68 hr; (b) for 50th unit \$11,494, for 200th unit \$8700; (c) $T_a = 317$, $T_u = 258$, $T_c = 3170$; (d) $\Phi = 93.75\%$, for 5th unit $T_a = 853$, $T_u = 781$, $T_c = 4265$; (e) $T_a = 114$, $T_u = 99.8$, $T_c = 11{,}478$.

CHAPTER 7

7.1 Lot = 2.125 hr, unit = 0.02833 hr.

7.3 Set-up = 2.9 hr, unit time = 5.16 min, 11.6 pieces/hr, 8.600 hr/100 pieces, 63.96 hr/lot.

7.5 Set-up = 1.26 hr, 0.922 min/piece, 65.1 pieces/hr, 1.537 hr/100 pieces, 16.4 hr/lot

7.7 (a) 10,238; (b) 15,000; (c) \$5905; (d) 0.5 year.

7.9 (a) \$1.65; (b) 3.19 min; (c) 30.13 min; (d) \$0.21; (e) \$0.32.

7.11 (a) 2.7 min; (b) 4.0 min.

7.13 (a) 11.25 in., 24.5 in.; (b) 16.7 in.; (c) 48.7 in.; (d) 2.26 in.

7.15 (a) 2.42 min, 108 RPM; (b) 0.74 min, 162 RPM, approx. 10 hr.

7.17 (a) 4.7 min; (b) 3.6 min.

7.19 67% more with WC tool.

7.21 Tool life at 100 fpm = 80 min; $n = 0.11$ and $K = 161.9$; $t_m = 11$ min, 197 RPM.

7.25 (a) 303 fpm, 125 RPM; (b) 5.75–5.77; (c) 4.86 min, 12.3 pieces/hr, 8.107 hr/100 pieces; (d) 347 fpm, 11.3 min.

7.27 \$39.63.

CHAPTER 8

8.1 \$168 / unit.

8.3 (a) \$0.54; (b) \$0.66; (c) \$0.68; (d) \$0.72.

8.5 For unit price of $4: (a) $1, 25%; (b) revenue $160,000, contribution $40,000.

8.7 (a) 5.1 years; (b) 0.569.

8.9 (a) Cost: $1109.85/lot, $6.34/unit; Price: $1,331.82/lot, $7.61/unit; (b) Countersink, $0.37/unit, $64.90/lot.

8.11 (a) Light = $43.51/hr, heavy = $113.72/hr, assembly = $37.87, finish = $44.56/hr, (b) $9455.40

8.13 $s = -0.1041$, 667 hr.

8.15 (a) 61.97; (b) $s = -0.1504$, 149,044 hr/lot, 60.59 hr avg./unit.

8.17 Total ECO = 479,756 hr, units 51–100 = 460,981 hr.

8.19 (a) $816,250; (b) $887,500.

8.21 (a) $925,000, $775,000; (b) $970,000, $730,000; (c) $500,000.

8.23 For 500 units: $14,431, $28.86/unit.

8.25 (a) 79.44 hr/unit average, 23,832 hrs total; (b) 4005 hr.

8.27 $43,795 bid, 20 hr error.

8.29 Price: U.S. $3.46, foreign $2.18, cost difference $0.83.

CHAPTER 9

9.1 Cumulative cash flow year 4 = $0, year 5 = $20,000.

9.3 Income $515,000, taxes $175,100.

9.5 Approximately 40%.

9.7 Positive cash flow does not occur during the first 5 years; year 5 it is −$8800.

9.9 Deflation in price, −10.37%, with associated inflation, −12.5%.

9.11 B.E. = 5000 units; profit for 10,000 units = $30,000.

9.13 1400 units.

9.15 Up to 6666, purchase; above that up to 13,889, batch, then above that, flow line.

9.17 120,000 m^3/yr.

9.19 Average cost: for 15 units = $3.33, for 30 units = $2.33.

9.25 Break-even points are 53 and 188 units; maximum profit of $11,200 at 110 units.

9.27 Minimum marginal cost at 3 units, minimum average total cost at 7 units, at the minimum point of the average variable cost curve.

9.29 Bidders 1 and 3 are "tied" at $139 & $140 respectively.

9.31 Supplier "C" at $369 million.

9.33 $162,816.

9.35 (a) $C(T) = 100n - 3 \times 10^{-3}n^2 + 10^{-7}n^3 + 10^5$; (c) average cost = $150/$yd^3$; (d) marginal price = $250.

CHAPTER 10

10.1 (a) 10%, 20%; (b) 22.5%.

10.3 (a) $90; (b) 4%; (c) $90,000 simple, $107,946 compound; (d) $565; (e) $8572; (f) $422; (g) 22.5; (h) 1 year, 6.67%, 2 years, 3.28%.

10.5 Enterprise B is profitable in year 3.

10.7 \$308.

10.9 (a) PW = \$18, FW = \$24, EAW = \$6.97.

10.11 (a) NPW = −\$358, NEAW = −\$58, NFW = −\$929; (b) NPW = −\$260, NEAW = −\$62, NFW = −\$3,321.

10.13 $i \approx 7.2\%, i = 7.18\%, n \approx 6$ yr.

10.15 Nominal, 12%, APR 12.68%.

10.17 \$1492 continuous, \$1469 annual.

10.19 (a) \$1768; (b) \$1775; (c) \$436,480.

10.21 Refrigeration and domestic heating is less at \$135,500.

10.23 (a) First A at \$338,533, then C then B; (b) A at \$80,755, then C then B; (c) A at \$2,096,031, then C then B.

10.25 (a) Alternative B at \$95,908; (b) alternative B at \$17,950.

10.27 Annual cost: large furnace \$981,176, new and current \$1,007,568.

10.29 \$120,623, buy it.

10.31 \$28.33 / unit.

10.33 (a) fixed equipment, NPW \$100,798, NEAW \$58,080; mobile equipment, NPW \$155,381, NEAW \$89,531.

CHAPTER 11

11.3 Make = \$90.00; buy = \$87.50; buy it.

11.5 Before taxes: unit operating cost \$8.27, profit \$222,400, yield 70%; after taxes: unit operating cost \$10.70, profit \$182,368, yield 57%.

11.7

Number of units	4000	8000	12000
Total production cost (\$)	113,400	164,800	216,200
Cost (\$/unit)	28.35	20.60	18.20

11.9 For 25 years \$504; for 15 years \$528.

11.11 PW = \$1000, price = \$1120; don't buy.

11.13 Stock \$960,000, bonds \$309,000.

11.15 (a) 5.5%; (b) 5.76%; (c) 17.6%; (d) preferred 7%, common 5.4%.

11.17 After-tax cost: bonds \$540,000, preferred stock \$900,000, common stock \$450,000.

11.19

Percent of capacity	50	75	100
After-tax profit (\$)	28,463	73,631	118,800
After-tax yield (%)	14.6	37.8	60.9

Bibliography

Ahuja, Hidra N., *Estimating: From Concept to Completion.* Prentice Hall: Upper Saddle River, NJ, 1987.

Ansari, Shahid L. and Jan E. Bell, *Target Costing.* Irwin: Chicago, IL, 1997.

Blanchard, Benjamin S., *Design and Manage to Life Cycle Cost.* Dilithium Press: 1978.

Bowman, Michael S., *Applied Economic Analysis for Technologists, Engineers, and Managers.* Prentice Hall: Upper Saddle River, NJ, 1999.

Bullinger, Clarence, *Engineering Economy*, 3rd ed. McGraw-Hill: New York, 1958.

Clark, Forrest D., A. B. Lorenzoni, and Michael Jimenez, *Applied Cost Engineering.* Marcel Dekker: New York, 1997.

Cooper, Robin Robert S. Kaplan, and Lawrence S. Maisel, et al., *Implementing Activity-Based Cost Management: Moving from Analysis to Action.* Institute of Management Accounts: Montvale, NJ, 1992.

Cooper, Robin, and Regine Slagmulder, *Target Costing and Value Engineering.* Productivity Press: Portland, OR 1997.

Creese, Robert, M. Adithan, B.S. Pabla, *Estimating and Costing for the Metal Manufacturing Industries.* Marcel Dekker: New York, 1992.

DeGarmo, E. Paul, William G. Sullivan, and James A. Bontadelli, *Engineering Economy*, 9th ed. Prentice Hall: Upper Saddle River, NJ, 1996.

Dhillon, B. S., *Life Cycle Costing: Techniques, Models, and Applications.* Gordon and Breach Science: New York, 1989.

Dudick, Thomas S., *Handbook of Product Cost Estimating and Pricing.* Prentice Hall: Upper Saddle River, NJ, 1991.

Freedman, Russell, *Kids at Work: Lewis Hine and the Crusade Against Child Labor.* Clarion Books: New York, 1994.

Gonen, Turan, *Engineering Economy for Engineering Managers.* Wiley: New York, 1990.

International Journal of Production Economics, various years.

Madachy, Raymond J., and Barry W. Boehm, *Software Process Modeling with System Dynamics.* 2003.

Michaels, Jack W., and William P. Wood, *Design to Cost*. Wiley: New York, 1989.

Newnan, Donald G., and Bruce Johnson, *Engineering Economic Analysis*, 5th ed. Engineering Press: San Jose, CA, 1995.

Ostwald, Phillip F., *AM Cost Estimator*, 4th ed. McGraw-Hill: New York, 1982.

Ostwald, Phillip F., *Construction Cost Analysis and Estimating*. Prentice Hall: Upper Saddle River, NJ, 2001.

Ostwald, Phillip F., and Jairo Munoz, *Manufacturing Processes and Systems*, 9th ed. Wiley: New York, 1997.

Park, Chan S., *Contemporary Engineering Economics*, 2nd ed. Addison-Wesley: Menlo Park, CA, 1997.

Peters, Max S., Klans D. Timmerhaus, and Ronald E. West, *Plant Design and Economics*, 5th ed. McGraw Hill: New York, 2003.

Rautenstrach, Walter, *The Economics of Business Enterprise*. Wiley: New York, 1939.

Riggs, James L., *Engineering Economics*, 4th ed. McGraw-Hill: New York, 1996.

Schweitzer, Marcell, Earnst Trossmann, and Gerald H. Lawson, *Break-Even Analysis*. Wiley: New York, 1992.

Stewart, Rodney D., *Cost Estimating*, 2nd ed. Wiley: New York, 1991.

Taylor, George A., *Managerial and Engineering Economy*. Van Nostrand: Princeton, NJ, 1964.

Tucker, Spencer A., *Profit Planning Decisions with the Break-Even System*. Thomos Press: New York, 1980.

ENTERPRISE AND COSTING SOFTWARE COMPANIES

ABC Technologies
ABIS-PRO
Agiltech—Metcapp
Bom.com
Boothroyd and Dewhurst
Cognition
Costimator
Cost Vision
GA Seer
Galorath
Hyperion
I4cast
Integrated Cost Management Solutions
MetalSoft
NDS Systems
Price
Savantage/Foresight Systems
Tecnomatix
Wright, Williams, and Kelly

Index

Note: Page numbers in italic refer to Picture Lessons.

USEFUL EQUATIONS *(continued)*

Title	Equation	Number
Opportunity margin	$M = P_o e^{-k_m t} - P_f e^{-k_f t}$	(8.17)
Actual to real dollars	$D_r = D_a\left(\frac{1}{1+f}\right)^{n-k}$	(9.3)
Cash flow with taxes	$F_c = (G - D_c - C)(1-t) + D_c$	(9.4)
Break-even	Total costs = net sales revenue	(9.5)
Break-even—cost side	$C_T = C_f + nC_v$	(9.6)
Break-even—revenue side	$R_T = nR_v$	(9.8)
Break-even for quantity	$n_{BE} = \frac{C_f}{(R_v - C_v)}$	(9.9)
Return on investment	$\text{Return} = \frac{\text{average earnings} - (\text{total investment} \div \text{economic life})}{\text{average investment}} \times 100, \%$	(10.2)
Compound-amount factor	$F = P(1+i)^n$	(10.7)
Present-worth factor	$P = F(1+i)^{-n}$	(10.8)
Uniform payments capital-recovery factor	$A = P\left[\frac{i(1+i)^n}{(1+i)^n - 1}\right]$	(10.10)
Uniform payments compound-amount factor	$F = A\left[\frac{(1+i)^n - 1}{i}\right]$	(10.11)
Uniform payments present-worth factor	$P = A\left[\frac{(1+i)^n - 1}{i(1+i)^n}\right]$	(10.12)
Equivalent annual cost or worth	$EAC = (P - F_s)(A/P, i\%, n) + F_s i$ $EAC = P(A/P, i\%, n) - F_s(A/F_s\ i\%, n)$	(10.13)
Present value bond	$P_0 = I(P/A, i_b, n) + M(P/F, i_b, n)$	(11.1)